Algebraic Geometric Codes: Basic Notions

Mathematical
Surveys
and
Monographs

Volume 139

Algebraic Geometric Codes: Basic Notions

Michael Tsfasman
Serge Vlăduţ
Dmitry Nogin

American Mathematical Society

EDITORIAL COMMITTEE

Jerry L. Bona   Peter S. Landweber
Michael G. Eastwood   Michael P. Loss
J. T. Stafford, Chair

Our research while working on this book was supported by the French National Scientific Research Center (CNRS), in particular by the Institut de Mathématiques de Luminy and the French-Russian Poncelet Laboratory, by the Institute for Information Transmission Problems, and by the Independent University of Moscow. It was also supported in part by the Russian Foundation for Basic Research, projects 99-01-01204, 02-01-01041, and 02-01-22005, and by the program Jumelage en Mathématiques.

2000 *Mathematics Subject Classification.* Primary 14Hxx, 94Bxx, 14G15, 11R58; Secondary 11T23, 11T71.

For additional information and updates on this book, visit
www.ams.org/bookpages/surv-139

Library of Congress Cataloging-in-Publication Data
Tsfasman, M. A. (Michael A.), 1954–
   Algebraic geometry codes : basic notions / Michael Tsfasman, Serge Vladut, Dmitry Nogin.
     p. cm. — (Mathematical surveys and monographs, ISSN 0076-5376 ; v. 139)
   Includes bibliographical references and index.
   ISBN 978-0-8218-4306-2 (alk. paper)
   1. Coding theory. 2. Number theory. 3. Geometry, Algebraic. I. Vladut, S. G. (Serge G.), 1954– II. Nogin, Dmitry, 1966– III. Title.

QA268.T754 2007
003'.54—dc22   2007061731

Copying and reprinting. Individual readers of this publication, and nonprofit libraries acting for them, are permitted to make fair use of the material, such as to copy a chapter for use in teaching or research. Permission is granted to quote brief passages from this publication in reviews, provided the customary acknowledgment of the source is given.

Republication, systematic copying, or multiple reproduction of any material in this publication is permitted only under license from the American Mathematical Society. Requests for such permission should be addressed to the Acquisitions Department, American Mathematical Society, 201 Charles Street, Providence, Rhode Island 02904-2294, USA. Requests can also be made by e-mail to reprint-permission@ams.org.

© 2007 by the American Mathematical Society. All rights reserved.
The American Mathematical Society retains all rights
except those granted to the United States Government.
Printed in the United States of America.
∞ The paper used in this book is acid-free and falls within the guidelines
established to ensure permanence and durability.
Visit the AMS home page at http://www.ams.org/
10 9 8 7 6 5 4 3 2 1   12 11 10 09 08 07

Contents

Preface ix

Advice to the Reader xvii

Chapter 1. Codes 1
  1.1. Codes and Their Parameters 1
    1.1.1. Definition of a Code 1
    1.1.2. $[n,k,d]_q$ Systems 3
    1.1.3. Spectra and Duality 7
    1.1.4. Bounds 15
    1.1.5. Bounds for Higher Weights 21
    1.1.6. Duality for Generalized Spectra 29
  1.2. Examples and Constructions 33
    1.2.1. Codes of Genus Zero 33
    1.2.2. Code Families 36
    1.2.3. Constructions 44
  1.3. Asymptotic Problems 49
    1.3.1. Main Asymptotic Problem 49
    1.3.2. Asymptotic Bounds 51
    1.3.3. Asymptotic Bounds for Higher Weights 57
    1.3.4. Polynomiality 60
    1.3.5. Other Asymptotics 64
  Historical and Bibliographic Notes 67

Chapter 2. Curves 69
  2.1. Algebraic Curves 69
    2.1.1. Quasiprojective Varieties 70
    2.1.2. Quasiprojective Curves 77
    2.1.3. Divisors 79
    2.1.4. Jacobians 85
    2.1.5. Riemann Surfaces 88
  2.2. Riemann–Roch Theorem 91
    2.2.1. Differential Forms 91
    2.2.2. Riemann–Roch Theorem 95
    2.2.3. Hurwitz Formula 100
    2.2.4. Special Divisors 102
    2.2.5. Cartier Operator 104
  2.3. Singular Curves 106
    2.3.1. Normalization 106

		2.3.2. Double-Point Divisor	107
		2.3.3. Plain Curves	108
	2.4.	Elliptic Curves	111
		2.4.1. Group Law	111
		2.4.2. Isomorphisms and the j-invariant	114
		2.4.3. Isogenies	115
		2.4.4. Complex Elliptic Curves	118
	2.5.	Curves over Nonclosed Fields	120
		2.5.1. Function Fields	120
		2.5.2. Places of a Function Field	122
		2.5.3. Divisors	125
		2.5.4. Function Fields and Algebraic Curves	126
	Historical and Bibliographic Notes		131

Chapter 3. Curves over Finite Fields — 133

3.1.	Zeta Function		133
	3.1.1.	Definition and Rationality	134
	3.1.2.	Functional Equation	138
	3.1.3.	Weil Theorem and Its Corollaries	141
	3.1.4.	Explicit Formula	143
	3.1.5.	Pellikaan's Two-Variable Zeta Function	144
3.2.	Asymptotics		146
	3.2.1.	Drinfeld–Vlăduţ Theorem	146
	3.2.2.	Lower Asymptotic Bounds	147
	3.2.3.	Points of Higher Degrees	147
	3.2.4.	Asymptotics for the Jacobian	149
	3.2.5.	Asymptotically Exact Families	151
3.3.	Elliptic Curves over Finite Fields		158
	3.3.1.	Isomorphism Classes	158
	3.3.2.	Isogeny Classes	161
	3.3.3.	Endomorphism Ring and the Zeta Function	162
	3.3.4.	Structure of $E(\mathbb{F}_q)$	163
3.4.	Remarkable Examples		165
	3.4.1.	Hermitian, Sub-Hermitian, and Maximal Curves	165
	3.4.2.	Kummer and Artin–Schreier Covers	167
	3.4.3.	García–Stichtenoth Towers	177
	3.4.4.	Curves of Small Genera	178
3.5.	Connection with Exponential Sums		180
	3.5.1.	Number of Points on Fermat Curves	180
	3.5.2.	L-functions of Characters	181
	3.5.3.	Estimates for Exponential Sums	183
Historical and Bibliographic Notes			187

Chapter 4. Algebraic Geometry Codes — 191

4.1. Constructions and Properties		191
	4.1.1. Basic Algebraic Geometry Constructions and Their Parameters	192
	4.1.2. Duality and Spectra	198
	4.1.3. Decoding Problem	202

4.2.	Additional Bounds and Constructions	206
	4.2.1. Extra Bounds	206
	4.2.2. Variants of the Basic Construction	216
	4.2.3. Partial Algebraic Geometry Codes	219
4.3.	Characterization of Algebraic Geometry Codes	225
	4.3.1. Three AG Levels	225
	4.3.2. All Linear Codes Are Weakly AG	226
	4.3.3. Criteria	228
4.4.	Examples	232
	4.4.1. Codes of Small Genera	232
	4.4.2. Elliptic Codes	234
	4.4.3. Hermitian Codes	241
	4.4.4. Other Examples	245
	4.4.5. Generalized Algebraic Geometry Codes	247
4.5.	Asymptotic Results	250
	4.5.1. The Basic Algebraic Geometry Bound and Its Variants	250
	4.5.2. Expurgation Bound and Codes with Many Light Vectors	253
	4.5.3. Constructive Bounds	263
	4.5.4. Other Bounds	266
4.6.	Nonlinear Algebraic Geometry Constructions	271
	4.6.1. Elkies Codes	271
	4.6.2. Xing Codes	280
Historical and Bibliographic Notes		284

Appendix. Summary of Results and Tables 287

A.1.	Codes of Finite Length	287
	A.1.1. Bounds	287
	A.1.2. Parameters of Some Codes	288
	A.1.3. Parameters of Some Constructions	289
A.2.	Asymptotic Bounds	292
	A.2.1. List of Bounds	292
	A.2.2. Comparison Diagrams	295
	A.2.3. Behaviour at the Endpoints	299
	A.2.4. Numerical Values	299
A.3.	Additional Bounds	306
	A.3.1. Constant-Weight Codes	306
	A.3.2. Self-dual Codes	306
	A.3.3. Bounds for Higher Weights	307

Bibliography 309

List of Names 329

Index 331

Preface

The book you hold in your hands is devoted to algebraic geometry codes, a comparatively young domain which emerged in the early 1980s at the meeting-point of several fields of mathematics. On the one side we see such respectable, well-developed, and difficult areas as algebraic geometry and algebraic number theory; on the other is a twentieth-century creation, information transmission theory (with its algebraic daughter, the theory of error-correcting codes), as well as combinatorics, finite geometries, dense sphere packings, and so on.

The relation between the two groups of domains, a priori distant from each other, was discovered by Valery Goppa. At the beginning of 1981 he presented his discovery at the algebra seminar of the Moscow State University; among the listeners there was Yuri Manin, who (being maybe the only one to understand), in turn, related the story at his seminar. One of the authors of this book (Tsfasman) was there and had the first chance in his life to hear the words *error-correcting code*. A couple of months later he, another author (Vlăduţ), and Thomas Zink wrote a small paper improving the asymptotic Gilbert–Varshamov bound. To the great astonishment of the authors, who were in doubt whether it was worth publishing or not, the world mathematical community, represented by both coding theorists and algebraic geometers, expressed lively interest. (All of a sudden, the authors found out that for a quarter of a century the asymptotic Gilbert–Varshamov bound had been attacked in vain, which had led to a widely spread—though not unanimous—opinion of its tightness.) This gave birth to a new domain, the algebraic geometry codes.

We (Tsfasman and Vlăduţ) were asked to write a book that could be of use to both coding specialists and algebraic geometers. The book [TV91] was published in 1991 by Kluwer Academic Publishers, and the Russian edition was scheduled and prepared for publication the same year by a Russian publisher, Nauka. The destiny of Russia was, however, different, which we by no means regret. The next time we were asked to publish the book in Russian came ten years later. By that time the book had become somewhat obsolete—the theory of algebraic geometry codes becoming twice as old—and we decided instead to write a new book in both English and Russian. The number of authors has increased, and the book has become two different books, being divided into *Basic Notions*, which you are reading now, and *Advanced Chapters*, yet to be written.

In the introduction to [TV91] we wrote that "the next decade will witness many new interesting results, methods, and problems in this fruitful area" and hoped for the book "to be of some use for those who choose this area for their own." We dare say that the prediction was correct and that the same words could be applied to this new book and to the decade to come.

To make the reading less tedious, we provide many exercises and problems.[1]

*****

The main interests of the authors of this book lie where algebraic geometry meets number theory. This leads to a point of view on coding theory different from that of an insider. This part of the introduction is to explain this view. (We restrict ourselves to *block codes*, called *codes* from now on.)

A finite-dimensional vector space over a normed field ($\mathbb{C}$, $\mathbb{R}$, $\mathbb{Q}_p$, etc.) possesses a natural metric. Vector spaces over $\mathbb{Q}$ and other algebraic number fields, as well as free $\mathbb{Z}$-modules of finite rank, have different possible metrics linked to different embeddings of the corresponding field or ring in a normed one. The same for global fields in finite characteristic: $\mathbb{F}_q(T)$ and its finite algebraic extensions.

How can we introduce a metric space structure on $\mathbb{F}_q^n$, a finite-dimensional space over a finite field? We know no metric that is more natural than the *Hamming metric*:

$$d(x,y) = \big|\{i \mid x_i \neq y_i\}\big|;$$

i.e., the distance equals the number of different coordinates, the corresponding norm being

$$\|x\| = \big|\{i \mid x_i \neq 0\}\big|.$$

This norm has a serious disadvantage: it depends on the choice of a basis, but this is, alas, unavoidable.

Then, $\mathbb{F}_q^n$ being a metric space, we make a direct analogy with the classical problem to construct dense packings of equal spheres in $\mathbb{R}^n$ and get the problem to place in $\mathbb{F}_q^n$ as many balls of a given diameter d as possible so that the packing density be as large as possible (the volume on $\mathbb{F}_q^n$ is much more natural than the metric, the volume of a set being nothing more than its cardinality). An almost equivalent problem is to construct the largest subset $C \subseteq \mathbb{F}_q^n$ such that the distance between any two distinct elements is at least d. Such a subset C is called an $[n,k,d]_q$ *code*,[2] where $k = \log_q |C|$. The question of finding the largest code for given q, n, and d looks like a natural combinatorial problem.

Among all the codes, there is an organic subset of *linear codes*, i.e., codes $C \subseteq \mathbb{F}_q^n$ that are linear subspaces. There are at least three reasons to study linear codes. First, they are natural analogues of lattice sphere packings in $\mathbb{R}^n$. Next, they are easier to construct and give many good examples. The third reason—most important for us—is that linear $[n,k,d]$ codes correspond to *projective systems*, i.e., systems $\mathcal{P}$ of n points defined over $\mathbb{F}_q$ in the $(k-1)$-dimensional projective space $\mathbb{P}^{k-1}$ over $\mathbb{F}_q$. Moreover, $n - d$ equals the maximal number of points in $\mathcal{P}$ lying in a hyperplane (details are given in Sec. 1.1.2; such $\mathcal{P}$ are called *projective*

---

[1] We tried to follow the distinction, calling a question that we know the answer to an EXERCISE, sometimes giving hints to more difficult ones, and reserving the word PROBLEM for research problems and open questions. A rather instructive story happened with Exercise 1.3.23. This exercise—and its analogue for sphere packings—was given in [TV91]. It happened to be slightly more difficult than we thought it to be. Being asked how to solve it, one of the authors took some time to understand that probabilistic arguments do give the necessary result. The attempt to do the same exercise for sphere packing was also successful, several years of reflection easily providing the solution ([ST01]; see also [Bli05]).

[2] See Remark 1.1.2, explaining why it is called a code.

$[n,k,d]_q$-*systems*). We have thus arrived at another problem: how to place n points in $\mathbb{P}^{k-1}$ so that there are many of them outside any hyperplane (this is a general position type condition). Note that here we have gotten rid of both the basis in the ambient space and the metric we were not quite content with.

Having read the last paragraph, each algebraic geometer will ask, what if for $\mathcal{P}$ we take (a part of) $\mathbb{F}_q$-points of some algebraic variety W? If we take a curve, the answer is almost obvious. Let V be an algebraic curve defined over $\mathbb{F}_q$ together with its projective embedding $V \hookrightarrow \mathbb{P}^{k-1}$; then n is at most $|V(\mathbb{F}_q)|$, the number of $\mathbb{F}_q$-points of the curve V, and $n-d$ is at most the degree of the curve $V \subset \mathbb{P}^{k-1}$. The question of possible relations between n, k, and d becomes that of sheer algebraic geometry. The most important part of this book treats further properties of such codes.

The latter remark determines our view of mathematical coding theory. Codes over any finite field $\mathbb{F}_q$ are equally interesting for us (in spite of the fact that for applications one mostly uses binary codes, i.e., those over the field $\mathbb{F}_2$ or at least over its extension). Linear codes are of special interest. The main problem for us is that of finding possible parameters of a code (ideally, not only should one find them out but also produce a code having these particular parameters). This problem has several main forms. Here is the list (the field $\mathbb{F}_q$ is always fixed).

Let the code length n be fixed. For a given d, find the maximal $k = K(n,d)$ such that there exists an $[n,k,d]_q$ code (respectively, a linear $[n,k,d]_q$ code). A close problem is to find the maximal $d = D(n,k)$ for a fixed k. Note that, given an $[n,k,d]_q$ code, we easily construct $[n+1,k,d]_q$, $[n,k-1,d]_q$, and $[n,k,d-1]_q$ codes, so that solving these problems, we get not just the best possible parameters but also all possible parameters. The third problem of the same kind is to find the minimal $n = N(k,d)$ for given k and d.

Being unable to solve these problems completely, we cut the problem into two (for simplicity, let us consider the problem, say, to find $K(n,d)$). On the one hand, it is useful to look for some conditions to restrict parameters of any code. Such conditions would give an *upper bound* for the cardinality of a code, $K(n,d) \leq K_{\text{up}}(n,d)$; they are also called *possibility bounds*. On the other hand, it would be nice to produce explicit codes enjoying good parameters. Each of such codes (with given n and d) gives a *lower bound* $K_{\text{low}}(n,d) \leq K(n,d)$, also called an *existence bound*; it is natural that codes are usually constructed not individually but by vast classes. If for given n and d the upper bound equals the lower one, the problem is solved.

The main disadvantage of putting the problem this way is that we have not one problem but an infinite series of problems, and experience shows that we are extremely far from the general solution. As a partial answer for small values of n, one has tables of values of $K_{\text{up}}(n,d)$ and $K_{\text{low}}(n,d)$ (see [Bro] for online versions of such tables, maintained and regularly updated by A.E. Brouwer). These tables are good to compare new methods of code constriction with existing ones.

Another type of problem arises when we start to get interested in the behaviour of code parameters as n grows. Consider a family of codes with $n \to \infty$; how can k and d vary? A reasonable statement of an asymptotic problem always depends on certain knowledge of the actual behaviour of parameters at infinity. In our case, at least three asymptotic problems look natural (see precise statements in Sec. 1.3).

First: what is the asymptotic behaviour of k with respect to n for a fixed d? Here, the character of the answer is known. Put

$$\varkappa_q(d) = \liminf_{n \to \infty} \left(\frac{n - K(n,d)}{\log_q n} \right);$$

then $0 < \varkappa_q(d) < \infty$, and the question of the exact value of or bounds for $\varkappa_q(d)$ is quite legitimate. It so happens that $\varkappa_q(d)/(d-1)$ is bounded from above and below by constants independent of d (roughly speaking, $1/2 \le \varkappa_q(d)/(d-1) \le (q-1)/q$ for any d).

Second: what is the behaviour of d with respect to n for a fixed k? The character of the answer is again known; moreover, the answer itself is known. Let

$$\delta_q(k) = \limsup_{n \to \infty} \left(\frac{D(n,k)}{n} \right);$$

then

$$\delta_q(k) = \frac{q^{k-1}(q-1)}{q^k - 1}.$$

The third problem is, we feel, the most interesting. We call it the *main asymptotic problem*. Let $n, k, d \to \infty$, $k/n \to R$, $d/n \to \delta$; what is the dependence between R and δ? The question can be stated rigorously and happens to be highly nontrivial. The core of the book treats this problem. There are other asymptotic questions to ask, which, however, look more artificial.

This ends the first and most important (as we see it) subject of mathematical coding theory, the *parameter problem*.

We approach the second subject equipped with the experience of the first one. We know several upper bounds; do there exist codes whose parameters reach these bounds? Usually, these codes are characterized by a nice property, such as, for example, perfect and equidistant codes. Then we know many classes of codes, and we can ask what are possible parameters of codes in a given class (good examples are cyclic codes, self-dual ones, MDS codes, etc.). Properties of specific codes are also of interest: their weight spectra (see Sec. 1.1.3), their automorphism groups, their behaviour under a natural duality, and so on. This subject can be called the *property and structure problem*; questions here are various and very important.

We would like not only to be able to find out the best parameters but also to construct codes with given parameters more or less explicitly. To construct a code for a fixed n is to produce a construction algorithm (for example, for a linear code this means to produce its generator matrix). Of course, there always exists a "stupid" universal algorithm: sorting out all subsets or all linear subspaces of $\mathbb{F}_q^n$; we would like to exclude such solutions. A way to do it in the asymptotic setting is given by the algorithm complexity theory: let us demand the construction algorithms be polynomial in n (see Sec. 1.3.3 for details). As of today, it is the only rigorous setting of the problem of explicit construction of codes, and we stick to it. This is the third subject, the *constructiveness problem*.

There is a fourth subject, almost totally ignored in this book: codes good from the practical point of view. Here we should admit that codes are able to correct errors that occur while transmitting information distorted by random noise in a transmission channel of a certain type. Codes for practical use should mostly be binary or at least 2^m-ary for a small m; they need to be rather long, but not too much; they must have not just a polynomial but a really simple and fast construction algorithm; and, not the least, they must enjoy a fast and simple decoding algorithm. This is a specific *problem of practical use*. This subject also puts forth interesting algebraic geometry problems; algebraic geometry codes have already led to significant progress in this direction.

And last, there is a subject that we may call the *problem of analogues*. We consider this problem to be the cornerstone for the role of coding theory in the whole building of modern mathematics. There is a beautiful analogy between linear codes and lattices in Euclidean spaces, corresponding to what is maybe the most fundamental analogy in mathematics, that between algebraic curves and number fields. Our interest in coding theory is due to a hope we can clarify this analogy.

The theory of algebraic geometry codes gave birth to many new problems in mother domains, including algebraic geometry and number theory. We hope to discuss all this in *Advanced Chapters*.

We believe that the possibilities of algebraic geometry codes are far from being exhausted and that this book will help to attract new forces to it.

$$* \; * \; * \; * \; *$$

The book is designed for mathematicians. The reader may or may not be acquainted with the main notions of coding theory and algebraic geometry.

The first two chapters are introductory. Chapter 1 is devoted to the basics of coding theory; algebraic geometry does not appear here explicitly, though in fact algebraic geometry codes determine the choice of material and results. There are few new results with respect to classical textbooks in coding theory. The main feature is that, first, we always work over an arbitrary finite field. Next, we profess the geometric approach whenever possible. To this end, we introduce projective systems, discuss spectra expressions motivated by these systems, and so on. This immediately attracts our attention to higher weights of codes, which are natural invariants of point configurations.

The second chapter provides an introduction to algebraic curves; codes are never mentioned, but the choice of topics is often determined by needs of coding theory. In this chapter we mostly work over an algebraically closed field. Starting from the very basics, we go as far as the Riemann–Roch theorem, discussing in some detail the theory of elliptic curves. At the end we pass to a nonclosed field and introduce the language of function fields.

Geometry over a finite field is the subject of the third chapter. We start to work with the zeta function from the very beginning and do not hide our interest in asymptotic problems. We give lots of examples. Relatively new are asymptotic bounds for the number of points on a curve and on its Jacobian, calculation of the number of divisors with prescribed properties, a theorem on the structure of the group of points of an elliptic curve over a finite field, and towers of curves with many points. We postpone the theory of infinite global fields and that of the asymptotic zeta functions to *Advanced Chapters*.

In the fourth chapter we discuss constructions of algebraic geometry codes, their spectra, many examples, codes of small genera, duality and self-duality, and so forth. We also consider the problem of characterization of algebraic geometry codes. Then we discuss in detail asymptotic lower bounds of algebraic geometry origin; these bounds are most prominent to show the power of algebraic geometry methods.

As an appendix, we give equations and tables of asymptotic bounds, tables of parameters of different classes of codes and of various coding constructions, and bounds for some classes of linear codes.

* * * * *

Now let us give a brief list of the most interesting topics you are not going to find in this book. In *Advanced Chapters* we would like to discuss the following topics, many of which have emerged quite recently.

First of all, we cannot abstain from briefly touching on the decoding of algebraic geometry codes, a topic not only very important for applications but also attracting our attention to new algebraic geometry questions. We shall also discuss relations of algebraic geometry codes to other types of error-correcting codes such as LDPC (low-density parity-check) codes and expander codes.

Next, we shall consider in more detail curves with many points, including modular curves and Deligne–Lusztig curves.

Third, we would like to discuss other problems linked with algebraic geometry codes by the same ideology. Such are the Rosenbloom–Tsfasman metrics and their applications to experimental design and uniform sequences, fast multiplication algorithms in finite fields using modular curves, authentication codes, quantum codes, codes over rings, graphs without short cycles, and many others.

Fourth, there is a very important analogue of codes, lattices, and sphere packings in Euclidean spaces. Here one should relate typical properties of random lattices, additive and multiplicative constructions of dense packings from algebraic number fields, nonlinear Lenstra codes, and Elkies–Shioda constructions of packings related to elliptic curves over global fields.

The fifth topic is multi-dimensional varieties over finite fields and related codes. Here it would be natural to discuss higher weights, generalized Reed–Muller codes, and codes from Grassmann and Schubert varieties. There is a very interesting question of the number of points on a surface over a finite field, both bounds and examples. Then there is the theory of abelian varieties over a finite field, especially such topics as the structure of the set of their points, statistics for the number of points, and behaviour of the eigenvalues of the Frobenius operator.

And the last topic, dear to us, is the asymptotic theory of number and function fields, towers of global fields and infinite extensions of $\mathbb{Q}$ and $\mathbb{F}_q(t)$, their zeta functions, the generalized Brauer–Siegel theorem, and other "infinite" properties.

It is clear that, even in the best circumstances, the attempt to include all the above in *Advanced Chapters* is bound to fail. We shall be happy if we manage to discuss at least some of these subjects.

One should not forget that while we are writing these books, many mathematicians—to start with you, dear reader—are getting or preparing to get new, beautiful results. There is no hope that we can catch up with you.

* * * * *

We (Serge Vlăduţ and Michael Tsfasman) are deeply grateful to our teacher Yuri Ivanovich Manin (who attracted our attention to the topic); Vladimir Drinfeld (for many interesting discussions of elliptic modules and modular curves); Gregory Katsman (who taught us coding theory); Leonid Bassalygo (who explained to us many coding subtleties); Gilles Lachaud (for many years of fruitful cooperation and hospitality, which helped us to write many chapters of the book); Alexander Barg (for many fruitful remarks and for the first version of tables of asymptotic bounds); Sergei Gelfand (for inducing us to publish our first paper improving the Gilbert–Varshamov bound); Gregory Kabatiansky (who made lots of valuable remarks reading the text of this book); Simon Litsyn (who attracted our attention to sphere packings in $\mathbb{R}^n$); Michael Rosenbloom (who worked with us on the problem of analogues); Alexei Skorobogatov (for many discussions); Andries Brouwer, Gerard van der Geer, and Marcel van der Vlught (who wrote appendices to the Russian edition of this book); members of the coding theory seminar and our colleagues from the Institute for Information Transmission Problems; and many other mathematicians for their friendship and help.

Dmitry Nogin thanks not only the above-named mathematicians but also his co-authors, who attracted him to undertake this work.

We are deeply grateful to our parents for their care and to our wives for their tender love.

Advice to the Reader

The original idea of this book was to convey to the reader what we know about algebraic geometry codes, starting from the very beginning and up to the most recent results. By no means is this goal achieved. The twenty-five-year domain of several hundred published papers can hardly be fully explored in a single book. That is why we have restricted ourselves to a textbook for a reader already working in or planning to plough this or adjacent fields. We could not expect the reader to be acquainted with either algebraic geometry or coding theory. In the first two chapters we briefly explain necessary results. These chapters are not supposed to be textbooks in those fields, but we hope that a diligent reader seeing one or both of these theories for the first time may still get an adequate feel of what they look like and will continue his way in the chosen direction.

We should stress the role of exercises dispersed everywhere in the book. They are numerous and constitute an organic part of the exposition. Their formulations are to be read and studied together with other statements of the book. If the reader has no desire to solve an exercise, he should just consider it to be one of the propositions given without proof. It is, however, highly recommended that you solve a certain number of exercises, this being important to acquire firm knowledge of the material. A large number of them are quite accessible even for an inexperienced reader.

The authors cherish the hope that they are able to solve all the exercises. Those we cannot solve are also presented but called problems (see, however, the footnote on p. x of the Preface).

The book is designed for several categories of readers.

If you are a specialist in codes interested mostly in fast knowledge of what algebraic geometry codes are, then we advise you glance through Secs. 1.1 and 1.2, just to rub shoulders with our terminology, which often differs from the standard one. Sec. 1.1.2 should be studied in more detail. Read Secs. 2.1.1–2.1.3, 2.2.1, 2.2.2, and Sec. 2.5. Then try to read at least a part of Secs. 2.4 and 3.3, then Secs. 3.4.1 and 3.4.2. After that you are ready to pass to the fourth chapter, to algebraic geometry codes themselves. You will surely be able to read Sec. 4.1 (except maybe for what concerns self-dual codes in Sec. 4.1.2) and Sec. 4.5 (except for Sec. 4.5.2). To understand what is written about self-dual codes in Sec. 4.1.2, you need Secs. 2.1.4; to understand Sec. 4.4.2, you need Sec. 2.4.1 and 2.4.2 and a part of Sec. 3.3; before you read Sec. 4.5.2, you need to study Secs. 2.1.4 and 3.2.4. This way gives you a rather thorough acquaintance with algebraic geometry codes at little expense. Then you can briefly glance through the rest of the book.

If you are a specialist in codes interested first of all in asymptotic problems of coding theory, then it is reasonable to start by looking through Secs. 1.1 and 1.2 and a more intense reading of Sec. 1.3, containing what is somewhat hidden in most books on coding theory. Then read Secs. 2.1.1–2.1.3 and 2.2.1, then Secs. 2.2.2 and 3.2. Now you are ready to read Sec. 4.5 (to understand Sec. 4.5.2, you also need Secs. 2.1.4 and 3.2.4). All this gives some insight into the asymptotic possibilities of algebraic geometry codes. Then we advise you to take a look at the rest of the book.

For an algebraic geometer interested in algebraic geometry codes as a new area to which to apply the algebraic geometry, we advise reading thoroughly Chapter 1 and then glancing at Secs. 2.1.1–2.1.3, 2.2.1, and 2.2.2 and studying those where you feel yourself less confident. Then read Chapter 4 starting from Sec. 4.1.1; after that, just read this chapter choosing one of the scenarios described above according to your interests.

If you are an algebraic geometer looking for new problems in your domain generated by algebraic geometry codes, then after Chapter 1 it is advisable to look through Chapter 2, concentrating on the material you are less familiar with. Then study closely Chapter 3. Then read Secs. 4.1 and 4.3 and pass to Secs. 4.5 and 4.6.

If codes are of no interest to you at all, after glancing through Chapter 2 you can then concentrate on Chapter 3. In this way the book can serve you as a textbook on algebraic geometry over a finite field.

Now we address the largest audience. If you are not a specialist in either codes or algebraic geometry and you wish to go to their meeting-point as soon as possible, we recommend the following course: Secs. 1.1.1, 1.1.2, and 1.2.1; then Secs. 2.1.1–2.1.3, 2.2.1, 2.2.2, Secs. 3.1 and 3.4. In the fourth chapter first read Secs. 4.1.1, 4.1.2, 4.3.1, 4.4.2, and 4.4.3. Now you are acquainted with algebraic geometry codes. Then read the rest of Chapter 4, turning when necessary to those parts of Chapters 1 and 2 you need.

And last, if you are already a specialist in algebraic geometry codes, start with the contents and then follow your own choice. Do not forget about the historical and bibliographic notes at the end of each chapter.

We heartily advise all readers to try to solve at least some of the exercises.

We also advise you not to forget about the tables and diagrams in the Appendix and about the list of names and index.

* * * * *

Chapter relation diagram.

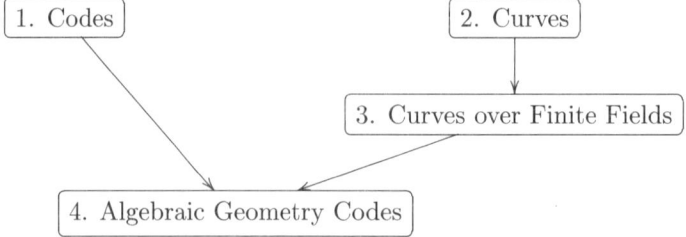

Section relation diagram.

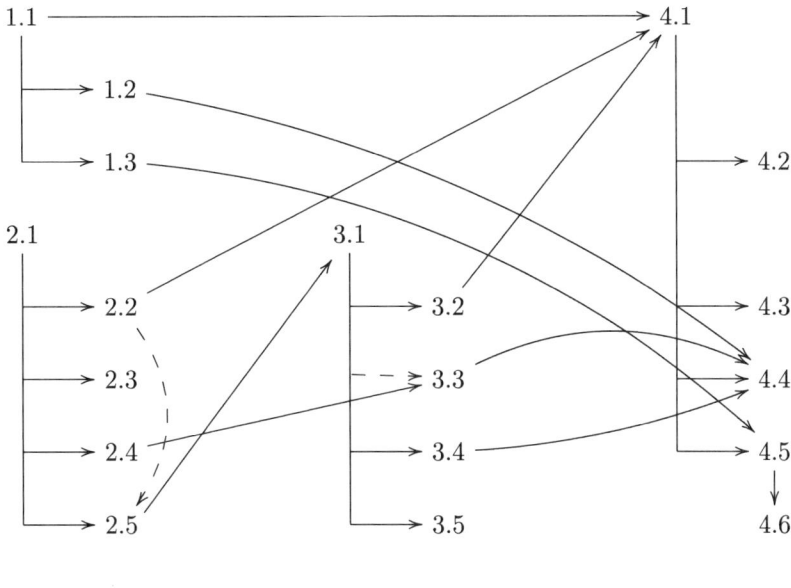

* * * * *

To avoid misunderstanding, we would like to point out that some terms and notation we use here differ somewhat from common use. In particular, in what follows:

- The notation $A \subset B$ means that A is a *proper* subset of B, i.e., $A \neq B$; when we do not exclude the possibility $A = B$, we write $A \subseteq B$. The set difference is denoted $A \setminus B$.
- An $[n,k,d]_q$ code may be linear or nonlinear. In the last case, k is not necessarily an integer; see Sec. 1.1.1.
- For *algebraic geometry codes* we use the notation $(X,\mathcal{P},D)_L$, $(X,\mathcal{P},D)_\Omega$, etc., while in many papers they are denoted, respectively, $C(G,D)$ and $C^*(G,D)$; see Sec. 4.1.1.
- By $\Omega(D)$ we denote the space
$$\Omega(D) = \{\omega \in \Omega(X)^* \mid (\omega) + D \geq 0\} \cup \{0\},$$
which in many papers may be denoted by $\Omega(-D)$; see Sec. 2.1.1.
- When describing code families and constructions, we mostly keep the usual meaning of the term, sometimes differing from it in details; see Secs. 1.2.2 and 1.2.3.
- $\lfloor x \rfloor$ means the integer part of $x \in \mathbb{R}$, i.e., $\lfloor x \rfloor \in \mathbb{Z}$, $\lfloor x \rfloor \leq x < \lfloor x \rfloor + 1$; and $\lceil x \rceil$ is the least integer such that $x \leq \lceil x \rceil < x + 1$.

CHAPTER 1

Codes

This chapter is an introduction to the theory of error-correcting codes. Presenting principal notions and examples, we always keep in mind the "algebraic geometry philosophy", which will explicitly come out in Chapter 4, but here we do not touch any results that use algebraic geometry codes.

The first section contains principal definitions and properties of codes, as well as some motivations. We also discuss in detail two "nonclassical" subjects, namely, projective systems and higher weights. The second section is devoted to particular examples (or, to be precise, interesting classes) of codes and various constructions of new codes starting with those already known. In the third section we discuss various asymptotic problems (i.e., a rigorous way to put questions about codes of large length), including the algorithmic point of view. We end this chapter with some notes on history and bibliography.

1.1. Codes and Their Parameters

In this section we define basic notions of the theory of error-correcting codes (coding theory). In Sec. 1.1.1 we define $[n,k,d]_q$ codes (both linear and nonlinear) and explain why they are called error-correcting. Section 1.1.2 is devoted to another (more invariant) definition of a linear code as a system of points in a linear or projective space; we believe that this approach is natural and fruitful. In Sec. 1.1.3 we analyze spectra of linear codes and introduce notions of dual and self-dual codes. Section 1.1.4 is devoted to the question of how good the parameters of a code can be, i.e., to bounds on the parameters. In Sec. 1.1.5 the same question for higher weights is considered. We end (in Sec. 1.1.6) with discussing spectra of higher weights.

1.1.1. Definition of a Code

Let A be a finite set, called an *alphabet*. The set $A^n = A \times \ldots \times A$ is equipped with the *Hamming metric*: the distance $d(a,b)$ is the number of coordinates at which a and b differ, i.e.,

$$d((a_1,\ldots,a_n),(b_1,\ldots,b_n)) = |\{i \mid a_i \neq b_i\}|.$$

By $q = |A|$ we denote the cardinality of the alphabet.

A nonempty subset $C \subseteq A^n$ is called a *q-ary code of length n*. The *cardinality* of a code is $M = |C| \in \mathbb{N}$, and the *log-cardinality* is $k = \log_q |C| \in \mathbb{R}$. The *minimum distance* of C is the nonnegative integer

$$d = \min\{d(a,b) \mid a,b \in C,\ a \neq b\}.$$

A code with these parameters is called an $[n,k,d]_q$ *code* (a coding theorist would call it a block error-correcting $[n,k,d]_q$ code). Elements of C are called *code vectors*, or *codewords*; their components are called *coordinates*, or *positions*.

Sometimes (especially in asymptotic problems; see Sec. 1.3), the following relative parameters are convenient to use: the *code rate* (*transmission rate*, or simply *rate*) $R = k/n$ and the *relative minimum distance* $\delta = d/n$. Of course, $0 \leq R \leq 1$ and $0 \leq \delta \leq 1$.

REMARK 1.1.1. In coding theory there is a good tradition assigning the symbols C, n, k, d, q, R, M, and δ to the defined notions. We try to follow this tradition wherever possible.

REMARK 1.1.2. Let us briefly explain why such a subset is called the *error-correcting code*. The notion appeared in information theory. Let us start with a given long enough message written in the alphabet A. When this message is transmitted over a channel, it is distorted by random fluctuations (noise).[1] Here is a way out. Take an $[n,k,d]_q$ code C. Let (for simplicity) k be an integer. Cut the message into pieces of length k each. Map every word of length k (in total, there are q^k possibilities) to an element of C—i.e., fix an embedding $E \colon A^k \hookrightarrow A^n$, $E(A^k) = C$—and, instead of a length-k piece a of the message, transmit the corresponding word $E(a)$ of length n (we encode the message; the parameter $n-k$ is called the *redundancy* of a code). The transmission is now $1/R$ times slower, which justifies the term *rate* for R. At the other end of the channel, we obtain a distorted word $E(a)' \in A^n$ and transform it into the nearest word $E(a)'' \in C$; this transformation (*maximum likelihood decoding*) is a mapping of sets $D \colon A^n \to C$. If the number of distortions is at most $\left\lfloor \dfrac{d-1}{2} \right\rfloor$, then $E(a)'' = E(a)$; i.e., the decoding is correct. The maximum likelihood decoding is almost an ideal that is out of reach. Usually we just have a map $D \colon U \to C$, where $U \subseteq A^n$ is the set of words that are at distance at most $t \leq \left\lfloor \dfrac{d-1}{2} \right\rfloor$ from C; i.e., U is the union of all balls of radius t centered in the elements of C. In this case we speak about *decoding up to t* (or correcting t errors). Usually $t = \left\lfloor \dfrac{d-1}{2} \right\rfloor$, but sometimes it is less. Set $\tau = t/n$. If the error probability per transmitted symbol is $p < \tau$ and the length n is large enough, then the probability of correct decoding is large enough (since the probability that the number of errors in n transmitted symbols is at most t is large enough).

Therefore, a "good" code is an $[n,k,d]_q$ code with large n and with R and δ as large as possible.

Both for constructing good codes and for designing algorithms that realize coding and decoding procedures, the notion of a code over an arbitrary alphabet is too poor in structure. It is possible to enrich this structure by introducing the notion of a linear code.

Now let $A = \mathbb{F}_q$ be a finite field, $q = p^m$. A *linear code* C of length n is a linear subspace in $\mathbb{F}_q^n$. For a linear code we have

$$k = \dim C, \qquad d = \min\{\|a\| \mid a \in C,\ a \neq 0\},$$

---

[1] Note that a communication channel can be either spatial (say, a telephone line) or temporal (say, hard-disk information storage). In the first case, noise is interference in the channel; in the second case, it is data contamination in the course of time: ink loses colour, paper yellows, the disk is exposed to cosmic rays, etc.

where $\|a\| = |\{i \mid a_i \neq 0\}|$ is the *weight* (*Hamming weight*, or *norm*) of a vector from $\mathbb{F}_q^n$. For the weight, the notation $\mathrm{wt}(a)$ is also used. In this case k is said to be the *dimension* of the linear code.

A choice of basis in C yields an embedding $C\colon \mathbb{F}_q^k \to \mathbb{F}_q^n$, which we usually denote by the same symbol as the code itself. The matrix of this map, usually denoted by G, is called a *generator matrix* of the code; linear combinations of its rows are precisely the code vectors. The map C is included into a short exact sequence
$$0 \longrightarrow \mathbb{F}_q^k \xrightarrow{C} \mathbb{F}_q^n \xrightarrow{H} \mathbb{F}_q^{n-k} \longrightarrow 0$$
(i.e., C is an injection, H is a surjection, and the kernel of H coincides with the image of C). It would be even more correct to say that we have $H\colon \mathbb{F}_q^n \to V$, where $\dim_{\mathbb{F}_q} V = n-k$; to define H as a matrix, that is, to write $H\colon \mathbb{F}_q^n \to \mathbb{F}_q^{n-k}$, we should fix a basis in V.

The matrix of the map H (which is also denoted by H) is called a *parity-check matrix* of C (since the condition $x \in C$ is equivalent to the equality $Hx^T = 0$).

EXERCISE 1.1.3. Let H be a parity-check matrix of a linear $[n,k,d]_q$ code C. Show that any $d-1$ columns of H are linearly independent (as vectors in $\mathbb{F}_q^{n-k}$) but there exist d linearly dependent columns.

According to our definition, H has n columns and $n-k$ linearly independent rows. Sometimes, by abuse of language, any matrix H' such that $H'x^T = 0$ if and only if $x \in C$ is also called a parity-check matrix. Such a matrix H' has $r \geq n-k$ rows, of which only $n-k$ are independent.

Equivalence and automorphisms. Denote by $\mathcal{A}$ the subgroup in the group of linear automorphisms of $\mathbb{F}_q^n$ generated by permutations of coordinates and multiplications of coordinates x_i by elements $a_i \in \mathbb{F}_q^*$. The group $\mathcal{A}$ acts on subsets of $\mathbb{F}_q^n$, i.e., on codes. Two codes, C and C', are called *equivalent* if $C' = A(C)$ for some $A \in \mathcal{A}$. The subgroup $\mathrm{Aut}\, C \subseteq \mathcal{A}$ consisting of elements that fix the code is called the *automorphism group* of the code. If is natural to consider codes up to equivalence; in many cases, speaking of a code, we rather mean its equivalence class. The group $\mathcal{A}$ is represented by monomial matrices (i.e., matrices in which each row and each column contain exactly one nonzero element) and is isomorphic to the semi-direct product of the groups $(\mathbb{F}_q^*)^n$ and S_n; we have $|\mathcal{A}| = (q-1)^n n!$.

If C is a linear $[n,k,d]_q$ code, it can be defined by its generator matrix G. A choice of G corresponds to a choice of a basis $\{e_1, \ldots, e_k\}$ in a k-dimensional linear space. The group $\mathrm{GL}(k, \mathbb{F}_q)$ acts on the set of such bases; two matrices, G and G', define the same code if and only if $G' = BG$ for some $B \in \mathrm{GL}(k, \mathbb{F}_q)$.

1.1.2. $[n,k,d]_q$ Systems

The notion of a linear code can be reformulated in the following elegant way.

Let V be a linear space over a finite field $\mathbb{F}_q$. By an $[n,k,d]_q$ *system* we call an ordered finite family $\mathcal{P}$ of points in V (in general, points of $\mathcal{P}$ need not be distinct) such that $\mathcal{P}$ does not lie in a hyperplane; obviously, $|\mathcal{P}| \geq \dim V$. The parameters of the system are defined as a triple of integers $[n,k,d]$ with
$$n = |\mathcal{P}|, \qquad k = \dim V, \qquad d = n - \max_H |\mathcal{P} \cap H| \geq 1$$

(the maximum is taken over all hyperplanes $H \subset V$; points are counted with their multiplicities).

Two $[n,k,d]_q$ systems, $\mathcal{P}$ and $\mathcal{P}'$, in V and V', respectively, are called *equivalent* if there is an isomorphism $V \xrightarrow{\sim} V'$ taking $\mathcal{P}$ to $\mathcal{P}'$. A natural object to consider is rather a class of equivalent systems.

PROPOSITION 1.1.4. *There is a one-to-one correspondence between the set of classes of $[n,k,d]_q$ systems and the set of linear $[n,k,d]_q$ codes.*

SKETCH OF THE PROOF. Let V^* be the space of linear forms on the linear space V, $(V^*)^* = V$. Set $\mathcal{P} = (P_1, \ldots, P_n)$. Consider the map $\varphi \colon V^* \to \mathbb{F}_q^n$ defined by $\varphi(Q) = (\varphi_1(Q), \ldots, \varphi_n(Q))$, $\varphi_i(Q) = Q(P_i)$. Then φ is injective; put $C = \operatorname{Im} \varphi$. Conversely, if $C \subseteq \mathbb{F}_q^n$, coordinate forms define elements of $V = C^*$ (columns of a generator matrix of the code). □

EXERCISE 1.1.5. Give a complete proof, checking in particular that the parameters do coincide.

Projective systems. Let us introduce a more invariant object. Let $\mathbb{P} = \mathbb{P}(V)$ be a projective space (for a definition, see Sec. 2.1.1) over $\mathbb{F}_q$. A *projective $[n,k,d]_q$ system* is a finite unordered family $\mathcal{P}$ of points in $\mathbb{P}$ that does not lie in any (projective) hyperplane H (note that $|\mathcal{P}| \geq \dim \mathbb{P} + 1 = \dim V$). By abuse of notation, we write $\mathcal{P} \subset \mathbb{P}$ (though, in general, there are multiplicities). The parameters $[n,k,d]$ are defined as follows:

$$n = |\mathcal{P}|, \qquad k = \dim \mathbb{P} + 1, \qquad d = n - \max_H |\mathcal{P} \cap H| \geq 1.$$

Just as for $[n,k,d]_q$ systems, two projective systems, $\mathcal{P} \subset \mathbb{P}$ and $\mathcal{P}' \subset \mathbb{P}'$, are called *equivalent* if there is a (projective) isomorphism $\mathbb{P} \cong \mathbb{P}'$ that takes $\mathcal{P}$ to $\mathcal{P}'$.

We call a linear code $C \subset \mathbb{F}_q^n$ *degenerate* if $C \subseteq \mathbb{F}_q^{n-1} \subset \mathbb{F}_q^n$, where $\mathbb{F}_q^{n-1}$ is the subspace of vectors with 0 at some fixed position. Otherwise, we say that the code is *nondegenerate*.

THEOREM 1.1.6. *Let $k \geq 1$ and $d \geq 1$. There is a one-to-one correspondence between the set of equivalence classes of nondegenerate linear $[n,k,d]_q$ codes and the set of equivalence classes of projective $[n,k,d]_q$ systems.*

SKETCH OF THE PROOF. Given a nondegenerate linear $[n,k]$ code C, we can construct a projective system as follows. Let us consider the coordinate functions $x_i \colon C \to \mathbb{F}_q$ such that

$$x_i \colon (c_1, \ldots, c_n) \mapsto c_i.$$

The restrictions $x_i|_C$ are linear functions on C, i.e., elements of the dual linear space C^*. Since C is nondegenerate, all the $x_i|_C$ are nonzero, i.e., correspond to n points P_i in the projective space $\mathbb{P}C^* = \mathbb{P}^{k-1}$.

Here, a nonzero codeword $c \in C$ corresponds to a hyperplane $H(c) \subset \mathbb{P}^{k-1}$, and vice versa. The weight of a codeword is the number of coordinate forms that do not vanish on this word, i.e.,

$$\|c\| = |\{P_i \mid P_i \notin H(c)\}|.$$

Note that, since $\|c\| > 0$ for $c \neq 0$, the points P_i do not lie in a hyperplane.

Conversely, given an arbitrary nondegenerate projective system $\{P_1, \ldots, P_n\}$ in $\mathbb{P}^{k-1} = \mathbb{P}V$, we can construct a corresponding linear code. To this end, we lift the

points $P_i \in \mathbb{P}V$ to vectors $v_i \in V$ (in an arbitrary way) and consider the map
$$(v_1,\ldots,v_n)\colon V^* \to \mathbb{F}_q^n.$$
The image of this map is a linear $[n,k]$ code. □

EXERCISE 1.1.7. Give a complete proof.

Projective systems have an advantage of dispensing with a choice of some particular code in its equivalence class; also, they look more natural than codes since no choice of basis in $\mathbb{F}_q^n$ is involved. Besides, the problem of possible parameters of a projective system looks quite natural, being just a question of how general a position of n points in $\mathbb{P}^{k-1}$ can be. Of course, this question can be posed over an arbitrary field. If the field is infinite, the answer is as follows: in $\mathbb{P}^{k-1}$ there always are n points in the general position, which means that any k of them are vertices of a nondegenerate simplex (EXERCISE!).

REMARK 1.1.8. Sometimes, it is indeed necessary to consider systems with multiplicities (for instance, when constructing codes whose parameters meet the Griesmer bound; cf. Theorem 1.1.43.)

The following research problem looks rather important and interesting.

PROBLEM 1.1.9. Rewrite existing books on coding theory in terms of projective systems (you may start with this chapter).

Dual systems. Let us present another approach. A *dual* $[n,k,d]_q$ *system* is a finite ordered family $\mathcal{Q}$ of points in a linear space W (multiplicities are again allowed) which does not lie in any hyperplane $H \subset W$. The parameters are as follows: $n = |\mathcal{Q}|$, $k = n - \dim W$ (note that $k \geq 0$), and d is the minimum number of linearly dependent vectors of $\mathcal{Q}$ (in particular, if $\mathcal{Q}$ includes multiplicities, then $d \leq 2$).

EXERCISE 1.1.10. Define a *dual projective* $[n,k,d]_q$ *system* and a proper notion of equivalence. Then prove the following theorem.

THEOREM 1.1.11. *There is a one-to-one correspondence between the set of equivalence classes of dual $[n,k,d]_q$ systems and the set of linear $[n,k,d]_q$ codes, and also between the classes of dual projective $[n,k,d]_q$ systems and the set of equivalence classes of nondegenerate linear $[n,k,d]_q$ codes.* □

n-sets. The language of projective systems (or dual projective systems) immediately shows that $k + d \leq n + 1$ (this inequality is known as the *Singleton bound*; cf. Proposition 1.1.41). Indeed, for any set of $k-1$ points in $\mathbb{P}^{k-1}$ there is a hyperplane passing through them; hence, $n - d \geq k - 1$ (an argument for dual systems: no $n - k + 1$ points in $\mathbb{P}^{n-k-1}$ can be in the general position). A projective $[n, k, n-k+1]_q$ system is called an *n-set*. The corresponding code C is called a *maximum distance separable code* (an *MDS code*).

In Secs. 1.2.1 and 4.4.1 we present some examples of such codes, namely, the Reed–Solomon codes and algebraic geometry codes on a projective line; their length, n, is at most $q + 1$.

We get a very interesting problem: for given k and q, find the maximum possible length $m_q(k)$ for which an $[m_q(k), k, m_q(k) + 1 - k]_q$ code exists.

Let us present some known results (omitting proofs):

- There always exist $[n,1,n]_q$, $[n,n,1]_q$, and $[n,n-1,2]_q$ codes (cf. Sec. 1.2.1); therefore,
$$m_q(1) = \infty, \qquad m_q(k) \geq k+1.$$
- $m_q(k) \geq q+1$.
- If q is even, then
$$m_q(3) \geq q+2, \qquad m_q(q-1) \geq q+2.$$
- If $k \geq 2$, then $m_q(k) \leq q+k-1$.
- If $k \geq 3$ and q is odd, then $m_q(k) \leq q+k-2$.
- If $k \geq q$, then $m_q(k) = k+1$.

PROBLEM 1.1.12 (the main conjecture on MDS codes). Prove that
$$m_q(k) = q+1$$
for $2 \leq k < q$, except for the case $m_q(3) = m_q(q-1) = q+2$ for q even.

This conjecture is proved whenever $q \leq 11$ or $k \leq 5$. It is also proved in a number of cases where k is less than a certain function of q, for example, for q prime and $k < \dfrac{q}{45} + \dfrac{37}{9}$, or $q = p^m$, $p \geq 5$, and $k < \sqrt{q}/2$, etc. These results are obtained by the theory of finite geometries. The main conjecture is very beautiful, but it seems that we are rather far from its proof.

Let $g \geq 0$. Denote by $m_q(g,k)$ the maximum length for which there exists an $[m_q(g,k), k, m_q(g,k)+1-k-g]_q$ code. In Sec. 4.4 we construct algebraic geometry codes which give some lower bounds on $m_q(g,k)$ for small g. The question on the precise value of $m_q(g,k)$ is most likely very difficult; it would be interesting to find some good bounds.

Generalized weights. Generalized Hamming weights (also called higher weights, support weights, effective lengths, or Wei weights), to be defined below, are another family of parameters of a linear code, not less natural than the minimum distance d.

Let $D \subset \mathbb{F}_q^n$ be a linear subspace of dimension r. The Hamming weight, defined on vectors in $\mathbb{F}_q^n$, is naturally generalized to the set of subspaces, namely,
$$\|D\| = |\mathrm{Supp}(D)|, \quad \mathrm{Supp}(D) = \{i : \exists\, v \in D,\ v_i \neq 0\}. \tag{1.1.1}$$

For a linear code C, define its r-th generalized weight d_r, $r = 1, \ldots, k$, as
$$d_r = d_r(C) = \min\{\|D\| :\ D \subset C,\ \dim D = r\}. \tag{1.1.2}$$

Obviously, $d_1 = d$ is the minimum distance of the code.

In geometric terms, the definition of higher weights $d_r = d_r(\mathcal{P})$ for projective systems is even more natural:
$$n - d_r = \max\{|\mathcal{P} \cap \Pi^r| :\ \Pi^r \text{ is a projective subspace of codimension } r \text{ in } \mathbb{P}\}$$
(cf. the definition of d in the beginning of this subsection).

If, from the mathematical point of view, the parameter d of projective systems deserves study, so does d_r. This is proved by both existing theorems and open problems.

There is a simple relation between the weight $\|D\|$ of a subspace and weights of vectors of this subspace. If $\dim D = r$, then

$$\|D\| = \frac{1}{q^r - q^{r-1}} \sum_{v \in D} \|v\|. \tag{1.1.3}$$

Indeed, among n coordinate forms x_i there are $\|D\|$ forms such that the map $x_i \colon D \to \mathbb{F}_q$ is surjective. Since the map is linear, the proportion of nonzero symbols at the ith position of vectors from D is $(q-1)/q$.

PROPOSITION 1.1.13. *For a nondegenerate projective system (respectively, for a code with $d_1 \geq 1$), we have*

$$0 < d_1 < d_2 < \ldots < d_k = n.$$

PROOF. Let $\mathcal{P}$ be a nondegenerate projective system. Then $d_1 \geq 1$ since outside any hyperplane there is a point of $\mathcal{P}$. Next, there is a subspace Π^r of codimension r such that there are d_r points of $\mathcal{P}$ outside it. If we pass a subspace Π^{r-1} through Π^r and one extra point of $\mathcal{P}$, we see that $d_r \leq d_{r-1} - 1$. Finally, since a subspace of codimension k is empty, we have $d_k = n$. □

The following theorem states an equivalence between the two languages.

THEOREM 1.1.14. *There is a natural one-to-one correspondence between the set of equivalence classes of nondegenerate projective systems and the set of equivalence classes of nondegenerate linear codes. The correspondence preserves the parameters $n, k, d_1, \ldots, d_k$.*

SKETCH OF THE PROOF. The correspondence is described above (see Theorem 1.1.6); let us check that the parameters match. For n and k this is immediate. Let us check this for d_i. A hyperplane in $\mathbb{P}$ is given by a linear form and therefore corresponds to a nonzero codeword of C. Hence, a subspace of codimension r in $\mathbb{P}$ corresponds to a subcode D of dimension r in C. The weight of D is the number of coordinate forms nonvanishing on D, i.e., the number of points of $\mathcal{P}$ that the space does not pass through. □

The collection of numbers $d_1, d_2, \ldots, d_k$ is called the (*weight*) *hierarchy* of the code.

1.1.3. Spectra and Duality

An important invariant of a code is its *weight enumerator*, or *spectrum*. We are going to study spectra of linear codes.

Let C be a linear $[n, k, d]_q$ code. Let $A_r = A_r(C)$ be the number of vectors of weight r in C. Obviously, $A_0 = 1$, $A_r \geq 0$, $A_r = 0$ for $0 < r < d$, and $\sum_{r=0}^{n} A_r = q^k$.

The *enumerator* is a homogeneous polynomial

$$W_C(x:y) = \sum_{r=0}^{n} A_r x^{n-r} y^r.$$

It is easily seen that $W_C(x:y) = \sum_{v \in C} x^{n-\|v\|} y^{\|v\|}$.

Sometimes it is more convenient to pass to nonhomogeneous coordinates and consider the polynomials
$$W_C(x) = \sum_{r=0}^{n} A_r x^{n-r} \quad \text{or} \quad W'_C(y) = \sum_{r=0}^{n} A_r y^r.$$

A code C has exactly one vector of weight 0 and no other vectors of weights less than d. Therefore, $A_0 = 1$, $A_1 = \ldots = A_{d-1} = 0$, and $A_d \neq 0$; i.e.,
$$W_C(x:y) = x^n + y^d \sum_{i=0}^{n-d} A_{d+i} y^i x^{n-d-i}.$$

Since in many cases we do not know the precise value of d but only know some lower estimate for it, the following form is rather convenient. Let a be some integer such that $d \geq n - a$. Then
$$W_C(x) = x^n + \sum_{i=0}^{a} A_{n-i} x^i.$$

EXERCISE 1.1.15. Show that if $d \geq n - a$, then
$$W_C(x) = x^n + \sum_{i=0}^{a} B_i (x-1)^i,$$

where
$$B_i = \sum_{j=n-a}^{n-i} \binom{n-j}{i} A_j \geq 0, \quad A_i = \sum_{j=n-i}^{a} (-1)^{n+i+j} \binom{j}{n-i} B_j.$$

Below (Exercise 1.1.29) we explain the conceptual meaning of B_i in terms of linear systems, and in Sec. 4.1.2 we give an interpretation of these numbers in the case of algebraic geometry codes.

For a linear $[n, k, d]_q$ code C, its *dual* code is defined as
$$C^\perp = \{x \in \mathbb{F}_q^n \mid xy = 0 \text{ for all } y \in C\}$$
(here, xy denotes the standard inner product, $xy = \sum_{i=1}^{n} x_i y_i$).

REMARK 1.1.16. One can also consider duality with respect to the *Hermitian* product $xy = \sum x_i \bar{y}_i$, where $\bar{a}$ for $a \in \mathbb{F}_q$, $q = p^e$, is defined (for even e) as $\bar{a} = a^{p^{e/2}}$; sometimes, $xy = \sum x_i y_i^p$ is also considered. The choice of a definition does not influence the results presented below. The same holds for the *twisted* inner product $xy = \sum x_i y_i \alpha_i$, where $\alpha_i \in \mathbb{F}_q^*$.

The parameters of the dual code $C^\perp$ are $[n, k^\perp = n-k, d^\perp]_q$; a generator matrix of the code is a parity-check matrix of its dual and vice versa; the dual distance $d^\perp$ depends (not in a simple way) on the equivalence class of the code C but not only on its parameters $[n, k, d]_q$. It can be computed if we know the enumerator W_C.

Moreover, there is a beautiful relation between the spectrum of a code and that of the dual one:

THEOREM 1.1.17 (MacWilliams identity).
$$W_{C^\perp}(x:y) = q^{-k} W_C(x + (q-1)y : x - y).$$

We prove this theorem later; now let us introduce one more notion, motivated, as will further be seen, by algebraic geometry constructions. Let g be a nonnegative integer.

A code C is called a *code of genus at most g* if the relations
$$k+d \geq n+1-g,$$
$$(n-k)+d^\perp \geq n+1-g$$
hold (Exercise 1.1.3 shows that for any linear code we have $k+d \leq n+1$ [the Singleton bound]; therefore, we consider only nonnegative values of g).

Of course, in this case $C^\perp$ is also a code of genus at most g.

THEOREM 1.1.18. *Let C be an $[n,k,d]_q$ code of genus at most g. Let $a = k+g-1$. Then the coefficients B_i in the representation*
$$W_C(x) = x^n + \sum_{i=0}^{a} B_i(x-1)^i$$
satisfy the inequalities
$$\binom{n}{i}(q^{a-i+1}-1) \geq B_i \geq \max\left\{0, \binom{n}{i}(q^{a-i-g+1}-1)\right\}$$
for $a-2g+2 \leq i \leq a$, and
$$B_i = \binom{n}{i}(q^{a-i-g+1}-1)$$
for $i \leq a-2g+1$.

PROOF. Apply Theorem 1.1.17:
$$W_{C^\perp}(x:y) = q^{-k} W_C(x+(q-1)y : x-y)$$
$$= q^{-k}\left[(x+(q-1)y)^n + \sum_{i=0}^{a} B_i(qy)^i(x-y)^{n-i}\right].$$

Passing to nonhomogeneous coordinates and setting $a^\perp = n-k+g-1$ and
$$W_{C^\perp}(x) = x^n + \sum_{i=0}^{a^\perp} B'_i(x-1)^i,$$
we obtain
$$x^n + \sum_{i=0}^{a^\perp} B'_i(x-1)^i = q^{-k}\left[(x+q-1)^n + \sum_{i=0}^{a} B_i q^i(x-1)^{n-i}\right].$$

Expanding in powers of $z = x-1$ yields
$$\sum_{i=0}^{a^\perp}\left(B'_i + \binom{n}{i}\right)z^i + \sum_{i=a^\perp+1}^{n}\binom{n}{i}z^i$$
$$= \sum_{i=0}^{n-a-1}\binom{n}{i}q^{n-k-i}z^i + \sum_{i=n-a}^{n}\left(B_{n-i} + \binom{n}{i}\right)q^{n-k-i}z^i.$$

Hence, for $n - a \leq i \leq n$ we have

$$B_{n-i} = \begin{cases} \binom{n}{i}(q^{-n+k+i} - 1) & \text{if } i \geq a^\perp + 1, \\ \binom{n}{i}(q^{-n+k+i} - 1) + B'_i q^{-n+k+i} & \text{if } i \leq a^\perp; \end{cases}$$

that is (we put $j = n - i$ and substitute $k = a - g + 1$ and $a^\perp = n - k + g - 1 = n - a + 2g - 2$),

$$B_j = \begin{cases} \binom{n}{j}(q^{a-j-g+1} - 1) & \text{if } j \leq a - 2g + 1, \\ \binom{n}{j}(q^{a-j-g+1} - 1) + B'_{n-j} q^{a-j-g+1} & \text{if } a - 2g + 2 \leq a. \end{cases}$$

This proves the desired equality and the lower bound. The upper bound will be proved below after we interpret the parameters B_i in terms of linear $[n, k, d]_q$ systems (see Theorem 1.1.28 and Exercise 1.1.29). $\square$

REMARK 1.1.19. It is easily seen that $B_k = B'_{n-k}$.

EXERCISE 1.1.20. Show that if $k + d \geq n + 1$, then we have $k + d = n + 1$ and $(n - k) + d^\perp = n + 1$; i.e., the code is of genus 0 (an MDS code).

EXAMPLE 1.1.21. Consider the $[4, 2, 1]_2$ code C with generator matrix

$$\begin{pmatrix} 1 & 0 & 0 & 0 \\ 0 & 1 & 0 & 0 \end{pmatrix}.$$

The dual code has the same parameters. The spectrum is $A_0 = 1$, $A_1 = 2$, $A_2 = 1$, and $A_3 = A_4 = 0$. This is a code of genus at most 2; $a = 3$; the computation of B_i gives $B_3 = A_1 = 2$, $B_2 = 3A_1 + A_2 = 7$, $B_1 = 3A_1 + 2A_2 + A_3 = 8$, and $B_0 = A_1 + A_2 + A_3 + A_4 = 3$. Note that in this case we have $B_i > \binom{n}{i}(q^{\lfloor (a-i)/2 \rfloor + 1} - 1)$ for $i = 2$; this observation will be used in Sec. 4.1.2.

EXAMPLE 1.1.22. Consider an $[n, k_0, d_0]_q$ code C_0 of genus zero. Let H_0 be its parity-check matrix. Let us add to H_0 one more row of weight w which is not a linear combination of rows of H_0. We obtain a matrix H such that any $d_0 - 1$ columns of H are linearly independent (this holds already for H_0). The parameters of the corresponding $[n, k, d]_q$ code C with the parity-check matrix H are $k = k_0 - 1$ and $d \geq d_0$; i.e., $k + d \geq n$. Now let $w < n - d$; then $d^\perp \leq w < n - d$ (since the added row is a code vector of $C^\perp$), and therefore $d^\perp + k^\perp = d^\perp + n - k < 2n - d - k \leq n$. If we assume that $k + d > n$, we get $k + d = k^\perp + d^\perp = n + 1$. Hence, we have constructed a code C with $k + d = n$ which is not a code of genus at most 1 (since $d^\perp + k^\perp < n$).

EXERCISE 1.1.23. Consider a $[6, 4, 2]_2$ code C with the generator matrix

$$\begin{pmatrix} 1 & 1 & 0 & 0 & 0 & 0 \\ 0 & 1 & 1 & 0 & 0 & 0 \\ 0 & 0 & 0 & 1 & 1 & 0 \\ 0 & 0 & 0 & 0 & 1 & 1 \end{pmatrix}.$$

Compute the spectra of C and $C^\perp$ in terms of B_i. (Answer: $B_0 = 15$, $B_1 = 42$, $B_2 = 45$, $B_3 = 24$, $B_4 = 6$; $B'_0 = 3$, $B'_1 = B'_2 = 6$, $B'_3 = 2$.)

Before proceeding to the proof of Theorem 1.1.17, let us recall that an *additive character* of $\mathbb{F}_q$ is a homomorphism ψ from the additive group $\mathbb{F}_q$ to the multiplicative group $\mathbb{C}^*$. It is easily seen that the image of such a homomorphism lies in the group μ_p of pth roots of unity.

EXERCISE 1.1.24. Prove that the group of additive characters of $\mathbb{F}_q$ is isomorphic to $\mathbb{F}_q$ and $0 \in \mathbb{F}_q$ corresponds to the trivial character $\psi_0(a) \equiv 1$.

LEMMA 1.1.25. *For any nontrivial additive character* $\psi \colon \mathbb{F}_q \to \mu_p$, *we have*

$$\sum_{a \in \mathbb{F}_q} \psi(ab) = \begin{cases} 0 & \text{for } b \neq 0, \\ q & \text{for } b = 0. \end{cases}$$

PROOF. Let $b_0 \neq 0$; choose a_0 such that $\psi(a_0 b) \neq 1$. Then

$$\psi(a_0 b) \sum_{a \in \mathbb{F}_q} \psi(ab) = \sum_{a \in \mathbb{F}_q} \psi((a_0 + a)b) = \sum_{a' \in \mathbb{F}_q} \psi(a' b)$$

since shifting by a_0 bijectively maps $\mathbb{F}_q$ onto itself. If $b = 0$, then $\psi(ab) = 1$ for any a. □

Fix a nontrivial character ψ_1. For $u, v \in \mathbb{F}_q^n$, let $uv \in \mathbb{F}_q$ be their inner product. Define

$$\psi_u(v) = \psi_v(u) = \psi_1(uv);$$

then $\psi_u \colon \mathbb{F}_q^n \to \mu_p$ is an additive character of the group $\mathbb{F}_q^n$.

Let A be an arbitrary $\mathbb{Z}[\mu_p]$-module. For a function $f \colon \mathbb{F}_q^n \to A$, define the transform

$$\widehat{f}(u) = \sum_{v \in \mathbb{F}_q^n} \psi_u(v) f(v).$$

LEMMA 1.1.26. *For any linear subspace* $C \subseteq \mathbb{F}_q^n$ *we have*

$$\sum_{u \in C^\perp} f(u) = \frac{1}{|C|} \sum_{u \in C} \widehat{f}(u).$$

PROOF. Indeed,

$$\sum_{u \in C} \widehat{f}(u) = \sum_{v \in \mathbb{F}_q^n} \sum_{u \in C} \psi_u(v) f(v)$$

$$= \sum_{v \in C^\perp} f(v) \sum_{u \in C} \psi_v(u) + \sum_{v \notin C^\perp} f(v) \sum_{u \in C} \psi_v(u). \qquad (1.1.4)$$

If $v \in C^\perp$, then $uv = 0$ for all $u \in C$; thus,

$$\sum_{u \in C} \psi_v(u) = \sum_{u \in C} \psi_1(0) = |C|.$$

Let v be such that there exists $u_0 \in C$ with $\psi_v(u_0) \neq 1$. Then

$$\psi_v(u_0) \sum_{u \in C} \psi_v(u) = \sum_{u \in C} \psi_v(u_0 + u) = \sum_{u \in C} \psi_v(u)$$

since $u_0 + C = C$; hence, $\sum_{u \in C} \psi_v(u) = 0$.

It remains to prove that if $\psi_v(u) = 1$ for any $u \in C$, then $v \in C^\perp$. Indeed, in this case $\psi_v(\alpha u) = \psi_1(\alpha(uv)) = 1$ for any $\alpha \in \mathbb{F}_q$, but there is $\beta \in \mathbb{F}_q^*$ such that

$\psi_1(\beta) \neq 1$; setting $\alpha = \beta(uv)^{-1}$, we get a contradiction, i.e., $uv = 0$. Vanishing of the second term in (1.1.4) proves the lemma. $\square$

In the course of the proof we have also established the following fact.

COROLLARY 1.1.27. *For any linear subspace* $C \subseteq \mathbb{F}_q^n$,
$$\sum_{u \in C} \psi_v(u) = \begin{cases} 0 & \text{for } v \notin C^\perp, \\ |C| & \text{for } v \in C^\perp. \end{cases}$$
$\square$

PROOF OF THEOREM 1.1.17. Consider $f \colon \mathbb{F}_q^n \to A$, where
$$A = \mathbb{C}[x,y], \qquad f(u) = x^{n-\|u\|} y^{\|u\|}.$$
Then the left-hand side of the identity of Lemma 1.1.26 equals $W_{C^\perp}(x:y)$. Let us calculate $\widehat{f}(u)$. Let $u = (u_1, \ldots, u_n)$ and $v = (v_1, \ldots, v_n)$; using the fact that $v_i^{q-1} = 1$ for $v_i \neq 0$ and Lemma 1.1.26, we get

$$\widehat{f}(u) = \sum_{v \in \mathbb{F}_q^n} \psi_u(v) x^{n-\|v\|} y^{\|v\|} = \sum_{v \in \mathbb{F}_q^n} \psi_1(u_1 v_1 + \ldots + u_n v_n) x^{n-\|v\|} y^{\|v\|}$$

$$= \sum_{v \in \mathbb{F}_q^n} \prod_{i=1}^n \psi_1(u_i v_i) x^{1-v_i^{q-1}} y^{v_i^{q-1}}$$

$$= \sum_{v_1 \in \mathbb{F}_q} \psi_1(u_1 v_1) x^{1-v_1^{q-1}} y^{v_1^{q-1}} \sum_{v_2 \in \mathbb{F}_q} \cdots$$

$$= \prod_{i=1}^n \sum_{v_i \in \mathbb{F}_q} \psi_1(u_i v_i) x^{1-v_i^{q-1}} y^{v_i^{q-1}} = \prod_{i=1}^n \left(x + y \sum_{v_i \in \mathbb{F}_q^*} \psi_1(u_i v_i) \right)$$

$$= \prod_{i=1}^n \left(x - y + y \sum_{\alpha \in \mathbb{F}_q} \psi_1(u_i \alpha) \right)$$

$$= \prod_{i=1}^n \left((x-y)^{u_i^{q-1}} (x+(q-1)y)^{1-u_i^{q-1}} \right)$$

$$= (x + (q-1)y)^{n-\|u\|} (x-y)^{\|u\|},$$

which proves the theorem. $\square$

Here is a strengthening of Theorem 1.1.18 (for an arbitrary linear code).

THEOREM 1.1.28. *Let C be a linear $[n, k, \geq d]_q$ code, and let the minimum distance of the dual code be at least $d^\perp$. Then*
$$W_C(x) = x^n + \sum_{i=0}^{n-d} B_i (x-1)^i,$$
where for $d^\perp - 1 \geq i \geq 0$ we have
$$B_i = \binom{n}{i}(q^{k-i} - 1),$$
and for $n - d \geq i \geq d^\perp$,
$$\binom{n}{i}\left(q^{\min\{n-d-i+1, k-d^\perp+1\}} - 1\right) \geq B_i \geq \max\left\{0, \binom{n}{i}(q^{k-i} - 1)\right\}.$$
$\square$

EXERCISE 1.1.29. Check the following interpretation of B_i in terms of linear systems. Let $\mathcal{P} = \{P_1,\ldots,P_n\}$ be a linear $[n,k,d]_q$ system, $P_i \in V$. Denote by H_i the hyperplane in V^* corresponding to the vector P_i. For any subset $\mathcal{R} \subseteq \mathcal{P}$, let $\ell(\mathcal{R}) = \dim\left(\bigcap_{P_i \in \mathcal{R}} H_i\right)$. Then

$$B_i = \sum_{\substack{\mathcal{R} \subset \mathcal{P} \\ |\mathcal{R}|=i}} \left(q^{\ell(\mathcal{R})} - 1\right).$$

Using this interpretation, prove Theorem 1.1.28.

Self-dual codes. A linear code C is called *self-dual* if $C = C^\perp$. A code C is called *quasi-self-dual* if there exists a vector $y = (y_1,\ldots,y_n) \in \mathbb{F}_q^n$, $y_i \neq 0$ for all $i = 1,\ldots,n$, such that $y \cdot C = C^\perp$. Here,

$$y \cdot C = \{y \cdot c = (y_1 c_1,\ldots,y_n c_n) \mid c \in C\}.$$

In this case, C is also said to be *y-self-dual*. Finally, a code C is called *formally self-dual* if $W_C = W_{C^\perp}$. Of course, any self-dual code is quasi-self-dual, and any quasi-self-dual code is formally self-dual. If $q = 2$, then any quasi-self-dual code is self-dual.

The following fact is quite easy to prove.

EXERCISE 1.1.30. Let $q = 2$ or $q = 3$. Show that weights of all code vectors of a self-dual q-ary code are divisible by q.

Note that this is not the case for formally self-dual codes: a $[2,1,1]_2$ code spanned by the vector $(1,0) \in \mathbb{F}_2^2$ is formally self-dual.

If an $[n,k,d]_q$ code C is (at least formally) self-dual, then $n = 2k$ since $q^k = W_C(1:1)$ is uniquely determined by the enumerator.

In the following, when we consider any questions concerning self-duality, we restrict ourselves to $[n,k,d]_q$ codes with $n = 2k$ only.

The most beautiful theorem on formally self-dual codes is proved with the help of the classical invariant theory. We have to content ourselves with its statement only.

THEOREM 1.1.31. *For any formally self-dual code C there exists a homogeneous polynomial $P(x:y)$ such that*

$$W_C(x:y) = P(x^2 + (q-1)y^2 : y(x-y)). \qquad \square$$

REMARK 1.1.32. Of course, not every polynomial of this form is actually an enumerator of some linear code. For example, the polynomial $y(x-y)$ cannot be an enumerator of a linear code since it lacks the term x^n, corresponding to the zero code vector. Another example of the same kind is $2x^2 + 2(q-1)y^2$. For $q = 2$, the enumerator of any self-dual code must be an even function in x (see Exercise 1.1.30). The question of which polynomials in $x^2 + (q-1)y^2$ and $y(x-y)$ are actually enumerators of self-dual codes (or even of any codes) is very subtle.

Let us present some other results (without proofs).

THEOREM 1.1.33. *The enumerator of a binary self-dual code is a polynomial in $x^2 + y^2$ and $x^2 y^2 (x^2 - y^2)^2$ (the same for a formally self-dual code with all weights even). The enumerator of a ternary self-dual code is a polynomial in $x^4 + 8xy^3$ and*

$y^3(x^3 - y^3)^3$ (the same for a formally self-dual code with all weights divisible by 3). The enumerator of a quaternary formally self-dual code with all weights even is a polynomial in $x^2 + 3y^2$ and $y^2(x^2 - y^2)^2$. The enumerator of a binary formally self-dual code with all weights divisible by 4 is a polynomial in $x^8 + 14x^4y^4 + y^8$ and $x^4y^4(x^4 - y^4)^4$. □

THEOREM 1.1.34. *Assume that all weights of a formally self-dual q-ary code C are divisible by an integer $t > 1$. Then either C is an $[n, n/2, 2]_q$ code and $W_C = ((q-1)x^2 + y^2)^{n/2}$ or $(q,t) = (2,2), (2,4), (3,3),$ or $(4,2)$.* □

EXERCISE 1.1.35. Check that if C is quasi-self-dual with respect to some $y \in (\mathbb{F}_q^*)^n$ and for each i there exists $x_i \in \mathbb{F}_q^*$ such that $y_i = x_i^2$, then $x \cdot C$ is a self-dual code. Check that if $q = 2^m$, then for any quasi-self-dual code C there exists a self-dual code equivalent to C. (*Hint:* Each element in a finite field of characteristic 2 is a square.)

Generalized spectra. The definition of a generalized spectrum is quite similar to the definition of a usual spectrum and can be given in either of the two languages. For a code $C \subseteq \mathbb{F}_q^n$, define
$$A_i^r = |\{D \subseteq C : \dim D = r, \|D\| = i\}|.$$
For a projective system $\mathcal{P} \subseteq \mathbb{P}$, let
$$A_i^r = |\{\Pi \subseteq \mathbb{P} : \operatorname{codim} \Pi = r, |\Pi \cap \mathcal{P}| = n - i\}|.$$

Note that for $i > 0$ we have $A_i^1 = A_i/(q-1)$, where A_i is a usual spectrum since there are $q - 1$ nonzero vectors in a one-dimensional linear space over $\mathbb{F}_q$.

In view of these definitions, Theorem 1.1.14 has the following generalization.

THEOREM 1.1.36. *There is a natural one-to-one correspondence between the set of equivalence classes of nondegenerate projective systems and the set of equivalence classes of nondegenerate linear codes. The correspondence preserves the parameters n, k, $d_1, \ldots, d_k$, and A_i^r.* □

Dual hierarchy. Similarly, for a linear code C (or a projective system $\mathcal{P}$) we get a set of parameters $d_1^\perp, \ldots, d_{n-k}^\perp$, i.e., higher weights of a dual code (dual system), which can as well be characterized in terms of the system $\mathcal{P}$.

THEOREM 1.1.37. *For a projective system $\mathcal{P}$, we have*
$$d_r^\perp = \min\{|\mathcal{Q}| : \mathcal{Q} \subset \mathcal{P}, |\mathcal{Q}| - \dim \operatorname{lin}\langle \mathcal{Q} \rangle = r\},$$
where $\langle \mathcal{Q} \rangle$ is the linear span of $\mathcal{Q}$, i.e., the minimal projective space containing $\mathcal{Q}$, and $\dim \operatorname{lin}\langle \mathcal{Q} \rangle$ is its linear dimension (which is greater by 1 than the projective dimension of $\langle \mathcal{Q} \rangle$).

PROOF. Let $D \subseteq C^\perp$; then elements of D are linear relations between elements of $\mathcal{P}$. Furthermore, they contain precisely $r = \dim D$ independent relations between elements of $\mathcal{Q} = \operatorname{Supp} D \subseteq \mathcal{P}$. Hence, $|\mathcal{Q}| - \dim \operatorname{lin}\langle \mathcal{Q} \rangle = r$. □

Now let us pass to a beautiful relation between the sequences $d_1, \ldots, d_k$ and $d_1^\perp, \ldots, d_{n-k}^\perp$.

THEOREM 1.1.38. *For a projective system (linear code), let $U = \{d_1, \ldots, d_k\}$ and $V = \{n + 1 - d_1^\perp, \ldots, n + 1 - d_{n-k}^\perp\}$. Then*
$$U \cap V = \varnothing \quad \text{and} \quad U \cup V = \{1, 2, \ldots, n\}.$$

PROOF. It suffices to prove that $U \cap V = \emptyset$. For an arbitrary r, consider the system $\mathcal{Q} \subseteq \mathcal{P}$ such that $|\mathcal{Q}| = d_r^\perp$ and $\dim \mathrm{lin} \langle \mathcal{Q} \rangle = d_r^\perp - r$ (see Theorem 1.1.37 above). Let $t = k + r - d_r^\perp$. Since $\Pi^t = \langle \mathcal{Q} \rangle \supset \mathcal{Q}$, we see that $d_t \leq n - d_r^\perp$.

We have to prove that $n + 1 - d_r^\perp$ is not contained in U. Assume the contrary; then, for some $\Delta \geq 0$, we have $d_{t+\Delta} = n - d_r^\perp + 1$ (since $d_t \leq n - d_r^\perp$); i.e., there exists a subspace $\Pi^{t+\Delta} \supset Z$ with $|Z| = d_r^\perp - 1$ and $\dim \mathrm{lin} \langle Z \rangle \leq k - t - \Delta = d_r^\perp - r - \Delta$. Then $|Z| - \dim \mathrm{lin} \langle Z \rangle \geq r + \Delta - 1 \geq r$, and Theorem 1.1.37 yields $|Z| \geq d_r^\perp$, a contradiction. □

We discuss relations between generalized spectra of a code and its dual in Sec. 1.1.6.

1.1.4. Bounds

We have already explained that a good code should have large k and d for a given n. Let q be fixed. For which n, k, and d does there exist a linear $[n,k,d]_q$ code (or any $[n,k,d]_q$ code)? Of course, $0 \leq k \leq n$ and $1 \leq d \leq n$.

We start with a rather strange (though quite useful) statement that, given a good code, we can get a lot of worse ones.

LEMMA 1.1.39 (spoiling lemma). *Assume that there exists a nondegenerate linear $[n,k,d]_q$ code C. Then we can construct a nondegenerate linear code with parameters*

(a) $\qquad\qquad\qquad [n+1, k, d]_q$

and, if $k \geq 1$ and $n > d \geq 2$, then also linear codes with parameters

(b) $\qquad\qquad\qquad [n-1, k-1, d]_q$,

(c) $\qquad\qquad\qquad [n-1, k, d-1]_q$,

(d) $\qquad\qquad\qquad [n, k-1, d]_q$,

(e) $\qquad\qquad\qquad [n, k, d-1]_q$.

PROOF. Let $\mathcal{P} \subset \mathbb{P}^{k-1}$ and $\mathcal{Q} \subset \mathbb{P}^{n-k-1}$ be the projective and dual projective systems corresponding to C (see Sec. 1.1.2).

(a) Choose a hyperplane $H_0 \subset \mathbb{P}^{k-1}$ such that $|H_0 \cap \mathcal{P}| = \max_H |H_0 \cap \mathcal{P}|$. Add to $\mathcal{P}$ one more point from H_0 (it does not matter whether it already belongs to $\mathcal{P}$ or not).

(b) Let $\mathcal{Q}_0 \subset \mathcal{Q}$ be a linearly dependent set of d vectors. Exclude from $\mathcal{Q}$ any vector that does not belong to $\mathcal{Q}_0$ (this is possible since $d < n$). Then n becomes less by 1, and d and $n - k$ do not change.

(c) Choose H_0 as in (a) and exclude from $\mathcal{P}$ any point that does not belong to H_0. Since $d \geq 2$, we again obtain a system (the remaining points cannot all lie in a hyperplane); k and $n - d$ do not change.

(d) Apply (b) and (a).

(e) Apply (c) and (a). □

EXERCISE 1.1.40. Prove the spoiling lemma for degenerate linear codes. Formulate and prove the spoiling lemma for nonlinear codes. (*Hint*: In this case, $k - 1$ is changed to $\log_q \lceil q^{k-1} \rceil$.)

In many cases it is difficult to calculate precise values of the code parameters, but it is possible to bound them. The spoiling lemma allows one to pass from a

$[\leq n, \geq k, \geq d]_q$ code C to an $[n,k,d]_q$ code. In this situation we say that C is an $[n,k,d]_q$ code *up to spoiling*.

Thus, we can always spoil parameters, but certainly we cannot always make them better. Here are some restrictions.

PROPOSITION 1.1.41 (Singleton bound). *For any linear $[n,k,d]_q$ code, we have*
$$n \geq k+d-1.$$

PROOF. Let us argue in terms of $[n,k,d]_q$ systems. Any set of $k-1$ vectors in a k-dimensional linear space V lies in a hyperplane; hence, $n-d = \max|\mathcal{P} \cap H| \geq k-1$. □

EXERCISE 1.1.42. Prove that the Singleton bound is also valid for nonlinear codes. (*Hint:* Shifts of C by different vectors of the form $(v_1, \ldots, v_{d-1}, 0, \ldots, 0)$ are disjoint.)

In the next statement, linearity is essential.

THEOREM 1.1.43 (Griesmer bound). *For a linear $[n,k,d]_q$ code, we have*
$$n \geq \sum_{i=0}^{k-1} \left\lceil \frac{d}{q^i} \right\rceil.$$

PROOF. Consider the corresponding projective system $\mathcal{P} \subset \mathbb{P}^{k-1}$. Let
$$|\mathcal{P} \cap H_0| = \max_H |\mathcal{P} \cap H| = n-d.$$
Set $\mathbb{P}' = H_0$ and $\mathcal{P}' = \mathcal{P} \cap H_0 \subset \mathbb{P}'$. This is a projective $[n',k',d']_q$ system with $n' = n-d$ and $k' = k-1$. Let H' be a hyperplane (of dimension $k-3$) in $\mathbb{P}'$ such that $|\mathcal{P}' \cap H'| = n'-d'$. In $\mathbb{P}^{k-1}$ there are $q+1$ hyperplanes H_i passing through H', and $|H_i \cap \mathcal{P}| \leq n-d$. Therefore,
$$(q+1)(n-d) \geq \sum_{i=1}^{q+1} |H_i \cap \mathcal{P}| = |V \cap \mathcal{P}| + q|H' \cap \mathcal{P}| = n + q(n-d-d')$$
since $\bigcup H_i = V$ and $\bigcap H_i = H'$. Hence we get $d' \geq \lceil d/q \rceil$. Iterating this operation k times, we obtain an $[n^{(k)}, 0, d^{(k)}]_q$ system with
$$n^{(k)} = n - d - d' - \ldots \leq n - \sum_{i=0}^{k-1} \left\lceil \frac{d}{q^i} \right\rceil.$$
The condition $n^{(k)} \geq 0$ proves the theorem. □

EXERCISE 1.1.44. Give an example of a nonlinear code that does not satisfy the Griesmer bound.

The following bounds are proved for arbitrary (not only linear) codes.

THEOREM 1.1.45 (Plotkin bound). *For any $[n,k,d]_q$ code C, we have*
$$d \leq \frac{nq^k(q-1)}{(q^k-1)q}.$$

PROOF. The minimum distance d cannot exceed the average pairwise distance between elements of C:
$$d \le \frac{1}{q^k(q^k-1)} \sum_{x,y \in C} d(x,y).$$

Set $X_{a,i} = |\{x \in C \mid x_i = a\}|$; then $\sum_{a \in \mathbb{F}_q} X_{a,i} = q^k$ for any i. Let $\delta_{a,b}$ be the Kronecker delta. Then
$$\sum_{x,y \in C} d(x,y) = \sum_{i=1}^{n} \sum_{x,y \in C} (1 - \delta_{x_i,y_i}) = \sum_{i=1}^{n} \sum_{a,b \in \mathbb{F}_q} (1 - \delta_{a,b}) X_{a,i} X_{b,i} \le n \max_Z Q(Z),$$

where Q is a quadratic form with matrix $(1 - \delta_{a,b})$; Z is a vector with coordinates $(Z_a)_{a \in \mathbb{F}_q}$, $Z_a \in \mathbb{R}$; and the maximum is taken over Z subject to the conditions $Z_a \ge 0$ and $\sum_{a \in \mathbb{F}_q} Z_a = q^k$. Now the theorem is implied by the following fact.

EXERCISE 1.1.46. Prove that this maximum equals $q^{2k-1}(q-1)$ and is attained when all Z_a are equal. □

THEOREM 1.1.47 (sphere packing bound or the Hamming bound). *For any $[n,k,d]_q$ code C, we have*
$$n - k \ge \log_q \sum_{i=0}^{\lfloor \frac{d-1}{2} \rfloor} \binom{n}{i} (q-1)^i.$$

PROOF. Consider spheres in $\mathbb{F}_q^n$ of radius $t = \lfloor (d-1)/2 \rfloor$ centered at the code vectors (by definition, a sphere of radius t centered at a is the set
$$B_t(a) = \{x \in \mathbb{F}_q^n \mid \|x - a\| \le t\}).$$
These spheres are disjoint; therefore, the product of $|C| = q^k$ and the volume (number of elements) of such a sphere is at most $|\mathbb{F}_q^n| = q^n$. Now the theorem is implied by the following statement.

LEMMA 1.1.48. *The volume of a sphere of radius t in $\mathbb{F}_q^n$ is*
$$|B_t(a)| = \sum_{i=0}^{t} \binom{n}{i} (q-1)^i.$$
□

EXERCISE 1.1.49. Prove the lemma.

The following theorem is obtained by combining the averaging procedure with sphere-packing arguments.

THEOREM 1.1.50 (Bassalygo–Elias bound). *For any $[n,k,d]_q$ code and any integer w such that $1 \le w \le n$ and*
$$A = d - 2w + \frac{qw^2}{(q-1)n} > 0,$$
we have the following inequality:
$$n - k \ge \log_q \binom{n}{w} + w \log_q (q-1) - \log_q d + \log_q A.$$

SKETCH OF THE PROOF. The idea is as follows: first we establish a relation between the possible cardinality of a code in $\mathbb{F}_q^n$ and that of a *spherical* (or *constant-weight*) code, i.e., a code on a sphere $S^n(w) = \{x \in \mathbb{F}_q^n \mid \|x\| = w\}$. Then we use the Plotkin method to estimate the cardinality of the spherical code. Let us start with the following mean value inequality.

LEMMA 1.1.51 (Bassalygo lemma). *Let $A(n,d) = q^{\max\{k\}}$, where the maximum is taken over all $[n,k,d]_q$ codes C with given n and d. Let $L \subset \mathbb{F}_q^n$, and let $A_L(n,d)$ be the maximum possible number of vectors from L such that the distance between any two of them is at least d. Then*

$$\frac{A(n,d)}{q^n} \leq \frac{A_L(n,d)}{|L|}.$$
□

EXERCISE 1.1.52. Prove the lemma. (*Hint*: Consider all possible shifts $L + v$ of the set L by vectors $v \in \mathbb{F}_q$, and use the fact that for any code C the volume of its intersection with some set $L + v$ is not less than the average volume of its intersection with such sets. Note that Exercise 1.1.42 is a particular case of using the Bassalygo lemma.)

For $L = S^n(w)$, we obtain the following result.

COROLLARY 1.1.53. *Let $A(n,d,w)$ be the maximum possible number of vectors of weight w in $\mathbb{F}_q^n$ such that the distance between any two of them is at least d. Then for any w we have*

$$\frac{A(n,d,w)}{\binom{n}{w}(q-1)^w} \geq \frac{A(n,d)}{q^n}.$$
□

LEMMA 1.1.54. *We have*

$$A(n,d,w) \leq \left\lfloor \frac{d}{d - 2w + \frac{qw^2}{(q-1)n}} \right\rfloor$$

whenever the denominator is positive.
□

EXERCISE 1.1.55. Prove the lemma. (*Hint*: Estimate the sum of all pairwise distances in a spherical code, as was done in the proof of the Plotkin bound [Theorem 1.1.45].)

EXERCISE 1.1.56. Derive Theorem 1.1.50 using the results of Corollary 1.1.53 and Lemma 1.1.54.
□

Let us present one more simple result, which in a number of cases allows one to improve upper bounds on $A(n,d)$:

THEOREM 1.1.57. *For any n and d, $0 \leq d \leq n$, we have*

$$A(n,d) \leq \frac{q^m}{V_{m,t}} A(n-m, d-2t),$$

where t and m are arbitrary integers satisfying the conditions

$$0 \leq d - 2t \leq n - m, \qquad 0 \leq t \leq m,$$

and $V_{m,t}$ is the volume of a sphere of radius t in $\mathbb{F}_q^m$.

PROOF. Consider an $[n,k,d]_q$ code C with the maximum number M of words, i.e., $M = q^k = A(n,d)$. Choose m arbitrary coordinates and fix a sphere B_t of radius t in the corresponding coordinate space $\mathbb{F}_q^m$. Consider the natural projections $\pi \colon \mathbb{F}_q^n \to \mathbb{F}_q^m$ and $\pi' \colon \mathbb{F}_q^n \to \mathbb{F}_q^{n-m}$ onto coordinate spaces, where $\mathbb{F}_q^{n-m}$ is generated by the remaining $n-m$ coordinates. Now consider the subset $C' = C \cap \pi^{-1}(B_t) \subseteq C$ (the subset consisting of words whose projections onto $\mathbb{F}_q^m$ belong to the sphere that we have chosen). Then, set $C'' = \pi'(C') \subseteq \mathbb{F}_q^{n-m}$ (in all words of C' delete symbols at these m positions). We thus obtain a code C'' of length $n-m$ with minimum distance at least $d-2t$ since deleted parts of words differ at no more than $2t$ positions. Hence, $|C''| \leq A(n-m, d-2t)$.

Let us show that if we appropriately choose a center of a sphere of radius t, we can get a code C'' with at least $\dfrac{M}{q^k}V_{m,t}$ words. This will immediately prove the theorem.

Let $y_1, \ldots, y_M$ be (not necessarily distinct) images of words of C under the projection π. Then $|C''| = |C'|$ is the number of words y_i that get into the sphere. Let a be the center of a sphere $B_t(a) \subseteq \mathbb{F}_q^m$. Consider the indicator function

$$\chi_a(y) = \begin{cases} 1 & \text{if } y \in B_t(a), \\ 0 & \text{otherwise.} \end{cases}$$

The number of words that get into the sphere can be expressed as $\sum\limits_{i=1}^{M} \chi_a(y_i)$. The average over all spheres is

$$\frac{1}{q^m} \sum_{a \in \mathbb{F}_q^m} \sum_{i=1}^{M} \chi_a(y_i) = \frac{1}{q^m} \sum_{i=1}^{M} \sum_{a \in \mathbb{F}_q^m} \chi_a(y_i) = \frac{M}{q^m} V_{m,t},$$

as required. $\square$

The next idea is to use spectra. Let C be a linear code. By Theorem 1.1.17, we know that

$$W_{C^\perp}(x : y) = q^{-k} W_C(x + (q-1)y : x - y)$$

or, in terms of coefficients,

$$A'_i = q^{-k} \sum_{j=0}^{n} A_j P_i(j),$$

where $P_i(x)$ is the *Krawtchouk polynomial*, defined as

$$P_i(x) = \sum_{j=0}^{i} (-1)^j (q-1)^{i-j} \binom{x}{j} \binom{n-x}{i-j}.$$

Note that $P_i(0) = \binom{n}{i}(q-1)^i$. The generator function of the polynomials $P_i(x)$ is

$$(1+(q-1)z)^{n-x}(1-z)^x = \sum_{i=0}^{\infty} P_i(x) z^i.$$

Since A'_i are coefficients of $W_{C^\perp}(x : y)$, they are nonnegative integers; i.e., for any i we have

$$\sum_{j=0}^{n} A_j P_i(j) \geq 0.$$

We want to get an upper bound for $q^k = 1 + \sum_{i=d}^{n} A_i$, i.e., to solve the following linear programming problem (note that $\sum_{j=0}^{n} A_j P_0(j) \geq 0$ for $A_j \geq 0$):

$$M = 1 + \sum_{i=d}^{n} x_i \longrightarrow \max$$

under the conditions

$$x_i \geq 0, \quad i = d, d+1, \ldots, n,$$

$$\binom{n}{j}(q-1)^j + \sum_{i=d}^{n} P_j(i) x_i \geq 0, \quad j = 1, \ldots, n.$$

Passing to the dual linear programming problem, we obtain the following result.

THEOREM 1.1.58 (linear programming bound). *Let there be given a set of non-negative real numbers $a_1, \ldots, a_n$ such that for any $j = d, d+1, \ldots, n$ we have*

$$1 + \sum_{i=1}^{n} a_i P_i(j) \geq 0.$$

Then for any $[n, k, d]_q$ code C we have

$$q^k \geq 1 + \sum_{i=1}^{n} a_i \binom{n}{i}(q-1)^i. \qquad \square$$

Note that this result can be strengthened if we apply the linear programming method to constant-weight (spherical) codes and then use Corollary 1.1.53.

Finally, we move on to an existence theorem.

THEOREM 1.1.59 (Gilbert–Varshamov bound). *Whenever*

$$q^{n-k} > \sum_{i=0}^{d-2} \binom{n-1}{i}(q-1)^i,$$

there always exists an $[n, k, d]_q$ code C.

PROOF. Let us construct a dual $[n, k, d]_q$ system $\mathcal{Q} \subset W$, where $\dim W = n - k$. For Q_1, take an arbitrary nonzero vector. Assume that we have already constructed a system of i vectors $Q_1, \ldots, Q_i$ such that any $d-1$ of them are linearly independent. Consider the set S_i consisting of vectors that are linear combinations of at most $d-2$ elements from $\{Q_1, \ldots, Q_i\}$; then

$$|S_i| \leq M_i = \sum_{j=0}^{d-2} \binom{i}{j}(q-1)^j.$$

If $M_i < |W|$, we can choose $Q_{i+1} \notin S_i$. Then any $d-1$ vectors from $\{Q_1, \ldots, Q_{i+1}\}$ are linearly independent. Proceeding in the same way, we obtain a desired system $\mathcal{Q} = \{Q_1, \ldots, Q_n\}$ of cardinality n. $\square$

REMARK 1.1.60. It is easily seen that there are many codes with parameters satisfying the inequality of the theorem (because of a large possibility of choices at each step). In a certain sense (see Remark 1.3.22), we can say that parameters of almost every code are quite near to those predicted by Theorem 1.1.59.

EXERCISE 1.1.61 (Varshamov procedure). Assume that the parameters of an $[n_0, k_0, d_0]_q$ code C_0 satisfy the inequality of Theorem 1.1.59. Prove that there exists an $[n_0, k_1, d_0]_q$ code $C_1 \supseteq C_0$ such that its parameters also satisfy the inequality and the triple $[n_0, k_1+1, d_0]$ does not satisfy it.

REMARK 1.1.62 (Gilbert bound). It is easier to prove a slightly weaker result: if

$$q^{n-k} > \sum_{i=0}^{d-1} \binom{n-1}{i}(q-1)^i,$$

then there exists a (nonlinear) $[n, k, d]_q$ code. This can be shown by choosing code vectors in $\mathbb{F}_q^n$ one by one so that each point does not lie in the union of spheres of radius $d-1$ centered at the preceding points.

The results exposed in Sec. 1.1.4 somehow bound possible values of the parameters. Still, the problem of actually finding out which values are possible and which are not is very difficult. This problem is unlikely to be solvable in its general setting.

1.1.5. Bounds for Higher Weights

First, let us fix some notation. As above, a linear subspace of codimension r (in both V and $\mathbb{P}$) is denoted by Π^r. We denote by $\begin{bmatrix} k \\ r \end{bmatrix} = \begin{bmatrix} k \\ r \end{bmatrix}_q$ the number of linear subspaces Π^r in $\mathbb{P}^{k-1}$ or, equivalently, the number of subspaces Π^r in a k-dimensional linear space V such that $\mathbb{P}^{k-1} = \mathbb{P}(V)$.

LEMMA 1.1.63. *The number of subspaces of codimension r in a k-dimensional linear space V equals the number of r-dimensional subspaces in V and equals*

$$\begin{bmatrix} k \\ r \end{bmatrix} = \frac{(q^k-1)(q^k-q)\ldots(q^k-q^{r-1})}{(q^r-1)(q^r-q)\ldots(q^r-q^{r-1})}$$
$$= \frac{(q^k-1)(q^{k-1}-1)\ldots(q^{k-r+1}-1)}{(q^r-1)(q^{r-1}-1)\ldots(q-1)}.$$

PROOF. The numerator of the first expression is the number of ways to choose r linearly independent vectors in V, and the denominator is the number of different bases in a given linear space of dimension r. Thus, this expression gives the number of r-dimensional subspaces. Furthermore, a subspace $\Pi^r \subseteq V$ canonically corresponds to an r-dimensional subspace

$$\Pi_r^* = \{v^* \mid v^*(v) = 0, \ \forall v \in \Pi^r\}$$

in V^*. □

It is easy to check that the numbers $\begin{bmatrix} k \\ r \end{bmatrix}$, called the *Gaussian binomial coefficients*, satisfy certain analogues of most standard relations for usual binomial coefficients.

LEMMA 1.1.64. *Let $r \leq s \leq k$. The number of subspaces Π^r in $\mathbb{P}^{k-1}$ (or in V, $\dim V = k$) passing through a fixed subspace Π_{fix}^s equals $\begin{bmatrix} s \\ r \end{bmatrix}$.*

PROOF. Consider the coset space of Π_{fix}^s in V. In this space W of dimension s, each Π^r becomes $\Pi^r / \Pi_{\text{fix}}^s = \Pi^{s-r}$. It remains to note that $\begin{bmatrix} s \\ s-r \end{bmatrix} = \begin{bmatrix} s \\ r \end{bmatrix}$. □

Upper bounds. Proposition 1.1.13 immediately implies the following fact.

COROLLARY 1.1.65. *For any linear code, we have*
$$r \leq d_r \leq n - k + r.$$ □

REMARK 1.1.66. The upper bound here generalizes the Singleton bound.

If, for a given r, a code has $d_r = n - k + r$, it is called an *r-th rank MDS code* or simply an *r-MDS code*. Any code is kth rank MDS.

COROLLARY 1.1.67. *Any r-th rank MDS code is also an s-th rank MDS for all $s \geq r$.* □

The following theorem generalizes the Plotkin bound.

THEOREM 1.1.68. *For any linear code, we have*
$$d_r \leq \left\lfloor \frac{n(q^r - 1)q^{k-r}}{q^k - 1} \right\rfloor.$$

PROOF. Denote by N the number of pairs (Π^r, x), where Π^r is a subspace of dimension r and $x \in \mathcal{P} \cap \Pi^r$. The total number of such subspaces is $\begin{bmatrix} k \\ r \end{bmatrix}$, and $\begin{bmatrix} k-1 \\ r \end{bmatrix}$ of them pass through any fixed point. Therefore, $N = n \begin{bmatrix} k-1 \\ r \end{bmatrix}$. Since each subspace Π^r contains at most $n - d_r$ points of $\mathcal{P}$, we get $N \leq (n - d_r) \begin{bmatrix} k \\ r \end{bmatrix}$. Hence,
$$n - d_r \geq n \begin{bmatrix} k-1 \\ r \end{bmatrix} / \begin{bmatrix} k \\ r \end{bmatrix} = n \frac{q^{k-r} - 1}{q^k - 1}.$$

Therefore,
$$d_r \leq n \left(1 - \frac{q^{k-r} - 1}{q^k - 1}\right) = n \frac{q^k - q^{k-r}}{q^k - 1}.$$ □

Consider now the following question. Let $\mathcal{P}$ be a projective system (code) with parameters $[n, k, d_1, \ldots, d_{k-1}]$. What systems (codes) can we construct using the existing one?

Deleting a point, we get (if $d_1 \geq 2$)
$$[n - 1, k, \geq d_1 - 1, \ldots, \geq d_{k-1} - 1]. \tag{1.1.5}$$

Projecting from a point outside $\mathcal{P}$, we get (if $k \geq 2$)
$$[n, k - 1, \geq d_1, \ldots, \geq d_{k-1}]. \tag{1.1.6}$$

Projecting from a point of $\mathcal{P}$, we get (if $k \geq 2$)
$$[n - 1, k - 1, \geq d_1, \ldots, \geq d_{k-1}]. \tag{1.1.7}$$

Adding a point, we get
$$[n + 1, k, \geq d_1, \ldots, \geq d_{k-1}]. \tag{1.1.8}$$

The above procedures are called *spoiling of a code* (cf. Lemma 1.1.39). The following proposition is less trivial.

PROPOSITION 1.1.69. *Given an $[n, k, d_1, \ldots, d_{k-1}]$ system $\mathcal{P}$, we can construct a system $\mathcal{P}'$ with parameters*
$$[n - d_s, k - s, d'_1 \geq d_{s+1} - d_s, \ldots, d'_{k-s-1} \geq d_{k-1} - d_s]$$

and a system $\mathcal{P}''$ with parameters
$$[d_s, s, d_1'' \geq d_1, \ldots, d_{s-1}'' \geq d_{s-1}]$$
for any $s = 1, \ldots, k-1$.

PROOF. Let Π^s be a subspace of codimension s such that $|\Pi^s \cap \mathcal{P}| = n - d_s$. Put $\mathcal{P}' = \mathcal{P} \cap \Pi^s \subset \Pi^s$. To estimate d_r', note that a subspace Π^{r+s} of codimension r in Π^s has codimension $r+s$ in $\mathbb{P}^{k-1}$, and
$$|(\mathcal{P} \cap \Pi^s) - (\mathcal{P} \cap \Pi^{r+s})| \geq (n - d_s) - (n - d_{r+s}).$$
For $\mathcal{P}''$, take the projection of $\mathcal{P} \setminus (\mathcal{P} \cap \Pi^s)$ from Π^s; then $\mathcal{P}'' \subset \mathbb{P}^{s-1}$, and the inverse image of a codimension-r subspace has codimension r in $\mathbb{P}^{k-1}$. □

This gives the following "generalized Griesmer bound".

COROLLARY 1.1.70. *For any linear code, we have*
$$d_r \geq \sum_{i=0}^{r-1} \left\lceil \frac{d_1}{q^i} \right\rceil.$$

PROOF. It suffices to apply the usual Griesmer bound $n \geq \sum_{i=0}^{k-1} \lceil d_1/q^i \rceil$ (Theorem 1.1.43) to the system $\mathcal{P}''$. □

REMARK 1.1.71. In fact, any upper bound for $[n, k, d_1]$ gives a lower bound on d_s in terms of d_1 and k (apply this upper bound to the $[d_s, s, d_1'' \geq d_1, \ldots, d_{s-1}'' \geq d_{s-1}]$ system $\mathcal{P}''$ of Proposition 1.1.69).

COROLLARY 1.1.72. *For the system $\mathcal{P}'$ of Proposition 1.1.69, we have*
$$d_r' \geq d_{r+s} - d_s \geq \frac{q^s - 1}{q^s(q^r - 1)} d_s.$$

PROOF. The first inequality is contained in Proposition 1.1.69. The second one is straightforward from the generalized Plotkin bound (Theorem 1.1.68). □

COROLLARY 1.1.73. *For any $r \leq s$, we have*
$$d_s \geq d_r + \sum_{i=1}^{s-r} \left\lceil \frac{(q-1)d_r}{(q^r - 1)q^i} \right\rceil.$$
In particular, for any $r \leq k$, we have
$$n \geq d_r + \sum_{i=1}^{k-r} \left\lceil \frac{(q-1)d_r}{(q^r - 1)q^i} \right\rceil.$$

PROOF. First, consider the second construction of Proposition 1.1.69. The system $\mathcal{P}''$ has parameters $n'' = d_s$, $k'' = s$, and $d_r'' \geq d_r$. Now apply the first construction of Proposition 1.1.69. We get a system $(\mathcal{P}'')'$ with parameters $n' = d_s - d_r$, $k' = s - r$, and
$$d_1' \geq d_{r+1} - d_r \geq \left\lceil \frac{(q-1)d_r}{(q^r - 1)q} \right\rceil.$$
By the (usual) Griesmer bound, we have
$$n' \geq \sum_{i=0}^{k-1} \left\lceil \frac{d_1'}{q^i} \right\rceil,$$

whence
$$d_s - d_r = n' \geq \sum_{i=0}^{s-r-1} \left\lceil \frac{(q-1)d_r}{(q^r-1)q^{i+1}} \right\rceil.$$
□

COROLLARY 1.1.74. *Let $r \leq s$. Then*
$$d_r \leq \left\lfloor \frac{d_s(q^r-1)q^{s-r}}{q^s-1} \right\rfloor.$$

PROOF. Apply Theorem 1.1.68 to the system $\mathcal{P}''$ of Proposition 1.1.69. □

COROLLARY 1.1.75. *For any $\ell \leq s-r$, we have*
$$d_r \leq \left\lfloor \frac{(d_s - \ell)(q^r-1)q^{s-\ell-r}}{q^{s-\ell}-1} \right\rfloor.$$

PROOF. Put $s' = s - \ell$. It suffices to apply Corollary 1.1.74 and the fact that $d_{s'} \leq d_s - s + s' = d_s - \ell$. □

Corollaries 1.1.74 and 1.1.75 give upper bounds on d_r in terms of d_s, $s > r$. Here is a lower bound.

THEOREM 1.1.76. *Let $r \leq s < k$. Then*
$$d_r \geq n - \left\lfloor \frac{(q^{k-r}-1)(n-d_s)}{q^{k-s}-1} \right\rfloor.$$

PROOF. Fix a subspace Π_{fix}^r of codimension r such that $|\mathcal{P} \cap \Pi_{\text{fix}}^r| = n - d_r$. Let us count the number A of pairs (x, Π^s), where Π^s is a subspace of codimension s in $\mathbb{P}^{k-1}$, $\Pi^s \subset \Pi_{\text{fix}}^r$, and $x \in \Pi^s \cap \mathcal{P}$. One the one hand, $A = (n-d_r)B$, where B is the number of subspaces $\Pi^s \subset \Pi_{\text{fix}}^r$ passing through a given point x, which equals the number of subspaces Π^{s-r} of codimension $s-r$ in $\mathbb{P}^{k-1-r}$ passing through a fixed point; i.e.,
$$A = (n-d_r) \begin{bmatrix} k-r-1 \\ s-r \end{bmatrix}.$$

On the other hand, $A \leq (n-d_s)C$, where C is the number of subspaces $\Pi^s \subset \Pi_{\text{fix}}^r$, since each Π^s contains at most $n - d_s$ points of $\mathcal{P}$; i.e.,
$$A \leq (n-d_s) \begin{bmatrix} k-r \\ s-r \end{bmatrix}.$$

Hence,
$$\frac{n-d_r}{n-d_s} \leq \frac{\begin{bmatrix} k-r \\ s-r \end{bmatrix}}{\begin{bmatrix} k-r-1 \\ s-r \end{bmatrix}} = \frac{q^{k-r}-1}{q^{k-s}-1}.$$
□

Up to this point, we considered only generalized weights of linear codes and corresponding systems. Let us now pass to essentially nonlinear theory. We can no longer use definitions given in terms of projective systems, so we have first to give a definition of higher weights for nonlinear codes in the style of equations (1.1.1) and (1.1.2); see p. 6. Here is one.

For a (not necessarily linear) code $C \subset \mathbb{F}_q^n$, consider the set A of all $(r+1)$-tuples $a = (a^{(0)}, \ldots, a^{(r)})$, $a^{(i)} \in C$, such that all vectors $a^{(1)} - a^{(0)}$, $a^{(2)} - a^{(0)}, \ldots, a^{(r)} - a^{(0)}$

are linearly independent. Then let

$$d_r = d_r(C) = \min_{a \in A} \left| \bigcup_{i=1}^{r} \mathrm{Supp}\left(a^{(i)} - a^{(0)}\right) \right|. \qquad (1.1.9)$$

The definition is obviously compatible with that for linear codes and is invariant under shifts of a code: $d_r(C) = d_r(C+v)$ for any $v \in \mathbb{F}_q^n$. For a nonlinear code, we set $k = \log_q |C|$.

Here is a natural generalization of the Hamming bound.

THEOREM 1.1.77. *Let* $t = \lfloor (d_r - 1)/(r+1) \rfloor$. *Then for any code* C *we have*

$$\sum_{i=0}^{t} \binom{n}{i} (q-1)^i \leq q^{n-k+r-1}.$$

PROOF. Consider a sphere B_t of radius t in the usual Hamming metric in $\mathbb{F}_q^n$. Look at the intersections $C_{v,t} = B_t \cap (C+v)$ for all $v \in \mathbb{F}_q^n$. These are q^{n-k} distinct disjoint sets. For $a^{(0)}, \ldots, a^{(r)} \in C_{v,t}$, we have

$$\left| \bigcup_{i=1}^{r} \mathrm{Supp}\left(a^{(i)} - a^{(0)}\right) \right| \leq \sum_{i=0}^{r} \left| \mathrm{Supp}\, a^{(i)} \right| \leq (r+1)t \leq d_r - 1;$$

hence (by definition (1.1.9)), the vectors $a^{(1)} - a^{(0)}, a^{(2)} - a^{(0)}, \ldots, a^{(r)} - a^{(0)}$ are linearly dependent. Thus, we have $|C_{v,t}| \leq q^{r-1}$ and $|B_t| \leq q^{n-k+r-1}$. □

Let us pass on to a generalization of the Bassalygo–Elias bound.

THEOREM 1.1.78. *For any* $[n, k, d_1, \ldots]_q$ *code* C, *we have*

$$d_r \leq \min_w \left\{ n - w \prod_{i=0}^{r-1} \left(\frac{w}{n(q-1)} - \frac{q^{n-k+i}}{\binom{n}{w}(q-1)^w} \right) \right.$$
$$\left. - (n-w) \prod_{i=0}^{r-1} \left(\frac{n-w}{n} - \frac{q^{n-k+i}}{\binom{n}{w}(q-1)^w} \right) \right\},$$

where the minimum is taken over all integers w *such that* $0 < w < n$ *and*

$$\binom{n}{w}(q-1)^w \min\left\{ \frac{w}{n(q-1)}, \frac{n-w}{n} \right\} \geq q^{n-k+r-1}. \qquad (1.1.10)$$
□

We have an obvious corollary, which is only slightly worse and much easier to use than the above theorem.

COROLLARY 1.1.79. *Let* $\omega = w/n$ *and* $\delta_r = d_r/n$. *Then*

$$1 - \delta_r \geq \max_w \left\{ \omega \left(\frac{\omega}{q-1} - \frac{q^{n-k+r-1}}{\binom{n}{w}(q-1)^w} \right)^r + (1-\omega)\left(1 - \omega - \frac{q^{n-k+r-1}}{\binom{n}{w}(q-1)^w} \right)^r \right\},$$

where the maximum is taken over all integers w *such that* $0 < w < n$ *and*

$$\binom{n}{w}(q-1)^w \min\left\{ \frac{w}{n(q-1)}, \frac{n-w}{n} \right\} \geq q^{n-k+r-1}.$$
□

REMARK 1.1.80. Definition (1.1.9) of higher weights has a serious disadvantage: according to this definition, it sometimes happens that $d_r = d_{r-1}$, which is impossible for linear codes (provided that $d_1 \geq 1$). Let us give another definition, which is also compatible with definition (1.1.1) and is also shift-invariant, but for which we always have $d_r > d_{r-1}$.

For a set of vectors $A = \{a^{(1)}, \ldots, a^{(m)}\} \subset C$, define

$$\operatorname{Supp} A = \left\{ i \mid \exists\, j, \ell \text{ such that } a_i^{(j)} \neq a_i^{(\ell)} \right\}.$$

Let
$$f_C(m) = \min |\operatorname{Supp} A|,$$

where the minimum is taken over all subsets $A \subset C$ of cardinality $m+1$. Obviously, $f_C(1) = d_1(C)$ is the usual minimum distance. The function $f_C(m)$ is a monotone nondecreasing step function. Define $d_r(C)$ to be its rth value. For a linear code, we have $d_r(C) = f_C(q^{r-1})$.

REMARK 1.1.81. Let us give one more possible (information-theoretic) definition, which is also free of the above-mentioned disadvantage of definition (1.1.9) and is compatible with definition (1.1.1) in the case of linear codes.

Let C be a (nonlinear) code of length n, $I = \{1, 2, \ldots, n\}$, $J \subseteq I$ be a coordinate subset, $P_J(C)$ be the set of all projections of codewords onto J, and X_J be a random variable taking values on $P_J(C)$ with probabilities induced by the uniform distribution on codewords. Let

$$h_i(C) = \max_j \left\{ H(X_J \mid X_{I \setminus J}) : |J| = i \right\},$$

where H is the conditional entropy; put $h_0(C) = 0$ by definition. Then $h_i(C)$ is a monotone nondecreasing step function with $0 \leq h_{i+1}(C) - h_i(C) \leq 1$. Define $d_r(C)$ as the smallest index i for which $h_i(C)$ takes its rth nonzero value.

Lower bounds. Let us give some existence bounds. All such bounds are based on the same idea: namely, we count the total number of objects of a certain kind (codes, systems, matrices, etc.) and compare it with the number of those with "bad" parameters. We shall state several theorems of this type for higher weights as the following "scheme of theorems".

THEOREM 1.1.82. *The statement*

> *The ratio of the number of objects of type $T_i(n, k)$ with d_r strictly less than a given number d_r^0 to the total number of objects of type $T_i(n, k)$ is not greater than $A_i(n, k, r)$; in particular, if $A_i(n, k, r) < 1$, then there exists an object of type $T_i(n, k)$ with $d_r \geq d_r^0$*

is valid in each of the following cases:

(a) $T_1 = $ *ordered projective $[n, k]$ systems, and*

$$A_1(n, k, r) = \begin{bmatrix} k \\ r \end{bmatrix} (q^k - 1)^{-n} \sum_{t=0}^{d-1} \binom{n}{t} ((q^r - 1) q^{k-r})^t (q^{k-r} - 1)^{n-t};$$

(b) $T_2 = $ *ordered linear $[n, k]$ systems (i.e., $n \times k$ matrices), and*

$$A_2(n, k, r) = q^{-rn} \begin{bmatrix} k \\ r \end{bmatrix} \sum_{t=0}^{d-1} \binom{n}{t} (q^r - 1)^t;$$

(c) T_3 = *ordered linear $[n,k]$ systems (i.e., $n \times k$ matrices)*, and
$$A_3(n,k,r) = q^{-r(n-d+1)} \begin{bmatrix} k \\ r \end{bmatrix} \binom{n}{d-1};$$

(d) T_4 = *linear $[n,k]$ codes*, and
$$A_4(n,k,r) = \binom{n}{d} \begin{bmatrix} d \\ r \end{bmatrix} \begin{bmatrix} k \\ r \end{bmatrix} / \begin{bmatrix} n \\ r \end{bmatrix};$$

(e) T_5 = *linear $[n,k]$ codes*, and
$$A_5(n,k,r) = \left(\sum_{t=0}^{d-1} \binom{n}{t} (q^r - 1)^t \right) \begin{bmatrix} k \\ r \end{bmatrix} / \binom{n}{r};$$

(f) T_6 = *full-rank $n \times k$ matrices (i.e., nondegenerate ordered linear systems)*, and
$$A_6(n,k,r) = q^{-rn} \begin{bmatrix} k \\ r \end{bmatrix} \left(\sum_{t=0}^{d-1} \binom{n}{t} (q^r - 1)^t \right) \prod_{j=0}^{k-1} (1 - q^{j-n})^{-1}.$$

PROOF. (a) The total number of $[n,k]$ systems $\mathcal{P} \subset \mathbb{P}^{k-1}$ is $\left(\frac{q^k-1}{q-1}\right)^n$. The number of those with $d_r < d$ is at most the number of pairs $(\mathcal{P}, \Pi^r)$ such that $|\mathcal{P} \cap \Pi^r| = n-t$ for $t < d$. In total, there are $\begin{bmatrix} k \\ r \end{bmatrix}$ subspaces Π^r. Fix t positions out of n; this can be done in $\binom{n}{t}$ different ways. Thus, the ratio in question is at most
$$\left(\frac{q^k-1}{q-1}\right)^{-n} \begin{bmatrix} k \\ r \end{bmatrix} \sum_{t=0}^{d-1} \binom{n}{t} \left(\frac{q^{k-r}-1}{q-1}\right)^{n-t} \left(\frac{(q^r-1)q^{k-r}}{q-1}\right)^t.$$

(b) The total number of $[n,k]$ systems $\mathcal{Q} \subset V^k$ is precisely the number of all $k \times n$ matrices, i.e., q^{kn}. The number of those with $d_r < d$ is at most the number of pairs $(\mathcal{Q}, \Pi^r)$ such that $|\mathcal{Q} \cap \Pi^r| = n-t$ for $t < d$; here, $|\Pi^r| = q^{k-r}$. Thus, the ratio in question is at most
$$q^{-kn} \begin{bmatrix} k \\ r \end{bmatrix} \sum_{t=0}^{d-1} \binom{n}{t} q^{(k-r)(n-t)} (q^k - q^{k-r})^t.$$

(c) The total number is q^{kn}. The number of those with $d_r < d$ is at most the number of triples $(\mathcal{Q}, \Pi^r, I_{n-d+1})$, where $I_{n-d+1} \subset [1,n]$ is a set of $n-d+1$ positions, and the corresponding $n-d+1$ elements of $\mathcal{Q}$ lie in Π^r. The ratio is at most
$$q^{-kn} \begin{bmatrix} k \\ r \end{bmatrix} \binom{n}{d-1} q^{(k-r)(n-d+1)} q^{k(d-1)}.$$

(d) The total number of $[n,k]$ codes is $\begin{bmatrix} n \\ k \end{bmatrix}$; the number of those containing a given subcode of dimension r is $\begin{bmatrix} n \\ k \end{bmatrix} \begin{bmatrix} k \\ r \end{bmatrix} / \begin{bmatrix} n \\ r \end{bmatrix}$. For a fixed support I_d, the number of $[d,r]$ codes is $\begin{bmatrix} d \\ r \end{bmatrix}$.

(e) The argument is the same, only that the number of codes with support exactly equal to I_t is bounded by $(q^r - 1)^t$ since there are $q^r - 1$ possibilities for each column.

(f) The total number of ordered nondegenerate affine $[n,k]$ systems equals $\prod_{j=0}^{k-1} (q^n - q^j)$. In total, there are $\begin{bmatrix} k \\ r \end{bmatrix}$ subspaces Π^r; for each of them, the number of

systems containing t vectors of Π^r is
$$\binom{n}{t}(q^k - q^{k-r})^t q^{(k-r)(n-t)}. \qquad \square$$

REMARK 1.1.83. Statement (b) is strictly better than (f). Statement (a) is the best when r is small as compared with other parameters.

Let $D_r(n,k)$ be the maximum possible d_r for given n and k. Simple spoiling considerations (see (1.1.5)–(1.1.8), p. 22) show:

If $D_r(n,k) \geq r+1$, then
$$D_r(n-1,k) \geq D_r(n,k) - 1.$$

If $k \geq 2$, then
$$D_r(n,k-1) \geq D_r(n,k)$$

and
$$D_r(n-1,k-1) \geq D_r(n,k).$$

We also have
$$D_r(n+1,k) \geq D_r(n,k).$$

PROPOSITION 1.1.84. *We have*
$$D_{s+1}(n,k) \leq D_1(n,k) + D_s(n - D_1(n,k), k-1-s).$$

PROOF. Consider an $[n,k]$ system $\mathcal{P}$ such that $d_{s+1}(\mathcal{P}) = D_{s+1}(n,k)$. For the $[n - d_1(\mathcal{P}), k-1]$ system $\mathcal{P}'$ of Proposition 1.1.69 we have $D_{s+1}(n,k) = d_{s+1}(\mathcal{P}) \leq d_1(\mathcal{P}') + D_s(n - d_1(\mathcal{P}'), k-1)$. Let $\lambda = D_1(n,k) - d_1(\mathcal{P}') \leq D_s(n - d_1(\mathcal{P}'), k-1)$. Then we get
$$D_s(n - d_1(\mathcal{P}') - \lambda, k-1-s) \geq D_s(n - d_1(\mathcal{P}'), k-1) - \lambda$$
since, deleting less than d_s points, we can never decrease the dimension by more than s, whence the proposition follows. $\qquad \square$

COROLLARY 1.1.85. *Define the sequence of integers $P_i(n,k)$ recursively as follows:* $P_1(n,k) = D_1(n,k)$ *and*
$$P_{i+1}(n,k) = P_i(n,k) + D_1(n - P_i(n,k), k-i-1).$$

Then
$$D_r(n,k) \leq D_1(n,k) + \sum_{i=0}^{r-1} D_1(n - P_i(n,k), k-i-1). \qquad (1.1.11)$$

PROOF. Use induction, applying Proposition 1.1.84. $\qquad \square$

The main question of the theory, the "parameter problem", is as follows.

PROBLEM 1.1.86. Find $D_r(n,k)$ for given n, k, and r.

As well as in the classical case $r = 1$, we do not hope to find any definite answer. One may try to study the problem for small values of parameters and in the general case to look for upper and lower bounds and to investigate asymptotics.

In fact even this problem is too weak. We would prefer to answer a stronger question:

PROBLEM 1.1.87. For given $\{d_1, d_2, \ldots, d_k\}$, does there exist an $[n, k, d_1, \ldots, d_k]_q$ code?

One might hope to answer this question from the geometric point of view when k is very small, studying configurations of points on a line, in a plane, and so on. For $q = 2$ and $k \leq 3$, the answer is as follows.

PROPOSITION 1.1.88. (a) *An $[n, 1, d_1]_2$ code exists if and only if $0 < d_1 \leq n$.*
(b) *An $[n, 2, d_1, d_2]_2$ code exists if and only if $0 < 3d_1 \leq 2d_2 \leq 2n$.*
(c) *An $[n, 3, d_1, d_2, d_3]_2$ code exists if and only if $0 < 3d_1 \leq 2d_2$, $7d_2 \leq 6d_3 \leq 6n$, and $d_3 - d_1 \leq 3(d_3 - d_2)$.* □

REMARK 1.1.89. Necessary and sufficient conditions for an $[n, 4, d_1, d_2, d_3, d_4]_2$ code to exist are also known; however, the answer is rather complicated. One might also hope to find an answer either to Problem 1.1.86 or to Problem 1.1.87 in the dual case, i.e., when $k - r$ is very small. In the binary case, $D_r(n, k)$ are known for $k - r \leq 3$.

1.1.6. Duality for Generalized Spectra

Here we consider the question on the relation between the generalized spectrum A_i^r of a code and that of a dual code.

The numbers A_i^r are closely related to some other parameters. Let us first consider the extended code $C^{(s)} = C \otimes \mathbb{F}_{q^s}$, the corresponding projective system being $\mathcal{P} \subset \mathbb{P}(\mathbb{F}_{q^s})$, i.e., the same set of points but in the projective space over the larger field. Let
$$A_i^{(s)} = A_i^{(s)}(C) = A_i^1(C^{(s)}).$$

THEOREM 1.1.90. *The first spectrum of the extended code (system) and the complete spectrum of the original code (system) are related by*
$$A_i^{(s)} = \sum_{m=0}^{k} [s]_m A_i^m,$$
where
$$[s]_m = \prod_{j=0}^{m-1} (q^s - q^j).$$

PROOF. To a hyperplane H defined over $\mathbb{F}_{q^s}$, assign its maximal linear subspace $\Pi^m \subseteq H$ defined over $\mathbb{F}_q$. Then $H \cap \mathcal{P} = \Pi^m \cap \mathcal{P}$ since $\mathcal{P}$ consists of $\mathbb{F}_q$-points. The number of such hyperplanes $H \subset \mathbb{P}$ passing through a fixed subspace Π^m_{fix} and containing no Π^{m-1} defined over $\mathbb{F}_q$ equals the number of points x in the dual projective space $\mathbb{P}^*$ (whose points are hyperplanes of $\mathbb{P}$) such that x is defined over $\mathbb{F}_{q^s}$, lies in some $\mathbb{P}^{m-1} \subseteq \mathbb{P}^*$, and does not lie in any $\mathbb{P}^{m-2}$ defined over $\mathbb{F}_q$. This means that its coordinates $x_i \in \mathbb{F}_{q^s}$, $i = 1, \ldots, m$, are linearly independent over $\mathbb{F}_q$. The number of such x is precisely $[s]_m$. □

The theorem can be used to obtain an analogue of the MacWilliams identities for generalized spectra (see Theorem 1.1.95 below). Another approach is as follows.

Consider the notions of puncturing and shortening. Let $\mathcal{P}$ be a projective system corresponding to a code C. For a subsystem $\mathcal{Q} \subseteq \mathcal{P}$, let $\mathbb{F}_q^{\mathcal{Q}} \subseteq \mathbb{F}_q^{\mathcal{P}} = \mathbb{F}_q^n$ be the subspace consisting of vectors that have zeroes at positions from $\mathcal{P} \setminus \mathcal{Q}$. The *shortening* $C^{\mathcal{Q}}$ of C is defined as
$$C^{\mathcal{Q}} = C \cap \mathbb{F}_q^{\mathcal{Q}}.$$

Let $\pi_{\mathcal{Q}}\colon \mathbb{F}_q^{\mathcal{P}} \to \mathbb{F}_q^{\mathcal{Q}}$ be the natural projection. The *puncturing* $C_{\mathcal{Q}}$ of C is defined as
$$C_{\mathcal{Q}} = \pi_{\mathcal{Q}}(C).$$
Obviously, $C_{\mathcal{Q}} = C/C^{\mathcal{P}\setminus\mathcal{Q}} \subseteq \mathbb{F}_q^{\mathcal{Q}}$.

In terms of projective systems, the puncturing is simply
$$\mathcal{P}_{\mathcal{Q}} = \mathcal{Q} \subseteq \langle \mathcal{Q} \rangle,$$
where $\langle \mathcal{Q} \rangle$ is the linear span of $\mathcal{Q}$, and the shortening is the projection from $\langle \mathcal{P}\setminus\mathcal{Q} \rangle$, i.e.,
$$\mathcal{P}^{\mathcal{Q}} = \mathrm{Im}(\mathcal{Q}) \subseteq \mathbb{P}/\langle \mathcal{P}\setminus\mathcal{Q} \rangle;$$
here, strictly speaking, by $\mathbb{P}/\langle \mathcal{P}\setminus\mathcal{Q} \rangle$ we understand $\mathbb{P}(V/\pi^{-1}(\langle \mathcal{P}\setminus\mathcal{Q} \rangle))$, where $\mathbb{P} = \mathbb{P}(V)$ and $\pi\colon V \to \mathbb{P}$ is the natural projection.

Let $m = |\mathcal{Q}|$. It is clear that the length is $n(\mathcal{P}_{\mathcal{Q}}) = n(\mathcal{P}^{\mathcal{Q}}) = m$, and for the dimension we have $k(\mathcal{P}_{\mathcal{Q}}) + k(\mathcal{P}^{\mathcal{P}\setminus\mathcal{Q}}) = k(\mathcal{P})$.

Define
$$N_m^s = N_m^s(\mathcal{P}) = \left|\{ \mathcal{Q} \subset \mathcal{P} \mid k(\mathcal{P}^{\mathcal{Q}}) = \dim(C^{\mathcal{Q}}) = \mathrm{codim}(\langle \mathcal{P}\setminus\mathcal{Q} \rangle) = s,\ |\mathcal{Q}| = m\}\right|.$$

As we shall see below, the numbers N_m^r behave nicely with respect to duality.

First, note the following fact.

THEOREM 1.1.91. *The numbers N_m^s and A_i^r are related by*
$$\sum_{j=0}^{m} \binom{n-j}{n-m} A_j^r = \sum_{s=r}^{k} \begin{bmatrix} s \\ r \end{bmatrix} N_m^s. \tag{1.1.12}$$

PROOF. Consider the set of pairs
$$P = \{(\Pi^r, \mathcal{Q}) \mid \mathrm{codim}\,\Pi^r = r,\ |\mathcal{Q}| = m,\ \Pi^r \cap \mathcal{P} = \mathcal{P}\setminus\mathcal{Q}\},$$
and count its cardinality in two different ways. The number of Π^r containing $\mathcal{P}\setminus\mathcal{Q}$ (and hence $\langle \mathcal{P}\setminus\mathcal{Q} \rangle$) equals $\begin{bmatrix} s \\ r \end{bmatrix}$, where $s = \mathrm{codim}\langle \mathcal{P}\setminus\mathcal{Q} \rangle$. Thus, $|P| = \sum_{s=r}^{k} \begin{bmatrix} s \\ r \end{bmatrix} N_m^s$. On the other hand, if $|\Pi^r \cap \mathcal{P}| = n-j$, then there are $\binom{n-j}{n-m}$ ways to chose $\mathcal{P}\setminus\mathcal{Q}$ inside $\Pi^r \cap \mathcal{P}$, and $|P| = \sum_{j=0}^{m} \binom{n-j}{n-m} A_j^r$. $\square$

MacWilliams-type identities for generalized spectra. The principal result on duality for the first weight ($r = 1$) is of course the MacWilliams identities, relating the spectrum of a code to that of its dual.

Theorems 1.1.90 and 1.1.91 can be used to deduce a generalization for any r. What is worth mentioning is that to express A_i^r one needs to know not only the rth spectrum $\widetilde{A}_j^r$ of the dual code but also $\widetilde{A}_j^s$ for all $s \leq r$.

Let us use the shortening and puncturing approach.

PROPOSITION 1.1.92. *Shortening is dual to puncturing in the sense that*
$$(\mathcal{P}_{\mathcal{Q}})^\perp = (\mathcal{P}^\perp)^{\mathcal{Q}} \quad \text{and} \quad (\mathcal{P}^{\mathcal{Q}})^\perp = (\mathcal{P}^\perp)_{\mathcal{Q}}.$$

PROOF. This is easier seen on codes than on systems. We have $\mathbb{F}_q^{\mathcal{P}} = \mathbb{F}_q^{\mathcal{Q}} \oplus \mathbb{F}_q^{\mathcal{P}\setminus\mathcal{Q}}$. If $u \in C^{\mathcal{Q}} = C \cap \mathbb{F}_q^{\mathcal{Q}}$ and $v \in (C^\perp)_{\mathcal{Q}} = \pi_{\mathcal{Q}}(C^\perp) = C^\perp/(C^\perp \cap \mathbb{F}_q^{\mathcal{P}\setminus\mathcal{Q}})$, then let $w = v \oplus 0 \in C^\perp$; obviously, $(v,u) = (w,u) = 0$. On the other hand, if $(u,v) = 0$ for any $v \in (C^\perp)_{\mathcal{Q}}$, then $(u \oplus 0, w) = 0$ for all $w = v \oplus v'$ and hence for all $w \in C^\perp$. $\square$

Let
$$\dim \mathcal{Q} = \dim \text{lin}\langle \mathcal{Q} \rangle = \dim \langle \mathcal{Q} \rangle + 1,$$
where $\dim \text{lin}\langle \mathcal{Q} \rangle$ is the linear dimension, which is greater by 1 than the projective dimension $\dim \langle \mathcal{Q} \rangle$.

Let $N_a^b = N_a^b(\mathcal{P})$ and $\widetilde{N}_a^b = N_a^b(\mathcal{P}^\perp)$. Let $k^\perp = k(\mathcal{P}^\perp) = n - k$.

LEMMA 1.1.93. *We have*
$$N_m^r = \widetilde{N}_{n-m}^{k^\perp + r - m}. \tag{1.1.13}$$

PROOF. Map $\mathcal{P}^\mathcal{Q}$ to $(\mathcal{P}^\perp)^{\mathcal{P} \setminus \mathcal{Q}}$. If $\dim \mathcal{P}^\mathcal{Q} = r$ and $|\mathcal{P}^\mathcal{Q}| = |\mathcal{Q}| = m$, then $|(\mathcal{P}^\perp)^{\mathcal{P}\setminus\mathcal{Q}}| = |\mathcal{P} \setminus \mathcal{Q}| = n - m$ and $\dim(\mathcal{P}^\perp)^{\mathcal{P}\setminus\mathcal{Q}} = k^\perp + r - m$ since $(C^\perp)^{\mathcal{P}\setminus\mathcal{Q}} = (C_{\mathcal{P}\setminus\mathcal{Q}})^\perp = (C/C^\mathcal{Q})^\perp \subseteq \mathbb{F}_q^{\mathcal{P}\setminus\mathcal{Q}}$, where C and $C^\perp$ are the corresponding codes; hence, $\dim(C^\perp)^{\mathcal{P}\setminus\mathcal{Q}} = k^\perp + r - m$. □

Then we can state the duality theorem.

THEOREM 1.1.94. *The spectra $\{A_i^r\}$ of a linear code (projective system) and $\{\widetilde{A}_v^u\}$ of its dual are related as follows* ($r = 0, 1, \ldots, k$; $j = 0, 1, \ldots, n$):
$$\sum_{i=0}^{j} \binom{n-i}{n-j} A_i^r = \sum_{u=0}^{r+k-j} q^{u(j-k-r+u)} \begin{bmatrix} j-k \\ r-u \end{bmatrix} \sum_{v=0}^{n-j} \binom{n-v}{j} \widetilde{A}_v^u. \tag{1.1.14}$$

PROOF. Using (1.1.12) and (1.1.13), we obtain
$$\sum_{i=0}^{j} \binom{n-i}{n-j} A_i^r = \sum_{m=r}^{k} \begin{bmatrix} m \\ r \end{bmatrix} N_j^m = \sum_{m=r}^{k} \begin{bmatrix} m \\ r \end{bmatrix} \widetilde{N}_{n-j}^{k^\perp + m - j}$$
$$= \sum_{s=k^\perp + r - j}^{n-j} \begin{bmatrix} s + j - k^\perp \\ r \end{bmatrix} \widetilde{N}_{n-j}^s$$
$$= \sum_{s=k^\perp + r - j}^{n-j} \left(\sum_{u=0}^{r+k-j} q^{u(j-k^\perp - r + u)} \begin{bmatrix} j - k^\perp \\ r - u \end{bmatrix} \begin{bmatrix} s \\ u \end{bmatrix} \right) \widetilde{N}_{n-j}^s$$
$$= \sum_{u=0}^{r+k-j} q^{u(j - k^\perp - r + u)} \begin{bmatrix} j - k^\perp \\ r - u \end{bmatrix} \sum_{v=0}^{n-j} \binom{n-v}{j} \widetilde{A}_v^u$$

since (EXERCISE!)
$$\begin{bmatrix} x+y \\ r \end{bmatrix} = \sum_{u=0}^{r-x} q^{u(x-r+u)} \begin{bmatrix} x \\ r-u \end{bmatrix} \begin{bmatrix} y \\ u \end{bmatrix}.$$
□

Another approach uses Theorem 1.1.90. Let
$$A^r(Z) = \sum_{i=0}^{n} A_i^r Z^i$$
be the rth enumerator for the code (written in a nonhomogeneous form with $Z = y/x$; cf. Sec. 1.1.3), and let $\widetilde{A}^r(Z)$ be the corresponding enumerator for the dual code.

THEOREM 1.1.95. *For any $m \geq 0$, we have*

$$\sum_{r=0}^{s}[s]_r \widetilde{A}^r(Z) = q^{-sk}(1+(q^s-1)Z)^n \sum_{r=0}^{s}[s]_r A^r\left(\frac{1-Z}{1+(q^s-1)Z}\right). \quad (1.1.15)$$

PROOF. Let $A^{(s)}(Z) = \sum_{i=0}^{n} A_i^{(s)} Z^i$. The MacWilliams identity for $C^{(s)}$ is

$$\widetilde{A}^{(s)}(Z) = q^{-sk}(1+(q^s-1)Z)^n A^{(s)}\left(\frac{1-Z}{1+(q^s-1)Z}\right).$$

Theorem 1.1.90 states that

$$A^{(s)}(Z) = \sum_{r=0}^{s}[s]_r A^r(Z),$$

and we are done. □

Of course, one can express A_i^r from (1.1.14) or from (1.1.15) in terms of $\widetilde{A}_v^u$ for $u \leq r$.

This leads to an interesting unsolved problem. The conditions $\widetilde{A}_v^u \geq 0$ are restrictions on A_i^r. For $r = 1$ these restrictions lead to linear programming bounds on d_1 and asymptotically to the McElice–Rumsey–Rodemich–Welch bound (Theorem 1.3.12), which is often the best asymptotic upper bound presently known.

PROBLEM 1.1.96. What is a proper generalization of the linear programming method for higher weights, $r > 1$? (There does not exist a direct analogue of the linear programming bound for higher weights.)

1.2. Examples and Constructions

In this section we present several examples of codes. Each example is in fact a method to construct a certain family of codes with rather good (in one sense or another) parameters. Since in many cases these families are natural ancestors of algebraic geometry codes, we try to choose constructions that are easy to generalize in that direction.

Section 1.2.1 is devoted to an example most important for us: to Reed–Solomon codes, which, as we shall see in Chapter 4, are precisely algebraic geometry codes of genus zero. We discuss them in detail, calculate the spectra, and give a decoding algorithm. In Sec. 1.2.2 we briefly discuss other interesting families of codes. Section 1.2.3 is devoted to a number of rather simple constructions that produce new codes starting with codes we already know.

1.2.1. Codes of Genus Zero

Recall that a *code of genus zero* (or an *MDS code*) is an $[n,k,d]_q$ code C such that $k+d = n+1$. This yields $(n-k) + d^\perp = n+1$; i.e., $C^\perp$ is also an MDS code (see Exercise 1.1.20).

Trivial codes. For any n there exist three simple q-ary codes which it is quite natural to call *trivial*. These are the $[n,n,1]_q$ code $C_1 = \mathbb{F}_q^n$ (called the *nonredundant code*), $[n,n-1,2]_q$ code C_2 consisting of all vectors $v = (v_1, \ldots, v_n) \in \mathbb{F}_q^n$ such that $\sum v_i = 0$ (called the *parity-check code*), and the one-dimensional $[n,1,n]_q$ code C_3 consisting of vectors of the form $(\alpha, \alpha, \ldots, \alpha)$, $\alpha \in \mathbb{F}_q$ (called the *repetition code*).

Now we are passing to a more conceptual construction.

Reed–Solomon codes. Let $\mathcal{P} = \{P_1, \ldots, P_n\} \subseteq \mathbb{F}_q$ be a subset of cardinality n. Consider the linear space $L(a)$ of all polynomials in one variable of degree at most a with coefficients in $\mathbb{F}_q$; $\dim L(a) = a+1$. For $n > a$, a nonzero polynomial $f(x) \in L(a)$ cannot vanish at all points of $\mathcal{P}$; moreover, there are at least $n-a$ points of $\mathcal{P}$ where it has nonzero values. The *evaluation map*

$$\mathrm{Ev}_\mathcal{P} \colon L(a) \to \mathbb{F}_q^n, \quad f \mapsto (f(P_1), \ldots, f(P_n)),$$

is injective for $n > a$, and its image is an $[n, a+1, n-a]_q$ code, known as a *Reed–Solomon code of degree a* (traditionally, this name is reserved for codes of length $n = q-1$ only, and others are called extended or shortened Reed–Solomon codes; however, for us it is natural to use the same name for all of them). The parameters of such a code satisfy the condition $k + d = n+1$, which is very good (according to Proposition 1.1.41, it could not be better); moreover, $k = a+1$ is an arbitrary integer chosen between 1 and n. Furthermore, Reed–Solomon codes form an *embedded family*: $L(a) \subset L(a+1)$. Unfortunately, their length is limited: $n \leq q$. In Sec. 4.4.1 we shall see that these codes can be naturally extended to codes on a projective line, with $k + d = n+1$ and $n \leq q+1$.

Take the basis $\{1, x, x^2, \ldots, x^a\}$ in $L(a)$. In this basis, the generator matrix of C is of the form (P_i^j), $j = 0, 1, \ldots, a$, $i = 1, 2, \ldots, n$; we identify a point of the line with the corresponding element of $\mathbb{F}_q$.

Dual code. Let us start with the simplest case.

EXERCISE 1.2.1. Let $\mathcal{P} = \mathbb{F}_q$ or $\mathcal{P} = \mathbb{F}_q^*$. Prove that in these cases $C^\perp$ is also a Reed–Solomon code with $a^\perp = n - a - 2$ and with parameters $[n, n-a-1, a+2]_q$. (Hint: Use the fact that $\sum_{x \in \mathbb{F}_q} x^i = 0$ for $i < q - 1$.)

Now let us find the dual code for the case of an arbitrary $\mathcal{P} \subseteq \mathbb{F}_q$. Set

$$g_0(x) = \prod_{P_i \in \mathcal{P}} (x - P_i)^{-1}, \qquad g_\ell(x) = x^\ell g_0(x).$$

Consider the vector space $\Omega(a)$ spanned by the functions $g_\ell(x)$ for $0 \leq \ell \leq n - a - 2$. Consider a function of the form $F(x) = f(x)g_0(x)$, f being a polynomial. As usual, define the *residue of F at P_i*:

$$\mathrm{Res}_{P_i} F = f(P_i) \prod_{j \neq i} (P_i - P_j)^{-1}.$$

EXERCISE 1.2.2 (residue formula). Prove that for $\deg f \leq n - 2$ we have

$$\sum_{P_i \in \mathcal{P}} \mathrm{Res}_{P_i} F = 0.$$

PROPOSITION 1.2.3. *The dual code $C^\perp$ for a Reed–Solomon code C is the image of $\Omega(a)$ under the map*

$$\mathrm{Res}_\mathcal{P} : g \mapsto (\mathrm{Res}_{P_1} g, \ldots, \mathrm{Res}_{P_n} g).$$

The code $C^\perp$ is equivalent to some Reed–Solomon code.

PROOF. Let $f \in L(a)$ and $g \in \Omega(a)$. Then

$$\sum_{P_i \in \mathcal{P}} f(P_i) \mathrm{Res}_{P_i} g = \sum_{P_i \in \mathcal{P}} \mathrm{Res}_{P_i}(fg) = 0.$$

Since $\dim L(a) + \dim \Omega(a) = n$, we get $C^\perp = \mathrm{Res}_\mathcal{P}(\Omega(a))$.

Note now that if $g(x) = h(x)g_0(x)$, $h(x)$ being a polynomial, then

$$\mathrm{Res}_{P_i} g = y_i h(P_i),$$

where

$$y_i = \prod_{j \neq i} (P_i - P_j)^{-1};$$

i.e., $C^\perp$ is obtained from the Reed–Solomon code $C' = \mathrm{Ev}_\mathcal{P}(L(a^\perp))$, where $a^\perp = n - a - 2$, by multiplying the ith coordinate of each vector by $y_i \in \mathbb{F}_q^*$. We write $C^\perp = y \cdot C'$, $y = (y_1, \ldots, y_n) \in (\mathbb{F}_q^*)^n$; such codes are called *generalized Reed–Solomon codes*. They are equivalent to Reed–Solomon codes in the sense of the definition in Sec. 1.1.1. □

In the cases $\mathcal{P} = \mathbb{F}_q$ and $\mathcal{P} = \mathbb{F}_q^*$, multiplication by y is an automorphism of C'.

Note that for n even and $a = n/2 + 1$, the code C is quasi-self-dual: $C^\perp = y \cdot C$; if $\mathcal{P} = \mathbb{F}_q$ or $\mathcal{P} = \mathbb{F}_q^*$, it is self-dual: $C^\perp = C$.

Spectra. Applying Theorem 1.1.18 to codes of genus zero, we get a complete answer.

PROPOSITION 1.2.4. *If C is an $[n,k,d]_q$ code of genus zero, then*

$$W_C(x) = x^n + \sum_{i=0}^{k-1} \binom{n}{i}(q^{k-i}-1)(x-1)^i.$$ □

EXERCISE 1.2.5. Establish this formula for Reed–Solomon codes by a direct computation. Check that for $i \neq 0$ we have

$$A_i = \binom{n}{i}(q-1)\sum_{j=0}^{i-d}(-1)^j\binom{i-1}{j}q^{i-d-j}.$$

Decoding. Consider an $[n, n-a-1, a+2]_q$ code C dual to a Reed–Solomon code of degree a. Recall that to decode C up to $t = \lfloor \frac{d-1}{2} \rfloor$ means to find an algorithm that, starting with some $v \in \mathbb{F}_q^n$ lying at distance at most t from some code vector $u \in C$, outputs this u. Let $v - u = e$, $\|e\| \leq t$; the vector e is called an *error vector*.

So we are given some $v \in \mathbb{F}_q^n$. Start with the calculation of the so-called *syndromes*:

$$s_j = s_j(v) = \sum_{P_i \in \mathcal{P}} v_i P_i^j, \quad 0 \leq j \leq 2t-1.$$

Note that if $u \in C$, then the corresponding sum is zero by the definition of the code C (since (P_i^j) is a parity-check matrix of the code). Hence, we have $s_j = \sum_{i \in I} e_i P_i^j$, where $I = \{i \mid e_i \neq 0\}$ denotes the (unknown) set of *error locators* (erroneous coordinates).

The next step is to find the polynomial

$$g(x) = \sum_{\ell=0}^{t} y_\ell x^\ell = c\prod_{i \in I}(x - P_i).$$

To this end, we solve (in undeterminants y_ℓ) the system of equations

$$\sum_{\ell=0}^{t} y_\ell s_{j+\ell} = 0, \quad 0 \leq j \leq t-1.$$

The sought-for function g is a solution of the system since $g(P_i) = 0$; hence,

$$\sum_{\ell=0}^{t} y_\ell s_{j+\ell} = \sum_{\ell=0}^{t}\sum_{P_i \in \mathcal{P}} y_\ell e_i P_i^{j+\ell} = \sum_{i \in I} e_i P_i^j g(P_i) = 0.$$

On the other hand, if $\{y'_\ell\}$ is another solution and $g' = \sum_{\ell=0}^{t} y'_\ell x^\ell$, then, setting

$$F_j(x) = \prod_{\substack{i \in I \\ i \neq j}}(x - P_i) = \sum_{k=0}^{t-1} b_k x^k,$$

for any $j \in I$ we have

$$e_j F_j(P_j) g'(P_j) = \sum_{i \in I} e_i F_j(P_i) g'(P_i) = \sum_{i \in I} \sum_{k=0}^{t-1} e_i b_k P_i^k g'(P_i)$$

$$= \sum_{k=0}^{t-1} \sum_{\ell=0}^{t} b_k y'_\ell \sum_{i \in I} e_i P_i^{k+\ell} = \sum_{k=0}^{t-1} b_k \left(\sum_{\ell=0}^{t} y'_\ell s_{k+\ell} \right) = 0.$$

Therefore, $g'(P_j) = 0$ for any $j \in I$; i.e., g is the only solution (up to a constant factor).

Expanding g into factors, we find the set of error locators I. Then we solve (in undeterminants e_i) the system

$$\sum_{i \in I} e_i P_i^j = s_j, \quad 0 \leq j \leq a.$$

The sought-for values of e_i satisfy the system. If $\{e'\}$ is another solution, then $\sum_{i \in I}(e_i - e'_i) P_i^j = 0$; i.e., the vector $e - e'$ belongs to C, but its weight is at most $2t \leq d-1$. The obtained contradiction shows that e is found uniquely.

REMARK 1.2.6. Further development of this approach leads to the classical Berlekamp–Massey algorithm. Another interesting approach to the decoding problem, the Guruswami–Sudan algorithm, is based on quite different ideas [GS99b, GR06]. We hope to consider it in *Advanced Chapters*.

1.2.2. Code Families

Let us discuss some other interesting examples. (In our exposition, discussing some well-known families of codes, we sometimes diverge in minor details from traditional constructions and definitions.)

First-order Reed–Muller codes. Consider the linear space L_m of all polynomials in m variables of degree 0 and 1; dim $L_m = m+1$. Let $\mathcal{P} = \{P_1, \ldots, P_n\} \subseteq \mathbb{F}_q^m$ be a subset of cardinality n such that no nonzero polynomial vanishes at all the points $P_1, \ldots, P_n$ (this is a fortiori the case if $n > q^{m-1}$ since the number of zeroes of a nonzero linear polynomial in m variables is at most q^{m-1}). The image of the evaluation map

$$\mathrm{Ev}_\mathcal{P} \colon L_m \to \mathbb{F}_q^n, \quad f \mapsto (f(P_1), \ldots, f(P_n)),$$

is an $[n, m+1, n - q^{m-1}]_q$ code with $n \leq q^m$.

Let us generalize this construction as follows. Consider all homogeneous linear forms in $m+1$ variables. Together with the zero form, they constitute a linear $\mathbb{F}_q$-space L'_m of dimension $m+1$. Now let $\mathcal{P} \subset \mathbb{F}_q^{m+1}$ be such that $P_i \neq (0, \ldots, 0)$ and, for any $\alpha \in \mathbb{F}_q^* \setminus \{1\}$, $P \in \mathcal{P}$ implies $\alpha P \notin \mathcal{P}$. Again consider the evaluation map. A nonzero form $f \in L'_m$ has at most q^m zeroes in $\mathbb{F}_q^{m+1}$; recall that $0 \notin \mathcal{P}$, and if a form f vanishes at $P_i \in \mathcal{P}$, then it also vanishes at all $\alpha P_i \notin \mathcal{P}$, $\alpha \neq 0, 1$. Therefore, the number of zeroes of f in $\mathcal{P}$ is at most $\frac{q^m - 1}{q - 1}$. The maximum possible cardinality of $\mathcal{P}$ is $\frac{q^{m+1} - 1}{q - 1}$ (for instance, let $\mathcal{P}$ be the set of all nonzero elements of $\mathbb{F}_q^{m+1}$ whose first nonzero coordinate is 1). We obtain an $\left[n, m+1, n - \frac{q^m - 1}{q - 1} \right]_q$

code for $n \leq \frac{q^{m+1}-1}{q-1}$, in particular, a code C with parameters
$$\left[\frac{q^{m+1}-1}{q-1}, m+1, q^m\right]_q.$$

This is a very good code, the best possible for $d/n > (q-1)/q$ (this code meets the Plotkin bound; see Theorem 1.1.45).

EXERCISE 1.2.7. Show that all nonzero code vectors of C are of weight q^m; i.e.,
$$W_C(x) = x^n + (q^{m+1}-1)x^{q^m}.$$

Hamming codes. Consider the code $C_H = C^\perp$ dual to the Reed–Muller code just considered. Theorem 1.1.18 makes it possible to find the weight spectrum of this code.

EXERCISE 1.2.8. Compute the spectrum.

The computations show that $d \geq 3$. However, this can also be seen without the spectrum. Indeed, the parity-check matrix of C_H has no proportional columns (if, for two points $P_1, P_2 \in \mathbb{F}_q^{m+1}$, all linear forms take proportional values, i.e., $f(P_1) = \alpha f(P_2)$ for some $\alpha \in \mathbb{F}_q^*$, then $P_1 = \alpha P_2$). Therefore, any two columns are linearly independent, and hence $d \geq 3$ (see Exercise 1.1.3).

Thus, we have constructed codes with parameters
$$[n, n-m-1, 3]_q, \quad n \leq \frac{q^{m+1}-1}{q-1}$$

(moreover, for $n = \frac{q^{m+1}-1}{q-1}$ we know the spectrum of such a code). These codes are very good among codes with $d = 3$. For $n = \frac{q^{m+1}-1}{q-1}$, they meet the Hamming bound (see Theorem 1.1.47).

Reed–Muller codes of order r. Let $r < m(q-1)$. Consider the linear space $L_m(r)$ of all polynomials in m variables of degree at most r. Fix a subset $\mathcal{P} = \{P_1, \ldots, P_n\} \subseteq \mathbb{F}_q^m$ and consider the evaluation map
$$\mathrm{Ev}_\mathcal{P} \colon L_m(r) \to \mathbb{F}_q^n, \quad f \mapsto (f(P_1), \ldots, f(P_n)).$$

Define $C = \mathrm{Ev}_\mathcal{P}(L_m(r))$. Finding the parameters of C is rather difficult. Let us for simplicity assume that $\mathcal{P} = \mathbb{F}_q^m$, $n = q^m$. The map $\mathrm{Ev}_\mathcal{P}$ is not injective in general. Indeed, $\mathrm{Ev}_\mathcal{P}(f) = \mathrm{Ev}_\mathcal{P}(f^q)$ for any function f. Denote by $L'_m(r)$ the linear space spanned by monomials of degree at most r that are not divisible by any q-power (i.e., by monomials of the form $x_1^{\alpha_1} \ldots x_m^{\alpha_m}$, $0 \leq \alpha_i \leq q-1$, $\sum \alpha_i \leq r$).

EXERCISE 1.2.9. Let
$$n = q^m, \quad r = a(q-1) + b \leq m(q-1), \quad 1 \leq b \leq q-1.$$
Prove that
$$C = \mathrm{Ev}_\mathcal{P}(L'_m(r)), \quad \mathrm{Ev}_\mathcal{P} \colon L'_m(r) \hookrightarrow \mathbb{F}_q^n,$$
$$k = \dim L'_m(r) = \sum_{i=0}^{r} \sum_{j=0}^{\lfloor i/q \rfloor} (-1)^j \binom{m}{j} \binom{m-1+i-qj}{m-1},$$
$$d = (q-b)q^{m-a-1}.$$

In particular, for $q=2$ we obtain a $\left[2^m, \sum_{i=0}^{r} \binom{m}{i}, 2^{m-r}\right]_2$ code. (*Hint:* To compute k, calculate the number of ways to arrange i objects in m cells so that no cell contains more than j objects; then apply the inclusion–exclusion principle. To compute d, use induction on m.)

Cyclic codes. Here we briefly touch on this important subject, which is mostly beyond the scope of this book.

A code $C \subseteq \mathbb{F}_q^n$ is called *cyclic* if it is invariant under a cyclic shift of coordinates; i.e., the condition $(c_1,\ldots,c_{n-1},c_n) \in C$ implies $(c_n, c_1 \ldots, c_{n-1}) \in C$. Note that, unlike most code properties, cyclicity is not an invariant of the equivalence class of a code.

The cyclicity condition is so strong that it allows us (in a sense) to describe a cyclic code. Let us identify $\mathbb{F}_q^n$ with the ring $R_n = \mathbb{F}_q[x]/(x^n - 1)$. A code is cyclic if and only if it is invariant under multiplication by x (indeed, if $c = c_0 + c_1 x + \ldots + c_{n-1} x^{n-1} \in R_n$, then $xc = c_{n-1} + c_0 x + \ldots + c_{n-2} x^{n-1}$ is its cyclic shift). The condition $xC \subseteq C$ is equivalent to C being an ideal of R_n. Since R_n is a principal ideal domain, we have $C = (g) \subseteq R_n$. Take for g the least-degree representative $g(x)$ in $\mathbb{F}_q[x]$. It is easily seen that $g(x) \mid (x^n - 1)$. Let $r = \deg g$; then the dimension of the code is $k = n - r$. An element $c(x) \in R_n$ belongs to the code if and only if there exists $f(x) \in \mathbb{F}_q[x]$, $\deg f \leq k$, such that $c(x) = f(x)g(x)$. The polynomial $g(x)$ is called the *generator polynomial* of the cyclic code C, and $h(x) = (x^n - 1)/g(x)$ is its *parity-check polynomial*.

EXERCISE 1.2.10. Prove that Reed–Solomon codes with $\mathcal{P} = \mathbb{F}_q^*$ are cyclic. The same holds for $\mathcal{P} = \mathbb{F}_p$ (for a prime p).

Let us impose an additional restriction $(n, q) = 1$; in this case the theory of cyclic codes can be considered from another point of view. With this restriction, the polynomial $x^n - 1$ has n distinct roots in $\overline{\mathbb{F}}_q$; let $\mathbb{F}_{q^m}$ be the splitting field of this polynomial, i.e., the smallest extension of $\mathbb{F}_q$ where $x^n - 1$ decomposes into linear factors.

EXERCISE 1.2.11. Check that m is the smallest integer for which $n \mid (q^m - 1)$.

The group $\mathbb{F}_{q^m}^*$ is cyclic. Its subgroup μ_n of nth roots of unity is also cyclic; fix its generator α. Then

$$x^n - 1 = \prod_{i=0}^{n-1}(x - \alpha^i).$$

Since $g(x) \mid (x^n - 1)$, we have $g(x) = \prod_{i \in I}(x - \alpha^i)$ for a subset $I \subseteq \{0, 1, \ldots, n-1\}$.

EXERCISE 1.2.12. Prove that coefficients of $g(x)$ belong to $\mathbb{F}_q$ if and only if $qI \equiv I \pmod{n}$.

Now it is quite clear that the condition $c(x) \in C$ is equivalent to the condition that $c(\alpha^i) = 0$ for all $i \in I$; therefore, a parity-check matrix of C can be written as follows (here, $I = \{i_1, \ldots, i_r\}$):

$$H = \begin{pmatrix} 1 & \alpha^{i_1} & \alpha^{2i_1} & \ldots & \alpha^{(n-1)i_1} \\ \vdots & & & & \vdots \\ 1 & \alpha^{i_r} & \alpha^{2i_r} & \ldots & \alpha^{(n-1)i_r} \end{pmatrix}.$$

Attention: Here H is a parity-check matrix in a Pickwickian sense; its entries do not lie in $\mathbb{F}_q$. Strictly speaking, each entry $h_{ij} \in H$ should be changed to a column of its coordinates in a basis of $\mathbb{F}_{q^m}$ over $\mathbb{F}_q$; the matrix H' thus obtained is a proper parity-check matrix of size $mr \times n$.

We are in the situation of the subfield restriction, discussed below in Sec. 1.2.3. Let there be given a linear $[n,k,d]_{q^m}$ code C'. Choose code vectors of C' with all coordinates lying in $\mathbb{F}_q$; i.e., consider the q-ary code $C = C' \cap \mathbb{F}_q^n$. Its parameters are at worst $[n, n-m(n-k), d]_q$.

It can in fact happen that some rows of H' are linearly dependent. For example, if $i_1 = qi_2$, then $\mathbb{F}_q$-expansions of α^{i_1} and α^{i_2} are obtained from each other by a permutation; hence, $\operatorname{rk} H' \leq ms$, $s = |I_0|$, where I_0 is such that $I_0 \cup qI_0 \cup \ldots \cup q^{m-1}I_0 = I$.

Let us summarize the results.

THEOREM 1.2.13. *Let $(n,q) = 1$. Any cyclic $[n,k,d]_q$ code C is the subfield restriction of an $[n,k',d']_{q^m}$ code C' with the parity-check matrix*

$$H = \begin{pmatrix} 1 & \alpha^{i_1} & \ldots & \alpha^{(n-1)i_1} \\ \ldots\ldots\ldots\ldots\ldots\ldots\ldots \\ 1 & \alpha^{i_r} & \ldots & \alpha^{(n-1)i_r} \end{pmatrix},$$

where m is the smallest integer such that $n \mid (q^m - 1)$, $\langle \alpha \rangle = \mu_n \subseteq \mathbb{F}_{q^m}^*$, and the set $I = \{i_1 \ldots, i_r\}$ has the property $qI \equiv I \pmod{n}$, $r = n - k'$; here, $d \geq d'$ and $k \geq n - mr$. Moreover, $k \geq n - ms$, where $s = \min |I_0|$ taken over all $I_0 \subseteq I$ such that $I_0 \cup qI_0 \cup \ldots \cup q^{m-1}I_0 = I$. □

It would also be good to know how to estimate the minimum distance of a cyclic code. In general, this is a very subtle question. Here is an important example.

BCH codes. Let $n \mid (q-1)$; i.e., $m = 1$. Consider a cyclic code with $I = \{0, 1, \ldots, r-1\}$. This is a Reed–Solomon code, and its dual (which is also a Reed–Solomon code) is constructed from polynomials of degree at most $r - 1$ evaluated at $\mathcal{P} = \{1, \alpha, \alpha^2, \ldots, \alpha^{n-1}\}$. For such codes we have $k = n - r$, and we know the minimum distance: $d = n + 1 - k = r + 1$. Let us generalize this example.

A *BCH code* is a cyclic code C such that

$$I \supseteq \{b, b+1, \ldots, b + \Delta - 2\}$$

for some $b \geq 0$ and $\Delta \geq 2$. We call Δ the *designed distance* of C. A *narrow sense BCH code* is a BCH code with $b = 1$, and a *primitive BCH code* is a BCH code of length $n = q^m - 1$ (in this case, α is a primitive element of $\mathbb{F}_{q^m}$).

EXERCISE 1.2.14. Prove that $d \geq \Delta$ for any BCH code. (*Hint*: There are at least two ways to do this. You can explicitly write out the condition for $\Delta - 1$ columns of H to be linearly dependent and check that this is a Vandermonde determinant; or you can reduce everything to the case $b = 0$, consider an $[n, k', d']_{q^m}$ Reed–Solomon code C', and note that $d \geq d'$.)

Then let us estimate the dimension, k, of C. Recall that $I \equiv qI \pmod{n}$; by Theorem 1.2.13 we have $n - k \leq ms$, where $s = \min |I_0|$ taken over $I_0 \subseteq I$ such that $I = I_0 \cup qI_0 \cup \ldots \cup q^{m-1}I_0$.

In what follows we assume that C is a narrow-sense primitive BCH code (i.e., $b=1$ and $n=q^m-1$) and $m \geq 2$. Put
$$I_1 = \{1, 2, \ldots, \Delta - 1\}, \qquad I_0 = I_1 \setminus (qI_1),$$
and let $I \subseteq \mathbb{Z}/n\mathbb{Z}$ be the smallest set such that $qI \equiv I \pmod{n}$ and $I \supseteq I_1$; thus, $s \leq |I_0| = \Delta - 1 - \left\lfloor \frac{\Delta - 1}{q} \right\rfloor$. We get
$$n - k \leq m\left(d - 1 - \left\lfloor \frac{d-1}{q} \right\rfloor\right).$$
Therefore, we have proved the following result.

THEOREM 1.2.15. *For any $m \geq 2$ and any $n = q^m - 1$ there exists a narrow-sense primitive q-ary BCH code with parameters*
$$\left[n = q^m - 1, \; k \geq n - m\left(d - 1 - \left\lfloor \frac{d-1}{q} \right\rfloor\right), \; \geq d\right]_q, \quad d = 2, 3, \ldots, 1 + \frac{nq}{m(q-1)}.$$
In particular, for $d = q\ell + 1$ and $n = q^m - 1$ we have $n - k \leq m\ell(q - 1)$. □

Of course, the most significant improvement, as compared with the standard estimate for the subfield restriction, is gotten for $q = 2$. At the end of Sec. 4.2.1 we return to the idea of "eliminating the qth powers" in a more refined version.

REMARK 1.2.16. Parameters of the codes of Theorem 1.2.15 are not always optimal. Choosing $b = 0$, i.e., $I_1 = \{0, 1, \ldots, \Delta - 2\}$, often gives better results. For example, for $q = 3$ and $d = 5$ this gives a code with better parameters, which in addition is optimal.

There are many other interesting cyclic codes; for example, such are post factum the Golay codes, some Goppa codes, Reed–Muller codes with $\mathcal{P} = \mathbb{F}_q^m \setminus \{0\}$, etc. Unfortunately, we cannot answer the following important question.

PROBLEM 1.2.17. *Do there exist infinite families of cyclic codes with $n \to \infty$ and $kd/n^2 \geq \varepsilon > 0$? (Such families are called asymptotically good; see Sec. 1.3.1.)*

REMARK 1.2.18. If we somewhat weaken the condition of cyclicity, this problem has a positive answer: for a fixed $s \geq 2$ there exist asymptotically good (as $n \to \infty$) *quasi-cyclic codes*, i.e., codes that are invariant under the cyclic s-shift
$$(c_1, \ldots, c_n) \mapsto (c_{s+1}, \ldots, c_n, c_1, \ldots, c_s);$$
moreover, for a fixed $s \geq 2$ there exist quasi-cyclic codes that asymptotically meet the Gilbert–Varshamov bound.

Here is another example of cyclic codes.

Quadratic-residue codes. Let $q = p$ be a prime, and let an odd ℓ be another prime such that p is a square modulo ℓ. Let I be the set of quadratic residues modulo ℓ. Recall that a *quadratic residue* is an element $i \in \mathbb{Z}/\ell\mathbb{Z}$ such that $i \equiv a^2 \pmod{\ell}$ for some a; i is a quadratic residue if and only if $i^{(\ell-1)/2} \equiv 1 \pmod{\ell}$. Since p is a quadratic residue, we have $pI \equiv I \pmod{\ell}$.

EXERCISE 1.2.19. Prove that a cyclic code C corresponding to the set I of quadratic residues (see Theorem 1.2.13) is an $\left[\ell, \frac{\ell+1}{2}, \geq \lceil \sqrt{\ell} \rceil\right]_p$ code. Let us add to each vector $v = (v_1, \ldots, v_\ell) \in C$ one coordinate more, $v_{\ell+1} = -\sum_{i=1}^{\ell} v_i$, and thus lengthen C by one position. Prove that for $\ell \equiv -1 \pmod 4$ the new code C' is self-dual.

Alternant codes. Let C_0 be a generalized Reed–Solomon code over $\mathbb{F}_{q^m}$ (the definition of such codes is given above; see the proof of Proposition 1.2.3). The code $C = C_0^\perp \cap \mathbb{F}_q^n$ is called an *alternant code*.

EXERCISE 1.2.20. Prove that the parameters of the alternant code C are at worst $[n, n - m(d-1), d]_q$.

There are a lot of alternant codes, much more than BCH codes; that is why there are many good codes among them (e.g., among primitive BCH codes there is no infinite family with $kd/n^2 \geq \varepsilon > 0$, whereas such families of alternant codes exist). Here is an important subclass of alternant codes.

Goppa codes. Let $G(z)$ be a polynomial with coefficients in $\mathbb{F}_{q^m}$. Let a subset $\mathcal{P} = \{P_1, \ldots, P_n\} \subseteq \mathbb{F}_{q^m}$ be such that $G(P_i) \neq 0$. Then the alternant code constructed from a generalized Reed–Solomon code with zeroes in $\mathcal{P}$ and with $y = (G(P_1)^{-1}, \ldots, G(P_n)^{-1})$ is called a *Goppa code*.

Let us give another definition of this code. For an arbitrary $v = (v_1, \ldots, v_n) \in \mathbb{F}_q^n$, define
$$R_v(z) = \sum_{i=1}^n \frac{v_i}{z - P_i}.$$
The code
$$C = \{v \in \mathbb{F}_q^n \mid R_v(z) \equiv 0 \pmod{G(z)}\}$$
is called a *Goppa code*.

EXERCISE 1.2.21. Check that these definitions are equivalent and prove that the parameters of the Goppa code are at worst $[n, n - mr, r+1]_q$, where $r = \deg G(z)$.

EXERCISE 1.2.22. Let $q = 2$. Show that if $G(z)$ has no multiple roots, then $d \geq 2r+1$. (Hint: In this case, $R_v(z) = f_v'(z)/f_v(z)$ for some polynomial $f_v(z)$, and $f_v'(z)$ is a square for any polynomial $f_v(z)$.)

Justesen codes. Let C_0 be an $[n, k, n+1-k]_{q^m}$ Reed–Solomon code, and let $y = (1, \alpha, \alpha^2, \ldots, \alpha^{n-1})$, where α is a primitive element of $\mathbb{F}_{q^m}$. Let
$$C = C_0 \oplus y \cdot C_0 \subseteq \mathbb{F}_{q^m}^{2n} \cong \mathbb{F}_q^{2mn},$$
i.e.,
$$C = \{(c_1, \ldots, c_n; c_1, \alpha c_2, \ldots, \alpha^{n-1} c_n) \mid (c_1, \ldots, c_n) \in C_0\},$$
and consider C as a q-ary code.

EXERCISE 1.2.23. Prove that C is a
$$\left[2mn, mk, \sum_{i=1}^\ell i \binom{2m}{i}(q-1)^i\right]_q \text{ code,}$$

where ℓ is the largest integer with the property $\sum_{i=1}^{\ell}\binom{2m}{i}(q-1)^i \le n-k+1$. (*Hint:* A string of the form $(c_1, c_1; c_2, \alpha c_2; \ldots; c_n, \alpha^{n-1} c_n)$ includes at least $n+1-k$ different nonzero q-ary vectors $(c_i, \alpha^{i-1} c_i)$ of length $2m$; estimate their aggregate weight.)

Golay code. Speaking of interesting examples, we mention with pleasure such a phenomenon as Golay codes. These codes can be described in several different ways.

Consider the field $\mathbb{F}_{2^{11}}$. We have $2^{11} - 1 = 2047 = 23 \times 89$; hence, the 23rd roots of 1 lie in this field. Let $\alpha \in \mathbb{F}_{2^{11}}$ be such a root. Let

$$C_{23} = \left\{ x \in \mathbb{F}_2^{23} \,\Big|\, \sum_{i=1}^{23} x_i \alpha^i = 0 \right\}.$$

The code C_{23} is in fact a $[23, 12, 7]_2$ code. It is easy to check that this code is given by Theorem 1.2.13 with $q = 2$, $m = 11$, $n = 23$, and $I_0 = \{1\}$; in this case,

$$I = \{1, 2, 3, 4, 6, 8, 9, 12, 13, 16, 18\}.$$

Its minimum distance is much larger than it is natural to expect. This code is, of course, cyclic. Its generator polynomial can be taken as either $g_1(x) = 1 + x + x^5 + x^6 + x^7 + x^9 + x^{11}$ or $g_2(x) = 1 + x^2 + x^4 + x^5 + x^6 + x^{10} + x^{11}$, depending on the choice of α; in fact, $x^{23} - 1 = (x-1)g_1(x)g_2(x)$.

The automorphism group M_{23} of the $[23, 12, 7]_2$ Golay code C_{23} is of order $23 \cdot 22 \cdot 21 \cdot 20 \cdot 48 = 10\,200\,960$. Its spectrum is

$$W_{23}(x) = x^{23} + 253 x^{16} + 506 x^{15} + 1288 x^{12} + 1288 x^{11} + 506 x^8 + 253 x^7 + 1.$$

Note that this polynomial is reciprocal:

$$W_{23}(x) = x^{23} W_{23}(1/x).$$

This code is quadratic-residue: the roots of its generator polynomial are α^i, i being a quadratic residue modulo 23 (the set of quadratic residues modulo 23 is I). Also, this code is related to quadratic residues modulo 11. Consider the vector $v = (1, 0, 1, 0, 0, 0, 1, 1, 1, 0, 1)$, where $v_i = 1$ if and only if i is a quadratic residue modulo 11. Let A be a matrix whose rows are formed by all cyclic shifts of this vector. Adding to A a row of ones, we get a matrix B. Then C_{23} is defined by the parity-check matrix

$$G = \boxed{\;E\;\big|\;B\;},$$

where $E = E_{12}$ is the identity matrix.

Even more wonderful is the code C_{24} obtained from C_{23} by adding the over-all parity check, i.e., adding the column $(1, 1, \ldots, 1, 0)$ to G. Its parameters are $[24, 12, 8]_2$, and its automorphism group M_{24} is 24 times as large: $|M_{24}| = 24 \cdot 23 \cdot 22 \cdot 21 \cdot 20 \cdot 48 = 244\,823\,040$. The group M_{24} is 5-transitive. The spectrum has weights 24, 16, 12, 8, and 0 only:

$$W_{24}(x) = x^{24} + 759 x^{16} + 2576 x^{12} + 759 x^8 + 1 = x^{24} W_{24}(1/x).$$

We have a similar situation over $\mathbb{F}_3$. Consider the field $\mathbb{F}_{3^5}$; $3^5 - 1 = 242 = 11 \cdot 22$. Let $\alpha \in \mathbb{F}_{3^5}$ be a primitive 11th root of 1. The code

$$C_{11} = \left\{ x \in \mathbb{F}_3^{11} \,\Big|\, \sum_{i=1}^{11} x_i \alpha^i = 0 \right\}$$

is an $[11,6,5]_3$ code. For this code, $I_0 = \{1\}$; $I = \{1,2,4,5,9\}$ is the set of quadratic residues modulo 11. This is a cyclic and quadratic-residue code:
$$x^{11} - 1 = (x-1)(x^5 + x^4 - x^3 + x^2 - 1)(x^5 - x^3 + x^2 - x - 1).$$

The generator matrix of C_{11} can be obtained in a similar manner, starting with $(0, 1, -1, -1, 1)$, i.e., using quadratic residues modulo 5.

Adding the overall parity check (the column $(-1, -1, -1, -1, -1, 0)$) results in a code C_{12} with parameters $[12, 6, 6]_3$, with automorphism group M_{12} of order $12 \cdot 11 \cdot 10 \cdot 9 \cdot 8 = 95040$, and the spectrum
$$W_{12}(x) = x^{12} + 264x^6 + 440x^3 + 24.$$

EXERCISE 1.2.24. *Prove all the above statements.* (*Attention:* This is rather difficult; use books on coding theory.)

Perfect codes. A code C is called *perfect* if $d = 2t + 1$ and $\mathbb{F}_q^n$ is the union of spheres of radius t centered at elements of the code. It is easily seen that this property depends on the code parameters only.

THEOREM 1.2.25. *Let q be a power of a prime, and let an $[n, k, d]_q$ code C (linear or nonlinear) be perfect. Then either $k = 0$; or $k = n$; or $q = 2$, $k = 1$, and $n = d$ is odd; or the parameters of C are exactly those of a Hamming or Golay code, i.e., $\left[n = \dfrac{q^m - 1}{q - 1}, n - m, 3\right]_q$ or $[23, 12, 7]_2$ or $[11, 6, 5]_3$.* □

Group codes. The notion of a cyclic code can be generalized to the case of any group G.

The *group algebra* of a group G is the algebra of functions
$$\mathbb{F}_q[G] = \{f \colon G \to \mathbb{F}_q\}$$
with multiplication defined as the convolution of functions:
$$(f_1 * f_2)(g) = \sum_{h \in G} f_1(h) f_2(h^{-1} g).$$

The group G acts on $\mathbb{F}_q[G]$ from the right: the action of $g \in G$ is defined as $(fg)(h) = f(hg)$. For a subgroup $H \subseteq G$, consider the invariant space
$$\mathbb{F}_q[G/H] = \{f \in \mathbb{F}_q[G] \mid fg = f \text{ for all } g \in H\}.$$

Any module of the form $\mathbb{F}_q[G/H]$ has a natural basis $f_1, \ldots, f_\ell$, where $\ell = [G : H]$, and the functions $\{f_i\}$ have the property
$$f_i(g_j H) = \begin{cases} 1 & \text{if } i = j, \\ 0 & \text{if } i \neq j, \end{cases}$$
$G = g_1 H \cup \ldots \cup g_\ell H$ being the decomposition of G into disjoint right cosets of H.

Thus, any subset $C \subseteq \mathbb{F}_q[G/H]$ can naturally be considered as a linear code since fixing a basis $\{f_1, \ldots, f_\ell\}$ defines an identification $\mathbb{F}_q[G/H] \cong \mathbb{F}_q^\ell$. The same is valid for any permutational G-module $M = \mathbb{F}_q[G/H_1] \oplus \ldots \oplus \mathbb{F}_q[G/H_m]$, where $H_1, \ldots, H_m$ are arbitrary subgroups of G, so that any subspace $C \subseteq M$ is a linear code over $\mathbb{F}_q$.

Now let $G \subseteq S_n \cap \operatorname{Aut} C$ be a subgroup of the automorphism group of a code $C \subseteq \mathbb{F}_q^n$ acting by permutations of coordinates. Let $B = \{e_1, \ldots, e_n\}$ be a basis of $\mathbb{F}_q^n$. The group G acts on B; hence B is represented as a disjoint union of

G-orbits: $B = \mathcal{O}_1 \cup \ldots \cup \mathcal{O}_m$, where $\mathcal{O}_i = Gb_i$ is an orbit of some $b_i \in B$. Let H_i be the stabilizer of b_i, so that there is an isomorphism of G-sets $\mathcal{O}_i \cong G/H_i$. Then $\mathbb{F}_q^n$ is identified with the G-module $\mathbb{F}_q[G/H_1] \oplus \ldots \oplus \mathbb{F}_q[G/H_m]$. In particular, if G acts transitively, then $\mathbb{F}_q^n$ is identified with $\mathbb{F}_q[G/H]$ and C with a G-submodule in $\mathbb{F}_q[G/H]$. Since $\mathbb{F}_q[G/H]$ is a right ideal in the group algebra $\mathbb{F}_q[G]$, the code C is also a right ideal in $\mathbb{F}_q[G]$ in this case.

For instance, if $\operatorname{Aut} C$ contains a cyclic subgroup of order n which permutes $e_1, \ldots, e_n$, then C is embedded in the group algebra $R_n = \mathbb{F}_q[x]/(x^n - 1)$ as an ideal; in other words, C is a cyclic code.

1.2.3. Constructions

There are many ways to construct new codes starting with known ones. The simplest were used in the proof of Lemma 1.1.39.

EXERCISE 1.2.26. Given an $[n,k,d]_q$ code, construct $[n+\ell,k,d]_q$, $[n-\ell,k-\ell,d]_q$, and $[n-\ell,k,d-\ell]_q$ codes. (Attention: If $k \notin \mathbb{Z}$, then by $k-\ell$ we should understand $\log_q(\lceil q^{k-\ell} \rceil)$.) If the given code is linear, new codes must also be linear.

Here we present some more constructions, starting with those that do not change q.

Direct sum. Let there be given two linear codes, $C_1 \subseteq \mathbb{F}_q^{n_1}$ and $C_2 \subseteq \mathbb{F}_q^{n_2}$. Their *direct sum*
$$C = C_1 \oplus C_2 \subseteq \mathbb{F}_q^{n_1+n_2}$$
consists of vectors $v = (v_1, v_2)$ with $v_1 \in C_1$ and $v_2 \in C_2$. It is easily seen that C is a linear $[n_1+n_2, k_1+k_2, d]_q$ code with $d = \min\{d_1, d_2\}$. We may also consider direct sums of any finite number of codes. If all these codes are equal, we get a *power* C^ℓ of a given code C; if C is an $[n,k,d]_q$ code, we obtain $[\ell n, \ell k, d]_q$ codes for all $\ell = 1, 2, \ldots$.

Tensor product. Consider the *tensor (Kronecker) product* of codes $C_1 \subseteq \mathbb{F}_q^{n_1}$ and $C_2 \subseteq \mathbb{F}_q^{n_2}$, i.e., the subspace
$$C = C_1 \otimes C_2 \subseteq \mathbb{F}_q^{n_1 n_2}$$
consisting of matrices such that all their rows belong to C_1 and all columns to C_2 (the space $\mathbb{F}_q^{n_1 n_2}$ being identified with the space of $n_1 \times n_2$ matrices). The code C is also called an *iterative code*.

EXERCISE 1.2.27. Prove that the obtained code C is an $[n_1 n_2, k_1 k_2, d_1 d_2]_q$ code.

Of course, we may consider a tensor product of any finite number of codes, in particular, a *tensor power* $C^{\otimes \ell}$ of a code. It has the parameters $[n^\ell, k^\ell, d^\ell]_q$.

Pasting. Start with an $[n_1, k, d_1]_q$ code $C_1 \colon \mathbb{F}_q^k \to \mathbb{F}_q^{n_1}$ and an $[n_2, k, d_2]_q$ code $C_2 \colon \mathbb{F}_q^k \to \mathbb{F}_q^{n_2}$. It is natural to consider the diagonal map
$$\mathbb{F}_q^k \xrightarrow{(C_1, C_2)} \mathbb{F}_q^{n_1} \oplus \mathbb{F}_q^{n_2} = \mathbb{F}_q^{n_1+n_2}.$$
Its image $C = (C_1 \,|\, C_2)$ is called the *pasting* of C_1 and C_2. It is easy to see that C is an $[n_1+n_2, k, d_1+d_2]_q$ code. If we apply this operation to one and the same code ℓ times, we obtain an ℓ-time *repetition* of a code; its parameters are $[\ell n, k, \ell d]_q$.

Code from an embedded pair. Let there be given a pair of embedded codes $C_1 \supset C_2$ with parameters $[n,k,d]_q$ and $[n,k-1,d+1]_q$. Choose a vector $v \in C_1$, $v \notin C_2$. An arbitrary vector from C_1 is of the form $w + xv$, $w \in C_2$, $x \in \mathbb{F}_q$. Extend the code C_1 by adding to $w + xv$ the $(n+1)$st coordinate equal to x. We obtain an $[n+1,k,d+1]_q$ code.

$(u\,|\,u+v)$ construction. Let $C_1 \subseteq \mathbb{F}_q^n$ and $C_2 \subseteq \mathbb{F}_q^n$ lie in the same space. Then we can consider the subspace $C = (C_1 \,|\, C_1 + C_2)$ consisting of vectors of the form $(u, u+v)$ with $u \in C_1$ and $v \in C_2$. It is easy to check that C is a $[2n, k_1 + k_2, d]_q$ code with $d = \min\{2d_1, d_2\}$.

Shortening by distance. We have shown in the proof of Theorem 1.1.43 that, starting with a linear $[n,k,d]_q$ code C, we can get an
$$\left[n - d, k - 1, \geq \left\lceil \frac{d}{q} \right\rceil \right]_q \text{ code.}$$

Shortening by dual distance. Let there be given a linear $[n,k,d]_q$ code C such that its dual code has minimum distance $d^\perp \leq k$. Choose a parity-check matrix H in such a way that it has a row v of weight precisely $d^\perp$. Delete from H this row and $d^\perp$ columns where v has nonzero coordinates. We obtain an $[n - d^\perp, k - d^\perp + 1, \geq d]_q$ code.

Parity check. Let $q = 2$. Then we can lengthen an $[n,k,d]_2$ code with an odd d by appending to a vector the sum of its coordinates. Since all vectors of the new code are of even weight, we obtain an $[n+1, k, d+1]_2$ code.

EXERCISE 1.2.28. Which of these constructions can be applied to nonlinear codes?

Now we pass to constructions of another type, where we change q. We start with linear constructions. Let C be a linear $[n,k,d]_q$ code, and let $q = r^m$.

Subfield restriction. Let $C' = C \cap \mathbb{F}_r^n \subseteq \mathbb{F}_q^n$ (we assume that $\mathbb{F}_r$ is naturally embedded in $\mathbb{F}_q$ as a subfield); $C' \subseteq \mathbb{F}_r^n$ is a linear $\mathbb{F}_r$-subspace. This is an $[n, k', d']_r$ code, and obviously
$$d' \geq d, \qquad k' \geq n - m(n - k).$$

This construction is reasonable to be applied to high-rate codes (i.e., to those with $R = k/n$ not too far from 1).

Concatenation. Let there be given, in addition to an $[n,k,d]_q$ code C with $q = r^m$, a linear $[N, m, D]_r$ code C_0 (we call C_0 the *inner* code and C the *outer* code). Define a new linear code C' as the image of the composition of maps
$$\mathbb{F}_r^{mk} \xrightarrow{\sim} \mathbb{F}_q^k \xrightarrow{C} \mathbb{F}_q^n \xrightarrow{\sim} (\mathbb{F}_r^m)^n \xrightarrow{(C_0,\ldots,C_0)} \mathbb{F}_r^{Nn}.$$

The code C' is called the *concatenation* of C and C_0, or simply a *concatenated* code.

EXERCISE 1.2.29. Prove that the minimum distance of C' is at least dD.

Thus, we have constructed an $[Nn, km, Dd]_r$ code (up to spoiling).

If for the inner code we take a nonredundant $[m, m, 1]_r$ code, the construction is called the *field descent*; it gives an $[nm, km, d]_r$ code.

Furthermore, inner codes in this construction can be different. In addition to an $[n,k,d]_q$ code C with $q=r^m$, let there be given n more linear $[N,m]_r$ codes $C_1,\ldots,C_n$ such that $d(C_i) \geq D$ for each $i=1,\ldots,n$. The image C' of the composition of maps

$$\mathbb{F}_r^{mk} \xrightarrow{\sim} \mathbb{F}_q^k \xrightarrow{C} \mathbb{F}_q^n \xrightarrow{\sim} (\mathbb{F}_r^m)^n \xrightarrow{(C_1,\ldots,C_n)} \mathbb{F}_r^{Nn}$$

is called the *concatenation of C and $C_1,\ldots,C_n$* (or simply a *concatenated* code).

EXERCISE 1.2.30. Show that in this case C' is also an $[Nn, km, Dd]_r$ code up to spoiling.

EXERCISE 1.2.31. Show that Justesen codes can be obtained as concatenated.

EXERCISE 1.2.32. Consider the case where the outer code C originates from a code over $\mathbb{F}_r$; i.e., all coefficients of its generator matrix G lie in fact in $\mathbb{F}_r$. Let $C' \subseteq \mathbb{F}_r^n$ be the code generated by the matrix G over $\mathbb{F}_r$. Prove that in this case the concatenated code coincides with $C_0 \otimes C'$ (i.e., with the iterative code constructed from C_0 and C').

Generalized concatenation. We start with a family of $[n,k_i,d_i]_{q_i}$ codes C_i, $i=1,\ldots,\ell$, and an embedded family

$$\{0\} = C'_0 \subset C'_1 \subset \ldots \subset C'_\ell$$

of $[N, K_i, D_i]_r$ codes such that $K_i = \sum_{j=1}^{i} m_j$ and $q_i = r^{m_i}$. Represent C'_ℓ as a direct sum $C'_\ell = \bigoplus_{j=1}^{\ell} V_j$, so that $C'_i = \bigoplus_{j=1}^{i} V_j$. Let W_i be the concatenation of V_i and C_i (note that $\dim V_i = m_i$). Set $C = \sum_{i=1}^{\ell} W_i$.

EXERCISE 1.2.33. Show that the sum is direct; i.e., $C = \bigoplus_{i=1}^{\ell} W_i$. Show that C is an

$$\left[Nn, \sum_{i=1}^{\ell} m_i k_i, \geq \min_{1 \leq i \leq \ell}\{d_i D_i\}\right]_r \text{ code.}$$

REMARK 1.2.34. It is possible to generalize both the concatenation and generalized concatenation to the case of nonlinear codes. The same is true for the subfield restriction (alphabet restriction), but instead of C' we should take "the best" of its shifts.

As we see from Exercise 1.2.32, the iterative construction (tensor product) is a particular case of concatenation. A similar particular case for the generalized concatenation is as follows.

Generalized iterative construction. Let there be given a linear $[N,k,d]_q$ code C' and a set of k linear $[n,k_i,d_i]_q$ codes C_i of the same length, $i=1,\ldots,k$. Let us represent codewords of C' as rows and codewords of C_i as columns.

Fix a basis $(c_1,\ldots,c_k)$ of C'. Denote by G the generator matrix of C with rows $c_1,\ldots,c_k$. For each i, $1 \leq i \leq k$, denote by C'_i the linear span of $c_1,\ldots,c_i$. Thus, C'_i is a code of length N and dimension i, and

$$C'_1 \subset C'_2 \subset \ldots \subset C'_k = C'.$$

Denote by D_i the minimum distance of the code C'_i. Note that in the above notation, all $m_i = 1$ and each V_i, $i = 1, \ldots, k$, is a one-dimensional $\mathbb{F}_q$-space spanned by the vector c_i.

Consider the set M consisting of all $n \times k$ matrices whose ith column lies in C_i for any i, $1 \leq i \leq k$. Obviously, M is a linear space of dimension

$$\dim M = \sum_{i=1}^{k} k_i.$$

Finally, consider a linear code

$$C = \{AG \mid A \in M\}.$$

The following statement corresponds to the result of Exercise 1.2.33.

PROPOSITION 1.2.35. *The obtained code C is an $[Nn, K, D]_q$ code with*

$$K = \sum_{i=1}^{k} k_i, \qquad D \geq \min\{d_i D_i \mid 1 \leq i \leq k\}. \qquad \square$$

REMARK 1.2.36. Of course, the construction of C depends not only on the choice of codes $C', C_1, \ldots, C_k$ but also on the choice of a basis $(c_1, \ldots, c_k)$ in C'.

REMARK 1.2.37. If we take for C_i one and the same code B, the construction yields the tensor product $C = B \otimes C'$, for which (Exercise 1.2.27) we have $D = d(B)d(C')$.

As an example of using this construction, consider the following exercise.

EXERCISE 1.2.38. Let C' be a Reed–Solomon code of length h and dimension $r + 1$ (where $2 \leq h \leq q$ and $1 \leq r < h$). Let the subcodes C'_i, $1 \leq i \leq r+1$, be Reed–Solomon codes of dimension i with minimum distance $D_i = h + 1 - i$. Let $C_1 = \ldots = C_{s+1}$ be one and the same $[n, k, d]$ code, and let $C_{s+2} = \ldots = C_h$ be the $[n, 1, n]$ repetition code. Check that the resulting code C has the parameters

$$N = hn, \qquad K = (s+1)k + r - s, \qquad D \geq \min\{d(h-s), n(h-r)\}.$$

Now we shall present two essentially nonlinear constructions. Let there be given an arbitrary $[n, k, d]_q$ code C, $k \in \mathbb{R}$, $q \in \mathbb{Z}$, $q \geq 2$. Let $M = q^k = |C|$.

Alphabet extension. Assume that $r \geq q$. Let us embed an alphabet A of cardinality q in an alphabet B of cardinality r. Then the embedding $C \hookrightarrow A^n \hookrightarrow B^n$ gives an $[n, k \log_r q, d]_r$ code.

Alphabet restriction. Now let $r \leq q$. Embed $B \subseteq A$ and introduce the structure of an abelian group on A (say, $A \cong \mathbb{Z}/q\mathbb{Z}$). Consider all the q^n shifts C_v of the code C by vectors $v \in A^n$, $C_v = \{v + c \mid c \in C\}$. In the totality of the sets C_v, each word of A^n occurs M times. Consider the intersections $B^n \cap C_v$; the total cardinality of these q^n sets is Mr^n. Hence, there exists a shift C_v such that $|B^n \cap C_v| \geq M\left(\dfrac{r}{q}\right)^n$.

Since shifting does not change the minimum distance of a code, we have obtained an $[n, \geq n - (n-k)\log_r q, \geq d]_r$ code.

Decoding. When we construct a code using another one or several others, there is a question how to decode the code we get, assuming that we know decoding algorithms for the codes that we start with.

For most of the above constructions, there is no problem finding a decoding algorithm; an interesting question is how to construct a "good" algorithm (from the point of view of complexity theory). For example, if there is a decoding algorithm for a q-ary code $C \subseteq \mathbb{F}_q^n$, the same algorithm decodes the field restriction $C' = C \cap \mathbb{F}_r^n \subseteq \mathbb{F}_r^n$.

With regard to the decoding of concatenated codes, the situation is more complicated. We need the following notion: an *erasure* is an error whose position is known. Assume that we have transmitted a vector $u \in C$ and received a vector $v \in \mathbb{F}_q^n$, $v = u + e + \varepsilon$, $I = \{i \mid e_i \neq 0\}$, $J = \{j \mid \varepsilon_j \neq 0\}$, where the set I is unknown to us (it is called the *set of error locators*) and J is known (it is called the *set of erasure locators*). The vector e is called an *error vector*, and ε an *erasure vector*. If there exists an algorithm which, starting with v, finds the nearest code vector u for any e and ε such that $|I| \leq t$ and $|J| \leq s$, the code is said to *correct t errors and s erasures*.

EXERCISE 1.2.39. Assume that we are given an algorithm which decodes any shortening $C' \subseteq \mathbb{F}_q^{n'}$ of a code C and corrects up to $t' \leq \left\lfloor \dfrac{d'-1}{2} \right\rfloor$ errors, where C is an $[n,k,d]$ code and $d' = d - (n - n')$. Construct a decoding algorithm for C correcting any t errors and s erasures for $2t + s \leq d - 1$.

Here is a result concerning the decoding of concatenated codes.

PROPOSITION 1.2.40. *Let C be a linear $[n,k,d]_q$ code, $q = r^m$, which possesses a decoding algorithm correcting t errors and s erasures for any t and s such that $2t + s \leq d' - 1$, $d' \leq d$. Let an $[N,m,D]_r$ code C_0 possess a decoding algorithm correcting any $T \leq \left\lfloor \dfrac{D'-1}{2} \right\rfloor$ errors, $D' \leq D$. Then the concatenated $[nN, km, dD]_r$ code C' possesses a decoding algorithm correcting any $u \leq \left\lfloor \dfrac{d'D'-1}{2} \right\rfloor$ errors.* □

1.3. Asymptotic Problems

Parameters of codes of large length n are of particular interest. As is often the case, in problems that have no simple solution for any specific n (in particular, in the problem of finding the best possible code parameters), passing to the limit as $n \to \infty$ smoothes out "random deviations" and clarifies the behaviour of parameters. In this section, we give precise settings of problems and discuss those results towards their solution that do not require using algebraic geometry codes. We shall again address asymptotic problems in Sec. 4.5, since such problems most strikingly illustrate the power and capabilities of algebraic geometry methods.

In Sec. 1.3.1 we state the main asymptotic problem (namely, $n, k, d \to \infty$; $k/n, d/n \to \text{const}$). In Sec. 1.3.2 we discuss known results towards its solution (except for those involving algebraic geometry), i.e., upper and lower asymptotic bounds. Section 1.3.3 deals with the problem of effective construction of asymptotically good codes; Sec. 1.3.4 is devoted to results in this area (polynomial bounds), again except for those involving algebraic geometry. In Sec. 1.3.5 we discuss other asymptotic settings (e.g., $n, k \to \infty$ and $d = \text{const}$) and present some results.

1.3.1. Main Asymptotic Problem

Fix the alphabet cardinality q. To a q-ary code C, assign its relative parameters $(\delta(C), R(C))$, i.e., a point in the unit square $[0,1]^2$ on the plane with coordinates (δ, R). All such points form a family of code points (code domain) V_q. Denote by U_q the set of limit points of this family. In other words, $(\delta, R) \in U_q$ if and only if there exists an infinite sequence of different codes C_i such that

$$\lim_{i \to \infty}(\delta(C_i), R(C_i)) = (\delta, R);$$

since there are only finitely many codes of any given length, for any sequence we have $n(C_i) \to \infty$. If $\delta > 0$ and $R > 0$, such a sequence (family) of codes C_i is called *asymptotically good*. In this situation, we sometimes just say "good codes".

THEOREM 1.3.1. *There exists a continuous function $\alpha_q(\delta)$, $\delta \in [0,1]$, such that*

$$U_q = \{(\delta, R) \mid 0 \leq R \leq \alpha_q(\delta)\}.$$

Moreover, $\alpha_q(0) = 1$, $\alpha_q(\delta) = 0$ for $(q-1)/q \leq \delta \leq 1$, and $\alpha_q(\delta)$ decreases on the segment $[0, (q-1)/q]$.

PROOF. Let $(\delta_0, R_0) \in U_q$. Consider the corresponding family of codes C_i and apply the spoiling lemma to each of them. For a given $[n, k, d]_q$ code, Lemma 1.1.39b yields an $[n-\ell, k-\ell, d]_q$ code for any $\ell \leq k$. The corresponding code points lie on the segment of the straight line $R = 1 - \delta\dfrac{n-k}{d}$, which passes through the points $\left(\dfrac{d}{n}, \dfrac{k}{n}\right)$ and $(0, 1)$, below and to the left of the point $\left(\dfrac{d}{n}, \dfrac{k}{n}\right)$. If $[n_i, k_i, d_i]_q$ are the parameters of C_i, $n_i \to \infty$, $k_i/n_i \to R_0$, and $d_i/n_i \to \delta_0$, then the points thus obtained fill the segment more and more densely (with step $1/n_i$); therefore, in the limit we obtain that any point of the straight line passing through (δ_0, R_0) and $(0, 1)$ belongs to U_q for $\delta \geq \delta_0$ (see Fig. 1.1).

Similarly, by Lemma 1.1.39c we get a straight line segment joining (δ_0, R_0) with $(1, 0)$ for $\delta \leq \delta_0$. The line is given by $R = R_0(1-\delta)/(1-\delta_0)$.

Put $\alpha_q(\delta) = \sup\{R \mid (\delta, R) \in U_q\}$; this function satisfies the first condition of the theorem. Let us check the continuity.

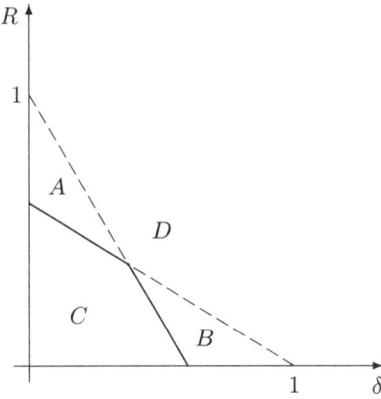

FIGURE 1.1.

Denote by $A = A(\delta_0, R_0)$ and $B = B(\delta_0, R_0)$, respectively, the lower and upper triangles between the straight lines $L_1 = L_1(\delta_0, R_0)$ and $L_2 = L_2(\delta_0, R_0)$ given by $R = 1 - (1 - R_0)\delta/\delta_0$ and $R = R_0(1 - \delta)/(1 - \delta_0)$ respectively (see Fig. 1.1). Let $R_0 = \alpha_q(\delta_0)$—i.e., (δ_0, R_0) is a point on the boundary of the code domain—and let $\delta_0 R_0 \neq 0$. Let us show that all other points of the boundary belong to the triangles A and B (this immediately implies the continuity of the boundary at the point (δ_0, R_0)). Indeed, if a point (δ_1, R_1) of the boundary lies in the area $C(\delta_0, R_0)$ (below both lines), then, since the point with the same δ_1 and larger R lying on the segments of $L_1(\delta_0, R_0)$ and $L_2(\delta_0, R_0)$ belongs to the code domain, we obtain a contradiction: (δ_1, R_1) cannot lie on the boundary. If a point (δ_1, R_1) of a boundary lies in the area $D(\delta_0, R_0)$ (above both lines), then the point (δ_0, R_0) lies in the area $C(\delta_1, R_1)$ and cannot be a boundary point. The fact that the boundary lies in sectors A and B proves also that the boundary is a decreasing function (up to the point where $\alpha_q(\delta)$ becomes zero).

Now let us show that $\alpha_q(\delta) = 0$ for $\delta \geq (q-1)/q$. Apply the Plotkin bound (Theorem 1.1.45): for any code, we have

$$\delta \leq \frac{q^k(q-1)}{(q^k-1)q}.$$

If (δ, R) is a code point with $R > 0$ and it is the limit point for a system of $[n_i, k_i, d_i]_q$ codes, then $k_i \to \infty$ and therefore $\delta \leq (q-1)/q$. Thus, for $\delta > (q-1)/q$ we have $\alpha_q(\delta) = 0$; by continuity, $\alpha_q((q-1)/q) = 0$ too.

It only remains to prove that $\alpha_q(\delta) > 0$ for $0 \leq \delta < (q-1)/q$. This follows from Theorem 1.3.20, which is proved below. □

We have given a detailed "geometrical" version of the proof; cf. Remark 1.3.4 below. An analytical version can be given as follows.

SKETCH OF THE PROOF. Let C be an arbitrary $[n, k, d]_q$ code. For a fixed $i = 1, \ldots, n$ there exists a subset consisting of q^{k-1} codewords with the same ith coordinates. Taking only these words and deleting the ith coordinate, we obtain a code with parameters $[n-1, \geq k-1, d]$. Repeating the procedure t times, we get

an $[n, \geq k-t]$ code with distance $\min(d, n-t)$. Thus,
$$R(\delta) \leq \tau + (1-\tau)R(\delta/(1-\tau)), \quad 0 \leq \tau \leq 1-\delta.$$
Now consider the variables $0 < \xi < \eta < 1/2$ such that $\xi/\eta = 1-\tau$. Then the above relation can be rewritten as
$$0 \leq R(\xi) - (1-\tau)R(\eta) \leq \tau.$$
To prove the continuity of $R(\delta)$, it remains to pass to the limit as $\tau \to 0$. □

Now, let q be a power of a prime.

Define code domains V_q^{lin} and U_q^{lin} for linear codes similarly to the definitions of V_q and U_q (consider only pairs (δ, R) that correspond to linear q-ary codes).

EXERCISE 1.3.2. Check that Theorem 1.3.1 is also valid for U_q^{lin} and α_q^{lin} (instead of U_q and α_q).

Obviously, $\alpha_q^{\text{lin}}(\delta) \leq \alpha_q(\delta)$.

Finding the functions α_q and α_q^{lin} is undoubtedly one of the central problems of coding theory. Presently, we are very far from its solution. Moreover, we do not even know any answers to the following questions:

PROBLEM 1.3.3. Are these functions differentiable in the interval $(0, (q-1)/q)$?

It is rather easy to check the differentiability at the points 0 and $(q-1)/q$ (EXERCISE!).

REMARK 1.3.4. If we were able to spoil a code "more smoothly", keeping its parameters not between the dotted and solid lines in Fig. 1.1 but between two smooth curves tangent to each other at the code point (say, two parabolas), we would not only prove that $\alpha_q(\delta)$ is differentiable but could also define it by a differential equation.

PROBLEM 1.3.5. Are these functions ∪-convex?

PROBLEM 1.3.6. Is it true that $\alpha_q(\delta) = \alpha_q^{\text{lin}}(\delta)$?

1.3.2. Asymptotic Bounds

Upper bounds. Since we have no answers to the above questions, we have to content ourselves with upper and lower bounds on α_q and α_q^{lin}. Let us see what the results of Sec. 1.1.4 look like as $n \to \infty$.

The Singleton bound gives
$$\alpha_q(\delta) \leq 1 - \delta.$$
The Griesmer bound yields
$$\alpha_q^{\text{lin}}(\delta) \leq 1 - \frac{q}{q-1}\delta.$$
The same inequality is also valid for nonlinear codes:

THEOREM 1.3.7 (asymptotic Plotkin bound). *We have*
$$\alpha_q(\delta) \leq R_P(\delta) = 1 - \frac{q}{q-1}\delta.$$
□

EXERCISE 1.3.8. Prove the theorem. (*Hint*: See the proof of Theorem 1.3.1; the Plotkin bound yields $\alpha_q\left(\frac{q-1}{q}\right) = 0$, and the rest follows from the spoiling lemma.)

Introduce the *q-ary entropy function*
$$H_q(x) = x \log_q(q-1) - x \log_q x - (1-x) \log_q(1-x)$$
(this is the entropy of a symmetric q-ary channel with error probability x, i.e., the entropy of the distribution $p_0 = 1-x$, $p_1 = \ldots = p_{q-1} = x/(q-1)$).

THEOREM 1.3.9 (Hamming bound). *We have*
$$\alpha_q(\delta) \leq R_H(\delta) = 1 - H_q(\delta/2).$$
□

EXERCISE 1.3.10. Prove that
$$\frac{1}{n} \log_q \left(\sum_{i=0}^{t} \binom{n}{i} (q-1)^i \right) = H_q\left(\frac{t}{n}\right) + o(1)$$
as $n \to \infty$, $t/n \to$ const. Using this fact, derive Theorem 1.3.9 from Theorem 1.1.47. Prove that
$$\frac{1}{n} \log_q \left(\binom{n}{t} (q-1)^t \right) = H_q\left(\frac{t}{n}\right) + o(1).$$
(*Hint*: Use the Stirling formula.)

THEOREM 1.3.11 (Bassalygo–Elias bound). *We have*
$$\alpha_q(\delta) \leq R_{BE}(\delta) = 1 - H_q\left(\frac{q-1}{q} - \frac{q-1}{q}\sqrt{1 - \frac{q\delta}{q-1}}\right).$$

PROOF. By Theorem 1.1.50 (as $n \to \infty$, $d/n \to \delta$, $\frac{1}{n}\log_q A(n,d) = k/n \to R$, $w/n \to \omega$), we have
$$R \leq 1 - H_q(\omega) + \frac{1}{n} \log_q \left[\frac{\delta}{\delta - 2\omega + \frac{q}{q-1}\omega^2} \right] \to 1 - H_q(\omega)$$
provided that
$$\omega^2 - 2\frac{q-1}{q}\omega + \frac{q-1}{q}\delta \geq \varepsilon > 0;$$
letting ε tend to zero and choosing the largest $\omega \leq 1$ with this property, i.e.,
$$\omega = \frac{q-1}{q} - \sqrt{\left(\frac{q-1}{q}\right)^2 - \frac{q-1}{q}\delta},$$
we obtain the theorem. □

The linear programming bound (Theorem 1.1.58) can also be used to get upper bounds, but this involves rather subtle technique, which is beyond the scope of our book. Here is the result:

THEOREM 1.3.12 (McEliece–Rodemich–Rumsey–Welch bound). *We have*
$$\alpha_q(\delta) \leq R_4(\delta) = H_q\left(\frac{(q-1) - \delta(q-2) - 2\sqrt{(q-1)\delta(1-\delta)}}{q}\right).$$
□

Linear programming applied to constant-weight codes (see the proof of Theorem 1.1.50) leads for $q=2$ to the following result.

THEOREM 1.3.13 (the second McEliece–Rodemich–Rumsey–Welch bound). *We have*
$$\alpha_2(\delta) \le R_{4(2)}(\delta) = \min_{0<u\le 1-2\delta} \left(1 + h(u^2) - h(u^2 + 2\delta u + 2\delta)\right),$$
where
$$h(x) = H_2\left(\frac{1-\sqrt{1-x}}{2}\right). \qquad \Box$$

Sometimes the function R_4 is called for short the MRRW bound. Note that for $q=2$ the bound $R_{4(2)}$ coincides with R_4 for $\delta \ge 0.273$ (since under this restriction the minimum is attained for $u = 1-2\delta$).

For $q=2$ all the above-mentioned bounds are $\cup$-convex, and $R_{4(2)}$ is the best known bound for all $\delta \in (0, 1/2)$. The behaviour of upper bounds for $q=2$ is shown in Fig 1.2a below.

For all $q \ge 3$, the Bassalygo–Elias bound and R_4 (see Fig. 1.2b,c) are no longer $\cup$-convex and (except for the case R_{BE} for $q=3$) no longer better than the Plotkin bound on the whole interval $(0, 1-1/q)$, which considerably complicates the situation with upper bounds.

Let us present known results in this direction.

THEOREM 1.3.14. *Let $q>2$. Then*
$$\alpha_q(\delta) \le R'_A(\delta) = 1 - \frac{\delta q}{q-2}\log_q(q-1)$$
for all $\delta \in \left[0, \left(\frac{q-2}{q}\right)^2\right]$. $\qquad\Box$

EXERCISE 1.3.15. Prove the theorem. (*Hint*: Apply the spoiling lemma to R_4.)

Linear programming applied to constant-weight codes yields for $q>2$ the following result (an analogue of the $R_{4(2)}$ bound):

THEOREM 1.3.16 (Aaltonen bound). *For $q>2$, we have*
$$\alpha_q(\delta) \le R_A(\delta) = \min\{1 - H_q(\omega) + f_q(\xi, \eta)\},$$
where
$$f_q(\xi,\eta) = H_q(\eta) + (1-\eta)H_q\left(\frac{\xi-\eta}{1-\eta}\right) - \xi\log_q(q-1) + \eta\log_q(q-2)$$
$$= H_q(\xi) + \xi H_q(\eta/\xi) - (\xi+\eta)\log_q(q-1) + \eta\log_q(q-2),$$
and the minimum is taken over ω, η, and ξ satisfying the conditions
$$0 \le \omega \le 1, \qquad 0 \le \eta \le (q-2)\omega/(q-1),$$
$$0 \le \xi - \eta \le \min\{\omega - \eta, 1-\omega\},$$
$$\beta = (1-\eta)h\left(\frac{\omega-\eta}{1-\eta}, \frac{\xi-\eta}{1-\eta}\right) \le \omega - \frac{q-1}{q-2}\eta,$$
$$\delta \ge 2\gamma + \left(2\beta + (\omega-\beta)K_{q-1}\left(\frac{\eta}{\omega-\beta}\right)\right)(1-\tau),$$

where
$$h(x,y) = \frac{x(1-x) - y(1-y)}{1 + 2\sqrt{y(1-y)}}, \quad 0 \leq x, y \leq 1,$$
$$K_{q-1}(x) = \frac{q-2}{q-1} - \frac{q-3}{q-1}x - \frac{2}{q-1}\sqrt{(q-2)x(1-x)}, \quad 0 \leq x \leq 1. \qquad \square$$

Note that for $\xi = \eta = 0$ the minimized expression $1 - H_q(\omega) + f_q(\xi, \eta)$ (under the given restrictions on the parameters) turns into the Bassalygo–Elias bound R_{BE}, and for
$$\omega = \frac{q-1}{q}, \qquad \eta = \frac{q-2}{q-1}\xi$$
into the R_4 bound. Thus, the Aaltonen bound improves both (see Fig. 1.2b,c).

Applying Theorem 1.1.57 yields the following result.

THEOREM 1.3.17. *Let $\delta \in (0, 1 - 1/q)$, and let μ and τ be arbitrary quantities satisfying the conditions*
$$0 \leq \delta - 2\tau \leq 1 - \mu, \qquad 0 \leq \tau \leq \frac{q-1}{q}\mu.$$
Denote $x = \dfrac{2\tau}{\mu}$ and $y = \dfrac{\delta - 2\tau}{1 - \mu}$. Then for $x \neq \delta$ we have
$$\alpha_q(\delta) \leq R(y) + (1 - H_q(x/2) - R(y))\frac{\delta - y}{x - y},$$
where $R(y)$ is any upper bound on α_q at the point y. $\qquad \square$

EXERCISE 1.3.18. Prove the theorem. (*Hint*: Use Theorem 1.1.57 combined with the asymptotics for $|B_t|$, Exercise 1.3.10.)

Since $1 - H_q(x/2) = R_H(x)$ (the Hamming bound), taking $(x_0, R_H(x_0)) = P$ and $(y_0, R(y_0)) = Q$ to be the tangency points of the common tangent to the graphs of the Hamming bound and any lower bound $R(\delta)$, we see that the straight line segment PQ is an upper bound for $\alpha_q(\delta)$. In particular, if for $R(\delta)$ we choose $R_4(\delta)$, we obtain the *Laihonen–Litsyn upper bound*, R_{LL}, which for large q improves (on some intervals of δ) the Aaltonen bound (see Fig. 1.2c).

Comparison of bounds for various q is presented in diagrams in Sec. A.2.2 of the Appendix.

Here we finish discussing upper (i.e., possibility) bounds and pass to existence bounds.

Lower bounds. Let
$$R_{GV}(\delta) = 1 - H_q(\delta).$$
This curve is known as the *Gilbert–Varshamov curve*.

EXERCISE 1.3.19. Check the following facts. On the segment $[0, (q-1)/q]$, the curve $R_{GV}(\delta)$ is (infinitely) differentiable and ∪-convex; also, $R_{GV}(0) = 1$ and $R_{GV}((q-1)/q) = 0$. For $\delta \to 0$ we have the asymptotics
$$R_{GV}(\delta) = 1 + \delta \log_q \delta + o(\delta \log_q \delta).$$

1.3. ASYMPTOTIC PROBLEMS

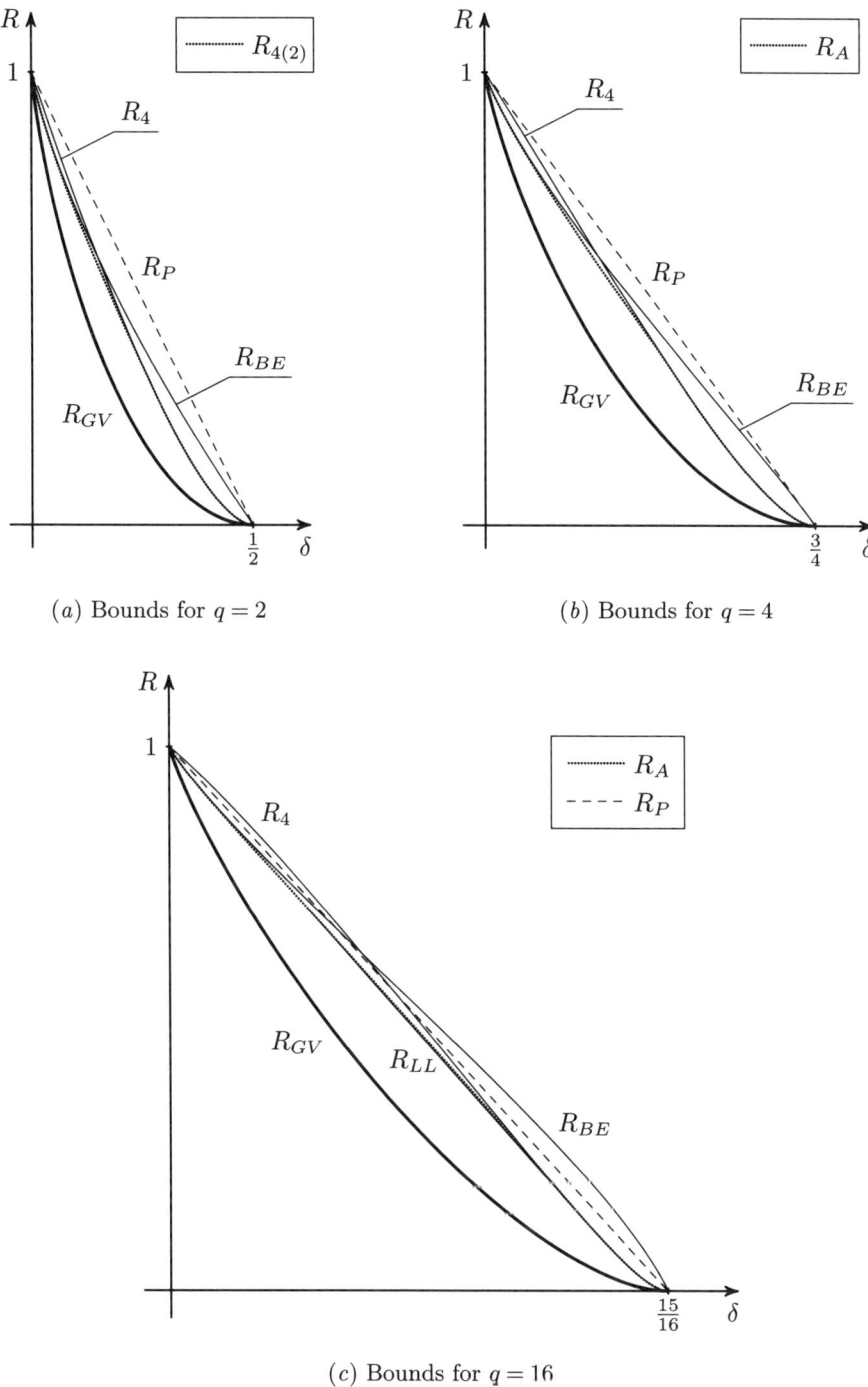

(a) Bounds for $q = 2$

(b) Bounds for $q = 4$

(c) Bounds for $q = 16$

FIGURE 1.2. Behaviour of asymptotic upper bounds for various q. For comparison, the lower bound R_{GV} is given.

In particular, the tangent at zero is vertical. For $\delta \to (q-1)/q$ we have the asymptotics
$$R_{GV}\left(\frac{q-1}{q} - x\right) = \frac{q^2}{2(q-1)\ln q}x^2 + o(x^2),$$
where $x = (q-1)/q - \delta$. The tangent at $(q-1)/q$ is horizontal, and the order of contact is two. Tangent lines to $R_{GV}(\delta)$ are of the form
$$R_t(\delta) = 1 - \left(\log_q(q^t + q - 1) - t\right) - t\delta, \quad 0 \leq t \leq \infty.$$
Each $R_t(\delta)$ is tangent to $R_{GV}(\delta)$ at the point
$$\delta_0 = \frac{q-1}{q^t + q - 1},$$
and
$$R(\delta_0) = 1 + \frac{tq^t}{q^t + q - 1} - \log_q(q^t + q - 1).$$

Now it is time to explain the role of R_{GV} in coding theory.

THEOREM 1.3.20 (Gilbert–Varshamov bound). *We have*
$$\alpha_q^{\text{lin}}(\delta) \geq R_{GV}(\delta) = 1 - H_q(\delta). \qquad \square$$

EXERCISE 1.3.21. Prove the theorem using Exercise 1.3.10 and Theorem 1.1.59.

REMARK 1.3.22. The Gilbert–Varshamov bound has remarkable statistical properties. We have the following facts, which can easily be stated rigorously:

(a) Almost all linear codes (i.e., points of V_q^{lin}) lie on the curve R_{GV}.

(b) Almost all nonlinear codes lie on the curve $R = R_{GV}/2$.

(c) Let us "adjust" a nonlinear $[n,k,d]_q$ code by deleting at most half of its code vectors. Then almost every code can be adjusted so that its parameters lie on the curve R_{GV}. Note that the rate of the code does not change asymptotically.

EXERCISE 1.3.23. For the above remark, formulate rigorous statements and prove them. (*Hint*: See the footnote on p. x of the Preface.)

Thus, codes whose parameters lie above the Gilbert–Varshamov bound are quite rare. "Bad" codes are also quite rare. Nevertheless, it is more than easy to construct a bad code and very difficult to construct a good one. It would be nice to understand the reason. Here is an ad homini argument: if it were quite the contrary, the mathematical coding theory would not exist.

Connections between different q. Let us derive some corollaries from the constructions of Sec. 1.2.3.

THEOREM 1.3.24. *We have*
$$\alpha_q(\delta) \geq \max\left\{\max_{q' \geq q}\left\{1 - (1 - \alpha_{q'}(\delta))\log_q q'\right\}, \max_{q' \leq q}\left\{\alpha_{q'}(\delta)\log_q q'\right\}\right\}. \qquad \square$$

EXERCISE 1.3.25. Prove the theorem. (*Hint*: Use alphabet restriction and alphabet extension; see Sec. 1.2.3.)

Concatenation helps us to establish the following important fact.

THEOREM 1.3.26. *If there exists an $[n,k,d]_q$ code C_0, q being a power of a prime and k being an integer, then*

$$\alpha_q(\delta) \geq \frac{k}{n}\alpha_{q^k}\left(\frac{d}{n}\delta\right).$$

If C_0 is linear, the same is valid for α^{lin}. □

COROLLARY 1.3.27. *We have*

$$\alpha_q(\delta) \geq \max_C \left(\frac{k}{n}\alpha_{q^k}\left(\frac{n}{d}\delta\right)\right),$$

where the maximum is taken over all $[n,k,d]_q$ codes with integer k. The same holds for linear codes. □

EXERCISE 1.3.28. Prove the theorem and the corollary.

Other problems. Besides the basic functions α_q and α_q^{lin}, it is reasonable to consider problems with certain algorithmic constraints (as is done below; see Sec. 1.3.4).

Self-dual problem. We can also consider an asymptotic problem for self-dual codes. For such codes we always have $R = 1/2$. Set

$$\delta_q^{\mathrm{sd}} = \limsup \frac{d}{n},$$

where the limit is taken over all self-dual codes (and, similarly, δ_q^{qsd} for quasi-self-dual ones). There is a "Gilbert–Varshamov bound" in this case:

THEOREM 1.3.29. *We have $\delta_q^{\mathrm{qsd}} \geq H_q^{-1}(1/2)$, where $H_q^{-1}(x)$ is the inverse function to the q-ary entropy. Moreover, if $q = 2^m$, the same inequality holds for δ_q^{sd}.* □

1.3.3. Asymptotic Bounds for Higher Weights

Just as for the usual weights, the bounds on parameters for higher weights are best to contemplate in an asymptotic setting.

We first study the main asymptotic setting, that is, $r = \mathrm{const}$, $n \to \infty$, $k/n \to R$, and $d_r/n \to \delta_r$; at the end of this subsection we explain why the case of growing r is much simpler.

Let $V_q^{(r)}$ be the family of points (δ_r, R) corresponding to all linear codes (projective systems). Let $U_q^{(r)}$ be the set of limit points. Just as for $r = 1$ (see Sec. 1.3.1), there exist continuous functions $\alpha^{(r)}(\delta) = \alpha_q^{(r)}(\delta)$ and $\delta_r(R) = \delta_{r,q}(R)$ such that

$$U_q^{(r)} = \{(\delta, R) \mid 0 \leq R \leq \alpha^{(r)}(\delta)\} = \{(\delta, R) \mid 0 \leq \delta \leq \delta_r(R)\} \cup \{R = 0\}.$$

The functions $\alpha^{(r)}(\delta)$ and $\delta_r(R)$, inverse to each other, are the unknown "true" bounds, which we are always seeking to estimate. It is also clear that $\delta_r(R)$ and $\alpha^{(r)}(\delta)$ are decreasing for $\delta \leq 1 - q^{-r}$.

Relations. There are some relations between the functions $\alpha_q^{(r)}$ for different r and q. For instance, a trivial one is $\alpha_q^{(r)}(\delta) \leq \alpha_q^{(s)}(\delta)$ for $r \leq s$. In the following, if q is fixed, we omit the subscript; for $r = 1$, we also omit the superscript.

THEOREM 1.3.30. *We have*

(a) $$\delta^{(r)}(R) \leq \frac{1-q^{-r}}{1-q^{-s}} \delta^{(s)}(R) \qquad \text{for } r \leq s,$$

(b) $$\delta^{(r)}(R) \geq \frac{1-q^{-r}}{1-q^{-1}} \delta(R),$$

(c) $$\delta^{(r)}(R) \geq 1 - q^{s-r}\left(1 - \delta^{(s)}(R)\right) \qquad \text{for } r \leq s,$$

(d) $$\delta^{(r)}(R) \leq 1 - q^{-r+1} + q^{-r+1}\delta(R),$$

(e) $$\alpha^{(r)}(\delta) \leq \alpha^{(s)}\left(\frac{1-q^{-s}}{1-q^{-r}}\delta\right) \qquad \text{for } r \leq s,$$

(f) $$\alpha^{(r)}(\delta) \geq \alpha\left(\frac{1-q^{-1}}{1-q^{-r}}\delta\right),$$

(g) $$\alpha^{(r)}(\delta) \geq \alpha^{(s)}\left(1 - q^{s-r}(1-\delta)\right) \qquad \text{for } r \leq s,$$

(h) $$\alpha^{(r)}(\delta) \leq \alpha\left(1 - q^{-r+1} + q^{-r+1}\delta\right),$$

(j) $$\alpha_q^{(r)}(\delta) \leq r\alpha_{q^r}(\delta).$$

PROOF. Bounds (a) and (c) follow from Corollary 1.1.74 and Theorem 1.1.76 respectively; bounds (b) and (d) are their particular cases. Inequalities (e)–(h) are the inverse statements. It remains to prove (j). Given an $[n,k,d_r]_q$ system $\mathcal{P}$, we can spoil it several times using (1.1.6) (see p. 22) in order to get an $[n,k',d_r]_q$ system $\mathcal{P}' \subset \mathbb{P}(V)$ with $r \mid k'$. Take an arbitrary lifting $\mathcal{Q} \subset V$ of $\mathcal{P}'$ and an arbitrary isomorphism $V \cong W$, where W is a linear space of dimension k'/r over $\mathbb{F}_{q^r}$. The corresponding system $Z \subset \mathbb{P}(W)$ is an $[n,k'/r,d_1]_{q^r}$ system with $d_1(W) \geq d_r(V)$ since any $\mathbb{F}_{q^r}$-hyperplane in W corresponds to a plane in V of codimension r. Passing to the asymptotics, we get (j). □

Upper bounds. We start with the simplest bound. From Corollary 1.1.65 we obtain the following result.

THEOREM 1.3.31. *We have*
$$\alpha^{(r)}(\delta_r) \leq 1 - \delta_r. \qquad \square$$

Hereafter, we assume that $\delta_r > 0$ and $R > 0$. It follows from Theorem 1.1.68 that
$$\delta_r(R) \leq 1 - q^{-r}.$$

THEOREM 1.3.32. *We have*
$$\alpha^{(r)}(\delta_r) \leq 1 - \frac{\delta_r}{1-q^{-r}}.$$

PROOF. Put $\ell = s = k$ in Corollary 1.1.75. □

This is a Plotkin-type bound. Theorem 1.1.77 immediately yields a Hamming-type bound. As usual, let $H_q(x) = x\log_q(q-1) - x\log_q x - (1-x)\log_q(1-x)$.

THEOREM 1.3.33. *We have*
$$\alpha^{(r)}(\delta_r) \le 1 - H_q\left(\frac{\delta_r}{r+1}\right).$$
□

A bound of the Bassalygo–Elias type is implied by Theorem 1.1.78 and is as follows.

THEOREM 1.3.34. *We have*
$$\alpha^{(r)}(\delta_r) \le 1 - H_q(\lambda), \tag{1.3.1}$$
where λ is the smallest root of
$$x^{r+1}(q-1)^{-r} + (1-x)^{r+1} = 1 - \delta_r.$$

PROOF. Assume that $\alpha^{(r)}(\delta_r) > 1 - H_q(\lambda)$. Then condition (1.1.10) of Theorem 1.1.78 is satisfied for $w = \omega n$, where $\omega = \lambda - \varepsilon$ is such that
$$\omega^{r+1}(q-1)^{-r} + (1-\omega)^{r+1} > 1 - \delta_r,$$
i.e.,
$$d_r > n - w\left(\frac{w}{n(q-1)}\right)^r - (n-w)\left(\frac{n-w}{n}\right)^r,$$
which contradicts Theorem 1.1.78. □

We can use the relations of Theorem 1.3.30 to obtain some other bounds. Let $\overline{R}_q(\delta)$ be the best known upper bound for the first weight, for codes over $\mathbb{F}_q$.

THEOREM 1.3.35. *We have*

(a) $$\alpha_q^{(r)}(\delta) \le r\overline{R}_{q^r}(\delta), \tag{1.3.2}$$

(b) $$\alpha_q^{(r)}(\delta) \le \overline{R}_q(1 - q^{-r+1} + q^{-r+1}\delta), \tag{1.3.3}$$

(c) $$\alpha_q^{(r)}(\delta) \le \overline{R}_q(1 - (1-\delta)^{1/r}). \tag{1.3.4}$$

PROOF. Inequality (a) follows from Theorem 1.3.30j, (b) follows from Theorem 1.3.30h, (c) is a consequence of Corollary 1.1.85: divide (1.1.11) by n and let n tend to infinity. Taking into account that δ_1 decreases as a function of R, we get
$$\delta_r \le \delta_1 + (1-\delta_1)\delta_1 + (1-\delta_1(1-\delta_1))\delta_1 + \ldots = 1 - (1-\delta_1)^r;$$
thus, $\delta_1 \ge 1 - (1-\delta_r)^{1/r}$. □

REMARK 1.3.36. We have written out several upper bounds; their minimum coincides with the minimum of (1.3.1) and (1.3.4). Therefore, for any q and r there exists $\mu_r(q)$ such that on the interval $(0, \mu_r(q))$ the minimum coincides with (1.3.1), and on $(\mu_r(q), 1-q^{-r})$, with (1.3.4). It is easily verified that $\mu_2(2) \approx 0.535$. Furthermore, one can further improve (1.3.4), not insisting on having an explicit formula for the bound. For simplicity, here we give a result only for $r = 2$.

For $r = 2$, formula (1.1.11) reads
$$D_2(n,k) \le D_1(n,k) + D_1(n - D_1(n,k), k-2).$$
Dividing by n and passing to the limit, we obtain
$$\delta_2 \le \delta_1 + (1-\delta_1)\delta_1\left(\frac{R}{1-\delta_1}\right),$$

where $\delta_1(R/(1-\delta_1))$ is the value of δ_1 corresponding to the rate $R' = R/(1-\delta_1)$ (not to R itself). Hence,

$$\delta_2 \leq \delta_1 + (1-\delta_1)\overline{R}_q^{-1}\left(\frac{R}{1-\delta_1}\right),$$

where $\overline{R}_q^{-1}$ is the inverse function to $\overline{R}_q$. This bound is better than (1.3.4).

Lower bounds. On the other hand, we can obtain a lower bound analogous to the Gilbert–Varshamov bound.

THEOREM 1.3.37. *We have*

$$\alpha_q^{(r)}(\delta_r) \geq 1 - H_{q^r}(\delta_r). \tag{1.3.5}$$

PROOF. This is the asymptotic form of Theorem 1.1.90 (of any of inequalities (a), (b), (e), or (f)). □

REMARK 1.3.38. We can use Theorem 1.3.30f together with the usual Gilbert–Varshamov bound to get the lower bound

$$\alpha_q^{(r)}(\delta_r) \geq 1 - H_q\left(\frac{1-q^{-1}}{1-q^{-r}}\delta\right).$$

However, this is always worse than (1.3.5).

Up to this point we have considered the main asymptotic setting. Now we would like to look at the case of growing r, i.e., $r \to \infty$, $n \to \infty$, $r/n \to \rho$, $k/n \to R$, and $d_r/n \to \delta_{"r"}$. Here we allow $\rho = 0$; also, we have to write $\delta_{"r"}$ instead of δ_r since at the limit r is no longer finite.

The simplest bound of Corollary 1.1.65 shows that

$$\rho \leq \delta_{"r"} \leq 1 - R + \rho.$$

In fact, one can show that each of such values of $\delta_{"r"}$ can be attained.

1.3.4. Polynomiality

Effectiveness problems play an important role in coding theory. It would be nice to be able to explicitly (effectively) construct good codes. The only rigorous notion of effectiveness we know is given by the computation complexity theory. Results of this theory are mostly asymptotic in themselves; i.e., sequences of codes of growing length should be considered rather than separate code with given parameters. There are three main questions about a code (or a class of codes): construction, encoding, and decoding. Therefore, we arrive at three corresponding questions on the *construction complexity*, *encoding complexity*, and *decoding complexity* of some class of codes. We first consider the question of the construction complexity and then briefly discuss the other two. It should be noted that for all codes considered in this book there is no real problem in encoding since most of them are linear codes, for which the encoding procedure (if we know the generator matrix) is trivial, and all the considered nonlinear codes differ from linear ones "by a finite factor" and are also easily encoded.

Polynomial families of codes. Let $\{C_i\}$ be a family of q-ary codes of growing length. This means that for any $i = 1, 2, \ldots$ there is an $[n_i, k_i, d_i]_q$ code C_i (the alphabet cardinality $q = |A|$ is always fixed); we also assume that the family is ordered in such a way that $n_{i+1} \geq n_i$ for any i and that each two codes in the family are different (as subsets in A^n). The latter condition immediately yields $n_i \to \infty$ as $i \to \infty$. We also assume for simplicity that $q = p^a$, where p is a prime, $A = \mathbb{F}_q$, and all codes C_i are linear (later on, we comment on the nonlinear case as well). Let $G_i \in \text{Mat}(n_i \times k_i, \mathbb{F}_q)$ be a generator matrix of C_i. The family C_i is called *polynomial* (or having a *polynomial construction complexity*) if there exists an algorithm $\mathfrak{A}$ for constructing matrices G_i whose complexity (the number of elementary operations in $\mathbb{F}_q$) is bounded by a polynomial in the length of a code. "Algorithm" here means any reasonable definition of this notion, since a particular computational model does not influence the class of codes thereby specified.

In general, a construction algorithm for a family of codes $\{C_i\}$ is an algorithm $\mathfrak{A}$ which for each i produces an encoding algorithm A_i for C_i (more precisely, a text $\bar{A}_i$ of this algorithm written in some language). The produced algorithm A_i, applied to a vector $x \in \mathbb{F}_q^k$ (for simplicity, we consider the case of $k \in \mathbb{Z}$; otherwise, we should consider $x \in \{1, 2, \ldots, q^k\}$ instead of $x \in \mathbb{F}_q^k$), outputs an element $A_i(x) \in C_i \subseteq \mathbb{F}_q^n$, so that to different values of x there correspond different $A_i(x)$. It is clear that a generator matrix G_i of a linear code (together with the rule of multiplying a vector by a matrix) is a particular case of such an algorithm A_i.

Saying that a family of codes $\{C_i\}$ is polynomial, we mean that not only the construction algorithm $\mathfrak{A}$ is polynomial in the code length n_i but also that the encoding algorithms A_i are polynomial, in the sense that, for any given x, computation of $A_i(x)$ needs a polynomial in n_i number of operations in $\mathbb{F}_q$.

Families of codes having a polynomial decoding procedure are defined in a similar way. Decoding of a code C up to $t \leq \lfloor (d-1)/2 \rfloor$ is given by a map $D: U \to C$, U being the set of vectors in $\mathbb{F}_q^n$ whose distance to the nearest point of C is at most t, such that $D(x) = x$ for $x \in C$ and $D(x)$ is the vector of C nearest to x in the general case. Let there be given a family $\{C_i\}$ of codes equipped by decoding algorithms $\{B_i\}$ defining a family of maps $D_i: U_i \to C_i$ such that $x \in U_i$ is taken to the nearest point of C_i. Then we call $(\{C_i\}, \{B_i\})$ a *polynomially decodable family* (or a family with a *polynomial decoding algorithm*) if there exists a universal algorithm $\mathfrak{B}$ generating all B_i, this algorithm is polynomial in n_i, and the number of operations in $\mathbb{F}_q$ required to realize D_i with the help of B_i is also polynomial in n_i. If $t_i/n_i \to \tau$, we say that the family is polynomially decodable up to τ.

Polynomial bounds. Now we are going to expose some definitions and results that are polynomial analogues of those of Sec. 1.3.1. They are all proved by a thorough analysis of the corresponding proofs in Sec. 1.3.1 and 1.1.4; therefore, we mostly abstain from giving any proofs here. Of course, not every result can be transferred to the polynomial situation; we shall point this out each time we come upon such a case.

Define $U_q^{\text{pol}} \subset [0, 1]^2$ as the set of limit points of relative parameters for polynomial families of codes; i.e., $(\delta, R) \in U_q^{\text{pol}}$ if and only if there exists a polynomial (in $n_i \to \infty$) family of $[n_i, k_i, d_i]_q$ codes with $d_i/n_i \to \delta$ and $k_i/n_i \to R$. Similarly, define $U_q^{\text{pol,lin}}$ for linear codes.

THEOREM 1.3.39. *There exist continuous functions $\alpha_q^{\mathrm{pol}}(\delta)$ and $\alpha_q^{\mathrm{pol,lin}}(\delta)$, $\delta \in [0,1]$, such that*
$$U_q^* = \{(\delta, R) \mid 0 \leq R \leq \alpha_q^*(\delta)\}.$$
Moreover, we have $\alpha_q^*(0) = 1$, $\alpha_q^*(\delta) = 0$ for $(q-1)/q \leq \delta \leq 1$, and on the segment $[0, (q-1)/q]$ the functions $\alpha_q^*(\delta)$ are decreasing (here the superscript $*$ means either pol *or* pol,lin*).* □

EXERCISE 1.3.40. Prove the theorem, assuming that we already know that $\alpha_q^{\mathrm{pol,lin}}(\delta) > 0$ for $0 \leq \delta < (q-1)/q$ (this is actually the case; see Theorem 1.3.45 and Exercise 1.3.48 below).

Of course,
$$\alpha_q^{\mathrm{pol,lin}}(\delta) \leq \alpha_q^{\mathrm{pol}}(\delta) \leq \alpha_q(\delta), \qquad \alpha_q^{\mathrm{pol,lin}}(\delta) \leq \alpha_q^{\mathrm{lin}}(\delta).$$

The main problems about polynomially constructable codes are as follows.

PROBLEM 1.3.41. Find $\alpha_q^{\mathrm{pol}}(\delta)$ and $\alpha_q^{\mathrm{pol,lin}}(\delta)$.

PROBLEM 1.3.42. Is it true that $\alpha_q^{\mathrm{pol}}(\delta) = \alpha_q(\delta)$ and $\alpha_q^{\mathrm{pol,lin}}(\delta) = \alpha_q^{\mathrm{lin}}(\delta)$?

PROBLEM 1.3.43. Is this at least true for the asymptotic behaviour of these bounds as $\delta \to 0$ and $\delta \to (q-1)/q$?

We do not know a single specifically polynomial upper bound. All known upper bounds are those for $\alpha_q(\delta)$. Codes attaining the Gilbert–Varshamov bound are constructed by an essentially nonpolynomial method involving an exponential (in n) search.

Concatenation is polynomial in the following sense.

EXERCISE 1.3.44. Start with two polynomial families of codes with parameters $[N_i, K_i, D_i]_{Q_i}$ and $[n_i, k_i, d_i]_q$, $Q_i = q^{k_i}$. Prove that their pairwise concatenations form a polynomial family of $[N_i n_i, K_i k_i, D_i d_i]_q$ codes. Moreover, this is also true if the family of $[n_i, k_i, d_i]_q$ codes is polynomial in N_i only (it can be exponential in n_i).

The latter fact is used to prove the following bound.

THEOREM 1.3.45 (Zyablov bound). *We have*
$$\alpha_q^{\mathrm{pol,lin}}(\delta) \geq R_Z(\delta) = \max_{\delta \leq \delta_0 \leq (q-1)/q} \left\{ \left(1 - \frac{\delta}{\delta_0}\right)(1 - H_q(\delta_0)) \right\}.$$

PROOF. Consider a family of Reed–Solomon $[N_i, K_i, D_i]_{Q_i}$ codes, $Q_i = q^{k_i}$, such that $k_i \to \infty$, $D_i/N_i \to \delta_1$, $K_i/N_i \to R_1$, $\delta_1 + R_1 = 1$, $N_i = Q_i$, and a family of Gilbert–Varshamov $[n_i, k_i, d_i]_q$ codes with $n \to \infty$, $d_i/n_i \to \delta_0$, $k_i/n_i \to R_0$, $R_0 = 1 - H_q(\delta_0)$. Concatenation yields $[n_i N_i, k_i K_i, d_i D_i]_q$ codes with $R = R_0 R_1 = (1 - H_q(\delta_0))(1 - \delta_1)$, $\delta = \delta_0 \delta_1$. We get the straight line $R = (1 - \delta/\delta_0)(1 - H_q(\delta_0))$, where the parameters of concatenated codes lie (asymptotically). Since we can choose any δ_0 so that $\delta \leq \delta_0 \leq (q-1)/q$ (then $R_0 \geq 0$) and $0 \leq \delta_1 = \delta/\delta_0 \leq 1$ (then $R_1 \geq 0$), we get the right-hand side of the inequality. We could choose the codes to be linear, so it only remains to prove that the construction is polynomial in the length $n_i N_i$ of the obtained codes. Indeed, the Gilbert–Varshamov codes are exponential in n_i, but $n_i \sim \log_q(n_i N_i)$, and all the rest is polynomial. □

REMARK 1.3.46. Justesen codes considered in Sec. 1.2.2 yield a bit worse bound (for $q=2$):
$$\alpha_2^{\text{pol,lin}}(\delta) \geq \frac{1}{2}\left(1 - \frac{\delta}{H_2^{-1}(1/2)}\right),$$
where H_2^{-1} is the inverse function to $H_2(\delta)$; $H_2^{-1}(1/2) \approx 0.110$. By certain elaboration of this construction (using the so-called punctured Justesen codes), this inequality can be sharpened a little.

REMARK 1.3.47 (Blokh–Zyablov bound). Using generalized concatenation, one can do better than in Theorem 1.3.45. The obtained bound is
$$\alpha_q^{\text{pol,lin}}(\delta) \geq R_{BZ}(\delta) = R_{GV}(\delta) - \delta \int_0^{R_{GV}(\delta)} \frac{dx}{R_{GV}^{-1}(x)},$$
where
$$R_{GV}(\delta) = 1 - H_q(\delta) = 1 - \delta\log_q(q-1) + \delta\log_q\delta + (1-\delta)\log_q(1-\delta),$$
and R_{GV}^{-1} is the inverse function to R_{GV}. This is the best known lower polynomial bound that does not involve algebraic geometry codes.

EXERCISE 1.3.48. Prove that both Zyablov and Bloch–Zyablov bounds are smooth curves passing through $(0;1)$ and $((q-1)/q;0)$. Compute their asymptotics at the endpoints. (*Answer*:
$$1 - R_{BZ}(\delta) \sim \frac{\ln q}{2}\delta(\log_q\delta)^2 \quad \text{as} \quad \delta \to 0,$$
$$R_{BZ}(x) \sim \frac{q^3}{6(q-1)^2\ln q}x^3 \quad \text{as} \quad x = \frac{q-1}{q} - \delta \to 0.)$$

Theorem 1.3.45 uses concatenation with outer codes over growing alphabets. Let us see what can be done over a fixed field $\mathbb{F}_{q^k}$.

THEOREM 1.3.49. *If there exists an $[n,k,d]_q$ code C_0, then*
$$\alpha_q^{\text{pol}}(\delta) \geq \frac{k}{n}\alpha_{q^k}^{\text{pol}}\left(\frac{d}{n}\delta\right);$$
if C_0 is linear, then
$$\alpha^{\text{pol,lin}}(\delta) \geq \frac{k}{n}\alpha_{q^k}^{\text{pol,lin}}\left(\frac{d}{n}\delta\right). \qquad \Box$$

COROLLARY 1.3.50. *We have*
$$\alpha_q^{\text{pol}}(\delta) \geq \max_C\left\{\frac{k}{n}\alpha_{q^k}^{\text{pol}}\left(\frac{d}{n}\delta\right)\right\},$$
where the maximum is taken over all q-ary codes C (here $[n,k,d]_q$ are the parameters of C). The same holds for $\alpha_q^{\text{pol,lin}}$ (the maximum being taken over all linear codes). $\qquad \Box$

EXERCISE 1.3.51. Prove the theorem and corollary.

For nonlinear codes (and arbitrary alphabets) we have an analogue of the second part of Theorem 1.3.24:

EXERCISE 1.3.52. Show that
$$\alpha_q^{\text{pol}}(\delta) \geq \max_{q' \leq q}\left\{\alpha_{q'}^{\text{pol}}(\delta)\log_q q'\right\}.$$

REMARK 1.3.53. It should be noted that the alphabet restriction, on the contrary, involves an exponential search over different shifts of the code and thus cannot be used to bound α_q^{pol}.

We return to bounds on α_q^{pol} in Sec. 4.5.3 after studying basic properties of algebraic geometry codes.

1.3.5. Other Asymptotics

The problem we have studied up to this point can be called the "main asymptotic problem of coding theory": $n,k,d \to \infty$, $k/n \to R$, and $d/n \to \delta$; find the dependence between R and δ. There are several other natural asymptotic settings. Note that $n \to \infty$ in any asymptotic problem; for k and d this is not necessarily so. In all problems q is assumed to be fixed. (If $q \to \infty$, many problems become trivial; nevertheless, this asymptotics can be of interest in studying some decoding algorithms that we hope to consider in *Advanced Chapters*.)

Fix the minimum distance d, and let $n \to \infty$. How can one express the relation between n and the largest possible k? Define
$$\varkappa_q(d) = \liminf_n \left(\inf \frac{n-k}{\log_q n}\right),$$
the infimum being taken over all $[n,k,d]_q$ codes with fixed n, q, and d. Similarly, for linear codes, one can define $\varkappa_q^{\text{lin}}(d)$.

THEOREM 1.3.54. *We have*
$$\left\lfloor \frac{d-1}{2} \right\rfloor \leq \varkappa_q(d) \leq \varkappa_q^{\text{lin}}(d) \leq \left\lceil \frac{(d-2)(q-1)}{q} \right\rceil. \qquad \square$$

PROOF. The left-hand inequality here is merely the Hamming bound (Theorem 1.1.47).

To prove the upper bound (the existence bound), consider the BCH codes with parameters
$$\left[n = q^m - 1, k = n - m\left(d - 1 - \left\lfloor \frac{d-2}{q} \right\rfloor\right) - 1, d\right]_q$$
(see Remark 1.2.16). Using $m = \log_q(n+1) \sim \log_q n$, we get
$$\varkappa_q^{\text{lin}}(d) \leq \left\lceil \frac{(d-2)(q-1)}{q} \right\rceil. \qquad \square$$

REMARK 1.3.55. For $q = 2$ and an odd d, the inequalities of Theorem 1.3.54 turn into equalities:
$$\varkappa_q(d) = \varkappa_q^{\text{lin}}(d) = \frac{d-1}{2}.$$

For $q = 2$ and an even d, by adding the parity-check position to the BCH code with parameters
$$\left[n = 2^m - 1, k = n - m\frac{d-2}{2}, d-1\right]_2,$$

we obtain an
$$\left[n = 2^m, k = n - m\frac{d-2}{2}, d\right]_2 \text{ code,}$$
whence we get $\varkappa_q(d) = (d-2)/2 = \lfloor(d-1)/2\rfloor$.

Note also that for any q and $d = 3$, using the Hamming $\left[n = \frac{q^m-1}{q-1}, n-m, 3\right]_q$ codes, we get $\varkappa_q(3) = 1$.

There are also some nice improvements for $d = 4, 5, 6$.

PROBLEM 1.3.56. Find the precise value of $\varkappa_q(d)$ for $q > 2$.

Consider another natural asymptotic setting. Define
$$\delta_q(k) = \limsup_n \left(\sup \frac{d}{n}\right),$$
the supremum being taken over all $[n, k, d]_q$ codes with fixed n, q, and k; similarly, one can define $\delta_q^{\text{lin}}(k)$.

THEOREM 1.3.57. *We have*
$$\delta_q(k) = \delta_q^{\text{lin}}(k) = \frac{q^k}{q^k - 1}\frac{q-1}{q}.$$

PROOF. The inequality $\delta_q(k) \leq \frac{q^k}{q^k-1}\frac{q-1}{q}$ is merely the Plotkin bound (see Theorem 1.1.45). The opposite inequality (existence bound) is immediately given by the first-order Reed–Muller codes with parameters
$$\left[\frac{q^m-1}{q-1}, m, q^{m-1}\right]_q. \qquad \square$$

There are other possible asymptotic settings. Since always $n \to \infty$ and we have already considered the cases $k = \text{const}$ and $d = \text{const}$, the only cases remaining are those with $n, k, d \to \infty$ and either $k/n \to 0$ or $d/n \to 0$ (if neither, we arrive at the main asymptotic problem). Here are some examples.

EXERCISE 1.3.58. Let $\delta_0 > (q-1)/q$. Prove that in the class of codes with $d/n \to \delta_0$ the dimension is bounded:
$$\limsup(k) \leq \left\lfloor\log_q\left(\frac{q\delta_0}{q\delta_0 - q + 1}\right)\right\rfloor.$$

Thus, there are two classes of asymptotic problems remaining: either $\delta \to 0$ and $R \to 1$ or $R \to 0$ and $\delta \to (q-1)/q$. Possible settings are as follows. Let $\varphi(n)$ be an increasing function with $\varphi(n)/n \to 0$.

Set
$$\rho_\varphi(n) = \inf(n - k),$$
the infimum being taken over all $[n, k, d]_q$ codes with a given n and $d \geq \varphi(n)$ (q is fixed). What is the asymptotic behaviour of $\rho_\varphi(n)$ as $n \to \infty$?

Another problem is to find the asymptotic behaviour of the best possible distance
$$d_\varphi(n) = \sup(d),$$
the supremum being taken over $[n, k, d]_q$ codes with $k \geq \varphi(n)$.

Sometimes we can answer these two questions using results on the main asymptotic problem and those about $\varkappa_q(d)$ and $\delta_q(k)$.

EXERCISE 1.3.59. Prove that for $\varphi = n^\alpha$, $0 < \alpha < 1$, we have
$$\frac{1-\alpha}{2} \lesssim \frac{\rho_\varphi(n)}{n^\alpha \log_q n} \lesssim \min\left\{1-\alpha, \frac{q-1}{q}\right\}.$$
(*Hint*: The lower estimate is given by the Hamming bound, and the upper one by the Gilbert–Varshamov bound and BCH codes; cf. Theorems 1.1.47, 1.1.59, and 1.2.15.)

EXERCISE 1.3.60. Prove that for $\varphi = n^\alpha$, $0 < \alpha < 1$, we have
$$d_\varphi(n) \sim \frac{q-1}{q} n.$$
(*Hint*: Use bounds for the main asymptotic problem.) What can be said about the second asymptotic term for $d_\varphi(n)$?

REMARK 1.3.61. Up to now, when studying existence bounds in all asymptotics, we considered limits over subsequences $\{n_i\} \subseteq \mathbb{N}$ for which the parameters are the best. However, other settings also give rather interesting problems: what can we say about any infinite subsequence $\{n_i\} \subseteq \mathbb{N}$? For example, set
$$\varkappa'_q(d) = \limsup_n \left(\inf \frac{n-k}{\log_q n}\right),$$
the infimum being taken over all codes of length n with distance d (q is fixed). What can be said about $\varkappa'_q(d)$? The same kind of question is also interesting for the main asymptotic problem and for various other asymptotics.

To end this section, here is another question which is of particular interest from the standpoint of algebraic geometry codes.

Let us call a family of $[n, k_i, d_i]_q$ codes, $i = 0, 1, \ldots, n+1-g$, a *g-family* for some natural g if
$$k_i \geq i,$$
$$k_i + d_i \geq n+1-g,$$
$$(n - k_i) + d_i^\perp \geq n+1-g.$$

PROBLEM 1.3.62. Set
$$\nu_q = \limsup_{g \to \infty} \frac{n_g}{g},$$
where n_g is the maximum possible length of a q-ary g-family. Find or estimate ν_q.

Historical and Bibliographic Notes

The theory of error-correcting codes emerged about half a century ago. Though it is still rather young, we may say that the main part of what we discuss in this chapter is quite classical. We have no intention of describing here the history of coding theory, and for the first part of its development we refer the reader to the books [MS77, KTII78, PW72, Ber68, vL99] and to the collection of papers [Ber74]. Our exposition of classical results follows in many respects the famous book by F.J. MacWilliams and N.J.A. Sloane [MS77], the main difference being that we always try to work over an arbitrary finite field, not just concentrate on binary codes.

Let us list the authors of major achievements. The notion of an error-correcting code was discovered by R.V. Hamming. Linear codes were independently discovered by D. Slepian and R.R. Varshamov. Cyclic codes are the discovery of E. Prange, I.S. Reed, and G. Solomon; BCH codes are due to R.K. Bose, D.K. Ray-Chaudhuri, and A. Hocquenghem. The spectrum of the dual code was calculated by F.J. MacWilliams. Classes of codes and asymptotic bounds mostly bear the name of their discoverers. Note that the Gilbert–Varshamov bound was first established by E.N. Gilbert for nonlinear codes and then by R.R. Varshamov for linear ones. The Bassalygo–Elias bound was found independently by P. Elias and L.A. Bassalygo. The generalization of the first bound of R.J. McEliece, E.R. Rodemich, H.C. Rumsey, Jr., and L.R. Welch to the q-ary case is due to M.J. Aaltonen. The concatenated construction is due to G.D. Forney, Jr.; the generalized concatenation, to E.L. Blokh, V.V. Zyablov, and V.A. Zinoviev. The first asymptotically good codes with polynomial construction complexity were discovered by V.V. Zyablov; almost immediately after that, J. Justesen constructed his codes having the same property. The completeness of the known list of perfect codes was established by A. Tietäväinen (based on previous work of S.P. Lloyd, J.H. van Lint, V.A. Zinoviev, and V.K. Leontiev). A.M. Gleason applied the invariant theory to self-dual codes. For complexity problems in coding theory, see the review [Bar98].

Let us now point out sources of the nonclassical content of this chapter, which mostly issued from the "algebraic geometry ideology" in coding theory.

Right after V.D. Goppa discovered the first algebraic geometry construction and the authors (M.A. Tsfasman and S.G. Vlăduţ) first encountered the notion of the code, we wanted to understand the existing results of coding theory from the general mathematics point of view. To do this, in [TV91] we introduced an easy and natural notion of an $[n,k,d]_q$ system (see Sec. 1.1.2, especially Proposition 1.1.4 and Theorems 1.1.6 and 1.1.11) and tried to use it in proofs. In fact, different versions of this notion appeared much earlier. Most likely, the first version is *modular representations* of Slepian [Sle56]; see also [PW72]. A particular case of projective systems was used (under another name) to study codes with two nonzero weights; see [CK86]. Another version was introduced in [HT78]; there projective systems were called mini-hypers (or maxi-hypers). This was used to construct codes reaching the Griesmer bound [Ham93]. The first step towards the solution of Problem 1.1.9 (rewriting coding theory in terms of projective systems) was done by S.M. Dodunekov and J. Simonis [DS98].

The current state of Conjecture 1.1.12 on MDS codes is described in the reviews [Hir83, HS98, HS01].

Results on spectra and duality (Exercise 1.1.15 and Theorems 1.1.18 and 1.1.28) are taken from the paper of G.L. Katsman and M.A. Tsfasman [KT87] (we have slightly changed the statements and proofs). On this topic, see also [AB99, ABL01a, ABL01b, CKL03].

Generalized weights were introduced by V.K. Wei [Wei91] because of a cryptography problem (codes being used in so-called "wire-tap channels II"). They also have other applications: to t-resilient functions, trellis complexity of codes, code properties with respect to shortening and puncturing, and so on. Similar properties of codes were earlier considered by T. Helleseth, T. Kløve, and J. Mykkeltveit [HKM77, Klø78]. In terms of projective systems, the notion of higher weights was rediscovered by one of the authors (M.A. Tsfasman [Tsf92b]) in an attempt to find natural geometric invariants of projective systems. This resulted in [HTV94]. In [TV95] the geometry of projective systems is regularly used to study higher weights. Our exposition mostly follows this paper.

Formula (1.1.3) is due to G. van der Geer and M. van der Vlugt [vdGvdV93]. Theorem 1.1.14 is proved in [TV91, HTV94]. The beautiful relation between the weight hierarchy of a code and its dual (Theorem 1.1.38) is noticed by V. Wei [Wei91]. The generalized Griesmer bound (Corollary 1.1.70) is also due to him.

Different definitions of higher weights for nonlinear codes were considered by G. Cohen, L. Huguet, G. Zémor, S. Litsyn, L.A. Bassalygo, I. Reuven, and Y. Be'ery [CHZ94, CLZ94, Bas94, RB99]. Bounds for nonlinear codes (Theorems 1.1.77 and 1.1.78) are established in [CHZ94, CLZ94].

As far as bounds for higher weights are concerned, part (a) of Theorem 1.1.82 is taken from [Tsf92b] (see [TV95]); parts (d) and (e) from [Wei91]; part (f) from the paper by T. Helleseth, T. Kløve, V.I. Levenshtein, and Ø. Ytrehus [HKLY95]. Proposition 1.1.84 and Corollary 1.1.85 are taken from the paper of S.G. Vlăduţ [Vlă96]. Proposition 1.1.88 comes from the paper by T. Helleseth, T. Kløve, and Ø. Ytrehus [HKY92], as well as part (j) of Theorem 1.3.30. Remark 1.1.89 comes from this paper and that of T. Kløve [Klø93]. Theorems 1.1.90 and 1.1.95 are taken from the paper of T. Kløve [Klø92]; in fact, they were already proved in [Klø78]. Theorems 1.1.91 and 1.1.94 and Proposition 1.1.92 are due to J. Simonis [Sim94]. In the direction of Problem 1.1.96, the following result is obtained in [ABL99]: from linear programming bounds for "pairs of vectors" there follows a bound for d_2 which slightly improves on Remark 1.3.36.

The example of Remark 1.2.16 is due to V.D. Goppa. Remark 1.2.18 is based on the paper by V.V. Chepyzhov [Che92]. Generalized iterative construction was independently discovered in [XNL99a]; see also [NX00, ÖS02]. Its relation with generalized concatenation was explained by G.A. Kabatiansky.

The importance of asymptotic settings was understood from the very birth of coding theory. Our statements of asymptotic problems (Theorem 1.3.1 and Exercise 1.3.2) are due to Yu.I. Manin (see [Man81, VM84]); they were also independently proposed by M.J. Aaltonen [Aal84].

Theorems 1.1.57 and 1.3.17 are due to T. Laihonen and S.N. Litsyn [LL98]; Theorems 1.3.14 and 1.3.16 to M.J. Aaltonen [Aal87, Aal90]. Theorem 1.3.26 and Corollary 1.3.27, and also Theorem 1.3.49 and Corollary 1.3.50, are taken from the papers of G.L. Katsman, M.A. Tsfasman, and S.G. Vlăduţ [KTV84, VKT84]; results of Theorem 1.3.24 and Exercise 1.3.52 are taken from the note of S.N. Litsyn and M.A. Tsfasman [LT86]. Concerning Remark 1.3.22, see [Bar02, BF02].

CHAPTER 2

Curves

In this chapter we expose necessary results from algebraic geometry. Since most constructions discussed in the book use only algebraic curves, here we mostly discuss the theory of curves. Multidimensional algebraic objects are used only when needed for curves (this is the case, e.g., for Jacobians). We try to minimize the number of technical tools and to work with intuitively lucid geometric objects. In spite of our attempt to restrict ourselves to more or less elementary means, the necessity to expose vast material may lead to some difficulties for the reader with no knowledge of algebraic geometry. We hope however that—maybe with the help of other books cited in the Historical and Bibliographic Notes to this chapter—even an inexperienced reader will be able to acquire knowledge permitting him to read the rest of the book.

In the first section we give definitions of most working tools for the theory of curves (including Jacobians); for curves over the field of complex numbers, we also explain the relation between curves and Riemann surfaces. Section 2 is concentrated around the theory of differential forms and the Riemann–Roch theorem; we also present the Hurwitz formula and the main properties of the Cartier operator. Section 3 is devoted to the first nontrivial case, the theory of elliptic curves; we describe many beautiful and important results specific to such curves. The last section gives a sketch of a purely algebraic approach to the theory, the main object being a function field; this theory is equivalent to that of algebraic curves. This language makes it easier to work with curves over algebraically nonclosed fields.

2.1. Algebraic Curves

Almost all objects we study in the book are, in one way or other, related to algebraic curves. This section is devoted to algebraic curves, their maps, and objects related to them: divisors, linear systems, line bundles, and Jacobians. The exposition is geometric, though we show how to study curves from the algebraic point of view and (for curves over the complex field) also from that of analysis and topology. The arithmetic of algebraic curves is beyond the scope of this chapter; that is why the ground field $\Bbbk$ is supposed to be algebraically closed.

Section 2.1.1 contains the definition and elementary properties of quasiprojective varieties. Section 2.1.2 is devoted to quasiprojective curves. Section 2.1.3 contains definitions and properties of divisors, linear systems, and line bundles on curves. Section 2.1.4 is devoted to Jacobians. In Sec. 2.1.5 we describe relations between complex algebraic curves and Riemann surfaces.

2.1.1. Quasiprojective Varieties

Throughout this section, we fix an algebraically closed field $\Bbbk$ (of an arbitrary characteristic), called in what follows the ground field. We denote by $\mathbb{A}^n$ the n-dimensional *affine space* over $\Bbbk$; its points are n-tuples $P = (x_1, \ldots, x_n)$, $x_i \in \Bbbk$. By $\mathbb{P}^n$ we denote the n-dimensional *projective space* over $\Bbbk$; its points are equivalence classes of $(n+1)$-tuples $Q = (y_0 : y_1 : \ldots : y_n)$, $y_i \in \Bbbk$, where not all of the y_i are zero, and $(y_0 : y_1 : \ldots : y_n)$ is equivalent to $(\lambda y_0 : \lambda y_1 : \ldots : \lambda y_n)$ for $\lambda \in \Bbbk$, $\lambda \neq 0$. Note that there is a natural embedding $\mathbb{A}^n \hookrightarrow \mathbb{P}^n$ given by

$$(x_1, \ldots, x_n) \mapsto (1 : x_1 : \ldots : x_n);$$

in what follows, unless otherwise specified, we use this embedding of $\mathbb{A}^n$ in $\mathbb{P}^n$. If we need to emphasize the dependence of $\mathbb{A}^n$ and $\mathbb{P}^n$ on $\Bbbk$, we write $\mathbb{A}^n(\Bbbk)$ and $\mathbb{P}^n(\Bbbk)$.

Sometimes it is convenient to use a coordinate-free definition of a projective space. Let V be a vector space of a finite dimension $n+1$ over $\Bbbk$, where $n \geq 0$. By $\mathbb{P}(V)$ we denote the set of lines in V, i.e., the set of equivalence classes $\{v\}$, $v \in V \setminus \{0\}$, where v is equivalent to λv for $\lambda \in \Bbbk \setminus \{0\}$. Fixing a basis in V defines identifications $V \cong \mathbb{A}^{n+1}$ and $\mathbb{P}(V) \cong \mathbb{P}^n$.

Closed sets. A *closed set* X in $\mathbb{A}^n$ (an *affine closed set*) is the set of common zeroes of a finite number of polynomials $F_1, \ldots, F_s \in \Bbbk[T_1, \ldots, T_n]$:

$$X = \{P = (x_1, \ldots, x_n) \in \mathbb{A}^n \mid F_1(x_1, \ldots, x_n) = \ldots = F_s(x_1, \ldots, x_n) = 0\}$$

(note that if $s = 0$, then $X = \mathbb{A}^n$). A subset U in $\mathbb{A}^n$ is *open* if its complement $\mathbb{A}^n \setminus U$ is closed in $\mathbb{A}^n$. A *closed* subset Y in $\mathbb{P}^n$ (a *projective closed set*) is the set of common zeroes of a finite number of homogeneous polynomials (*forms*) $G_1, \ldots, G_r \in \Bbbk[T_0 : \ldots : T_n]$:

$$Y = \{Q = (y_0 : \ldots : y_n) \in \mathbb{P}^n \mid G_1(y_0 : \ldots : y_n) = \ldots = G_r(y_0 : \ldots : y_n) = 0\}$$

(here we need forms, since the vanishing of G_i must depend on the equivalence class of Q only). A complement $\mathbb{P}^n \setminus Y$ of a closed set Y is called *open* in $\mathbb{P}^n$. To any closed affine set $X \subseteq \mathbb{A}^n$ we assign a projective closed set $\widetilde{X} \subseteq \mathbb{P}^n$ containing X, called the *projective closure* of X, which is obtained as follows. Let $F_1 = \ldots = F_s = 0$ be equations defining X, where $F_1, \ldots, F_s \in \Bbbk[T_1, \ldots, T_n]$; define the forms $\widetilde{F}_i \in \Bbbk[T_0, \ldots, T_n]$ by

$$\widetilde{F}_i(T_0, \ldots, T_n) = \sum_{j=0}^{\deg F_i} T_0^{\deg F_i - j} F_{ij}(T_1, \ldots, T_s), \quad i = 1, \ldots, s,$$

where F_{ij} are forms of degree j involved in the decomposition

$$F_i(T_1, \ldots, T_n) = \sum_{j=0}^{\deg F_i} F_{ij}(T_1, \ldots, T_n)$$

of the polynomial F_i into a sum of homogeneous polynomials. Then $\widetilde{X}$ is defined by

$$\widetilde{F}_1(T_0 : \ldots : T_n) = \ldots = \widetilde{F}_s(T_0 : \ldots : T_n) = 0.$$

Now let $S \subseteq \mathbb{P}^n$ be an arbitrary set. A subset $T \subseteq S$ is called *closed in* S if there exists a projective closed set $X \subseteq \mathbb{P}^n$ such that $T = S \cap X$. A subset $U \subseteq S$ is called *open in* S if its complement $S \setminus U$ is closed in S.

2.1. ALGEBRAIC CURVES

EXERCISE 2.1.1. Show that the above definitions of closed and open sets define a topology on S.

This topology is known as the *Zariski topology*.

EXERCISE 2.1.2. Show that $\mathbb{P}^n$ and $\mathbb{A}^n$ are irreducible in the Zariski topology (recall that a topological space X is called *irreducible* if any two nonempty open subsets $U, V \subseteq X$ have a nonempty intersection).

In particular, $\mathbb{P}^n$ and $\mathbb{A}^n$ are nonseparable in the Zariski topology.

Quasiprojective sets. An open subset U of a closed projective set X is called a *quasiprojective set*. In particular, any affine closed set is quasiprojective. A quasiprojective set $X \subseteq \mathbb{P}^n$ is called *reducible* if there exists a pair of nonempty proper closed subsets $X_1, X_2 \subset X$ such that $X = X_1 \cup X_2$. Otherwise, X is called *irreducible*. An irreducible quasiprojective set is referred to as a *quasiprojective variety*. Most often, we omit the term *quasiprojective* and just speak about a *variety*. If a variety is closed in $\mathbb{P}^n$, it is called a *projective* variety.

THEOREM 2.1.3. *Any quasiprojective set X can be uniquely represented as a union of finitely many nonempty varieties X_i such that X_i does not contain X_j for $i \neq j$.* □

The varieties X_i are called *irreducible components* of X.

EXERCISE 2.1.4. Show that $\mathbb{A}^1 \setminus \{0\}$ is a quasiprojective variety but is not projective.

EXERCISE 2.1.5. Show that the set $X = \{T_1 T_2 = 0\}$, where $(T_0 : T_1 : T_2)$ are homogeneous coordinates in $\mathbb{P}^2$, is a reducible closed projective set. Find its irreducible components.

Dimension. For a variety X, its *dimension* $n = \dim X$ is the largest integer n such that there exists a strictly descending chain of algebraic varieties

$$X = X_0 \supset \ldots \supset X_n \neq \varnothing$$

with X_i closed in X_{i-1} for $i = 1, \ldots, n$. If Y is a closed subvariety of X, we define the *codimension* of Y in X as

$$\operatorname{codim}_X Y = \dim X - \dim Y.$$

EXERCISE 2.1.6. Show that $\dim \mathbb{P}^n = \dim \mathbb{A}^n = n$.

EXERCISE 2.1.7. Let $X \subset Y$ be projective varieties. Show that $\dim X < \dim Y$. In particular, if a variety does not coincide with $\mathbb{P}^n$ and is closed in $\mathbb{P}^n$, then $\dim X < n$.

The dimension of a quasiprojective set is, by definition, the maximum of dimensions of its irreducible components, i.e., $\dim X = \max_i \dim X_i$. Varieties of dimension 1 are called *curves*; of dimension 2, *surfaces*; and of dimension 3, *threefolds*.

Rational functions. Let $X \subseteq \mathbb{P}^n$ be a quasiprojective variety. Let $F, G \in \Bbbk[T_0, \ldots, T_n]$ be forms of equal degree in homogeneous coordinates in $\mathbb{P}^n$, and let $G(P) \neq 0$ for at least one point $P \in X$. Then the fraction $F/G \in \Bbbk(T_0 : \ldots : T_n)$ is called a *rational function* on X. Fractions F/G and F'/G' define the same rational function if the form $F'G - FG'$ vanishes at all points $P \in X$. The set of rational functions on X is denoted by $\Bbbk(X)$.

EXERCISE 2.1.8. Show that $\Bbbk(X)$ is a field (addition and multiplication of rational functions is defined as usual:
$$\frac{F}{G}+\frac{F'}{G'}=\frac{FG'+F'G}{GG'}, \qquad \frac{F}{G}\cdot\frac{F'}{G'}=\frac{FF'}{GG'}).$$
The field $\Bbbk(X)$ is called the *field of rational function on X*; it is a fundamental invariant of a variety X.

A function $f \in \Bbbk(X)$ is called *regular* at $P \in X$ if it admits a representation $f = F/G$ with $G(P) \neq 0$. In this case, $f(P) = F(P)/G(P)$ is called the *value* of f at P.

EXERCISE 2.1.9. Show that the set U_f of points $P \in X$ at which a fixed rational function $f \in \Bbbk(X)$ is regular is open and nonempty.

A function $f \in \Bbbk(X)$ is called *regular on X* if it is regular at all points $P \in X$. The set $\Bbbk[X]$ of functions regular on X is a $\Bbbk$-algebra.

EXERCISE 2.1.10. Show that $\Bbbk[X]$ is an integral ring (i.e., has no nontrivial zero divisors).

EXERCISE 2.1.11. Show that $\Bbbk[\mathbb{P}^n] = \Bbbk$, and show that $\Bbbk[\mathbb{A}^n] = \Bbbk[T_1, \ldots, T_n]$ is a polynomial ring.

EXERCISE 2.1.12. Prove the following fact. Let $X \subseteq \mathbb{A}^n$ be an affine variety. Then the field of fractions of $\Bbbk[X]$ is $\Bbbk(X)$. (*Hint*: Consider restrictions of coordinate functions on $\mathbb{A}^n$ onto X.)

Exercises 2.1.11 and 2.1.12 reveal a profound distinction between properties of regular functions on an affine variety and of those on $\mathbb{P}^n$ (see also Corollary 2.1.24 below).

Rational maps. An n-tuple of rational functions $f = (f_1, \ldots, f_n)$, where $f_i \in \Bbbk(X)$, defines a *rational map* $f\colon X \to \mathbb{A}^n$ from a variety $X \subseteq \mathbb{P}^m$ to $\mathbb{A}^n$. A map f is called *regular at $P \in X$* if each f_i is regular at P. If a rational map is regular at each $P \in X$, it is called a *regular map from X to $\mathbb{A}^n$*.

A *rational map* f from $X \subseteq \mathbb{P}^m$ to $\mathbb{P}^n$ is given by an $(n+1)$-tuple $(F_0 : \ldots : F_n)$ of forms of the same degree in $m+1$ homogeneous coordinates in $\mathbb{P}^m$ such that at least one of the forms does not vanish at each point of X. Here, $(F_0 : \ldots : F_n)$ and $(F'_0 : \ldots : F'_n)$ define the same rational map if all the forms $F_i F'_j - F'_i F_j$ vanish on X. A rational map $f\colon X \to \mathbb{P}^n$ is called *regular* at $P \in X$ if f has a representation $(F_0 : \ldots : F_n)$ such that $F_i(P) \neq 0$ for at least one $i = 0, 1, \ldots, n$.

EXERCISE 2.1.13. Show that the set $U = U_f$ of points $P \in X$ where a given rational map $f\colon X \to \mathbb{P}^n$ is regular is open in X.

A rational map regular at all points $P \in X$ is called a *regular map* (or *morphism*) of varieties. Thus, a rational map of X is a regular map of an open subset of X. If there exists a variety $Y \subseteq \mathbb{P}^n$ such that $F(P) \in Y$ for any $P \in U_f$, we write $f\colon X \to Y$ and refer to f as a *rational map from X to Y*.

EXERCISE 2.1.14. Show that any regular map $f\colon X \to Y$ is continuous in the Zariski topology (i.e., that the set $f^{-1}(U)$ is open in X for any open subset $U \subseteq Y$).

EXERCISE 2.1.15. Show that a rational map $f\colon X \to \mathbb{A}^n$ defines a rational map $\bar{f}\colon X \to \mathbb{P}^n$; moreover, if f is regular, then $\bar{f}$ is also regular. The converse is not true: if $\bar{f}$ is regular, f is not necessarily regular (*give an example!*).

Let $f\colon X \to Y$ be a rational map. The set $\operatorname{Im} f = f(U_f)$ is called the *image* of f. If the image of f is dense in Y (i.e., the closure of $\operatorname{Im} f$ in Y coincides with Y), then f is called a *dominant* map. It is clear that any rational map $f\colon X \to \mathbb{P}^n$ can be considered as a dominant map $f\colon X \to Y$, Y being the projective closure of $\operatorname{Im} f$ in $\mathbb{P}^n$.

Let $f\colon X \to Y$ be a dominant rational map, and let $g \in \Bbbk(Y)$. Define a function $f^*(g)$ by
$$f^*(g)(P) = g(f(P)), \tag{2.1.1}$$
where $P \in X$ is a point such that $P \in U_f$ and $f(P) \in U_g$.

EXERCISE 2.1.16. Check that $f^*(g)$ is a rational function on X.

Thus, a rational dominant map $f\colon X \to Y$ defines a field embedding $f^*\colon \Bbbk(Y) \hookrightarrow \Bbbk(X)$, which is identical on $\Bbbk$.

EXERCISE 2.1.17. Check that this is indeed an embedding of fields.

The converse is also true:

EXERCISE 2.1.18. Let $\varphi\colon \Bbbk(Y) \hookrightarrow \Bbbk(X)$ be an arbitrary embedding of fields of rational functions identical on $\Bbbk$. Show that there exists a rational dominant map $f\colon X \to Y$ such that $\varphi = f^*$. (*Hint*: Reduce everything to the case of affine varieties and their regular maps; cf. Exercise 2.1.26 below.)

If $f\colon X \to Y$ and $f'\colon Y \to Z$ are rational maps and f is dominant, one can define the composition $f' \circ f$ in the usual way: $(f' \circ f)(P) = f'(f(P))$ for $P \in X$. Since f is dominant, we necessarily have $U_{f'} \cap \operatorname{Im} f \neq \varnothing$; i.e., the set of regular points of the map $f' \circ f$ is nonempty. Hence, $f' \circ f\colon X \to Z$ is a rational map.

EXERCISE 2.1.19. Show that $(f' \circ f)^* = f^* \circ f'^*$ if f and f' are dominant.

Now let $f\colon X \to Y$ be a regular map (not necessarily dominant). Then (2.1.1) defines a homomorphism of $\Bbbk$-algebras $f^*\colon \Bbbk[Y] \to \Bbbk[X]$.

EXERCISE 2.1.20. Show that if $f\colon X \to Y$ and $f'\colon Y \to Z$ are regular maps, then $(f' \circ f)^* = f^* \circ f'^*$.

If the fields $\Bbbk(X)$ and $\Bbbk(Y)$ of two varieties X and Y are isomorphic as $\Bbbk$-algebras, then X and Y are called *birationally isomorphic*. According to Exercises 2.1.18 and 2.1.19, this is equivalent to the existence of dominant rational maps $f\colon X \to Y$ and $f'\colon Y \to X$ such that $f' \circ f$ and $f \circ f'$ coincide (as rational maps) with the identical automorphisms of X and Y respectively. If f and f' can be chosen to be regular, then X and Y are called (*biregularly*) *isomorphic*. One may consider isomorphic varieties as one and the same variety, which is possibly differently embedded in projective spaces.

EXERCISE 2.1.21. Show that any variety is birationally isomorphic to each of its open nonempty subsets.

EXERCISE 2.1.22. Let X be a hyperbola in $\mathbb{A}^2$ given by $x_1 x_2 = 1$, and let $Y = \mathbb{A}^1 \setminus \{0\}$. Show that the projection onto the first coordinate $f\colon X \to Y$ defines an isomorphism from X to Y.

A variety birationally isomorphic to $\mathbb{A}^n$ (or, equivalently, to $\mathbb{P}^n$) is called *rational*; thus, Exercise 2.1.22 shows that hyperbola is a rational curve. A variety isomorphic to a closed subset in $\mathbb{A}^n$ is called an *affine* variety. Note that Exercise 2.1.22 shows that an affine variety need not be closed in the ambient affine space. For projective varieties, the situation is completely different.

THEOREM 2.1.23. *The image of a projective variety X under a regular map $f\colon X \to \mathbb{P}^n$ is closed in $\mathbb{P}^n$.* □

Applying this theorem to a *function f*, i.e., to a regular map $f\colon X \to \mathbb{A}^1$, we obtain the following fact.

COROLLARY 2.1.24. *For any projective variety X, we have $\Bbbk[X] = \Bbbk$.* □

COROLLARY 2.1.25. *An affine variety of positive dimension is never isomorphic to a projective variety.* □

In contrast to a projective variety (which, according to Corollary 2.1.24, has no nonconstant regular functions), an affine variety is completely determined by its algebra of regular functions.

EXERCISE 2.1.26. Let X and Y be affine varieties. Show that for any morphism of $\Bbbk$-algebras $\varphi\colon \Bbbk[X] \to \Bbbk[Y]$ there exists a unique regular map $f\colon Y \to X$ such that $\varphi = f^*$. (*Hint:* $\Bbbk[X] = \Bbbk[\mathbb{A}^n]/I_X$, where I_X is the ideal in $\Bbbk[\mathbb{A}^n]$ consisting of functions that vanish on X.)

EXERCISE 2.1.27. Deduce from the previous exercise that affine varieties X and Y are isomorphic if and only if the $\Bbbk$-algebras $\Bbbk[X]$ and $\Bbbk[Y]$ are isomorphic.

Local ring of a point. Let P be a point on a variety X. The set of rational functions that are regular at P is denoted by $\mathcal{O}_P$; obviously, $\mathcal{O}_P$ is a ring. When we want to emphasize the dependence of this set on X, we denote it by $\mathcal{O}_{X,P}$.

EXERCISE 2.1.28. Check that $\mathcal{O}_P$ has a unique maximal ideal $\mathfrak{m}_P$, which consists of functions $f \in \mathcal{O}_P$ such that $f(P) = 0$ (recall that a commutative ring with a unique maximal ideal is called a *local* ring).

The ring $\mathcal{O}_P$ is called the *local ring of P*.

EXERCISE 2.1.29. Show that $\mathcal{O}_P/\mathfrak{m}_P = \Bbbk$. (*Hint:* Show first that $\mathcal{O}_P/\mathfrak{m}_P$ is an algebraic extension of $\Bbbk$.)

The quotient group $\mathfrak{m}_P/\mathfrak{m}_P^2$ is an $\mathcal{O}_P/\mathfrak{m}_P$-module; i.e., it is a vector space over $\Bbbk$. The space $\theta_P = (\mathfrak{m}_P/\mathfrak{m}_P^2)^*$ dual to $\mathfrak{m}_P/\mathfrak{m}_P^2$ is called the *tangent space* to X at P (and $\mathfrak{m}_P/\mathfrak{m}_P^2$ is called the *cotangent space* to X at P).

EXERCISE 2.1.30. Explain why θ_P is called the tangent space.

EXERCISE 2.1.31. Let $X \subseteq \mathbb{P}^n$. Show that for any $P \in X$ we have
$$\dim X \le \dim_\Bbbk \theta_P \le n.$$
(*Hint*: The left-hand inequality is easily proved by induction on the dimension of X; the right-hand one follows from the fact that $\mathfrak{m}_P/\mathfrak{m}_P^2$ is generated by $x_1 - x_1(P), \ldots, x_n - x_n(P)$, where $x_1, \ldots, x_n$ are coordinates in $\mathbb{A}^n \ni P$.)

Smooth and singular points. A point P is called *nonsingular* (*smooth*, *regular*, or *simple*) if $\dim(\mathfrak{m}_P/\mathfrak{m}_P^2) = \dim X$. Otherwise, P is called *singular*.

EXERCISE 2.1.32. Show that $P = (0,0,0)$ is a singular point on the surface $X \subset \mathbb{A}^3$ given by $x_1 x_2 = x_3^2$, and all the other points $Q \in X$ are smooth.

If all points of a variety are smooth, it is called a *smooth* (or *nonsingular*) variety.

EXERCISE 2.1.33. Show that $\mathbb{P}^n$ and $\mathbb{A}^n$ are smooth.

PROPOSITION 2.1.34. *The set X_{smooth} of smooth points of a variety X is open in X and nonempty.* $\square$

Product of varieties. Let $X \subseteq \mathbb{A}^r$ and $Y \subseteq \mathbb{A}^s$. Then the set $X \times Y$ of pairs of the form $(P,Q) \in \mathbb{A}^{r+s}$ with $P \in X$ and $Q \in Y$ is called the *product* of X and Y.

EXERCISE 2.1.35. Check that if X and Y are closed sets, then $X \times Y$ is also closed in $\mathbb{A}^{r+s}$.

EXERCISE 2.1.36. Show that if X and Y are varieties, then $X \times Y$ is also a variety.

Now, let $X \subseteq \mathbb{P}^n$ and $Y \subseteq \mathbb{P}^n$ be quasiprojective varieties. Denote by $X \times Y$ the set of pairs (P,Q) with $P \in X$ and $Q \in Y$. We would like to find an embedding $\varphi \colon X \times Y \to \mathbb{P}^N$ (for some N) such that $\varphi(X \times Y)$ is a quasiprojective subvariety of $\mathbb{P}^N$. To this end, it suffices to consider the case $X = \mathbb{P}^n$ and $Y = \mathbb{P}^m$. Indeed, if an embedding $\psi \colon \mathbb{P}^n \times \mathbb{P}^m \to \mathbb{P}^N$ is already constructed, then $\varphi \colon X \times Y \to \mathbb{P}^N$ is obtained by restriction of ψ onto $X \times Y \subseteq \mathbb{P}^n \times \mathbb{P}^m$.

To construct $\psi \colon \mathbb{P}^n \times \mathbb{P}^m \to \mathbb{P}^N$, set $N = (m+1)(n+1) - 1 = mn + m + n$. Let homogeneous coordinates T_{ij} in $\mathbb{P}^N$ be indexed by pairs (i,j) with $0 \leq i \leq n$ and $0 \leq j \leq m$. For $P = (x_0 : \ldots : x_m) \in \mathbb{P}^n$ and $Q = (y_0 : \ldots : y_n) \in \mathbb{P}^m$, set

$$\psi_{(i,j)}((P,Q)) = x_i y_j.$$

This embedding is known as the *Segre embedding*.

EXERCISE 2.1.37. Show that $\psi(\mathbb{P}^n \times \mathbb{P}^m)$ is closed in $\mathbb{P}^N$. (*Hint:* Show that $\psi(\mathbb{P}^n \times \mathbb{P}^m)$ is defined by the equations $T_{ij} T_{kl} = T_{il} T_{kj}$.)

PROPOSITION 2.1.38. *For any quasiprojective varieties X and Y, we have*

$$\dim(X \times Y) = \dim X + \dim Y.$$ $\square$

Line bundles. Let X be a quasiprojective variety. A *family of vector spaces* on X is a regular map $p \colon \mathscr{E} \to X$ such that for any $P \in X$ its fibre $\overline{\mathscr{E}}_P = p^{-1}(P)$ (which is closed in $\mathscr{E}$ by Exercise 2.1.14) is a vector space over $\Bbbk$.

A *morphism* from a family $p \colon \mathscr{E} \to X$ to a family $p' \colon \mathscr{E}' \to X$ is a regular map $f \colon \mathscr{E} \to \mathscr{E}'$ such that $p = p' \circ f$ and the induced map of fibres $f_P \colon \overline{\mathscr{E}}_P \to \overline{\mathscr{E}}'_P$ is a morphism of vector spaces over $\Bbbk$. Isomorphism of families is defined in a natural way.

The simplest example of a family of vector spaces is the product $X \times \mathbb{A}^m$ with the natural projections $p_X \colon X \times \mathbb{A}^m \to X$. This family is called *trivial*. If a set U is open in X and $p \colon \mathscr{E} \to X$ is a family, then $p|_{p^{-1}(U)} \colon p^{-1}(U) \to U$ is also a family of vector spaces, called the *restriction of $\mathscr{E}$ onto U*; it is denoted by $\mathscr{E}|_U$.

In the following, we consider line families of vector spaces only (i.e., families with fibres of dimension 1). A line family $p\colon \mathscr{L} \to X$ is called a *line bundle* (*line vector bundle*) on X if for any P there exists an open set $U \subset X$, $U \ni P$, such that $\mathscr{E}|_U$ is isomorphic to the trivial family. We denote line bundles by $\mathscr{L}, \mathscr{M}, \mathscr{N}$, etc. The trivial line bundle is denoted by $\mathscr{O}$. A *section* of a line bundle $p\colon \mathscr{L} \to X$ is a regular map $s\colon X \to \mathscr{L}$ such that $p \circ s = \mathrm{id}_X$. In particular, any line bundle $\mathscr{L}$ has the zero section s_0 for which $s_0(P) = 0 \in \overline{\mathscr{L}}_P$. The set of sections $s\colon X \to \mathscr{L}$ of a bundle $\mathscr{L}$ is a vector space with operations

$$(s+s')(P) = s(P) + s'(P) \quad \text{and} \quad (\alpha s)(P) = \alpha \cdot s(P)$$

for $P \in X$ and $\alpha \in \Bbbk$. This vector space (space of sections) is denoted by $H^0(X, \mathscr{L})$. In particular, one easily checks that $H^0(X, \mathscr{O})$ coincides with the set $\Bbbk[X]$ of regular functions on X. Each section $s \in H^0(X, \mathscr{L})$ defines a morphism $\tilde{s}\colon \mathscr{O} \to \mathscr{L}$ uniquely determined by the condition $\tilde{s}(1) = s(P)$, where 1 is considered as an element in the fibre of $\mathscr{O}$. The subset $T_s = \{P \in X \mid s(P) = 0\}$ is called the *zero set* of s. Note that from the definition of a line bundle and Exercise 2.1.14 one can easily deduce that T_s is a closed subset of X.

Let $p\colon \mathscr{L} \to X$ be a line bundle, and let $X = \bigcup U_\alpha$ be an *open covering* of X (i.e., a covering of X by open sets) such that $\mathscr{L}|_{U_\alpha}$ is trivial for any α. Let $\varphi_\alpha\colon \mathscr{L}|_{U_\alpha} \xrightarrow{\sim} U_\alpha \times \mathbb{A}^1$ be the corresponding isomorphism. Consider the map

$$\varphi_\alpha \circ \varphi_\beta^{-1}\colon (U_\alpha \cap U_\beta) \times \mathbb{A}^1 \to (U_\alpha \cap U_\beta) \times \mathbb{A}^1.$$

It is an isomorphism of trivial families over $U_\alpha \cap U_\beta$. Let $P \in U_\alpha \cap U_\beta$; then $(\varphi_\alpha \circ \varphi_\beta^{-1})_P$ is a $\Bbbk$-automorphism of the one-dimensional fibre $\overline{\mathscr{L}}_P$, i.e., an element $\lambda_P \in \Bbbk^*$. Thus, $\varphi_\alpha \circ \varphi_\beta^{-1}$ defines a regular function $f_{\alpha\beta} \in \Bbbk[U_\alpha \cap U_\beta]$, which vanishes nowhere on $U_\alpha \cap U_\beta$. It is easily checked that $f_{\alpha\alpha} = \mathrm{id}$ for any α and that $f_{\alpha\gamma} = f_{\alpha\beta} f_{\beta\gamma}$ on $U_\alpha \cap U_\beta \cap U_\gamma$ for any triple (α, β, γ). Conversely, any family of functions $f_{\alpha\beta} \in \Bbbk[U_\alpha \cap U_\beta]$ satisfying these relations defines a line bundle.

Let $\mathscr{L}$ and $\mathscr{M}$ be line bundles given by families $\{f_{\alpha\beta}\}$ and $\{g_{\alpha\beta}\}$ respectively, where $f_{\alpha\beta}, g_{\alpha\beta} \in \Bbbk[U_\alpha \cap U_\beta]$ (one can easily show the existence of an open covering $\{U_\alpha\}$ that trivializes both $\mathscr{L}$ and $\mathscr{M}$ by refining the coverings on which $\mathscr{L}$ and $\mathscr{M}$ are defined). Then the system of functions $\{f_{\alpha\beta} g_{\alpha\beta}\}$ defines a line bundle $\mathscr{L} \otimes \mathscr{M}$, called the (*tensor*) *product* of $\mathscr{L}$ and $\mathscr{M}$. The system $f_{\alpha\beta}^{-1}$ defines a bundle $\mathscr{L}^{-1} = \mathscr{L}^{\otimes(-1)}$ such that $\mathscr{L}^{-1} \otimes \mathscr{L} \cong \mathscr{O}$. The tensor product of $m > 0$ copies of a bundle $\mathscr{L}$ is denoted by $\mathscr{L}^{\otimes m}$, or simply $\mathscr{L}^m$; for $m < 0$, we set $\mathscr{L}^m = \mathscr{L}^{\otimes m} = (\mathscr{L}^{-1})^{\otimes(-m)}$. By definition, $\mathscr{L}^0 = \mathscr{L}^{\otimes 0} = \mathscr{O}$. It is easily checked that the set $\mathrm{Pic}(X)$ of line bundles on X is a group, which is known as the *Picard group* of X.

Here is an example of a line bundle on $\mathbb{P}^n$, which plays a key role in various problems. Let V be a vector space of dimension $n+1$ such that $\mathbb{P}^n = \mathbb{P}(V)$. Consider the set

$$\mathscr{E} = \{(P, v) \in \mathbb{P}^n \times V \mid v \in \ell_P\},$$

where ℓ_P is the line in V corresponding to P. There is a natural projection $\mathscr{E} \to \mathbb{P}^n$, $(P, v) \mapsto P$, whose fibres are isomorphic to $\mathbb{A}^1$. Thus, one can check that $\mathscr{E}$ defines a line bundle over $\mathbb{P}^n$ (*tautological bundle*), which is denoted by $\mathscr{O}(-1)$. Set $\mathscr{O}(1) = \mathscr{O}(-1)^{\otimes -1}$ and $\mathscr{O}(m) = \mathscr{O}(1)^{\otimes m}$. For any section $s \in H^0(\mathbb{P}^n, \mathscr{O}(1))$, its zero set T_s is a hyperplane in $\mathbb{P}^n$, and vice versa, for any hyperplane $H \subset \mathbb{P}^n$ there is a section

$s \in H^0(\mathbb{P}^n, \mathscr{O}(1))$ such that $T_s = H$. Moreover, there is a canonical isomorphism $H^0(\mathbb{P}^n, \mathscr{O}(1)) = V^*$, where V^* is the space dual to V.

EXERCISE 2.1.39. Prove all the above statements about line bundles.

2.1.2. Quasiprojective Curves

A *quasiprojective curve* (*algebraic curve*, or simply a *curve*) is a quasiprojective variety of dimension 1. Since for any nonempty variety X each point $P \in X$ is closed in X, a curve can be defined as a variety whose proper closed subvarieties are only points. If a curve is closed in some $\mathbb{P}^n$, it is called *projective*, or *complete* (note that by Theorem 2.1.23 this property depends only on the isomorphism class of the curve).

EXAMPLE 2.1.40. The zero set of an irreducible homogeneous form F in $\mathbb{P}^2$ is a complete curve. The curve $\mathbb{A}^1$ is not complete.

Sometimes, a union $X = \bigcup X_i$ of several curves X_i is also called a (*reducible*) *curve*; in this case, the curves X_i are referred to as *irreducible components* of the reducible curve X.

Since curves are the simplest nontrivial varieties, many general properties of varieties described above can be specified more concretely in the case of curves.

Nonsingular points on a curve. It follows from Proposition 2.1.34 that a curve X has only finitely many singular points. Let $P \in X$ be an arbitrary point.

EXERCISE 2.1.41. Show that $P \in X$ is nonsingular if and only if the ideal $\mathfrak{m}_P$ in the local ring $\mathscr{O}_P$ is *principal* (i.e., $\mathfrak{m}_P = t_P \mathscr{O}_P$ for some $t_P \in \mathscr{O}_P$).

If P is a nonsingular point, then any function $t_P \in \mathscr{O}_P$ such that $\mathfrak{m}_P = t_P \mathscr{O}_P$ is called a *local parameter* at P.

To check whether a particular point P on a curve X is nonsingular, it is convenient to use the *differential criterion of nonsingularity*. Let $P \in X \subseteq \mathbb{P}^n$; without loss of generality, we may assume that $P \in X \cap \mathbb{A}^n$ (since $\mathbb{P}^n$ is a union of $n+1$ copies of $\mathbb{A}^n$). Let $(x_1, \ldots, x_n)$ be coordinates in $\mathbb{A}^n$, and let $F_1(x), \ldots, F_N(x) \in \Bbbk[x_1, \ldots, x_n]$ be a system of generators of the ideal $I_{X \cap \mathbb{A}^n}$. (Recall that $I_{X \cap \mathbb{A}^n}$ is the ideal consisting of polynomials that vanish on $X \cap \mathbb{A}^n$.) Thus, $F_1(x) = \ldots = F_N(x) = 0$ is a complete system of equations defining the affine curve $X \cap \mathbb{A}^n$. Consider the Jacobi matrix

$$\frac{\partial F}{\partial x}(P) = \left(\frac{\partial F_i}{\partial x_j}(P)\right)_{\substack{i=1,\ldots,N \\ j=1,\ldots,n}}.$$

PROPOSITION 2.1.42. *A point $P \in X$ is nonsingular if and only if the rank of $\dfrac{\partial F}{\partial x}(P)$ equals $n-1$.* □

In particular, if $N = 1$ and $n = 2$ (the case of a plane curve with affine equation $F(x_1, x_2) = 0$), then $P \in X$ is nonsingular if and only if at least one of the partial derivatives $\partial F/\partial x_1$ and $\partial F/\partial x_2$ does not vanish at P.

EXERCISE 2.1.43. Which of the points listed below are nonsingular?
(a) $P = (0,0)$ on the curve X given by $x_1^2 = x_2^3$;
(b) $P = (0,1)$, $X \colon \{x_1^2 + 1 = x_2^3\}$;
(c) $P = (0,0,0)$, $X \colon \{x_1^2 = x_2^2 + x_3,\ x_1^2 = x_3^3 + x_2\}$.

Power series expansion. Nonsingular points have another important property: a function regular in a neighbourhood of a nonsingular point P has a unique power series expansion in powers of an arbitrary local parameter t_P.

More precisely, let P be a nonsingular point of a curve X. Fix a local parameter t_P at P. For any formal power series $F(t) = \sum_{i=0}^{\infty} a_i t^i \in \Bbbk[[t]]$ and any integer $m > 0$, denote by $(F(t))_m$ the truncated series of degree m, i.e., the polynomial $\sum_{i=0}^{m} a_i t^i$.

PROPOSITION 2.1.44. *There exists a unique embedding $\tau_P \colon \mathcal{O}_P \to \Bbbk[[t_P]]$ of $\Bbbk$-algebras such that $f - (\tau_P(f))_m \in \mathfrak{m}_P^{m+1}$ for any $m \geq 0$.* □

Thus, any function regular at P has a unique expansion into a (formal) series in powers of a fixed local parameter t_P.

EXERCISE 2.1.45. Let $X = \mathbb{A}^1$, $P = \{0\}$, $t_P = t$, and $f = (1-t)^{-1}$. Find $\tau_P(f)$. (*Hint*: This is a geometric progression.)

Note that the power series expansion is directly extended to rational functions: any $f \in \Bbbk(X)$ has a unique expansion into a Laurent series at any nonsingular point $P \in X$, since there exists an integer m such that $t_P^m f$ is regular at P and we can set $\tau_P(f) = t_P^{-m} \tau_P(t_P^m f)$. Thus, we obtain a field embedding $\tau_P \colon \Bbbk(X) \hookrightarrow \Bbbk((t_P))$ for any nonsingular $P \in X$.

Smooth complete curves. Smooth complete curves are of the greatest importance. They possess the following useful property.

THEOREM 2.1.46. *Smooth complete curves are isomorphic if and only if they are birationally isomorphic.* □

EXERCISE 2.1.47. Let $X \subset \mathbb{P}^2$ be the curve defined by the homogeneous equation $x_0 x_1^2 = x_2^3$. Show that X is birationally isomorphic to $\mathbb{P}^1$ (i.e., X is a rational curve). Is X isomorphic to $\mathbb{P}^1$?

Since it is clear that $\mathbb{A}^1$ is birationally isomorphic (but not isomorphic) to $\mathbb{P}^1$, neither smoothness nor completeness in the condition of Theorem 2.1.46 can be omitted.

REMARK 2.1.48. For varieties of higher dimensions, Theorem 2.1.46 is also wrong. For example, smooth projective surfaces $\mathbb{P}^2$ and $\mathbb{P}^1 \times \mathbb{P}^1$ are birationally isomorphic but not isomorphic.

Fields $\mathbb{K}$ that occur as $\Bbbk(X)$ for smooth complete curves X have the following description.

PROPOSITION 2.1.49. *A field $\mathbb{K}$ containing $\Bbbk$ is isomorphic to $\Bbbk(X)$ for some smooth complete curve X if and only if $\mathbb{K}$ is of transcendence degree 1 over $\Bbbk$ and is finitely generated over $\Bbbk$.* □

These results allow us to use both equivalent languages—algebraic and geometric—when studying smooth complete curves, choosing the one more convenient for the matter in question. Below, in Sec. 2.5, we state some of the above notions and results using the purely algebraic language of functional fields.

Let as explain how one can algebraically express the notion of a point on a curve.

Let $V \subset \Bbbk(X)$ be an arbitrary proper subring in $\Bbbk(X)$, and let $V \supset \Bbbk$. A ring V is called a *valuation ring* if the condition that a function $f \in \Bbbk(X)$ does not belong to V implies $f^{-1} \in V$. An example of a valuation ring is the local ring $\mathscr{O}_P$ of a nonsingular point $P \in X$. To prove this, consider the following important notion. For $f \in \mathscr{O}_P$, $f \neq 0$, define

$$\operatorname{ord}_P(f) = \max\left\{ k \mid f \in \mathfrak{m}_P^k,\ f \notin \mathfrak{m}_P^{k+1} \right\}.$$

For an arbitrary rational function $f = g/h$ with $g, h \in \mathscr{O}_P$, define

$$\operatorname{ord}_P(f) = \operatorname{ord}_P(g) - \operatorname{ord}_P(h).$$

Since $\mathscr{O}_P \supset \Bbbk[U]$ for any open neighbourhood U of P, the field of fractions of $\mathscr{O}_P$ coincides with $\Bbbk(X)$, and $\operatorname{ord}_P(f)$ is well defined for any $f \in \Bbbk(X) \setminus \{0\}$. The number $\operatorname{ord}_P(f)$ is called the *(zero) order* of f at P. If $\operatorname{ord}_P(f) < 0$, then $|\operatorname{ord}_P(f)|$ is called the *pole order* of f at P. Thus, ord_P is a surjective group homomorphism $\operatorname{ord}_P : \Bbbk(X)^* \to \mathbb{Z}$; as is easily checked, it possesses the following property: for any $f, g \in \Bbbk(X)^*$,

$$\operatorname{ord}_P(fg) = \operatorname{ord}_P(f) + \operatorname{ord}_P(g),$$
$$\operatorname{ord}_P(f+g) \geq \min\bigl(\operatorname{ord}_P(f), \operatorname{ord}_P(g)\bigr).$$

Such a homomorphism is called a *discrete valuation* of $\Bbbk(X)$. Note that

$$\mathscr{O}_P = \{f \in \Bbbk(X)^* \mid \operatorname{ord}_P(f) \geq 0\} \cup \{0\}.$$

It is obvious from this description of $\mathscr{O}_P$ that if $f \in \Bbbk(X)$ and $f \notin \mathscr{O}_P$, then $f^{-1} \in \mathscr{O}_P$. Thus, $\mathscr{O}_P$ is a valuation ring. Moreover, we have the following important result.

THEOREM 2.1.50. *Let X be a smooth projective curve. Then the correspondence $P \mapsto \mathscr{O}_P$ is a bijection between the set of points $P \in X$ and the set of valuation rings in $\Bbbk(X)$.* □

Degree of a map. Note that if $f : X \to Y$ is a nonconstant rational map, then f is dominant. Hence, $f^* : \Bbbk(Y) \hookrightarrow \Bbbk(X)$ is a field embedding. Since the fields $\Bbbk(X)$ and $\Bbbk(Y)$ are of transcendence degree 1 over $\Bbbk$ and are finitely generated, the degree of extension $[\Bbbk(X) : f^*(\Bbbk(Y))]$ is finite; it is called the *degree of the map f* and is denoted by $\deg f$. In the next section we give another interpretation of this number.

Covers. If X and Y are smooth complete curves and $f : X \to Y$ is a rational nonconstant map, it can be shown that f is regular and surjective. In this case, f is called a *cover* of curves; the degree of f is referred to as the *covering degree*.

2.1.3. Divisors

Let X be a smooth complete curve over $\Bbbk$ (as before, we assume $\Bbbk$ to be algebraically closed and the curve X to be irreducible). A *divisor* D on X is a formal finite sum of the form $D = \sum a_P P$, where P are points of X and a_P are integers. The *support* $\operatorname{Supp} D$ of a divisor D is the set $\{P \in X \mid a_P \neq 0\}$. Sometimes we write $D = \sum a_i P_i$ instead of $D = \sum a_P P$, where $a_i = a_{P_i}$. The set of divisors on a curve X is denoted by $\operatorname{Div}(X)$. It is an abelian group with the operation of addition of divisors: if $D = \sum a_P P$ and $E = \sum b_P P$, then $D \pm E = \sum (a_P \pm b_P) P$. The *degree* $\deg P$ of a divisor $D = \sum a_P P$ is the integer $\sum a_P$. The degree map $\operatorname{Div}(X) \to \mathbb{Z}$ is surjective; its kernel, i.e., the subgroup of divisors of degree zero on X, is denoted by $\mathrm{Div}^0(X)$. If all a_P for a divisor $D = \sum a_P P$ are nonnegative, D is called an *effective* divisor; we write $D \geq 0$. If $D \neq 0$, the divisor is called *positive*. The set of effective divisors is denoted by $\mathrm{Div}^+(X)$. This definition induces a partial order on $\mathrm{Div}(X)$; namely, $D \geq F$ if $D - F \in \mathrm{Div}^+(X)$. Note that any divisor is a difference of two effective ones.

Now let $f \in \Bbbk(X)^*$ be an arbitrary nonzero rational function. Then to each f we assign the divisor
$$(f) = \sum \mathrm{ord}_P(f) P,$$
called the *divisor of the function* f. Note that this definition is correct, since any $f \in \Bbbk(X)^*$ has finitely many zeroes and poles on X and therefore $\mathrm{ord}_P(f) \neq 0$ for a finite number of points P only. Note also that
$$(f) = (f)_0 - (f)_\infty,$$
where
$$(f)_0 = \sum_{\mathrm{ord}_P(d) > 0} \mathrm{ord}_P(f) P \quad \text{and} \quad (f)_\infty = \sum_{\mathrm{ord}_P(f) < 0} (-\mathrm{ord}_P f) P$$
are effective divisors; $(f)_0$ is called the *divisor of zeroes* of f, and $(f)_\infty$ is the *divisor of poles*.

EXERCISE 2.1.51. Find the divisor (f) on $\mathbb{P}^1$ for an arbitrary polynomial $f = F(X) \in \Bbbk[X]$, where $x = T_1/T_0$ is a coordinate in $\mathbb{A}^1 \subset \mathbb{P}^1$.

Divisors of the form (f) are called *principal*. It is seen from the definition that $(fg^{\pm 1}) = (f) \pm (g)$; therefore, principal divisors form a subgroup $P(X)$ in $\mathrm{Div}(X)$. Divisors D_1 and D_2 such that $D_1 - D_2 \in P(X)$ are called *linearly equivalent* (or simply *equivalent*); we write $D_1 \sim D_2$.

THEOREM 2.1.52. *The degree of a principal divisor equals zero.* □

Thus, $P(X) \subseteq \mathrm{Div}^0(X)$. It follows from the theorem that all divisors in a given linear equivalence class are of the same degree. The quotient groups $\mathrm{Cl}(X) = \mathrm{Div}(X)/P(X)$ and $\mathrm{Cl}^0(X) = \mathrm{Div}^0(X)/P(X)$ play an important role in studying properties of curves. One can check that $\mathrm{Cl}(X)$ is canonically isomorphic to the group $\mathrm{Pic}(X)$ (we shall discuss this later; see Exercise 2.1.70). In the following, we use the notation $\mathrm{Pic}(X)$ and $\mathrm{Pic}^0(X)$ instead of $\mathrm{Cl}(X)$ and $\mathrm{Cl}^0(X)$, respectively.

Let D be a divisor on a curve X. Consider the set
$$L(D) = \{f \in \Bbbk(X)^* \mid (f) + D \geq 0\} \cup \{0\}.$$
Obviously, $L(D)$ is a vector space over $\Bbbk$. It is called the *space associated to the divisor* D; its dimension is denoted by $\ell(D)$.

THEOREM 2.1.53. *The dimension of $L(D)$ is finite for any $D \in \mathrm{Div}(X)$.*

PROOF. Let $D = D_1 - D_2$, where $D_1 \geq 0$ and $D_2 \geq 0$. Since $L(D) \subseteq L(D_1)$, we may assume D to be effective: let $D = \sum a_P P$, where all $a_P \geq 0$. Choose a local parameter t_P at some $P \in \mathrm{Supp}\, D$. Since $t_P^{a_P} f$ is regular at P for any $f \in L(D)$, we can define a linear functional $\varphi_P \colon L(D) \to \Bbbk$ by setting $\varphi_P(f) = (t_P^{a_P} f)(P)$. Obviously, the kernel of φ_P lies in $L(D')$, where $D' = D - P$ is an effective divisor with $\deg D' < \deg D$. Since $\ell(D) \leq \ell(D') + 1$, induction on $\deg D$ proves the theorem. □

COROLLARY 2.1.54. *For an effective divisor D, we have*
$$\ell(D) \leq \deg D + 1.$$
□

EXERCISE 2.1.55. Show that for any $D \in \mathrm{Div}(X)$ we have
$$\ell(D) \leq \max\{0, \deg D + 1\}.$$

EXERCISE 2.1.56. Let $X = \mathbb{P}^1$ and $D = n \cdot \infty$ with $n \geq 0$. Show that $L(D)$ consists of polynomials of degree at most n and therefore $\ell(D) = n + 1$.

PROPOSITION 2.1.57. *The value of $\ell(D)$ depends only on the linear equivalence class of D.*

PROOF. If $D_1 - D_2 = (f)$, where $f \in \Bbbk(X)^*$, then multiplication by f defines an isomorphism from $L(D_1)$ to $L(D_2)$. □

Linear systems. Let $M \neq \{0\}$ be a nonzero subspace of $L(D)$. A set of effective divisors of the form $(f) + D$, where f runs over $M \setminus \{0\}$, is called a *linear system* and is denoted by $|M|$. If $M = L(D)$, then $|M|$ is called a *compete linear system* and is denoted by $|D|$.

EXERCISE 2.1.58. Show that $|M| \cong \mathbb{P}(M)$.

The dimension of $\mathbb{P}(M)$ is denoted by $\dim|M|$. Thus, $\dim|M| = \dim M - 1$; in particular, $\dim|D| = \ell(D) - 1$.

There is a close relation between linear systems on a curve X and rational maps from X to projective spaces. Indeed, let $\varphi \colon X \to \mathbb{P}^n$ be a rational map,
$$\varphi \colon X \mapsto (f_0(P) : \ldots : f_n(P)), \qquad (2.1.2)$$
and assume that $\mathrm{Im}(\varphi)$ is contained in no hyperplane $H \subset \mathbb{P}^n$ (otherwise, φ can be considered as a map from X to $\mathbb{P}^m$ with $m < n$). Let
$$(f_i) = \sum a_{P,i} P, \quad i = 0, 1, \ldots, n,$$
and let
$$D = -\inf\{(f_0), \ldots, (f_n)\}$$
with respect to the partial order on $\mathrm{Div}(X)$ described above; i.e., $D = -\sum a_P P$ with $a_P = \min_{0 \leq i \leq n} a_{P,i}$. Obviously, $(f_i) + D \geq 0$; i.e., $f_i \in L(D)$. Let M_φ be the linear subspace in $L(D)$ generated by f_i. Thus, we have assigned to φ a linear system $|M_\varphi|$. Conversely, to any nonempty linear system $|M| \subseteq |D|$ there corresponds a rational map $\varphi \colon X \to \mathbb{P}^n$, $n = \dim|M|$, defined by (2.1.2), where $\{f_0, \ldots, f_n\}$ is a basis in M. In this case, the divisors $(f_i) + D$ can be regarded as inverse images, $\varphi^*(H_i)$, of the coordinate hyperplanes $H_i \colon x_i = 0$. More generally, if $\lambda = (\lambda_0 : \ldots : \lambda_n)$ and H_λ is a hyperplane given by $\sum \lambda_i x_i = 0$, then $f^*(H_\lambda) = (\sum \lambda_i f_i) + D$.

One can also describe φ in coordinate-free terms. To a point $P \in X \setminus \mathrm{Supp}\, D$, assign the functional φ_P on M defined as $\varphi_P(f) = f(P)$. Thus, we can define a map
$$\varphi = \varphi_M \colon X \setminus \mathrm{Supp}\, D \longrightarrow \mathbb{P}(M^*)$$
by $\varphi_M(P) = \varphi_P$ for $P \in X \setminus \mathrm{Supp}\, D$, i.e., a rational map from X to $\mathbb{P}(M^*)$. In the following, we use the notation φ_M for the map defined by a linear system $|M|$; if $|M| = |D|$, we write φ_D for φ_M. Using the connection between linear systems and rational maps, one can prove the following important fact.

THEOREM 2.1.59. *Any rational map of a smooth complete curve X to a protective space is regular.*

PROOF. Let $\varphi\colon X \to \mathbb{P}^n$ be defined by (2.1.2). Obviously, φ is regular at any point $P \notin \operatorname{Supp} D$. Let $Q \in \operatorname{Supp} D$. Choose a local parameter t_Q at Q and consider the map $\varphi'\colon X \to \mathbb{P}^n$ given by
$$\varphi'\colon P \mapsto (f'_0(P) : \ldots : f'_n(P)),$$
where $f'_i = t_Q^{-a_Q} f_i$. By the definition of D, all f'_i are regular at Q and there exists i_0 such that $f'_{i_0}(Q) \neq 0$. Hence, φ' is regular at Q. It remains to note that φ and φ' coincide as rational maps. □

COROLLARY 2.1.60. *Let $\varphi\colon X \to Y$ be a rational map, X and Y being smooth complete curves. Then φ is regular.* □

As a corollary of Theorem 2.1.59, one can also obtain Theorem 2.1.46.

Lattices. Now we shall explain how one can describe a divisor on X in terms of the function field $\Bbbk(X)$.

For each point $P \in X$, let there be given a *lattice* in $\Bbbk(X)$, i.e., a free $\mathcal{O}_P$-submodule $\mathscr{L}_P$ of rank 1 in $\Bbbk(X)$; $\mathscr{L}_P = \ell_P \mathcal{O}_P$ for $\ell_P \in \Bbbk(X)$. In this situation we say that we are given a *family of lattices* $(\mathscr{L}_P)_{P \in X}$. A family $(\mathscr{L}_P)_{P \in X}$ is called an *asymptotically standard lattice family* if, for all but a finite number of points $P \in X$, we have $\mathscr{L}_P = \mathcal{O}_P$.

With any divisor D we can associate an asymptotically standard lattice family as follows. Let $D = \sum a_P P$; then
$$\mathscr{L}_P = \begin{cases} \mathcal{O}_P & \text{if } P \notin \operatorname{Supp} D, \\ t_P^{-a_P} \mathcal{O}_P & \text{if } P \in \operatorname{Supp} D, \end{cases} \qquad (2.1.3)$$
where t_P is a local parameter at P. Note that $\mathscr{L}_P$ does not depend on the choice of a local parameter.

PROPOSITION 2.1.61. *Formula (2.1.3) defines a bijection of $\operatorname{Div}(X)$ onto the set of asymptotically standard lattice families on X.*

PROOF. Let us construct an inverse map $(\mathscr{L}_P) \mapsto D$. For any point $P \in X$, we have $\mathscr{L}_P = t_P^{a_P} \mathcal{O}_P$, where a_P is an integer; since $\mathscr{L}_P = \mathcal{O}_P$ for almost all P (except for a finite number of points), a divisor $D = \sum (-a_P) P$ is well defined. It is clear that the maps $(\mathscr{L}_P) \mapsto D$ and $D \mapsto (\mathscr{L}_P)_{P \in X}$ are inverse to each other. □

Let $\mathscr{L} = (\mathscr{L}_P)_{P \in X}$ be an asymptotically standard lattice family. The intersection $\bigcap_P \mathscr{L}_P \subset \Bbbk(X)$ is called the *space of sections* of $\mathscr{L}$. It is denoted by $H^0(\mathscr{L})$; its elements are called *sections* of $\mathscr{L}$. This definition immediately implies the following fact.

PROPOSITION 2.1.62. *Let $\mathscr{L} = (\mathscr{L}_P)_{P \in X}$ and $D \in \operatorname{Div}(X)$ correspond to each other. Then $H^0(\mathscr{L}) = L(D)$.* □

By a *fibre* of a family $(\mathscr{L}_P)$ at a point $P \in X$ we mean the one-dimensional vector space $\overline{\mathscr{L}}_P = \mathscr{L}_P / t_P \mathscr{L}_P$ over $\Bbbk = \mathcal{O}_P / t_P \mathcal{O}_P$. The image of a section $s \in H^0(\mathscr{L})$ in $\overline{\mathscr{L}}_P$ is called the *value* of s at P.

If $s \in H^0(\mathscr{L}) \setminus \{0\}$, then we define the *divisor of zeroes* of s as
$$D_s = \sum_P (\mathrm{ord}_P(s) - a_P) P,$$
where $\mathscr{L}_P = t_P^{a_P} \mathscr{O}_P$ for all P.

EXERCISE 2.1.63. Show that D_s is well defined and effective.

Asymptotically standard families $\mathscr{L} = (\mathscr{L}_P)$ and $\mathscr{M} = (\mathscr{M}_P)$ are called *linearly equivalent* if there exists $f \in \Bbbk(X)^*$ such that $\mathscr{L}_P = f \mathscr{M}_P$ for all $P \in X$.

EXERCISE 2.1.64. Show that asymptotically standard lattice families are linearly equivalent if and only if the corresponding divisors are.

Cartier divisors. One can give an alternative definition for a divisor (*Cartier divisor*), this definition being very useful in various situations. Let $D \in \mathrm{Div}(X)$, $D = \sum a_P P$, and let U be an open subset in X. By $D|_U$ we denote the following divisor on U:
$$D|_U = \sum_{P \in U} a_P P.$$

EXERCISE 2.1.65. Show that for any $D \in \mathrm{Div}(X)$ and any $P \in X$ there exists an open set $U = U(P, D)$ such that $P \in U$ and $D|_U$ is a principal divisor on U; i.e., $D|_U = (f)|_U$ for some $f \in \Bbbk(X)$.

EXERCISE 2.1.66. Deduce from the previous exercise that for any $D \in \mathrm{Div}(X)$ there exists a finite open covering $\{U_i\}$ of X such that $D|_{U_i} = (f_i)|_{U_i}$. Show that on the set $U_i \cap U_j$, the function $f_i f_j^{-1}$ is regular and has no zeroes. Conversely, show that for any system $(\{U_i\}, \{f_i\})$, where $\{U_i\}$ is a finite open covering of X, f_i are rational functions, and $f_i f_j^{-1} \in \Bbbk[U_i \cap U_j]^*$, there is a unique divisor $D \in \mathrm{Div}\, X$ with $D|_{U_i} = (f_i)|_{U_i}$.

A system $(\{U_i\}, \{f_i\})$ with these properties is called a *Cartier divisor*.

Functoriality. Using Cartier divisors, we can now define the *inverse image* $\varphi^*(D) \in \mathrm{Div}(X)$ of a divisor $D \in \mathrm{Div}(Y)$ under a regular map $\varphi \colon X \to Y$ of smooth irreducible complete curves. Indeed, let $(\{U_i\}, \{f_i\})$ be a Cartier divisor on Y, corresponding to $D \in \mathrm{Div}\, Y$. Then $\varphi^*(D)$ is defined on the covering $\{\varphi^{-1}(U_i)\}$ of X by the system of functions $\{\varphi^*(f_i)\}$, where $\varphi^*(f_i)(x) = f(\varphi(x))$. One can easily check that φ^* defines a group homomorphism $\varphi^* \colon \mathrm{Div}(Y) \to \mathrm{Div}(X)$; here, $\varphi^*(P(Y)) \subseteq P(X)$ since $\varphi^*((f)) = (\varphi^*(f))$ for any $f \in \Bbbk(Y)$. Thus, we obtain a homomorphism $\varphi^* \colon \mathrm{Pic}(Y) \to \mathrm{Pic}(X)$. Furthermore, one can easily show that $\varphi^*(\mathrm{Div}^0(Y)) \subseteq \mathrm{Div}^0(X)$ (cf. Proposition 2.1.74 below), and thus there is also a map $\varphi^* \colon \mathrm{Pic}^0(Y) \to \mathrm{Pic}^0(X)$.

EXERCISE 2.1.67. Explain why it would be incorrect to define the inverse image of a divisor $D = \sum a_P P$ on Y as $\sum a_P \left(\sum_Q Q \right)$, where the inner sum is taken over the inverse images $Q \in \varphi^{-1}(P)$.

Using Cartier divisors, for a curve $X \subset \mathbb{P}^n$ we can also define the *divisor* $(F) \in \mathrm{Div}\, X$ of a form F on $\mathbb{P}^n$. Indeed, put $U_i = \{T_i \neq 0\} \cap X$ and $f_i = F/T_i^s$, where $s = \deg F$ and $(T_0 : \ldots : T_n)$ are homogeneous coordinates in $\mathbb{P}^n$.

EXERCISE 2.1.68. Check that this system $(\{U_i\}, \{f_i\})$ defines a Cartier divisor.

In particular, if $F = L$ is a linear form, then $(F) = (L)$ is called a *hyperplane section divisor*; all such divisors are linearly equivalent. The degree $\deg(L)$ is called the *degree of X*. It is denoted by $\deg X$. If a map $f\colon X \to \mathbb{P}^m$ is given by a linear system $M \subseteq L(D)$, then effective divisors $D' \in |M|$ are precisely the inverse images of hyperplane section divisors $(L) \in \mathrm{Div}(f(X))$.

EXERCISE 2.1.69. Prove the last statement.

Connection with line bundles. There is a close connection between divisors and line bundles which is easy to explain using Cartier divisors. If $D = (\{U_\alpha\}, \{f_\alpha\})$ is a Cartier divisor on X, we set $f_{\alpha\beta} = f_\alpha f_\beta^{-1}\big|_{U_\alpha \cap U_\beta}$; since $f_{\alpha\beta} \in \Bbbk[U_\alpha \cap U_\beta]^*$, the functions $f_{\alpha\beta}$ define a line bundle on X, denoted by $\mathscr{O}(D)$.

EXERCISE 2.1.70. Show that the correspondence $D \mapsto \mathscr{O}(D)$ is a bijection of the set of linear equivalence classes of divisors on X onto the set of isomorphism classes of line bundles on X.

By virtue of Exercise 2.1.70, the degree $\deg \mathscr{L}$ of a line bundle $\mathscr{L}$ on X makes sense. If $\mathscr{L} \cong \mathscr{O}(D)$, the map φ_D is sometimes denoted by $\varphi_\mathscr{L}$.

EXERCISE 2.1.71. Let $\mathscr{L} \cong \mathscr{O}(D)$. Show that $H^0(X, \mathscr{L})$ and $L(D)$ are (canonically) isomorphic.

Let $s \in H^0(X, \mathscr{L})$, and let $X = \bigcup U_\alpha$ be an open covering such that all $\mathscr{L}|_{U_\alpha}$ are trivial. Since $\mathscr{L}|_{U_\alpha} \cong \mathscr{O}|_{U_\alpha}$, the restriction $s|_{U_\alpha}$ of s can be regarded as a regular function f_α on U_α. Let $D_{s,\alpha}$ be the divisor of zeroes of f_α on U_α, i.e., $D_{s,\alpha} = (f_\alpha)_0|_{U_\alpha}$.

EXERCISE 2.1.72. Show that there exists a unique divisor D_s such that $D_s|_{U_\alpha} = D_{s,\alpha}$ for all α.

The divisor D_s is called the *divisor of zeroes of s*. It is effective and is linearly equivalent to any divisor $D \in \mathrm{Div}(X)$ with $\mathscr{L} \cong \mathscr{O}(D)$.

EXERCISE 2.1.73. Prove this fact.

Inverse images of points. To study nonconstant maps of smooth projective curves $f\colon X \to Y$, it is important to consider divisors of the form $f^*(P)$, where P is a point of Y.

PROPOSITION 2.1.74. *For any point $P \in Y$, we have*
$$\deg f^*(P) = \deg f,$$
and for any divisor $D \in \mathrm{Div}\,Y$, we have
$$\deg f^*(D) = (\deg D)(\deg f). \qquad \square$$

EXERCISE 2.1.75. Deduce Theorem 2.1.52 from this proposition. (*Hint*: Apply Proposition 2.1.74 to the map $f\colon X \to \mathbb{P}^1$ and the divisor $D = 0 - \infty$.)

Ramification points. Divisors of the form $f^*(Q)$ allow us to define an important notion of ramification points of a map. Let $f\colon X \to Y$ be a cover (i.e., a surjective regular map) of smooth complete curves. Let $P \in X$ and $f(P) = Q$. It is clear that $P \in \mathrm{Supp}(f^*(Q))$. If $f^*(Q) = P + D'$, where $P \notin \mathrm{Supp}\,D'$, the map f is said to be *unramified* at $P \in X$. Otherwise, we have $f^*(Q) = e_P P + D'$, where

$e_P \geq 2$ and $P \notin \operatorname{Supp} D'$, and f is said to be *ramified* at P. Then P is called a *ramification point* of f, and e_P is called the *ramification index* of f at P. It easily follows from the definition of $f^*(Q)$ that, for local parameters t_P and t_Q at P and Q respectively, we have $f^*(t_Q) = t_P^{e_P} u$, where $u \in \mathscr{O}_P^*$. If, for $Q \in Y$ and for all $P \in f^{-1}(Q)$, the map f is unramified at P, we say that f is *unramified* at Q. Otherwise, Q is called a *ramification point* of f.

REMARK 2.1.76. Definitions of divisors and corresponding objects can as well be given for a higher-dimensional variety X (which we for simplicity assume to be smooth and projective). A *prime (simple) divisor* F on X is an (irreducible) subvariety in X of codimension 1 (not necessarily smooth). A *divisor* D on X is a finite linear combination
$$D = \sum n_i F_i, \quad n_i \in \mathbb{Z},$$
where F_i are prime divisors. The set $\operatorname{Div}(X)$ of divisors on X is an abelian group with respect to addition of divisors; the set $\operatorname{Div}^+(X)$ of effective divisors $D \geq 0$ (i.e., with all $n_i \geq 0$) is its subsemigroup.

If $f \in \Bbbk(X)^*$ and F is a prime divisor on X, one can define an integer $\operatorname{ord}_F(f)$ as follows. It can be shown that there exists an open affine set $U \subset X$ with $F \cap U \neq 0$ such that the subvariety $F \cap U$ in U is given by $g = 0$, where $g \in \Bbbk[U]$. Define
$$\operatorname{ord}_F(f) = \min \left\{ \ell \mid f = g^\ell h \text{ for } h \in \Bbbk[U] \right\}.$$
It is easily checked that $\operatorname{ord}_F(f)$ is well defined (i.e., does not depend on the choice of U); moreover,
$$\operatorname{ord}_F(f_1 f_2) = \operatorname{ord}_F(f_1) + \operatorname{ord}_F(f_2)$$
for any $f_1, f_2 \in \Bbbk(X)^*$, and
$$\operatorname{ord}_F(f_1 + f_2) \geq \min\{\operatorname{ord}_F(f_1), \operatorname{ord}_F(f_2)\}$$
if $f_1 + f_2 \neq 0$.

If $\operatorname{ord}_F(f) \geq 0$, we call it the *zero order* of f at F; if $\operatorname{ord}_F(f) < 0$, then $|\operatorname{ord}_F(f)|$ is called the *pole order* of f at F. For $f \in \Bbbk(X)^*$, similarly to the case of a curve, we define the *principal divisor*
$$(f) = \sum \operatorname{ord}_F(f) F.$$
For an arbitrary $D \in \operatorname{Div}(X)$, put
$$L(D) = \{f \in \Bbbk(X)^* \mid (f) + D \geq 0\} \cup \{0\}.$$

Using properties of $\operatorname{ord}_F(f)$, one can easily show that $L(D)$ is a vector subspace in $\Bbbk(X)$; one can also prove that it is of finite dimension for any D. The dimension of $L(D)$ is denoted by $\ell(D)$. The above-described connection between divisors and rational maps can be transferred word-for-word to the case of arbitrary smooth projective varieties. The theory of Cartier divisors and the connection between divisor classes and line bundles is also quite general.

2.1.4. Jacobians

Important information on an algebraic curve is provided by its Jacobian. The Jacobian of a curve is an algebraic group, i.e., an algebraic variety which is also a group such that the structure of the group and that of the algebraic variety are compatible.

Algebraic groups. Let G be a quasiprojective algebraic variety which is also a group. Then G is called an *algebraic group* if the maps
$$\psi\colon G\times G\to G, \quad \psi(g_1,g_2)=g_1g_2,$$
and
$$\varphi\colon G\to G, \quad \varphi(g)=g^{-1},$$
are regular.

EXERCISE 2.1.77. Check that the following varieties are algebraic groups:

(a) Affine line $\mathbb{A}^1$ with a group law given by the standard addition of coordinates of points. This algebraic group is called *additive* and is denoted by $\mathbb{G}_a$.

(b) Affine curve $\mathbb{A}^1\setminus\{0\}$ with a group law given by multiplication of coordinates. This algebraic group is called *multiplicative* and is denoted by $\mathbb{G}_m$.

(c) The set of nondegenerate $n\times n$ matrices over $\mathbb{k}$
$$\mathrm{GL}(n)=\mathrm{GL}(n,\mathbb{k})=\{A\mid \det A\neq 0\}$$
with the standard matrix multiplication law. This group is known as the *general linear group* (over $\mathbb{k}$). Note that $\mathrm{GL}(1)=\mathbb{G}_m$.

Note that the map $L_h\colon G\to G$, $L_h(g)=hg$, where h is an arbitrary fixed element of the group, is an isomorphism (of varieties but not of groups) since $(L_h)^{-1}=L_{h^{-1}}$.

PROPOSITION 2.1.78. *If G is an algebraic group, then G is a smooth variety.*

PROOF. There exists a nonsingular point $g\in G$ (Proposition 2.1.34). Applying L_h for all $h\in G$, we find that all points of G are nonsingular. □

Abelian varieties. If G is an algebraic group and a projective variety, then G is called an *abelian variety*. This name is justified by the following result.

THEOREM 2.1.79. *An abelian variety is abelian as a group.* □

We also have the following fact.

PROPOSITION 2.1.80. *Let $\psi\colon G\to H$ be a regular map from an abelian variety G to an algebraic group H. Then there exists a morphism of algebraic groups (i.e., a regular map which is a group homomorphism) $\varphi\colon G\to H$ such that $\psi=L_h\circ\varphi$, where $h=\psi(e)\in H$.* □

COROLLARY 2.1.81. *If A and B are isomorphic (as varieties) abelian varieties, then A and B are also isomorphic as groups.* □

Thus, for abelian varieties "all algebra is determined by geometry."

Jacobian of a curve. Now, let X be a smooth projective curve. Recall that $\mathrm{Pic}^0(X)$ denotes the subgroup in $\mathrm{Pic}(X)=\mathrm{Div}(X)/P(X)$ consisting of linear equivalence classes of divisors of degree zero.

THEOREM 2.1.82. *For each smooth projective curve there exists a unique abelian variety J_X such that*

(a) *J_X is isomorphic to $\mathrm{Pic}^0(X)$ as a group.*

(b) *The map*
$$i_{P_0}\colon X\to J_X, \quad P\mapsto P-P_0,$$

where P_0 is an arbitrary fixed point of X, is regular.

(c) For any regular map $\varphi\colon X \to A$ from X to an abelian variety A such that $\varphi(P_0)$ is the neutral element of A, there exists a morphism of abelian varieties $\lambda\colon J_X \to A$ with $\varphi = \lambda \circ i_{P_0}$. □

The abelian variety J_X is called the *Jacobian* of X.

The dimension of J_X is called the *genus*, $g(X)$, of X. One can show that this definition coincides with the definition of the genus of X in terms of differential forms given in Sec. 2.2.1; for $\Bbbk = \mathbb{C}$, it also coincides with the topological definition of the genus given in Sec. 2.1.5.

EXAMPLE 2.1.83. (a) In the case of $X = \mathbb{P}^1$, the Jacobian J_X is trivial since $\mathrm{Pic}^0(X) = \{0\}$. Therefore, $g(\mathbb{P}^1) = 0$.

(b) If X is a smooth plane cubic curve, then $J_X \cong X$ (see Sec. 2.4.1 below). Hence, $g(X) = 1$.

Functoriality. For a regular map $f\colon X \to Y$ of smooth projective curves, there are two regular maps of Jacobians, $f_*\colon J_X \to J_Y$ and $f^*\colon J_Y \to J_X$; for a regular map $g\colon Y \to Z$, one has

$$(f \circ g)_* = f_* \circ g_*, \qquad (f \circ g)^* = g^* \circ f^*.$$

EXERCISE 2.1.84. Deduce the existence of the map $f_*\colon J_X \to J_Y$ from Theorem 2.1.82c.

The existence of the regular map f^* is implied by the following result.

THEOREM 2.1.85. *Let $f\colon X \to Y$ be a regular map of smooth projective curves. Then the map $f^*\colon J_Y \to J_X$ defined by the inverse image of a divisor is a morphism of abelian varieties.* □

Embedding in the Jacobian. It should be noted that any curve which is not isomorphic to $\mathbb{P}^1$ can be embedded in its Jacobian:

PROPOSITION 2.1.86. *If $i_{P_0}\colon X \to J_X$ is not injective, then X is isomorphic to $\mathbb{P}^1$.*

PROOF. If i_{P_0} is not injective, then there are two distinct points, P and Q, such that $P = Q + (f)$ for some $f \in \Bbbk(X) \setminus \Bbbk$. Consider f as a regular map $f\colon X \to \mathbb{P}^1$. Since $(f) = P - Q$, we have $(f)_0 = P$; therefore, by Proposition 2.1.74 we have $\deg f = 1$. Hence, f is an isomorphism. □

Note that J_X can also be identified with the set of isomorphism classes of line bundles of degree zero. Then, as is easily verified, for any line bundle $\mathscr{L}$ of degree $a = \deg \mathscr{L}$, the map $\mathscr{M} \mapsto \mathscr{M} \otimes \mathscr{L}$ is a bijection from the set of line bundles of degree zero to the set of line bundles of degree a. Let us present this as a separate statement.

PROPOSITION 2.1.87. *There is a natural bijection between the set of isomorphism classes of line bundles of a fixed degree on a smooth projective curve X and the Jacobian J_X of this curve.* □

2.1.5. Riemann Surfaces

In the case where the ground field $\Bbbk$ is the complex field $\mathbb{C}$, any smooth algebraic curve can be considered as a Riemann surface; a projective curve defines a compact Riemann surface.

Recall that a *Riemann surface* is a connected complex-analytic manifold of complex dimension 1. More precisely, a Riemann surface is a connected Hausdorff topological space T, equipped with an atlas S. By an *atlas* we mean a family $S = \{(U_\alpha, p_\alpha) \mid \alpha \in A\}$, where A is an index set, $\{U_\alpha\}_{\alpha \in A}$ is an open covering of T, and $p_\alpha : U_\alpha \to V_\alpha$ is a homomorphism of U_α onto an open subset $V_\alpha \subseteq \mathbb{C}$ such that, for $U_\alpha \cap U_\beta \neq 0$, the map

$$p_\beta \circ p_\alpha^{-1} : p_\alpha(U_\alpha \cap U_\beta) \longrightarrow p_\beta(U_\alpha \cap U_\beta)$$

is holomorphic. Usually, one also requires S to be maximal among sets that satisfy these conditions. In fact, this requirement is not essential, since for any S there always exists a maximal set S' that contains S. In what follows, we do not require S to be maximal.

So let X be a smooth quasiprojective curve over the complex field $\mathbb{C}$. Let us define a structure of a (complex) topological space on X. First, we make some elementary remarks on the complex topology of affine and projective spaces. It is clear that $\mathbb{A}^N(\mathbb{C})$ is homeomorphic to $\mathbb{R}^{2N}$ in the topology of the product of N copies of $\mathbb{C}$. The *complex topology* on $\mathbb{P}^N(\mathbb{C})$ is defined as the quotient topology of $\mathbb{A}^{N+1}(\mathbb{C}) \setminus \{0\}$ under the canonical projection

$$\mathbb{A}^{N+1}(\mathbb{C}) \setminus \{0\} \longrightarrow \mathbb{P}^N(\mathbb{C}).$$

If we note that the restriction of this map onto the $(2N+1)$-sphere

$$S^{2N+1} = \left\{ z = (z_0, \ldots, z_N) \in \mathbb{A}^{N+1}(\mathbb{C}) \setminus \{0\} \,\Big|\, \sum_{i=0}^{N} |z_i|^2 = 1 \right\}$$

canonically embedded in $\mathbb{A}^{N+1} \setminus \{0\}$ yields a surjection $S^{2N+1} \to \mathbb{P}^N(\mathbb{C})$, we get that $\mathbb{P}^N(\mathbb{C})$ is compact. Now let us define the complex topology on a curve X, regarded as a subset of some $\mathbb{P}^N(\mathbb{C})$, and then prove that this topology does not depend on an embedding. Since X is a projective curve $\overline{X}$ without a finite number of points, it suffices to define the topology on $\overline{X}$ as the topology induced from $\mathbb{P}^N(\mathbb{C})$. The set $\overline{X}$ is closed in $\mathbb{P}^N(\mathbb{C})$ as the intersection of zero sets of continuous functions. This immediately implies that a projective curve is compact.

Now let us show that, for a smooth curve X, this complex topology can be defined intrinsically, independently of an embedding in a projective space. It suffices to define a fundamental basis of neighbourhoods for each point. Let $P \in X$ be an arbitrary point (by our assumption, it is a nonsingular point of X). Let U be an affine neighbourhood of P, $U \subset \mathbb{A}^N(\mathbb{C})$. Consider a local parameter t_P at P. Since t_P is a rational function regular at P, there exists $\varepsilon > 0$ such that the open set $U_\varepsilon = \{Q \in X \mid t_P(Q) < \varepsilon\}$ is continuously mapped to $\mathbb{C}$ by t_P. Consider the functions $t_1, \ldots, t_N$, restrictions of coordinate functions $T_1, \ldots, T_N$ on $\mathbb{A}^N$ onto X. By Proposition 2.1.44, they correspond to the formal series $\Phi_i(t) = \tau_P(t_i)$ at P. From the implicit function theorem, one can deduce the following fact.

LEMMA 2.1.88. *For each point P there exists $\varepsilon > 0$ such that the series $\Phi_i(t)$ converges in a disc of radius ε.* □

For a disk D_ε of radius ε in $\mathbb{C}$, define a map $u\colon D_\varepsilon \to \mathbb{A}^N(\mathbb{C})$ by $u(t) = (\Phi_1(t),\ldots,\Phi_N(t))$. One easily checks that the maps $t_P \circ u$ and $u \circ t_P$ are identical on some neighbourhoods of 0 and P (in $\mathbb{C}$ and U_ε respectively). Therefore, t_P defines a homeomorphism of a neighbourhood U_P of P in X onto a neighbourhood V_P of 0 in $\mathbb{C}$. Since this argument is valid for any choice of t_P, the complex topology on X depends on neither the choice of local parameter nor the choice of embedding $X \subset \mathbb{P}^N$. Moreover, Lemma 2.1.88 implies that the family (U_P, t_P) is an atlas on X. To check that we thus obtain a Riemann surface, we should prove the following statement.

THEOREM 2.1.89. *Any algebraic curve over $\mathbb{C}$ is connected in the complex topology.*

SKETCH OF THE PROOF. Let, for simplicity, X be a projective curve. We shall use here the following immediate corollary of the Riemann–Roch theorem (note that the proof of the Riemann–Roch theorem given in Sec. 2.2 does not employ results of this subsection): Let $P_0 \in X$; then there exists a nonconstant rational function $f \in \Bbbk(X)$ regular at all points of X except for P_0. Assume now that $X = X_1 \cup X_2$, where X_1 and X_2 are nonempty disjoint closed subsets of X. Let $P_0 \in X_1$; consider a nonconstant rational function f regular everywhere outside P_0. Since f is regular on X_2 and X_2 is compact, $f|_{X_2}$ is constant. Hence, f itself is constant (consider a neighbourhood of any point in X_2 isomorphic to a disk in $\mathbb{C}$), a contradiction. □

Since a Riemann surface can be considered as a two-dimensional orientable differentiable manifold, i.e., an orientable surface, any smooth complex algebraic curve X is an orientable surface. If X is projective, this surface is compact. Recall the following fact.

THEOREM 2.1.90. *A compact orientable connected surface S is diffeomorphic to a sphere with g handles, g being a nonnegative integer.* □

The integer g is called the *genus* of S (if S corresponds to an algebraic curve X, it is also called the *genus* of X).

In particular, for $g = 0$ the surface S is diffeomorphic to the sphere S^2; for $g = 1$, to the torus $T^2 = S^1 \times S^1$.

One can show that this topological definition of the genus coincides with both the above definition in terms of Jacobians and the one in terms of differential forms given in Sec. 2.2.1.

In the compact case it turns out that the construction of a Riemann surface from an algebraic curve can be inverted. This profound result is known as the *Riemann existence theorem*:

THEOREM 2.1.91. *Let Y be a compact Riemann surface. Then there exists a unique (up to an isomorphism) smooth projective algebraic curve X with $Y = X_{\mathrm{an}}$, where X_{an} is the Riemann surface associated with X.* □

In fact, one can also show that, for any complete curves X and X', the set of holomorphic maps from X_{an} to X'_{an} coincides with the set of regular maps from X to X'.

Therefore, the notion of a smooth projective complex curve and that of a compact Riemann surface are essentially equivalent. This makes it appropriate to apply the powerful technique of complex analysis to the study of complex algebraic curves, as well as to use (very carefully) our intuition gathered from differential geometry in the study of curves over other fields. We shall use the connection between curves and Riemann surfaces to construct some classes of algebraic curves that naturally arise as Riemann surfaces (the so-called modular curves; see *Advanced Chapters*).

2.2. Riemann–Roch Theorem

Since many important questions of the theory of algebraic curves are reduced to the calculation of the dimension of $L(D)$ for various divisors D, an explicit expression for $\ell(D)$ plays an essential role in the theory of curves. Such an expression is given by the Riemann–Roch theorem, which is a crucial result of the theory. To state it, one should consider differential forms on curves, which are also useful in many other questions.

In this section, we give an exposition of this subject to the minimum extent necessary. In Sec. 2.2.1, the definition and basic properties of differential forms on curves are given; Sec. 2.2.2 contains the statement and a scheme of a proof of the Riemann–Roch theorem, as well as of some of its corollaries. Section 2.2.3 is devoted to the Hurwitz formula, which describes the behaviour of genera under regular maps; this result often allows one to compute the genus of a particular curve. In Sec. 2.2.4 we describe basic properties of special divisors and Weierstrass points on curves. Section 2.2.5 is devoted to the Cartier operator on differential forms, which is specific for curves over fields of positive characteristics and plays an essential role in studying such curves.

In this section, by X we mean a smooth projective curve over an algebraically closed field $\Bbbk$.

2.2.1. Differential Forms

Let $P \in X$, and let f be a function regular at P, i.e., $f \in \mathscr{O}_P$. By $d_P f$ we denote the image of $f - f(P) \in \mathfrak{m}_P$ in the one-dimensional vector space $\mathfrak{m}_P/\mathfrak{m}_P^2$ over $\Bbbk$. The element $d_P f$ is called the *differential* of f at P.

EXERCISE 2.2.1. Check that the differential $d_P \colon \mathscr{O}_P \to \mathfrak{m}_P/\mathfrak{m}_P^2$ thus defined is a morphism of vector $\Bbbk$-spaces; moreover, the *"Leibniz rule"*

$$d_P(fg) = f(P)\,d_P g + g(P)\,d_P f$$

holds for any $f, g \in \mathscr{O}_P$.

Let $U \subset X$ be an open subset in X. Let $f \in \Bbbk[U]$ be a rational function on X regular on U. Next, let $\Phi[U]$ be the set of maps φ that send each point $P \in U$ to some element in $\mathfrak{m}_P/\mathfrak{m}_P^2$; obviously, $\Phi[U]$ is a module over the ring $\Bbbk[U]$. Any function $f \in \Bbbk[U]$ defines an element $df \in \Phi[U]$ by $(df)(P) = d_P f$. We call $\varphi \in \Phi[U]$ a *differential form regular on U* if for any $P \in U$ there exists a neighbourhood V of P such that the restriction $\varphi|_V$ lies in the $\Bbbk[V]$-submodule of $\Phi[V]$ generated by elements df, $f \in \Bbbk[V]$. Differential forms regular on U form a $\Bbbk[U]$-module, denoted by $\Omega[U]$. In particular, if $U = X$, we obtain a vector $\Bbbk$-space $\Omega[X]$. Sometimes, we simply write Ω for $\Omega[X]$.

THEOREM 2.2.2. *The space $\Omega[X]$ is of finite dimension over $\Bbbk$.*

We give the proof of this theorem a bit later (after the proof of Corollary 2.2.8). The dimension $\dim_\Bbbk \Omega[X]$ is denoted by $g = g(X)$ and is called the *genus* of X. Note once more that this definition coincides with those given in Secs. 2.1.4 and 2.1.5.

PROPOSITION 2.2.3. *Let $P \in X$, and let $t = t_P$ be a local parameter at P. Then there exists a neighbourhood U of P such that $\Omega[U] = \Bbbk[U]\,dt$.*

SKETCH OF THE PROOF. First, note that for any $F \in \Bbbk[T_1,\ldots,T_N]$ and any $f_1,\ldots,f_N \in \Bbbk[U]$ we have

$$d(F(f_1,\ldots,f_N)) = \sum_{i=1}^{N} \frac{\partial F}{\partial T_i}(f_1,\ldots,f_N)\, df_i,$$

which easily follows from the linearity of the operator d and from $d(fg) = f\,dg + g\,df$. Now let V be an open neighbourhood of P that is an affine curve. Let $V \subset \mathbb{A}^N$, and let the polynomials $F_1,\ldots,F_m \in \Bbbk[T_1,\ldots,T_N]$ be a basis of the ideal $I_V \subset \Bbbk[T_1,\ldots,T_N]$ formed by polynomials vanishing on V. Since $F_i|_V = 0$, we obtain

$$\sum_{j=1}^{N} \frac{\partial F_i}{\partial T_j} dt_j = 0 \quad \text{for} \quad i = 1,\ldots,m, \tag{2.2.1}$$

where $t_j = T_j|_V \in \Bbbk[V]$. Since P is a nonsingular point, the rank of the matrix $(\partial F_i/\partial T_j(P))$ is $N-1$; without loss of generality, we can assume $t = t_1$ to be a local parameter. Hence, we can express dt_j in terms of dt from (2.2.1):

$$dt_j = f_j\, dt, \quad j = 2,\ldots,N,$$

f_j being rational functions regular at P. If $U \subset V$ is an affine neighbourhood of P in which all the f_j are regular, one can easily show that $\Omega[U] = \Bbbk[U]\, dt$, since any form $\omega \in \Omega[U]$ can be expressed in terms of $dt_1,\ldots,dt_N$ and hence in terms of dt. □

COROLLARY 2.2.4. *Let $\omega \in \Omega[U]$. Then the zero set F_ω of ω (i.e., the set of points $P \in U$ with $\omega_P = 0$) is closed in U.* □

Let $\omega \in \Omega[U]$ be a differential form regular on an open subset U of X. Then ω is said to define a *rational differential form* on X; two forms, $\omega \in \Omega[U]$ and $\omega' \in \Omega[U']$, define the same rational differential form on X if the restrictions of ω and ω' onto $U \cap U'$ coincide. The set of rational differential forms is denoted by $\Omega(X)$. Obviously, $\Omega(X)$ is a vector space over the field $\Bbbk(X)$ of rational functions on X. Moreover, Proposition 2.2.3 implies the following fact.

PROPOSITION 2.2.5. *The dimension of $\Omega(X)$ over $\Bbbk(X)$ is 1.* □

Canonical class. From the definition of a differential form ω, it follows that any point P has a neighbourhood U such that $\omega = f\,dt$, $f \in \Bbbk(X)$, where $t - t(P)$ is a local parameter at any $P \in U$. Hence it follows that for $\omega \neq 0$ there exists a finite open covering $\{U_i\}$ of X such that $\omega|_{U_i} = f_i\, dt_i$. Since we have $f_i\, dt_i = f_j\, dt_j$ on $U_i \cap U_j$, where $t_i - t_i(Q)$ and $t_j - t_j(Q)$ are local parameters at any $Q \in U_i \cap U_j$, the functions f_i/f_j and f_j/f_i are regular and do not vanish on $U_i \cap U_j$. Therefore, the system $(\{U_i\}, \{f_i\})$ defines a Cartier divisor, denoted by (ω) and called the *divisor associated with* ω. Note that any other form $\omega' \in \Omega(X)$ by Proposition 2.2.5 can be represented as $\omega' = f\omega$ for $f \in \Bbbk(X)$, so that $(\omega') = (\omega) + (f)$. Hence, the linear equivalence class $K = K_X$ of (ω), $\omega \in \Omega(X)$, does not depend on the choice of ω; it is called the *canonical class* of X. Sometimes, by K_X we also denote an arbitrary divisor from the canonical class.

EXAMPLE 2.2.6. (a) Let $X = \mathbb{P}^1$, $\omega = dt$, where t is a coordinate on $\mathbb{P}^1$. Set $u = t^{-1}$,

$$U_0 = \{x \in \mathbb{P}^1 \mid u \neq 0\}, \qquad U_1 = \{x \in \mathbb{P}^1 \mid t \neq 0\}.$$

Then $\mathbb{P}^1 = U_0 \cup U_1$, the function $t - t(P)$ is a local parameter at any point $P \in U_0$, and $u - u(Q)$ is a local parameter at any $Q \in U_1$. In U_0 we have the representation $\omega = d(u^{-1}) = -u^{-2} du$. Hence, $(\omega) = -2\infty$, where ∞ is a point on $\mathbb{P}^1$ with $u(\infty) = 0$. Therefore, $K_{\mathbb{P}^1}$ consists of divisors of degree -2.

(b) Let X be the curve in $\mathbb{P}^2$ given by $x_0^3 + x_1^3 + x_2^3 = 0$, and let
$$U_{ij} = \{P \in X \mid x_i(P) x_j(P) \neq 0\}, \quad i,j = 0, 1, 2;$$
thus, $X = U_{01} \cup U_{02} \cup U_{12}$. Set $x = x_1/x_0$, $y = x_2/x_0$, and consider $\omega = dy/x^2$ in U_{01}. Then $\omega = -dx/y^2$ in U_{02} and $\omega = dv/u^2$ in U_{12}, where $v = x_0/x_1$ and $u = x_2/x_1$. Therefore, ω is a regular form on X, and one easily checks that $(\omega) = 0$. Hence, the canonical class of X is zero.

Smooth plane curves. Let $F(x_0 : x_1 : x_2) = 0$ be a homogeneous equation of degree m that defines a smooth curve X in $\mathbb{P}^2$. Let $x = x_1/x_0$ and $y = x_2/x_0$ be coordinates on $\mathbb{A}^2 \subset \mathbb{P}^2$, and let $G(x,y) = F(1 : x : y) = 0$ be the equation of the affine curve $X' = X \cap \mathbb{A}^2$. Consider rational differential forms on X given by
$$\omega = \frac{P \, dy}{G'_x}, \qquad (2.2.2)$$
where $G'_x = \partial G/\partial x$ and $P(x,y) \in \Bbbk[x,y]$.

PROPOSITION 2.2.7. *A form given by (2.2.2) is regular if and only if $\deg P \leq m - 3$. Conversely, any regular form on X can be represented in the form (2.2.2) with some $P \in \Bbbk[x,y]$, $\deg P \leq m - 3$. Thus,*
$$\Omega[X] = \left\{ \frac{P \, dy}{G'_x} \,\Big|\, P(x,y) \in \Bbbk[x,y], \; \deg P \leq m - 3 \right\}.$$
Moreover, $K_X = (m-3)L$, where L is the class of a hyperplane (linear) section.

PROOF. Consider the form $\omega_0 = dy/G'_x$. It is regular at any point $P = (x,y) \in X' = X \cap \mathbb{A}^2$ with $G'_x(x,y) \neq 0$. Furthermore, since $G(x,y) = 0$ on X, we have $\omega_0 = -\omega'_0$, where $\omega'_0 = dx/G'_y$; hence, ω_0 is also regular at any point of $X \cap \mathbb{A}^2$ with $G'_y(x,y) \neq 0$. Since X is smooth, for each point of X' at list one of the derivatives G'_x and G'_y does not vanish, and ω_0 is regular on X'. Moreover, it is clear that $(\omega_0)|_{X'} = 0$. To investigate the behaviour of ω_0 on the whole X, it remains to consider points outside $\mathbb{A}^2$. Let us consider the set $X'_1 = X \cap \mathbb{A}^2_1$, where $\mathbb{A}^2_1 = \{(x_0 : x_1 : x_2) \in \mathbb{P}^2 \mid x_1 \neq 0\}$. Coordinates on $\mathbb{A}^2_1$ are $u = 1/x$ and $v = y/x$; hence, we have $x = 1/u$, $y = v/u$, $dx = -du/u^2$, and $dy = (u\,dv - v\,du)/u^2$. The equation of X'_1 in $\mathbb{A}^2_1$ is the polynomial $H(u,v) = u^m G(1/u, v/u)$, so that $H'_v(u,v) = u^{m-1} G'_y(x,y)$, whence we get $\omega_0 = -u^{m-3} du/H'_v$. Therefore, the divisor of ω_0 on X'_1 is $(m-3)(u) = (m-3)(x_0)$. Thus, $K_X = (m-3)L$, where L is the class of the intersection of X with a line in $\mathbb{P}^2$.

From these arguments, it is clear that $\Omega[X'] = \{P\omega_0 \mid P \in \Bbbk[x,y]\}$. Indeed, the condition $(f\omega_0) \geq 0$ for $f \in \Bbbk(X)$ is equivalent to $(f)|_U \geq 0$; i.e., $(f) = D - m(x_0)$ for some $D \geq 0$ and $m \geq 0$. Now let $\omega = P\omega_0$ with $P \in \Bbbk[x,y]$, $\deg P = s$. Then on X'_1 we have $\omega = -\widetilde{P}(u,v) u^{m-3-s} du$, where $\widetilde{P}(u,v) = u^s P(1/u, v/u)$. This immediately implies that ω is regular in X'_1 if and only if $s \leq m - 3$. Applying the same reasoning to the other neighbourhood $X'_2 = X \cap \mathbb{A}^2_2$, where $\mathbb{A}^2_2 = \{(x_0 : x_1 : x_2) \in \mathbb{P}^2 \mid x_2 \neq 0\}$, we obtain the proposition. □

Thus, divisors from the canonical class K_X are divisors of the form (F), where F is a form of degree $m-3$. It is possible to say that divisors from the canonical class are intersections of X with curves given by equations $F = 0$ of degree $m-3$. Such curves are called *adjoint* for X. For $m \leq 3$, there are no adjoint curves.

COROLLARY 2.2.8. *The genus of a smooth plane curve X of degree m is*
$$g(X) = \frac{(m-1)(m-2)}{2}.$$

PROOF. Since the space of polynomials in two variables of degree at most $m-3$ is of dimension $(m-1)(m-2)/2$, it suffices to show that the forms $P\omega_0$ and $Q\omega_0$ are distinct if P and Q are distinct polynomials of degree at most $m-3$. Indeed, if $P\omega_0 = Q\omega_0$, then the difference $P - Q$ is divisible by G, which is possible only for $P = Q$. □

From the definition of the divisor (ω) one can easily see that a form ω is regular if and only if $(\omega) \geq 0$. Hence, $\Omega[X]$ is isomorphic to $L(K_X)$, where $K_X = (\omega)$, ω being an arbitrary nonzero rational differential form on X. In view of Theorem 2.1.53, this implies Theorem 2.2.2. Below we also need the space $\Omega(D)$, $D \in \mathrm{Div}(X)$, defined as follows:
$$\Omega(D) = \{\omega \in \Omega(X) \setminus \{0\} \mid (\omega) + D \geq 0\} \cup \{0\}.$$

From the definition of $\Omega(D)$ it is clear that $\Omega(D) \cong L(K_X + D)$; in particular, $\Omega(D)$ is of finite dimension for any $D \in \mathrm{Div}(X)$. The line bundle $\mathscr{O}(K_X + D)$ is sometimes denoted by $\Omega(D)$; thus, $\Omega(D) \cong H^0(\Omega(D))$.

Functoriality. Let $\varphi \colon X \to Y$ be a regular map of smooth projective curves, and let $\omega = \sum\limits_{i=1}^{m} f_i\, dg_i \in \Omega(Y)$. We set
$$\varphi^*(\omega) = \sum_{i=1}^{m} \varphi^*(f_i)\, d\varphi^*(g_i) \in \Omega(X).$$

EXERCISE 2.2.9. Show that $\varphi^*(\omega)$ is well defined, i.e., does not depend on the choice of a representation of the form $\omega = \sum f_i\, dg_i$. Show that if ω is regular at $Q \in Y$, then $\varphi^*(\omega)$ is also regular at any point $P \in \varphi^*(Q)$.

Thus, there exist $\Bbbk$-linear maps
$$\varphi^* \colon \Omega(Y) \to \Omega(X) \quad \text{and} \quad \varphi^* \colon \Omega[Y] \to \Omega[X].$$

EXERCISE 2.2.10. Show that if φ is a *separable map* (i.e., the corresponding field extension $\Bbbk(X)/\varphi^*(\Bbbk(Y))$ is separable), then φ^* is an injection. Otherwise, φ^* is trivial. Show that if $\varphi \colon X \to Y$ and $\psi \colon Y \to Z$ are regular maps, then $(\psi \circ \varphi)^* = \varphi^* \circ \psi^*$.

REMARK 2.2.11. Note that in the general case we have $\varphi^*((\omega)) \neq (\varphi^*(\omega))$ (consider any map $\varphi \colon X \to \mathbb{P}^1$ of degree greater than 1). Thus, the maps φ^* for divisors and differential forms, in general, do not commute. A precise commutation rule will be given in Sec. 2.2.4 (the Hurwitz formula).

Automorphisms. An isomorphism $g\colon X \to X$ of a smooth projective curve X onto itself is called an *automorphism*. The set of automorphisms of X is denoted by $\mathrm{Aut}(X)$, or $\mathrm{Aut}_{\Bbbk}(X)$ if it is necessary to indicate its dependence on a ground field. It is clear that $\mathrm{Aut}(X)$ is a group.

EXERCISE 2.2.12. Show that $\mathrm{Aut}(\mathbb{P}^1) = \mathrm{PGL}(2,\Bbbk)$, the quotient group of the group $\mathrm{GL}(2,\Bbbk)$ of nondegenerate 2×2 matrices modulo its center (the center of $\mathrm{GL}(2,\Bbbk)$ consists of matrices of the form $\begin{pmatrix} a & 0 \\ 0 & a \end{pmatrix}$, $a \in \Bbbk^*$).

Thus, $\mathrm{Aut}(\mathbb{P}^1)$ is an infinite group. Below, in Sec. 2.4.1, we shall show that if $g(X) = 1$, then X is an abelian variety (of dimension 1), and hence $\mathrm{Aut}(X)$ contains X as a subgroup; this subgroup is normal, and the quotient group $\mathrm{Aut}(X)/X$ is a finite group G of order 2, 4, or 6 (if p, the characteristic of $\Bbbk$, equals neither 2 nor 3; for $p = 3$, the order of G is a divisor of 12; for $p = 2$, it is a divisor of 24). Hence, for $g(X) = 1$, the group $\mathrm{Aut}(X)$ is also infinite. On the other hand, we have the following fact.

THEOREM 2.2.13. *If $g(X) \geq 2$, then $\mathrm{Aut}(X)$ is finite.* □

REMARK 2.2.14. If $p = \mathrm{char}\,\Bbbk = 0$, then
$$|\mathrm{Aut}(X)| \leq 84(g-1).$$

This is also valid for the case of a finite characteristic provided that $p \geq g+2$, with the only exception of the curve $y^2 = x^p - x$, $p \neq 2$, which has $g = (p-1)/2$ and $|\mathrm{Aut}(X)| = 2p^2(p-1)$. For $p \leq g+1$, as we shall see in Sec. 3.4, the group $\mathrm{Aut}(X)$ can be very large for some other curves too.

Now let $H \subseteq \mathrm{Aut}(X)$ be an arbitrary finite subgroup (the case $g(X) \leq 1$ is allowed). It is clear that H also acts on $\Bbbk(X)$. Consider the field $\mathbb{K} = \Bbbk(X)^H$ of invariants under this action. Since $\mathbb{K}$ is finitely generated over $\Bbbk$ and is of transcendence degree 1 over $\Bbbk$, by Proposition 2.1.49 and Theorem 2.1.46 there exists a unique smooth projective curve Y such that $\Bbbk(Y) = \mathbb{K}$. The embedding $\Bbbk(Y) \hookrightarrow \Bbbk(X)$ defines a regular map $X \to Y$ of degree $h = |H|$. In this situation, Y is denoted by X/H. In particular, if $H = \langle g \rangle$ is a cyclic subgroup in $\mathrm{Aut}(X)$ generated by $g \in \mathrm{Aut}(X)$, then X/H is denoted by $X/\langle g \rangle$ or by X^g.

EXERCISE 2.2.15. Show that there is the equality of sets $Y = X/H$.

EXERCISE 2.2.16. Let $Y = X/H$, let $f\colon X \to Y$ be the natural map, and let $D \in \mathrm{Div}(Y)$. Then H acts on $L(f^*D)$, and $L(f^*D)^H$ is canonically isomorphic to $L(D)$.

2.2.2. Riemann–Roch Theorem

THEOREM 2.2.17 (Reimann–Roch theorem). *Let X be a smooth projective curve. Then for any divisor $D \in \mathrm{Div}(X)$ we have*

$$\ell(D) - \ell(K - D) = \deg D - g + 1, \tag{2.2.3}$$

$K = K_X$ being the canonical class of X. □

We do not prove this profound result; we just give an insight into the main points of its proof. On the way, we introduce some important notions of the geometry of algebraic curves.

Note first of all that formula (2.2.3) is valid for $D=0$ since $\ell(0)=1$, $\deg(0)=0$, and $\ell(K)=g$ by the definition of g.

Let $R = R(X)$ be the *algebra of distributions* on X, i.e., of families $\{r_P\}_{P\in X}$, $r_P \in \Bbbk(X)$, such that $r_P \in \mathscr{O}_P$ for almost all $P \in X$ (i.e., for all but a finite number). If $D = \sum a_P P$ is a divisor on X, then by $R(D)$ we denote the set of distributions $r = \{r_P\}$ such that $\mathrm{ord}_P(r_P) \geq -a_P$ for any $P \in X$. Since for any $f \in \Bbbk(X)$ we have $f \in \mathscr{O}_P$ for almost all $P \in X$, the field $\Bbbk(X)$ is a $\Bbbk$-subalgebra of R. Consider the vector $\Bbbk$-space
$$I(D) = R/(R(D) + \Bbbk(X)).$$
Below, we shall see that $I(D)$ is of finite dimension for any D. Set
$$i(D) = \dim_\Bbbk I(D).$$
A primary version of the Riemann–Roch theorem was as follows.

THEOREM 2.2.18. *For any divisor $D \in \mathrm{Div}(X)$, we have*
$$\ell(D) - i(D) = \deg D - g + 1. \tag{2.2.4}$$

PROOF. To verify (2.2.4) for $D=0$, it is sufficient to show that $i(0) = g$. This follows from Theorem 2.2.22, stated below. Next, it is clear that, to prove (2.2.4), it suffices to demonstrate that (2.2.4) holds for a divisor D if and only if it holds for $D' = D + P$, where P is an arbitrary point of X. Indeed, any divisor can be obtained from the trivial one by successively adding or subtracting single points. We have
$$\deg D' - g + 1 = (\deg D - g + 1) + 1.$$
Therefore, it suffices to show that
$$\ell(D') - i(D') = (\ell(D) - i(D)) + 1.$$

In fact, we shall prove that there are two possibilities: either $\ell(D') = \ell(D)$ and $i(D') = i(D) - 1$, or $\ell(D') = \ell(D) + 1$ and $i(D') = i(D)$.

To begin with, note that arguments in the proof of Theorem 2.1.53 show that $\ell(D')$ equals either $\ell(D)$ or $\ell(D) + 1$.

Since $\dim R(D')/R(D) = 1$, it is clear that $i(D')$ equals either $i(D)$ or $i(D) - 1$. From the definitions of $I(D)$ and $I(D')$, it follows that the sequences
$$0 \to R(D) + \Bbbk(X) \to R \to I(D) \to 0,$$
$$0 \to R(D') + \Bbbk(X) \to R \to I(D') \to 0$$
are exact. (Recall that a sequence of abelian groups
$$0 \to A \xrightarrow{f} B \xrightarrow{g} C \to 0$$
is called *exact* if f is injective, g is surjective, and $\mathrm{Im}(f) = \mathrm{Ker}(g)$.) Hence we get exact sequences
$$0 \to (R(D) + \Bbbk(X))/\Bbbk(X) \to R/\Bbbk(X) \to I(D) \to 0,$$
$$0 \to (R(D') + \Bbbk(X))/\Bbbk(X) \to R/\Bbbk(X) \to I(D') \to 0.$$

Using the isomorphism $(A+B)/B \cong A/(A\cap B)$, which holds for any subgroups A and B of an abelian group C, we obtain exact sequences
$$0 \to R(D)/L(D) \to R/\Bbbk(X) \to I(D) \to 0,$$
$$0 \to R(D')/L(D') \to R/\Bbbk(X) \to I(D') \to 0$$
since $L(D) = \Bbbk(X) \cap R(D)$ for any $D \in \mathrm{Div}(X)$. Since $\dim R(D)/R(D') = 1$, the equality $L(D) = L(D')$ implies $i(D') = i(D) - 1$, and $\ell(D') = \ell(D) + 1$ implies $i(D) = i(D')$. □

Residues. To deduce Theorem 2.2.17 from Theorem 2.2.18, we need the following important notion. Let ω be a rational differential form on X. Let $t = t_P$ be a local parameter at $P \in X$, and let $\omega = f dt$ for some $f \in \Bbbk(X)$. Expanding f into a Laurent power series in t, we get
$$f = \sum_{i=-M}^{\infty} a_i t^i.$$
The coefficient a_{-1} is called the *residue of ω at P* and is denoted by $\mathrm{Res}_P(\omega)$. The basic properties of residues are as follows.

PROPOSITION 2.2.19. (a) $\mathrm{Res}_P(\omega)$ *does not depend on the choice of a local parameter* t_P;
(b) Res_P *is a $\Bbbk$-linear functional on* $\Omega(X)$;
(c) *if ω is regular at P, then* $\mathrm{Res}_P(\omega) = 0$;
(d) $\mathrm{Res}_P(df) = 0$ *for all* $f \in \Bbbk(X)^*$;
(e) $\mathrm{Res}_P(df/f) = \mathrm{ord}_P(f)$ *for all* $f \in \Bbbk(X)^*$.

PROOF. Properties (b)–(d) are obvious. Let us prove (e): If $f = t^n u$, where $\mathrm{ord}_P(u) = 0$, then $df/f = n\, dt/t + du/u$. Hence, $\mathrm{Res}_P(df/f) = n$ since the form du/u is regular at P. Here we shall prove (a) on the assumption that $\Bbbk$ is of characteristic zero; in fact, the general case can be reduced to this. Let us write
$$\omega = \sum a_n\, dt/t^n + \omega_0,$$
where ω_0 is regular at P. If we choose another local parameter u and consider the corresponding residue $\mathrm{Res}'_P(\omega)$, then properties (b) and (c) imply
$$\mathrm{Res}'_P(\omega) = \sum a_n \mathrm{Res}'_P(dt/t^n).$$
Now it suffices to note that the rational functions $g_n = t^{-(n-1)}/(n-1)$ are well defined for $n \geq 2$ since $\mathrm{char}\,\Bbbk = 0$; hence, $\mathrm{Res}'_P(dt/t^n) = \mathrm{Res}'_P(dg) = 0$ by (e). Therefore, $\mathrm{Res}'_P(\omega) = \mathrm{Res}_P(\omega)$. □

Here is the most important result on residues.

PROPOSITION 2.2.20 (residue formula).
$$\sum_{P \in X} \mathrm{Res}_P(\omega) = 0 \qquad (2.2.5)$$
for any differential form $\omega \in \Omega(X)$. □

Note that formula (2.2.5) makes sense since $\mathrm{Res}_P(\omega) \neq 0$ for finitely many points only. We do not prove this proposition. We only note that if $\Bbbk = \mathbb{C}$, then X has a natural structure of a Riemann surface, and one can prove that $\mathrm{Res}_P(\omega) = \dfrac{1}{2\pi i}\oint_P \omega$. Then (2.2.5) follows from the Stokes theorem.

With the help of residues, one can define a pairing between differential forms and distributions:
$$\langle \cdot, \cdot \rangle : \Omega(X) \times R \to \Bbbk, \quad \langle \omega, r \rangle \mapsto \sum_{P \in X} \mathrm{Res}_P(r_P \omega).$$

EXERCISE 2.2.21. Show the following properties:
(a) If $r \in \Bbbk(X) \subset R$, then $\langle \omega, r \rangle = 0$.
(b) For any $\omega \in \Omega(-D)$ and $r \in R(D)$, we have $\langle \omega, r \rangle = 0$.
(c) If $f \in \Bbbk(X)$, then $\langle f\omega, r \rangle = \langle \omega, fr \rangle$.

Let $\omega \in \Omega(-D)$. Then, according to Exercise 2.2.21b, the map $r \mapsto \langle \omega, r \rangle$ defines a linear functional $\theta(\omega)$ on $I(D) = R/(\Bbbk(X) + R(D))$.

THEOREM 2.2.22 (duality theorem). *The map $\omega \mapsto \theta(\omega)$ defines an isomorphism from $\Omega(-D)$ onto the space dual to $I(D)$.* □

From Theorem 2.2.22 in combination with Theorem 2.2.18, the Riemann–Roch theorem follows since $\Omega(-D) \cong L(K - D)$. Now let us present some corollaries of the Riemann–Roch theorem.

EXERCISE 2.2.23. Let X be a smooth projective curve, K_X be its canonical class, and $g = g(X)$ be the genus of X. Show that
(a) $\deg K_X = 2g - 2$,
(b) $\ell(K - D) = g - 1 - \deg D$ if $\deg D < 0$,
(c) $\ell(D) = \deg D - g + 1$ if $\deg D > 2g - 2$.
(*Hint*: Use the fact that $\ell(D) = 0$ for $\deg D < 0$.)

It is sometimes convenient to use the Riemann–Roch theorem in the language of line bundles. Let $\mathscr{L}$ be a line bundle on X. Then $\mathscr{L} = \mathcal{O}(D)$ for some divisor D, and thus
$$h^0(\mathscr{L}) = \dim_{\Bbbk} H^0(\mathscr{L}) = \ell(D).$$

THEOREM 2.2.24. *Let $\mathscr{L} = \mathcal{O}(D)$ be a line bundle on X, and let $\mathscr{K}_X = \mathcal{O}(K_X)$ be the canonical bundle. Then*
$$h^0(\mathscr{L}) - h^0(\mathscr{K}_X \otimes \mathscr{L}^{-1}) = \deg \mathscr{L} - g + 1.$$ □

COROLLARY 2.2.25. (a) $\deg \mathscr{K}_X = 2g - 2$;
(b) $h^0(\mathscr{L}) = \deg \mathscr{L} - g + 1$ *if* $\deg \mathscr{L} > 2g - 2$. □

As a corollary of the Riemann–Roch theorem, one easily obtains the uniqueness of a smooth projective curve X of genus $g(X) = 0$.

EXERCISE 2.2.26. Show that if $g(X) = 0$, then $X \cong \mathbb{P}^1$. (*Hint*: Apply the Riemann–Roch theorem to the divisor $D = P$, where P is a point of X.)

Embeddings in projective spaces. In many cases, the Riemann–Roch theorem makes it possible to prove that the map $\varphi_D \colon X \to \mathbb{P}^n$ defined by a complete linear system $|D|$ is an embedding, i.e., defines an isomorphism of X onto its image in $\mathbb{P}^n$. This can be done with the use of the following result:

PROPOSITION 2.2.27. *Let D be a divisor on X such that*

$$\ell(D - P - Q) = \ell(D) - 2 \tag{2.2.6}$$

for any points $P, Q \in X$. Then the map $\varphi_D \colon X \to \mathbb{P}^n$, $n = \dim |D|$, is an embedding.

SKETCH OF THE PROOF. For $P \neq Q$, equality (2.2.6) implies that $\varphi_D(P) \neq \varphi_D(Q)$; i.e., φ_D is injective. For $P = Q$, (2.2.6) implies that φ_D^* defines an isomorphism from $\mathfrak{m}_{P'}/\mathfrak{m}_{P'}^2$ onto $\mathfrak{m}_P/\mathfrak{m}_P^2$, where $P' = \varphi_D(P)$ and $\mathfrak{m}_{P'}$ is the maximal ideal of the local ring of P' on the image $Y = \varphi_D(X)$ of X.

Thus, φ_D is an injection that induces isomorphisms on all tangent spaces. Such an injection must be an embedding. Indeed, the injectivity of φ_D (taking into account that $\varphi_D^* \neq 0$) implies that $\deg \varphi_D = 1$; therefore, there exists an inverse rational map $\varphi_D^{-1} \colon Y \to X$. Since $\varphi_D^* \colon \mathfrak{m}_{P'}/\mathfrak{m}_{P'}^2 \to \mathfrak{m}_P/\mathfrak{m}_P^2$ is an isomorphism for any P, the curve Y is smooth. Hence, φ_D is an isomorphism. □

EXERCISE 2.2.28. Give a detailed proof of the proposition.

Applying Exercise 2.2.23c, we obtain the following result.

COROLLARY 2.2.29. *If $\deg D \geq 2g + 1$, then φ_D is an embedding of X in $\mathbb{P}^n$.* □

Of particular interest are maps φ_{mK} defined by multiplicities of the canonical class $K = K_X$. For these maps we get the following corollary.

COROLLARY 2.2.30. *If X is a curve of genus at least 2, then φ_{3K} is an embedding. If X is a curve of genus at least 3, then φ_{2K} is an embedding.* □

REMARK 2.2.31. The map φ_K for a curve of genus $g \geq 2$ need not always be an embedding. Curves for which φ_K is not an embedding are called *hyperelliptic*. Let us give another description of hyperelliptic curves. If φ_K is not an embedding, then there exist points P and Q such that (2.2.6) does not hold, i.e.,

$$\ell(K - P - Q) \geq \ell(K) - 1 = g - 1.$$

Since $\ell(K - P) \geq \ell(K - P - Q)$, the equality $\ell(K - P) = g$ implies by the Riemann–Roch theorem that

$$\ell(P) = 1 + 1 - g + \ell(K - P) = 2.$$

Therefore, there exists a nonconstant map $f \colon X \to \mathbb{P}^1$ such that $\deg f = \deg P = 1$; hence, $X = \mathbb{P}^1$ and $g(X) = 0$. Then

$$\ell(K - P - Q) = \ell(K - P) = g - 1$$

for some $P, Q \in X$. By the Riemann–Roch theorem we get

$$\ell(P + Q) = 2 - g + 1 + \ell(K - P - Q) = 2;$$

hence, there is a map $f \colon X \to \mathbb{P}^1$ of degree 2. Conversely, if $f \colon X \to \mathbb{P}^1$ is a regular map of degree $\deg f = 2$ and $g(X) > 0$, then $f^*(Q) = P + P'$, where Q is a point of $\mathbb{P}^1$ and P and P' are points of X (not necessarily distinct). Then, by the Riemann–Roch theorem,

$$\ell(K - P - P') = 2g - 2 + 2 + 1 - g + \ell(P + P') = g - 1,$$
$$\ell(K - P) = 2g - 2 - 1 + 1 - g + \ell(P) = g - 1 = \ell(K - P - P')$$

(here, $\ell(P) = 1$ since X is not rational, and $\ell(P+P') = 2$ since $P + P' = f^*(Q)$); therefore, φ_K is not an embedding. It can be shown that there are "few" hyperelliptic curves; they are (in some precise sense, which we do not specify here) rare exceptions among all curves.

Noether theorem. The canonical class of a nonhyperelliptic curve X has another important property, which can be proved using the canonical embedding $\varphi_K \colon X \hookrightarrow \mathbb{P}^{g-1}$. To formulate it, note that for any divisors D and D' on X there is a natural map

$$L(D) \otimes L(D') \to L(D + D'), \quad f \otimes f' \mapsto ff'.$$

In particular, for any $m \geq 1$ we have a map

$$L(D)^{\otimes m} \to L(mD).$$

Note that for an arbitrary permutation π of the set $\{1, \ldots, m\}$, the images of $f_1 \otimes \ldots \otimes f_m$ and $f_{\pi(1)} \otimes \ldots \otimes f_{\pi(m)}$ in $L(mD)$ coincide. Thus, we have a natural map from the mth symmetric power $S^m(L(D))$ to $L(mD)$.

THEOREM 2.2.32. *For any nonhyperelliptic curve X, the map $S^m(L(K)) \to L(mK)$ is surjective.* □

A curve being an image $\varphi_K(X)$ of some curve X under the canonical embedding in $\mathbb{P}^{g-1}$ is called a *canonical curve*.

EXERCISE 2.2.33. Show that Theorem 2.2.32 can be reformulated as follows: A canonical curve in $\mathbb{P}^{g-1}$ is *projectively normal*; i.e., for any divisor $D \in |mL|$ (where (L) is the hyperplane section divisor) there exists a form F of degree m such that $(F) = D$.

2.2.3. Hurwitz Formula

Let $f \colon X \to Y$ be a nonconstant regular map of smooth complete curves. A formula due to Hurwitz gives an expression for the genus of X in terms of the genus of Y and some parameters of f. To obtain the formula, we first reduce the problem of computing the genus to the case of a separable map f. Note that if the characteristic p of the ground field is zero, then f is a fortiori separable; thus, this reduction is required for $p > 0$ only.

Consider the field extension $\Bbbk(X)/f^*(\Bbbk(Y))$. Let $\mathbb{K}'$ be the maximal separable subextension, $f^*(\Bbbk(Y)) \subseteq \mathbb{K}' \subseteq \Bbbk(X)$. By Proposition 2.1.49, there exists a smooth projective curve X' such that $\mathbb{K}' = \Bbbk(X')$. Thus, the map $f \colon X \to Y$ can be factorized,

$$f \colon X \xrightarrow{f'} X' \xrightarrow{f''} Y,$$

where f' is purely inseparable and f'' is separable. Let us show that $g(X) = g(X')$. To this end, introduce the following definition. Let X be a smooth projective curve over a field $\Bbbk$ of characteristic $p > 0$. Let $\mathbb{K} = \Bbbk(X)$; consider the field $\mathbb{K}_p = \mathbb{K}^{1/p}$ generated by the pth roots of elements of $\mathbb{K}$:

$$\mathbb{K}_p = \{a \in \overline{\mathbb{K}} \mid a^p \in \mathbb{K}\}.$$

By Proposition 2.1.49, there exists a curve X_p such that $\mathbb{K}_p = \Bbbk(X_p)$; the inclusion $\mathbb{K} \subset \mathbb{K}_p$ defines a map $\varphi_p \colon X_p \to X$, which is known as the *Frobenius morphism over $\mathbb{F}_p$* (or simply the *Frobenius morphism*). Since the transcendence degree of $\mathbb{K}$ is 1, we have $\deg \varphi_p = p$.

PROPOSITION 2.2.34. *Let $f\colon X \to Y$ be a nonconstant purely inseparable morphism. Then $g(X) = g(Y)$.*

SKETCH OF THE PROOF. Let $\deg f = p^r$. Using induction, we may assume that $r = 1$. Then $\Bbbk(X) \supseteq f^*(\Bbbk(Y)^{1/p})$; since $[\Bbbk(X) : f^*(\Bbbk(Y))] = p$, the map f coincides with the Frobenius morphism over $\mathbb{F}_p$, and $X = Y_p$. It suffices to show that $\Omega[Y]$ and $\Omega[Y_p]$ are of the same dimension. This follows from the fact that if $\omega = f\,dg \in \Omega[Y]$, then the form $\omega' = f'dg'$ belongs to $\Omega[Y_p]$, where f' and g' are obtained from f and g by raising all their coefficients (which lie in $\Bbbk$) to the pth power; conversely, if $\omega' = f'dg' \in \Omega[Y_p]$, then $\omega = f\,dg$ lies in $\Omega[Y]$, where f and g are obtained from f' and g' by taking the pth roots of their coefficients. $\square$

Thus, Proposition 2.2.34 reduces the problem of computing the genus of X given that of Y to the case of a separable map $f\colon X \to Y$.

PROPOSITION 2.2.35. *For a separable map $f\colon X \to Y$, the set of its ramification points is finite.*

PROOF. Deleting finitely many points from X and Y, we may assume that X and Y are affine curves. Let $\Bbbk[X] = A$ and $\Bbbk[Y] = A'$. Let $a \in A$ be a primitive element of the extension $\Bbbk(X)/f^*(\Bbbk(Y))$, $F(T) \in A'[T]$ be its minimal polynomial, and $D(F) \in A'$ be its discriminant. Since f is separable, $D(F) \neq 0$, and hence $D(F)$ has only finitely many zeroes on Y. Outside these zeroes, f is unramified. $\square$

Let $f\colon X \to Y$ be a nonconstant separable map of smooth projective curves. Let $P \in X$, $f(P) = Q$, and let t_P and t_Q be local parameters at P and Q respectively. Then $f^*(dt_Q) = g\,dt_P$ for some $g \in \mathscr{O}_P$. Let $\operatorname{ord}_P(g) = b_P$; it is clear that $b_P \neq 0$ only for ramification points of f. Let

$$B = B_f = \sum b_P P \in \operatorname{Div}(X);$$

the divisor B_f is called the *ramification divisor* of f.

THEOREM 2.2.36. *Let $f\colon X \to Y$ be a nonconstant separable map of degree $n = \deg f$. Then*

$$2g(X) - 2 = n(2g(Y) - 2) + \deg B_f.$$

PROOF. By Exercise 2.2.23a and Proposition 2.1.74, it suffices to show that $K_X = f^*K_Y + B_f$. Let $\omega \in \Omega(Y)$ be an arbitrary nonzero rational differential form such that $\operatorname{Supp}(\omega)$ is disjoint from the finite set of points where f is ramified. Let us compute the divisor $K_X = (f^*(\omega))$. If Q is not a ramification point and $\omega = g\,dt_Q$, where t_Q is a local parameter at Q, then $f^*(t_Q)$ is a local parameter for any P such that $f(P) = Q$; hence, $K_X|_U = f^*K_Y|_U$, where $U = X \setminus \operatorname{Supp} B_f$. Let Q be a ramification point; then $\omega = h\,dt_Q$ for some function h. Since $\operatorname{Supp}(\omega)$ does not contain Q, we have $\operatorname{ord}_Q h = 0$. Let $f^*(dt_Q) = g\,dt_P$, where $f(P) = Q$. Then $(f^*(\omega))_P = (g)_P$, i.e., $(f^*(\omega))_P = b_P$. Therefore, $\operatorname{ord}_P(f^*(h)) = 0$. $\square$

A map $f\colon X \to Y$ is said to be *tamely ramified* if neither of the ramification indices e_P, $P \in X$, is divisible by $p = \operatorname{char}\Bbbk$; otherwise, we say that ramification is *wild*.

COROLLARY 2.2.37 (Hurwitz formula). *Let a nonconstant separable map* $f\colon X \to Y$ *of smooth projective curves be tamely ramified. Then*

$$2g(X) - 2 = n(2g(Y) - 2) + \sum_{P \in X} (e_P - 1). \qquad (2.2.7)$$

PROOF. It suffices to show that if the ramification is tame, then

$$\deg B_f = \sum_{P \in X} (e_P - 1).$$

This follows from the fact that if $f^*(t_Q) = h t_P^{e_P}$, $\mathrm{ord}_P h = 0$, then

$$f^*(dt_Q) = h e_P t_P^{e_P - 1}\, dt_P + t_P^{e_P}\, dh;$$

since $e_P \neq 0$ in $\Bbbk$, this implies the desired formula. □

The following particular case of the Hurwitz formula is of special interest.

COROLLARY 2.2.38. *Let* $f\colon X \to \mathbb{P}^1$ *be a nonconstant map from a smooth projective curve* X *to a projective line, and let* f *have tame ramification only. Then*

$$g(X) = 1 - n + \frac{1}{2} \sum_{P \in X} (e_P - 1), \qquad (2.2.8)$$

where $n = \deg f$. □

EXERCISE 2.2.39. Let $\operatorname{char}\Bbbk = 0$. Prove that any map of degree greater than 1 from a curve to a projective line has at least two ramification points. (Note that this is not true for $\operatorname{char}\Bbbk = p > 0$; try to give a counterexample.)

2.2.4. Special Divisors

A divisor D is called *special* if $\ell(K - D) > 0$.

EXERCISE 2.2.40. Show that D is a special divisor if and only if $D \sim K - D'$ for some effective divisor D'.

EXERCISE 2.2.41. Show that any divisor D of degree at least $2g - 1$ is nonspecial and any divisor D' of degree at most $g - 2$ is special.

EXERCISE 2.2.42. Show that for any divisors D and D' we have

$$\ell(D + D') \geq \ell(D) + \ell(D') - 1;$$

i.e.,

$$\dim|D| + \dim|D'| \leq \dim|D + D'|.$$

(*Hint*: Consider the kernel of the natural map

$$L(D) \otimes L(D') \to L(D + D').)$$

Using the result of this exercise, one can prove the following important result.

THEOREM 2.2.43 (Clifford theorem). *For any special effective divisor* D, *we have*

$$\dim|D| \leq \frac{1}{2} \deg D.$$

PROOF. According to Exercise 2.2.40, the divisor $D' = K - D$ is equivalent to an effective one. Applying the result of Exercise 2.2.42 to D and D', we get
$$\dim|D| + \dim|K - D| \leq \dim|K| = g - 1.$$
By the Riemann–Roch theorem we have
$$\dim|D| - \dim|K - D| = \deg D + 1 - g.$$
Adding these inequalities, we obtain the theorem. □

REMARK 2.2.44. Usually, the Clifford theorem is formulated with the following additional claim: If $\dim|D| = \frac{1}{2}\deg D$, then either $D = 0$ or $D = K$, or X is hyperelliptic. The proof of this fact is somewhat more complicated and is not given here since we do not need this fact in the book.

Weierstrass points. An elementary example of an effective divisor is a divisor of the form $D = aP$, $P \in X$. Analysis of the question of whether such a divisor is special leads to the definition of Weierstrass points, which play an important role in the study of geometry of curves of genus $g \geq 2$.

Let $P \in X$, $g = g(X) \geq 2$, and let $a \geq 1$ be an integer. We call a a *gap* at P if $\ell(aP) = \ell((a-1)P)$ and a *nongap* otherwise. In other words, a is a nongap at P if there exists a rational function $f \in \Bbbk(X)$ with a pole of order a at P and with no other poles on X. If such a function does not exist, then a is a gap at P.

EXERCISE 2.2.45. Show the following properties for any $P \in X$:
(a) If a and b are nongaps at P, then $a + b$ is also a nongap.
(b) 1 is a gap at P.
(c) The number of gaps at P equals g.
(d) If $a \geq 2g$, then a is a nongap at P.

We call $P \in X$ a *Weierstrass point* if the gap sequence $\{a_1, \ldots, a_g\}$ at P does not coincide with $\{1, \ldots, g\}$.

EXERCISE 2.2.46. Let P be an arbitrary point of X. Show that the conditions below are equivalent:
(a) P is a Weierstrass point.
(b) The divisor gP is special.
(c) The divisor aP for some $a \geq g$ is special.

Let $\{a_1, \ldots, a_g\}$, $1 \leq a_1 < a_2 < \ldots < a_g \leq 2g - 1$, be the gap sequence at $P \in X$. We define the (*Weierstrass*) *weight* of P as
$$w(P) = \sum_{i=1}^{g}(a_i - i).$$
In particular, $w(P) > 0$ if and only if P is a Weierstrass point.

We have the following *mass formula*:
$$\sum_{P \in X} w(P) = (g-1)g(g+1) + \alpha(X),$$
where $\alpha(X) \geq 0$. Moreover, $\alpha(X) = 0$ if $p = \text{char}\,\Bbbk = 0$; for curves in characteristic $p > 0$, the number $\alpha(X)$ can be positive for certain "exceptional" curves only.

Therefore, the number of Weierstrass points on X is finite and does not exceed $(g-1)g(g+1) + \alpha(X)$.

EXERCISE 2.2.47. Show that $w(P) \leq g(g-1)/2$ and that $w(P) = g(g-1)/2$ if and only if 2 is a nongap at P.

A point P with $w(P) = g(g-1)/2$ or, equivalently, a point for which 2 is a nongap is called a *hyperelliptic* point.

EXERCISE 2.2.48. Show that X has hyperelliptic points if and only if X is a hyperelliptic curve. Show that if the ground field has characteristic $p \neq 2$, then a hyperelliptic curve has $2(g+1)$ Weierstrass points, all of them being hyperelliptic.

To construct Weierstrass points on a particular curve, one can use the following result.

PROPOSITION 2.2.49. *Let $\varphi\colon X \to Y$ be a regular map of degree n of smooth projective curves. Let Q be a point on Y such that $\varphi^{-1}(Q)$ consists of a single point P, and let $g(Y) \leq \lfloor g(X)/n \rfloor - 1$. Then P is a Weierstrass point.*

PROOF. By the Riemann–Roch theorem, we have $\ell((g(Y)+1)Q) \geq 2$. Hence, there exists a nonconstant function $f \in \Bbbk(Y)$ with a pole at Q of order at most $g(Y)+1$ and with no other poles on Y. Consider the function $\varphi^*(f)$. Its unique pole is P; since the ramification index at P is n, for the pole order we have

$$\mathrm{ord}_P(\varphi^*(f)) = n\,\mathrm{ord}_Q(f) \leq n(g(Y)+1).$$

Since $n(g(Y)+1) \leq g(X)$, we have $\ell(g(X)P) \geq 2$; thus, P is a Weierstrass point. □

This result is especially convenient to use in the following case:

COROLLARY 2.2.50. *Let $h \in \mathrm{Aut}(X)$ be an element of period n such that P is a fixed point of h, i.e., $h(P) = P$, and let $g(X^h) \leq \lfloor g(X)/n \rfloor - 1$. Then P is a Weierstrass point.* □

2.2.5. Cartier Operator

In the case $\mathrm{char}\,\Bbbk = p > 0$, there is an operator on $\Omega(X)$ that plays an important role in many questions of the geometry of curves in positive characteristic. Let $\mathbb{K} = \Bbbk(X)$, where X is a smooth projective curve over $\Bbbk$, and let $t \in \mathbb{K}$, $t \notin \mathbb{K}^p$, so that $\mathbb{K} = \mathbb{K}^p(t)$. An arbitrary element $\omega \in \Omega(X)$ can be represented as $\omega = y\,dt$ or

$$\omega = (y_0^p + y_1^p t + \ldots + y_{p-1}^p t^{p-1})\,dt, \quad y_i \in \mathbb{K}. \tag{2.2.9}$$

Set
$$C(\omega) = y_{p-1}\,dt.$$

It can be shown that the definition of C does not depend on the choice of t; i.e., it does not change if we represent ω in the form $z\,dw$, $w \notin \mathbb{K}^p$. Formula (2.2.9) implies that any $\omega \in \Omega(X)$ can be uniquely represented in the form $\omega = df + g^p\,dt/t$ (the uniqueness is obvious). For this representation, one easily checks that

$$C(\omega) = g\,dt/t.$$

Let us list some basic properties of the Cartier operator.

EXERCISE 2.2.51. Let z be an arbitrary element of $\mathbb{K}$. Then we have the following relations:
(a) $C(z^p \omega) = z C(\omega)$;
(b) $C(dz) = 0$;

(c) $C(dz/z) = dz/z$;
(d) if $P \in X$ and ω is regular at P, then $C(\omega)$ is also regular at P;
(e) for any $P \in X$ we have
$$\bigl(\mathrm{Res}_P(C(\omega))\bigr)^p = \mathrm{Res}_P \omega.$$

Using the Cartier operator, we can give a characterization of *exact* (i.e., of the form $\omega = df$) and *logarithmic* (i.e., of the form $\omega = df/f$) differential forms:

PROPOSITION 2.2.52. *Let $\omega \in \Omega(X)$. Then*
(a) $C(\omega) = 0$ *if and only if ω is exact; i.e., $\omega = df$ for some $f \in \mathbb{K}$.*
(b) $C(\omega) = \omega$ *if and only if ω is logarithmic; i.e., $\omega = df/f$ for some $f \in \mathbb{K}$.* □

Property (a) is obvious; the fact that ω is logarithmic whenever $C(\omega) = \omega$ requires a more complicated proof.

The operator $C = C_p$ we have introduced is known as the *Cartier operator over $\mathbb{F}_p$*. If $q = p^m$ is a power of p, then by the *Cartier operator over $\mathbb{F}_q$* we call $C_q = C_p^m$.

Below, we shall need the following result.

EXERCISE 2.2.53. Let $D = P_1 + \ldots + P_n$ be a divisor with all the P_i being distinct points of X. Let $P \in X$ be a point different from any of the P_i, and let $\omega \in \Omega(D - aP)$. Then
$$C_q(\omega) \in \Omega\left(D - \left\lfloor \frac{a}{q} \right\rfloor P\right).$$

2.3. Singular Curves

Nonsingular curves and their properties play the main role in the theory of curves. Nevertheless, the study of singular curves is in a number of cases indispensable, since many smooth curves have singular models that are useful to prove some properties of the curves. (A *model of a field* $\mathbb{K}$, or a *model of an algebraic variety* X, with $\mathbb{K}$ being the field of rational functions on X, is any variety Y such that $\mathbb{K} = \Bbbk(Y)$.) Thus, for example, each curve has a plane (singular) model. In this section we briefly discuss some properties of singular curves and describe their interrelations with smooth curves.

Section 2.3.1 is devoted to normalization of a curve, which is the simplest way to obtain a smooth curve from a singular one. In Sec. 2.3.2 we define the double-point divisor, which plays an essential role in studying singular curves. In Sec. 2.5.31 we give some properties of plane singular curves; in particular, we describe differential forms on them.

In this section, the ground field $\Bbbk$ is assumed to be algebraically closed.

2.3.1. Normalization

Let A be an *integral domain* (or simply a *domain*), i.e., a commutative ring without divisors of zero. Let K be the field of fractions of this ring, and let L be an extension of K. An element $x \in L$ is called *integral* over A if there are elements $a_0, a_1, \ldots, a_{n-1} \in A$ such that $x^n + a_{n-1}x^{n-1} + \ldots + a_0 = 0$. The set $\bar{A}_L$ consisting of all elements $x \in L$ integral over A is called the *integral closure* of A in L. The set $\bar{A}_K$ is denoted by $\bar{A}$. If $\bar{A} = A$, the domain is said to be *integrally closed*.

EXERCISE 2.3.1. Show that, for any domain A and any extension L, the set $\bar{A}_L$ is an integrally closed domain.

A curve X is called *normal* if the local rings $\mathscr{O}_P$ are integrally closed for all points $P \in X$.

EXERCISE 2.3.2. Show that a curve X is normal if and only if the ring $\Bbbk[U]$ is integrally closed for any open subset $U \subset X$.

EXERCISE 2.3.3. Show that a smooth curve is normal.

Moreover, we have the following fact.

PROPOSITION 2.3.4. *Any normal curve is smooth.* □

Thus, the normality and nonsingularity conditions for curves are equivalent.

REMARK 2.3.5. These condition are equivalent over an arbitrary perfect field (recall that a field $\Bbbk$ is *perfect* if its algebraic closure $\bar{\Bbbk}$ is separable over $\Bbbk$). If the ground field is not perfect, there exist singular normal curves. Likewise, Proposition 2.3.4 becomes invalid for surfaces and higher-dimensional varieties; however, it is known that a smooth variety is always normal, and the subset of singular points of a normal variety over a perfect field is of codimension at least two.

EXERCISE 2.3.6. Let X be an affine curve, and let $A = \Bbbk[X]$. Show that there exists a unique affine curve X^ν such that $\Bbbk[X^\nu] = \bar{A}$.

The inclusion $\Bbbk[X] \subseteq \Bbbk[X^\nu]$ defines a regular surjective map $X^\nu \to X$.

THEOREM 2.3.7. *Let X be a quasiprojective curve. There exists a unique smooth curve X^ν such that there is a unique regular surjective morphism $\nu\colon X^\nu \to X$ which is a birational isomorphism. If X is a projective curve, then X^ν is also projective.* □

We call ν the *normalization map*; the curve X^ν is called the *normalization* of X.

EXERCISE 2.3.8. Let U be an affine open subset of X. Let $\nu\colon X^\nu \to X$ be the normalization map, and let $V = \nu^{-1}(U)$. Show that $V = U^\nu$.

2.3.2. Double-Point Divisor

Local invariants. In the study of singular curves, it is important to consider characteristics of a point P on a curve X that depend on the local ring $\mathscr{O}_P$ only. Such characteristics are called *local invariants* of P. Let $\nu\colon X^\nu \to X$ be the normalization map, and let $\mathscr{O}'_P$ be the integral closure of $\mathscr{O}_P$.

EXERCISE 2.3.9. Show that the set $S(P) = \nu^{-1}(P)$ is finite for any point $P \in X$ and that $\mathscr{O}'_P = \mathscr{O}_{S(P)}$, where $\mathscr{O}_{S(P)}$ is the ring of rational functions on X^ν regular at any point $Q \in S(P)$. Show that

$$\mathscr{O}_{S(P)} = \bigcap_{Q \in S(P)} \mathscr{O}_Q$$

and that $\mathscr{O}_{S(P)}$ is a *semi-local* ring, i.e., it has a finite number of maximal ideals (this number equals the cardinality of $S(P)$).

Let c_P be the *annihilator* of the $\mathscr{O}_P$-module $\mathscr{O}'_P/\mathscr{O}_P$, i.e.,

$$c_P = \{a \in \mathscr{O}_P \mid a\mathscr{O}'_P \subseteq \mathscr{O}_P\}.$$

It is clear that c_P is an ideal both in $\mathscr{O}_P$ and in $\mathscr{O}'_P$. It can be shown that c_P is the largest ideal in $\mathscr{O}_P$ among those that are also ideals in $\mathscr{O}'_P$. An ideal c_P is called the *conductor* of the extension $\mathscr{O}'_P/\mathscr{O}_P$.

EXERCISE 2.3.10. Show that the conductor c_P is a nontrivial ideal. (*Hint*: $\dim_\Bbbk \mathscr{O}'_P/\mathscr{O}_P < \infty$.)

Set
$$n_P = \dim_\Bbbk \mathscr{O}'_P/c_P, \qquad \delta_P = \dim_\Bbbk \mathscr{O}'_P/\mathscr{O}_P;$$

we call n_P and δ_P the *fundamental local invariants* of P.

EXERCISE 2.3.11. Show the equivalence of the following conditions:
- P is a nonsingular point,
- $\delta_P = 0$,
- $n_P = 0$.

Hence, δ_P and n_P are nonzero for singular points only.

EXERCISE 2.3.12. Show that if $\delta_P \neq 0$, then

$$\delta_P \leq n_P - 1.$$

(*Hint*: $\mathscr{O}_P \supseteq \Bbbk + c_P$.)

One can also prove the following fact:

PROPOSITION 2.3.13. *For any point $P \in X$, we have*
$$n_P \leq 2\delta_P.$$
□

EXERCISE 2.3.14. (a) Let $X \subset \mathbb{A}^2$ be defined by the equation $y^2 = x^2 + x^3$ for char$\mathbb{k} \neq 2$, and let $P = (0,0)$. Show that $X^\nu = \mathbb{A}^1$ and that the map ν can be given as $\nu(t) = (t^2 - 1, t(t^2 - 1))$, where t is a coordinate on $\mathbb{A}^1$. Show that $S(P) = \{Q, Q'\}$, where $Q, Q' \in \mathbb{A}^1$ are given by $t = 1$ and $t = -1$ respectively. Show that $\mathscr{O}'_P$ consists of fractions of the form $P(t)/Q(t)$, where $Q(t)$ is divisible by neither $t+1$ nor $t-1$, and show that $c_P = (t^2 - 1)\mathscr{O}'_P$, $\mathscr{O}_P = \mathbb{k} + c_P$, $n_P = 2$, and $\delta_P = 1$.

(b) Let $X \subset \mathbb{A}^2$ be defined by $y^2 = x^3$, and let $P = (0,0)$. Show that $X^\nu = \mathbb{A}^1$, $\nu(t) = (t^2, t^3)$, $S(P) = Q = \{t = 0\}$,
$$\mathscr{O}'_Q = \{P(t)/Q(t) \mid P, Q \in \mathbb{k}(t),\ Q(0) \neq 0\},$$
$c_P = t^2 \mathscr{O}'_P$, $\mathscr{O}_P = \mathbb{k} + c_P$, $n_P = 2$, and $\delta_P = 1$.

A point $P \in X$ is called a *simplest singularity* if $\delta_P = 1$ (and hence $n_P = 2$). For such a point, $S(P)$ consists of either one or two elements. If $S(P)$ has two elements, then P is called a *simplest double point* (*nodal point*, or simply a *node*). If $S(P)$ consists of a single point, then P is called a *simplest cuspidal point* (or simply a *cusp*). Thus, in Exercise 2.3.14a, P is a simplest double point; in Exercise 2.3.14b, P is a simplest cuspidal point.

Double-point divisor. Let X be a projective curve, $P \in X$, and $Q \in S(P)$. Since Q is a nonsingular point, we have $c_P \mathscr{O}_P = \mathfrak{m}_Q^{n_Q}$ for a nonnegative integer n_Q. The divisor
$$D = \sum_P \sum_{Q \in S(P)} n_Q Q$$
is called the *double-point divisor* (or the *conductor*) of X. Note that D is well-defined, since for a nonsingular point P we always have $|S(P)| = 1$, and $n_Q = 0$ for $Q \in S(P)$. Thus, D is a divisor on X^ν. It plays a crucial role in the study of the interrelationship between X and X^ν.

EXERCISE 2.3.15. Show that
$$n_P = \sum_{Q \in S(P)} n_Q.$$

EXERCISE 2.3.16. Find the double-point divisors for projective closures of curves from Exercise 2.3.14.

EXERCISE 2.3.17. Let $X \subset \mathbb{P}^2$ be a curve given by
$$x_0 x_2^4 = x_0 x_1^4 + x_1^5,$$
and let $P = (1:0:0)$. Compute n_P, δ_P, and D. Is P a simplest singularity?

2.3.3. Plain Curves

Note that for any smooth projective curve X there exists a projective plane curve X' birationally isomorphic to X. We call such a curve X' a *plane model* of X. Indeed, let $f \in \mathbb{k}(X) \setminus \mathbb{k}$ be a function such that $\mathbb{k}(X)/\mathbb{k}(f)$ is a separable extension (it is easily shown that f does exist). By the primitive element theorem there exists a function $g \in \mathbb{k}(X)$ such that $\mathbb{k}(X) = \mathbb{k}(f, g)$. It is clear that $F(f, g) = 0$ for some polynomial $F \in \Bbbk[T_1, T_2]$. Let G be a polynomial of the least possible degree such that $G(f,g) = 0$. Then G is irreducible, and we can define X' as the projective closure of the plane affine curve defined by $G(x,y) = 0$.

As a rule, X' cannot be chosen to be singular. Indeed, the genus of a smooth projective plane curve equals $(m-1)(m-2)/2$, where m is its degree (Corollary 2.2.8). Since the genus of a curve can be any nonnegative integer (for instance, for char $\Bbbk \ne 2$, the genus of the curve $y^2 = f(x)$, $f \in \Bbbk[x]$, equals $\lfloor (\deg f - 1)/2 \rfloor$), there are many smooth projective curves that are not isomorphic to smooth projective plane curves. (Even for curves of genus $g = (m-1)(m-2)/2 > 3$, the property of a curve to have a smooth plane model is an exception; on the contrary, all non-hyperelliptic curves of genus 3 possess this property, as well as all curves of genus 1.) Therefore, studying singular plane curves is indispensable. Let X be a smooth projective curve, X' be its plane model, and $f, g \in \Bbbk(X)$ be as above. Then the rational map $\varphi \colon X \to X'$ defined by $x \mapsto (f(x), g(x))$ is regular by Theorem 2.1.59. Since the normalization of a curve is unique, we see that $X = (X')^\nu$, and the map φ is the normalization map for X'. The main question is to express the fundamental invariants of X in terms of invariants of X' or rather in terms of its equation in $\mathbb{P}^2$.

Genus. By definition, the genus of a singular curve is the genus of its normalization.

THEOREM 2.3.18. *Let X' be a projective plane (singular) curve of degree m. Let X be its normalization, and let $g = g(X') = g(X)$ be its genus. Then*

$$g = \frac{(m-1)(m-2)}{2} - \sum_P \delta_P, \qquad (2.3.1)$$

where the sum is taken over all singular points of X'. □

The number $p_a(X') = (m-1)(m-2)/2$ is sometimes called the *arithmetic genus* of X'. Thus, for smooth plane curves (and only for them), we have

$$p_a(X') = g(X').$$

EXAMPLE 2.3.19. The curve $x_0 x_2^2 = x_1^3 - x_0 x_1^3$ (or $y^2 = x^3 - x^2$ in affine coordinates) is of arithmetic genus 1 (its genus being zero). The same is true for the curve $x_0 x_2^2 = x_1^3$.

Differential forms. In fact, $\Omega[X]$ can be described similarly to how it was done in Proposition 2.2.7 for smooth plane curves. Let $F(x_0 : x_1 : x_2) = 0$ be the homogeneous equation of X', and let $0 = G(x,y) = F(1 : x : y)$ be its affine equation. For arbitrary $P, Q \in \Bbbk[x,y]$, the expression $P(x,y) dy / Q(x,y)$ defines a rational differential form on X.

PROPOSITION 2.3.20. *We have*

$$\Omega[X] \subseteq \left\{ \frac{P(x,y) dy}{G'_x(x,y)} \,\Big|\, \deg P(x,y) \le m - 3 \right\}. \qquad \square$$

A polynomial $P(x,y)$ is called *adjoint* (for X') if $P(x,y) dy / G'_x(x,y) \in \Omega[X]$. In the case $X = X'$, any polynomial P of degree $\deg P \le m - 3$ is adjoint by Proposition 2.2.7. Let us give a description for adjoint polynomials in the general case. Note that to any polynomial $P(x,y)$ of degree s we may assign a form F_P of degree s:

$$F_P(x_0 : x_1 : x_2) = x_0^s P(x_1/x_0, x_2/x_0).$$

In Sec. 2.1.2 it was shown that any form F defines a Cartier divisor $(F) = (\{U_i, f_i\})$ on a smooth plane curve X. Actually, one can observe that this definition does not use nonsingularity of X and is therefore valid for any plane curve. The Cartier divisor $(\{\nu^{-1}(U_i), \nu^*(f_i)\})$ on X, where $\nu \colon X \to X'$ is the normalization map, is called the *divisor of F on X*; it is also denoted by (F). Recall that by D we denote the double-point divisor on X'. In this notation, we have the following fact.

PROPOSITION 2.3.21. *A polynomial $P \in \Bbbk[x,y]$ of degree $s \leq m-3$ is adjoint for X' if and only if $(F_P) \geq D$ in $\mathrm{Div}(X)$.* □

Thus, the curve $F_P(x_0 : x_1 : x_2) = 0$ for an adjoint polynomial P contains all singular points of X'. If X' has simplest singularities only, this is also a sufficient condition for P to be adjoint for X'; in this case, Theorem 2.3.18 is implied by Propositions 2.3.20 and 2.3.21.

Noether theorem. In the study of plane curves, an essential role is played by the following result, called the *Noether $A\varphi + B\psi$ theorem*.

THEOREM 2.3.22. *Let F be the homogeneous equation of X', and let G and H be forms in variables (x_0, x_1, x_2). Then the condition $(G) \geq (H) + D$ implies that there exist forms A and B such that*
$$G = AH + BF$$
in the ring $\Bbbk[x_0, x_1, x_2]$. □

In other words, the condition $(G) \geq (H) + D$ is sufficient for $\overline{H}$ to divide $\overline{G}$ in the homogeneous coordinate ring $R_X = \Bbbk[x_0, x_1, x_2]/(F)$ of X', where $\overline{G}$ and $\overline{H}$ are the images of G and H in R_X.

COROLLARY 2.3.23. *Let X' be a smooth complete plane curve. Then $(G) \geq (H)$ implies that $\overline{G}$ is divisible by $\overline{H}$ in R_X.* □

Auxiliary formulae. In the above notation, let G be of degree m as a polynomial of x and of degree n as a polynomial of y. Let its total degree be r, and let g be the genus of X'.

PROPOSITION 2.3.24. *We have*
$$\deg D = (r-1)(r-2) - 2g.$$
□

Expressions for the divisors (dx) and (G'_x) of the differential form dx and the function G'_x, respectively, are of use.

PROPOSITION 2.3.25. *Let B_x (respectively, B_y) be the ramification divisor of the map $X \to \mathbb{P}^1$ defined on X' by $(x,y) \mapsto x$ (respectively, by $(x,y) \mapsto y$). Then*
$$(dx) = B_x - 2(x)_\infty,$$
$$(G'_x) = B_y - (m-2)(x)_\infty - n(y)_\infty - D,$$
where, as usual, $(f)_\infty$ is the divisor of poles of $f \in \Bbbk(X)$. □

EXERCISE 2.3.26. Prove Proposition 2.3.25; deduce Proposition 2.3.24 from it.

2.4. Elliptic Curves

Curves of genus 1, called *elliptic*, are of special importance in the theory of curves. Elliptic curves possess a number of remarkable properties, the most important of them being the group structure on the set of points of an elliptic curve. In this section we give basic results concerning elliptic curves to the minimum necessary extent.

In Sec. 2.4.1 we describe the group law on an elliptic curve. Section 2.4.2 is devoted to isomorphism classes of elliptic curves and to the j-invariant, which classifies them. In Sec. 2.4.3 we consider morphisms of elliptic curves. Section 2.4.4 describes the connection between complex elliptic curves and classical elliptic functions.

All curves in this section are assumed to be smooth projective curves defined over an algebraically closed field $\Bbbk$ (in Sec. 2.4.4, we assume $\Bbbk = \mathbb{C}$).

2.4.1. Group Law

There are several ways to define the (unique!) group structure on an elliptic curve. We start with the most invariant (though the most abstract) way and then describe the obtained group law in more down-to-earth terms.

Let E be an elliptic curve, i.e., a curve of genus 1. Fix a point $P_0 \in E$ and consider the map (of sets) $\lambda_{P_0} \colon E \to \operatorname{Pic}^0(E)$, $\lambda_{P_0} \colon P \mapsto \{P - P_0\}$, where $\{P - P_0\}$ is the class of the divisor $P - P_0$ in $\operatorname{Pic}^0(E)$ (the group of linear equivalence classes of degree-0 divisors on E). In Sec. 2.1.4 we stated (without a proof) that $\operatorname{Pic}^0(E)$ is an abelian variety and that λ_{P_0} is an isomorphism. We shall not use these facts here; on the contrary, we shall obtain them as consequences of our arguments.

PROPOSITION 2.4.1. *The map $\lambda_{P_0} \colon E \to \operatorname{Pic}^0(E)$ is bijective.*

PROOF. Let D be an arbitrary divisor of degree 0 on E. Applying the Riemann–Roch theorem to $D + P_0$, we obtain

$$\ell(D + P_0) - \ell(K - D - P_0) = 1.$$

Note that, since $\deg K = 2g - 2 = 0$, we have $\deg(K - D - P_0) = -1$. Therefore, $\ell(K - D - P_0) = 0$, and thus $\ell(P_0 + D) = 1$. This means that there exists a unique effective divisor of degree 1 (i.e., a point $P \in E$) equivalent to $D + P_0$. Therefore, for any divisor $D \in \operatorname{Div}(E)$ there is a unique point P such that $D \sim P - P_0$, as required. □

Since $\operatorname{Pic}^0(E)$ is a group, a group structure on E arises, with P_0 being the neutral element. Let us represent the obtained group law in geometric terms. To this end, consider the divisor $3P_0$ and the corresponding map $\varphi_{3P_0} \colon E \to \mathbb{P}^2$ (note that $\ell(3P_0) = 3$ by the Riemann–Roch theorem). Then Corollary 2.2.29 yields the following result.

PROPOSITION 2.4.2. *The map φ_{3P_0} defines an embedding of E in $\mathbb{P}^2$.* □

Corollary 2.2.8 shows that a smooth irreducible plane cubic (i.e., a curve of degree 3) is of genus 1. Conversely, Proposition 2.4.2 states that any elliptic curve (up to an isomorphism) is a plane cubic; moreover, the curve $\varphi_{3P_0}(E)$ intersects the line of infinity at the point P_0 only (with multiplicity 3). In what follows, we assume that E is such a plane cubic. Let P, Q, and R be arbitrary points on E. The condition $P + Q + R = 0$ (in the sense of the group law on E) can be rewritten as the linear equivalence of divisors, $P+Q+R \sim 3P_0$. Since $3P_0 = (L)$, where L is a linear form, this means that P, Q, and R lie on a line ℓ (with equation $L=0$). If two of the points coincide, $P=Q$, this means that ℓ is tangent to E at P. If $P=Q=R$, then P has order 3 in E, and ℓ is tangent to E with multiplicity greater than or equal to 2; i.e., P is a flex point of E (the notions of a tangent line and a flex point are well known in the case where the ground field is real; for an arbitrary ground field, one should regard the above as their definitions).

Coordinatewise expression. To express the group law in coordinates, it is convenient to represent the equation of a plane cubic in a special form. Note that the Riemann–Roch theorem gives $\ell(nP_0) = n$ for $n \geq 1$. Let
$$x \in L(2P_0) \setminus \Bbbk, \qquad y \in L(3P_0) \setminus L(2P_0).$$
Then the seven functions 1, x, y, xy, x^2, y^2, and x^3 lie in $L(6P_0)$ and are therefore linearly dependent. Since only y^2 and x^3 have a pole of order 6 at P_0, their coefficients in this linear dependence are nonzero (otherwise, either the functions 1, x, y, xy, x^2, and y^2 or the functions 1, x, y, xy, x^2, and x^3 would be linearly dependent, and the curve would be rational). Multiplying x and y by appropriate elements of $\Bbbk$, we may assume that the coefficients are equal to 1. Thus, we have the relation
$$y^2 + a_1 xy + a_3 y = x^3 + a_2 x^2 + a_4 x + a_6 \tag{2.4.1}$$
with $a_i \in \Bbbk$. Assume for simplicity that $\operatorname{char} \Bbbk \neq 2$ (note however that similar results are valid for $\operatorname{char} \Bbbk = 2$ too). Then, making the substitution $y \mapsto y + (a_1 x + a_2)/2$, we obtain the equation
$$y^2 = (x-a)(x-b)(x-c),$$
where a, b, and c are distinct elements of $\Bbbk$ (if any two of them coincide, the curve is singular; *check this!*). Now, make the substitution $x \mapsto (x-a)/(b-a)$. We get an equation of the form
$$y^2 = x(x-1)(x-\lambda) \tag{2.4.2}$$
with $\lambda \in \Bbbk$; the parameter λ is known as the *Legendre parameter* of E.

In homogeneous coordinates, equation (2.4.2) can be rewritten as
$$x_0 x_2^2 = x_1 (x_1 - x_0)(x_1 - \lambda x_0),$$
where $x = x_1/x_0$ and $y = x_2/x_0$. The coordinates of P_0 are in this case $(0:0:1)$. Since the line $x_1 = ax_0$ intersects E at the points $Q = (1:a:y_0)$ and $Q' = (1:a:-y_0)$, where $y_0^2 = a(a-1)(a-\lambda)$, and also passes through P_0, the sum of Q and Q' is P_0. Thus, $Q' = -Q$. Now it is easy to interpret the addition law geometrically and to express it in coordinates. If $P_1 = (1:x_1:y_1)$ and $P_2 = (1:x_2:y_2)$, then $Q = P_1 + P_2$ is obtained as follows: Let $Q' = (1:x_3:-y_3)$ be the intersection point of E and the line passing through P_1 and P_2; reflect Q' about the x axis (see Fig. 2.1); then $Q = P_1 + P_2 = (1:x_3:y_3)$.

EXERCISE 2.4.3. Let $P_i = (x_i, y_i)$, $i = 1, 2, 3$, and $P_1 + P_2 = P_3$. Express x_3 and y_3 in terms of x_1, x_2, y_1, and y_2. (*Answer:*
$$\begin{aligned} x_3 &= -x_1 - x_2 - \left(\frac{y_1 - y_2}{x_1 - x_2}\right)^2 + \lambda + 1, \\ y_3 &= \frac{x_2 y_1 - x_1 y_2}{x_1 - x_2} - x_3 \left(\frac{y_1 - y_2}{x_1 - x_2}\right).) \end{aligned} \tag{2.4.3}$$

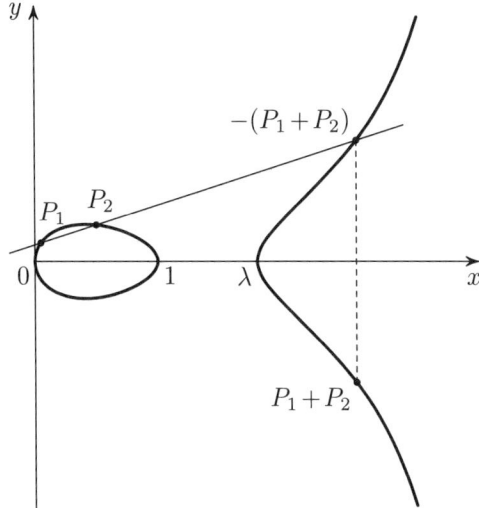

FIGURE 2.1. Addition law on an elliptic curve.

It is immediately seen from (2.4.3) that the map $\mu\colon E\times E\to E$, $\mu(P_1,P_2) = P_1+P_2$, is regular (note that (2.4.3) is valid in the case $x_1\ne x_2$ only, but it is easy to get a similar formula for $P_1 = P_2$ (EXERCISE!), and for $P_1\ne P_2$ and $x_1 = x_2$ we have $y_1+y_2 = 0$ and $P_1+P_2 = 0$). The map $P\mapsto -P$ is obviously regular. Thus, E is an abelian variety and can be identified with its Jacobian. Usually, the obtained group is denoted by $E(\Bbbk)$ to emphasize its dependence on $\Bbbk$ and is called the *group of points of E with values in* $\Bbbk$.

Structure of $E(\Bbbk)$. Here is an important property of this group.

PROPOSITION 2.4.4. *The group $E(\Bbbk)$ is divisible; i.e., for any positive integer n and any point $P\in E(\Bbbk)$ there is a point $Q\in E(\Bbbk)$ such that $nQ = P$ in $E(\Bbbk)$.*

SKETCH OF THE PROOF. We may assume that $P = (1:x:y)$, since otherwise we have $P = P_0$ and $Q = P_0$. Let $Q = (1:z:w)$, where (z,w) are (so far unknown) affine coordinates of Q. The group law makes it possible to express (x,y) in terms of (z,w). We obtain two equations of the form $F(z,w) = 0$ and $G(z,w) = 0$ in unknowns (z,w), where F and G are polynomials with coefficients depending on x, y, and λ. It can be shown that, since $\Bbbk$ is algebraically closed, the system has a solution. □

For $\Bbbk = \mathbb{C}$ there is a more precise description of $E(\Bbbk)$, which is given in Sec. 2.4.4 below. Let us note another useful property of elliptic curves.

Automorphisms. Since E is an abelian variety, we have the following statement.

PROPOSITION 2.4.5. *For any fixed point $P\in E$, the map $L_P\colon Q\mapsto Q+P$ is an automorphism of E (of an algebraic variety but not of a group).* □

COROLLARY 2.4.6. *The action of $\mathrm{Aut}(E)$ on E is transitive.* □

2.4.2. Isomorphisms and the j-invariant

Let us assume that char$\mathbb{k} \neq 2$. Note that the parameter λ in equation (2.4.2) is not an invariant of E; i.e., isomorphic curves may have different λ's. For instance, it can easily be shown that equations of the form (2.4.2) with λ and $\lambda' = 1/\lambda$ define isomorphic curves (EXERCISE!). An invariant of the isomorphism class of a curve E is the following expression, called the *j-invariant* (or *absolute invariant*) of E:

$$j = j(E) = 2^8 \frac{(\lambda^2 - \lambda + 1)^3}{\lambda^2(1-\lambda)^2}. \qquad (2.4.4)$$

The coefficient 2^8 is introduced in order that the j-invariant have integer coefficients as a series in some natural parameter q (we shall consider this question in *Advanced Chapters* when discussing modular curves).

The role of the j-invariant in the theory of elliptic curves is shown by the following result.

THEOREM 2.4.7. (a) *The value $j = j(\lambda) = j(E) \in \mathbb{k}$ is well defined, i.e., depends on the isomorphism class of E only.*

(b) *Two elliptic curves, E and E', are isomorphic if and only if $j(E) = j(E')$.*

(c) *For any $j \in \mathbb{k}$ there exists an elliptic curve E over $\mathbb{k}$ such that $j(E) = j$.*

Thus, the rule $E \mapsto j(E)$ defines a bijection from the set of isomorphism classes of elliptic curves over $\mathbb{k}$ onto $\mathbb{A}^1(\mathbb{k})$.

SKETCH OF THE PROOF. (a) Note first that $j(\lambda) = j(\lambda')$ for any $\lambda' \in \Lambda = \{\lambda, 1/\lambda, 1-\lambda, (1-\lambda)^{-1}, \lambda/(\lambda-1), (\lambda-1)/\lambda\}$, which is easily verified by a direct substitution. If we write the equation of E in the form (2.4.2), the projection $f(x,y) = x$ defines a morphism $f: E \to \mathbb{P}^1$ of degree 2, which has exactly four ramification points: $0, 1, \lambda$, and ∞. Let

$$y^2 = x(x-1)(x-\lambda')$$

be another representation of E; i.e., we have another morphism $f': E \to \mathbb{P}^1$ of degree 2 with ramification points $0, 1, \lambda'$, and ∞. Let P and P' be points of E such that $f(P) = \lambda$ and $f(P') = \lambda'$. By Corollary 2.4.6, there is an automorphism $\sigma \in \text{Aut}(E)$ that takes P to P'. Since f is defined by the linear system $|2P|$ and f' is defined by $|2P'|$, the morphisms f and $f' \circ \sigma$ are defined by the same liner system and hence differ by an automorphism of $\mathbb{P}^1$. It is easily checked that a linear fractional automorphism of $\mathbb{P}^1$ takes $(0, 1, \lambda, \infty)$ to $(0, 1, \lambda', \infty)$ if and only if $\lambda' \in \Lambda$, which proves (a).

(b) Let E and E' be two elliptic curves, let λ and λ' be the corresponding Legendre parameters, and let $j(\lambda) = j(\lambda')$. If we consider λ as a parameter and λ' as an unknown, we obtain an equation of degree 6 in λ', which vanishes on elements of Λ. Hence, it has no other roots, and therefore E and E' are isomorphic.

(c) Let $j \in \mathbb{k}$, and let λ be any root of the equation

$$2^8(\lambda^2 - \lambda + 1)^3 = j(1-\lambda)^2 \lambda^2.$$

Then (2.4.2) defines an elliptic curve E with $j(E) = j$. $\square$

REMARK 2.4.8. For char$\mathbb{k} \neq 2, 3$, it is often convenient to use another form of the equation of an elliptic curve E. Making the substitution $x \mapsto x - (\lambda+1)/3$, we reduce (2.4.2) to

$$y^2 = x^3 + ax + b. \qquad (2.4.5)$$

2.4. ELLIPTIC CURVES

Usually, one makes the substitution $y \mapsto 4y$, $x \mapsto 4x$ and writes (2.4.5) in the form
$$y^2 = 4x^3 - g_2 x - g_3, \qquad (2.4.6)$$
which is known as the *Weierstrass normal form* of E. Reasons to introduce this form are related to the theory of elliptic functions; they will be explained in Sec. 2.4.4 below. It is easy to verify that for a curve E with equation (2.4.6) we have
$$j(E) = 1728 g_2^3/(g_2^3 - 27 g_3^2).$$

REMARK 2.4.9. Theorem 2.4.7 is also valid if $\operatorname{char} \Bbbk = 2$. We do not give a proof for this case; let us only present a formula for the j-invariant in characteristic 2. Making the substitution $x \mapsto x + a_2$ in (2.4.2), we reduce the equation of E to the form
$$y^2 + a_1 xy + a_3 y = x^3 + a_4 x + a_6. \qquad (2.4.7)$$
Then $j(E) = a_1^{12}/\Delta$, where
$$\Delta = a_1^6 a_6 + a_1^5 a_3 a_4 + a_1^4 a_4^2 + a_3^4 + a_1^3 a_3^3.$$

EXERCISE 2.4.10. Let $F: E \to E^{(p)}$ be the Frobenius morphism of an elliptic curve E over a field of characteristic $p > 0$. Show that $j(E^{(p)}) = (j(E))^p$.

2.4.3. Isogenies

Let $f: E \to E'$ be a nonconstant morphism of elliptic curves. Let P_0 be the zero of the group law on E. If we choose the point $Q_0 = f(P_0)$ as the zero of the group law on E', then f is a group homomorphism due to the general properties of abelian varieties (Proposition 2.1.80). Such a morphism is called an *isogeny*. Since E and E' are curves, we may speak about the degree of f; it is referred to as the *degree of the isogeny*. Since $E \cong J_E$, any isogeny $f: E \to E'$ gives rise to the *dual isogeny*, defined as $f^t = f^*: J_{E'} \to J_E$.

EXERCISE 2.4.11. Show that $(f^t)^t = f$ for any isogeny f.

PROPOSITION 2.4.12. *Let $f: E \to E'$ be an isogeny of degree n. Then we have $\deg f^t = n$, $f \circ f^t = n_{E'}$, and $f^t \circ f = n_E$, where n_E and $n_{E'}$ are the morphisms of multiplication by n on E and E' respectively.* □

In the case of complex elliptic curves, this will be proved in the next subsection.

COROLLARY 2.4.13. *The degree of n_E equals n^2.* □

The subgroup $\operatorname{Ker}(n_E)$ is denoted by E_n or $E[n]$; it is called the *group of n-torsion points* of E.

EXERCISE 2.4.14. Let $p \nmid n$. Show that
$$E[n] \cong (\mathbb{Z}/n\mathbb{Z})^2.$$
(*Hint*: First consider the case of a prime n; then use induction on the number of divisors of n.)

EXERCISE 2.4.15. Let f be an isogeny of degree $n = n'n''$, where n' and n'' are coprime. Show that there exist isogenies f' and f'' such that $f = f' \circ f''$, $\deg f' = n'$, and $\deg f'' = n''$.

Now let us describe $E[p^a]$ for $a \geq 1$, where $p = \operatorname{char} \Bbbk$. First, we consider the case $a = 1$. For any point $P \in E$, the morphism p_E induces the zero map on the space $\mathfrak{m}_P/\mathfrak{m}_P^2$ since, as easily follows from the addition formulae, $f(nP) \equiv nf(P)$ (mod $\mathfrak{m}_Q^2$) if $f(P) \in \mathfrak{m}_Q$. Hence, p_E is an inseparable morphism since a separable morphism induces an isomorphism on the spaces $\mathfrak{m}_P/\mathfrak{m}_P^2$ for all P except for a finite number of its ramification points. Therefore, one of the morphisms, f or f', is purely inseparable. Since there is only one purely inseparable morphism of degree p on E, namely, the Frobenius morphism $F \colon E \to E^{(p)}$, we see that one of the morphisms f and f^t coincides with F and the other is $V = F^t$ (called the *Verschiebung morphism*). Now we have to consider two cases: (a) V is a separable morphism, (b) V is purely inseparable. In case (a), E is called an *ordinary* elliptic curve; in case (b), the curve is called *supersingular* (in this case, p_E is also purely inseparable).

PROPOSITION 2.4.16. *The kernel of the multiplication by p^a, $a \geq 1$, in the group $E(\Bbbk)$ is trivial for a supersingular curve E and is isomorphic to $\mathbb{Z}/p^a\mathbb{Z}$ for an ordinary E.*

SKETCH OF THE PROOF. It follows from the above argument that the proposition holds for $a = 1$. For an arbitrary $a \geq 1$, it can easily be deduced by induction on a. □

Supersingular curves. For a fixed p, there are only finitely many nonisomorphic supersingular curves over a field of characteristic p. Supersingular values of λ are described as follows.

THEOREM 2.4.17. *Let $\operatorname{char} \Bbbk = p > 2$, and let E be the curve given by $y^2 = x(x-1)(x-\lambda)$. Then E is supersingular if and only if*

$$\sum_{i=0}^{(p-1)/2} \binom{(p-1)/2}{i}^2 \lambda^i = 0. \qquad \square$$

Thus, all supersingular values of λ and of the j-invariant lie in some finite field. This field can be specified explicitly.

PROPOSITION 2.4.18. *Let $\operatorname{char} \Bbbk = p > 2$. Let $j = j(E)$ for a supersingular curve E in characteristic p. Then $j \in \mathbb{F}_{p^2}$.*

PROOF. Since Theorem 2.4.17 implies $j \in \overline{\mathbb{F}}_p$, it suffices to show that $j^{p^2} = j$.

Indeed, for a supersingular curve, the morphism $V = F^t$ is the Frobenius morphism on $E^{(p)}$, and $j(E^{(p)}) = j^p$. Hence, $(j^p)^p = j(E)$, as required. □

REMARK 2.4.19. Moreover, a supersingular curve can be defined by an equation with coefficients in $\mathbb{F}_{p^2}$.

REMARK 2.4.20. For $p = 2$ there exists precisely one supersingular value of the j-invariant, namely, $j = 0$. For example, the curve $y^2 + y = x^3$ with $j(E) = 0$ is supersingular.

The group $\operatorname{Hom}(E, E')$. Now let E and E' be two elliptic curves. The set of *morphisms of algebraic groups* $f \colon E \to E'$ (i.e., of regular maps that are group homomorphisms) is denoted by $\operatorname{Hom}(E, E')$; if $E = E'$, it is denoted by $\operatorname{End}(E)$. The set $\operatorname{Hom}(E, E')$ is an abelian group with the operation of addition of morphisms: $(f+g)(x) = f(x)+g(x)$; the set $\mathrm{End}(E)$ is a ring: multiplication is the composition of endomorphisms. It is clear that $\mathrm{Hom}(E, E')$ is a torsion-free group since, if $nf = 0$, then $f(E)$ is contained in the finite set $E[n]$ and hence is trivial. Analysis of the action of f on torsion points of E yields the following fact.

PROPOSITION 2.4.21. *The group* $\mathrm{Hom}(E, E')$ *is of rank* 0, 1, 2, *or* 4; *moreover, rank* 4 *is possible only if* E *and* E' *are supersingular elliptic curves (and hence the characteristic of the ground field is positive).* □

If $\mathrm{Hom}(E, E') \neq 0$, then the curves E and E' are said to be *isogeneous*. Consider the ring $\mathrm{End}^0(E) = \mathrm{End}(E) \otimes \mathbb{Q}$, which is obtained from $\mathrm{End}(E)$ by the adjunction of all elements of the form $1/n$, $n \in \mathbb{Z}$. Since the group $\mathrm{End}(E)$ is torsion-free, $\mathrm{End}(E)$ is embedded in $\mathrm{End}^0(E)$; by the relations $f^t \circ f = n$ and $f \circ f^t = n$, the ring $\mathrm{End}^0(E)$ is a division $\mathbb{Q}$-algebra. Possible types of such algebras are described by the following theorem.

THEOREM 2.4.22. *There are the following possibilities for* $\mathrm{End}^0(E)$:
(a) $\mathrm{End}^0(E) = \mathbb{Q}$;
(b) $\mathrm{End}^0(E)$ *is a complex quadratic field*;
(c) $\mathrm{End}^0(E)$ *is a quaternion algebra over* $\mathbb{Q}$ *ramified at* p *and at* ∞; *this is the case if* $p = \mathrm{char}\,\Bbbk > 0$ *and* E *is a supersingular curve over* $\Bbbk$. □

Recall that a *quaternion algebra over* $\mathbb{Q}$ *ramified at* p *and at* ∞ is a division algebra (i.e., a skew field) $H_{p,\infty}$ of degree 4 over $\mathbb{Q}$ such that for any prime $\ell \neq p$ there is an isomorphism of algebras $H_{p,\infty} \otimes \mathbb{Q}_\ell \cong M_2(\mathbb{Q}_\ell)$, where $M_2(\mathbb{Q}_\ell)$ is the algebra of 2×2 matrices over the ℓ-adic field $\mathbb{Q}_\ell$, whereas $H_{p,\infty} \otimes \mathbb{Q}_p$ and $H_{p,\infty} \otimes \mathbb{R}$ are division algebras over $\mathbb{Q}_p$ and $\mathbb{R}$ respectively. One can prove that $H_{p,\infty}$ exists for any prime p and is uniquely determined by these properties.

Thus, $\mathrm{End}(E)$ is an order in a division algebra of degree 1, 2, or 4 over $\mathbb{Q}$ (recall that an *order* in a division algebra A is a subring $\mathcal{O} \in A$ which is a free $\mathbb{Z}$-module such that $\mathcal{O} \otimes \mathbb{Q} = A$).

THEOREM 2.4.23. *Possible types of the rings* $\mathrm{End}(E)$ *are as follows*:
(a) $\mathrm{End}(E) = \mathbb{Z}$;
(b) $\mathrm{End}(E) = \mathbb{Z} + c\mathcal{O}_\Bbbk$, $c \in \mathbb{N}$, $p \nmid c$, *and* $\mathcal{O}_\Bbbk$ *is a maximal order in the complex quadratic field* $\Bbbk = \mathrm{End}^0(E)$ (*in this case, c is called the* conductor *of the order*);
(c) $\mathrm{End}(E)$ *is a maximal order in the quaternion algebra* $H_{p,\infty} = \mathrm{End}^0(E)$. □

Proofs of Proposition 2.4.21 and Theorems 2.4.22 and 2.4.23 are rather cumbersome. For $\Bbbk = \mathbb{C}$ there are simpler proofs, which are presented in Sec. 2.4.4 (note however that, for $\Bbbk = \mathbb{C}$, the most complicated case (c) of Theorems 2.4.22 and 2.4.23 does not occur).

Automorphism group. Theorem 2.4.23 yields a description of the group $\mathrm{Aut}_0(E)$ of automorphisms of the curve E viewed as an algebraic group (i.e., of automorphisms that preserve the zero point P_0). This group is isomorphic to the group of units $\mathrm{End}(E)^*$ of the ring $\mathrm{End}(E)$. The answer is as follows.

THEOREM 2.4.24. (a) *If* $j(E)$ *is neither* 0 *nor* 1728, *then*
$$\mathrm{Aut}_0(E) = \{\pm 1\}.$$

(b) *For* $p \neq 2, 3$,

$$\text{if } j(E) = 0, \quad \text{then } \text{Aut}_0(E) = \mu_6;$$
$$\text{if } j(E) = 1728, \quad \text{then } \text{Aut}_0(E) = \mu_4,$$

where $\mu_n = \{\zeta \in \mathbb{C}^* \mid \zeta^n = 1\}$ *is the group of n-th roots of unity.*

(c) *If* $p = 2$ *and* $j = 0 = 1728$, *then* $\text{Aut}_0(E) = \text{SL}(2, \mathbb{F}_3)$ *is a group of order* 24.

(d) *If* $p = 3$ *and* $j = 0 = 1728$, *then* $\text{Aut}_0(E)$ *is the semi-direct product of* $\mathbb{Z}/3\mathbb{Z}$ *and* $\mathbb{Z}/4\mathbb{Z}$, *hence a group of order* 12. □

REMARK 2.4.25. (a) Since it is clear that the group $\text{Aut}(E)$ of all automorphisms of E (as of an algebraic curve) is a semi-direct product of $E(\Bbbk)$ and $\text{Aut}_0(E)$, Theorem 2.4.24 also gives a description of $\text{Aut}(E)$.

(b) For supersingular curves, we have a noteworthy *mass formula*:

$$\sum |\text{Aut}(E)|^{-1} = \frac{p-1}{24}, \qquad (2.4.8)$$

where the sum is taken over the set of all nonisomorphic supersingular elliptic curves in characteristic $p > 0$.

EXERCISE 2.4.26. Find the number of nonisomorphic supersingular elliptic curves in characteristic p. Find supersingular values of the j-invariant for all $p \leq 13$. (*Hint*: Derive from (2.4.8) that the desired number is of the form $\lfloor p/12 \rfloor + \delta_p$, where $\delta_p \in \{0, 1, 2\}$ depends on the residue of p modulo 12. Then calculate δ_p.)

In more detail, elliptic curves over a finite field will be considered in Sec. 3.3.

2.4.4. Complex Elliptic Curves

The theory of elliptic curves over $\mathbb{C}$ is closely connected with the classical theory of elliptic functions. The latter makes it possible to obtain a number of profound results that are rather difficult to prove using purely algebraic methods.

Elliptic functions. Let Λ be a *lattice* on the complex plane $\mathbb{C}$, i.e., a free subgroup of $\mathbb{C}$ with two generators, ω_1 and ω_2, such that $\tau = \omega_2/\omega_1$ is not real. Changing the order of ω_1 and ω_2 if necessary, we assume hereinafter without loss of generality that $\text{Im}(\tau) > 0$.

An *elliptic function* f with the *period lattice* Λ is a function meromorphic on $\mathbb{C}$ and periodic with respect to Λ, i.e., a meromorphic function such that $f(z + w) = f(z)$ for any $z \in \mathbb{C}$ with $f(z) \neq \infty$ and for any $w \in \Lambda$. Note that any integral (i.e., analytical on the whole plane $\mathbb{C}$) elliptic function is constant by the Liouville theorem since it is continuous on the compact $\mathbb{C}/\Lambda$, hence is bounded on $\mathbb{C}/\Lambda$ and by periodicity is bounded on $\mathbb{C}$.

It is not obvious from the definition that nonconstant elliptic functions do exist. The existence follows from an explicit construction of the *Weierstrass $\mathcal{P}$-function*: let

$$\mathcal{P}(z) = \frac{1}{z^2} + \sum_{\omega \in \Lambda \setminus \{0\}} \left(\frac{1}{(z - \omega)^2} - \frac{1}{\omega^2} \right). \qquad (2.4.9)$$

EXERCISE 2.4.27. Show that (2.4.9) absolutely converges uniformly on any compact K such that $K \cap \Lambda = \varnothing$.

2.4. ELLIPTIC CURVES

EXERCISE 2.4.28. Show that $\mathcal{P}(z)$ is an even elliptic function with a double pole at any point $\omega \in \Lambda$ and no other poles. Show that
$$\mathcal{P}'(z) = -2 \sum_{\omega \in \Lambda} \frac{1}{(z-\omega)^3}$$
is an odd elliptic function with a triple pole at any point of Λ and no other poles.

THEOREM 2.4.29. *The function $\mathcal{P}(z)$ satisfies the differential equation*
$$\mathcal{P}'(z)^2 = 4\mathcal{P}(z)^3 - g_2 \mathcal{P}(z) - g_3,$$
where
$$g_2 = 60 \sum_{\omega \in \Lambda \setminus \{0\}} \omega^{-4}, \qquad g_3 = 140 \sum_{\omega \in \Lambda \setminus \{0\}} \omega^{-6}. \qquad \square$$

EXERCISE 2.4.30. Prove Theorem 2.4.29. (*Hint*: Consider the Laurent series expansions of $\mathcal{P}(z)$ and $\mathcal{P}'(z)$ at the origin.)

Uniformization. Theorem 2.4.29 shows that points of the form $(\mathcal{P}(z), \mathcal{P}'(z))$ lie on an elliptic curve with equation
$$y^2 = 4x^3 - g_2 x - g_3.$$

Furthermore, we have the following important fact:

THEOREM 2.4.31. *The map*
$$\varphi \colon \mathbb{C} \to \mathbb{P}^2(\mathbb{C}), \quad z \mapsto \begin{cases} (1 : \mathcal{P}(z) : \mathcal{P}'(z)) & \text{if } z \notin \Lambda, \\ (0:0:1) & \text{if } z \in \Lambda, \end{cases}$$
defines an isomorphism of the Riemann surfaces $\mathbb{C}/\Lambda$ and $E(\mathbb{C})$, where E is the elliptic curve with the homogeneous equation
$$x_0 x_2^2 = 4x_1^3 - g_2 x_0^2 x_2 - g_3 x_0^3,$$
which is also a group isomorphism. $\square$

REMARK 2.4.32. Of course, the complex-analytic map φ is not algebraic.

In view of Theorem 2.4.31, it is said that elliptic functions *uniformize* elliptic curves.

COROLLARY 2.4.33. *The group $E(\mathbb{C})$ is isomorphic to the torus $(\mathbb{R}/\mathbb{Z})^2$ as a real Lie group.* $\square$

Corollary 2.4.33 immediately implies Propositions 2.4.4 and 2.4.12 for the case of complex elliptic curves. Theorems 2.4.22 and 2.4.23 are also easily obtained from Theorem 2.4.31 using the following fact:

EXERCISE 2.4.34. Show that for any $f \in \mathrm{End}(E)$ there exists $\alpha_f \in \mathbb{C}$ such that $\alpha_f \Lambda \subseteq \Lambda$ and the endomorphism $\varphi^{-1} \circ f \circ \varphi$ of $\mathbb{C}/\Lambda$ coincides with the multiplication by α_f. (*Hint*: Consider the Taylor series expansion of $\varphi^{-1} \circ f \circ \varphi$ at the origin.)

EXERCISE 2.4.35. From the above exercise, deduce the parts of Theorems 2.4.22 and 2.4.23 that concern the case $\mathbb{k} = \mathbb{C}$.

2.5. Curves over Nonclosed Fields

So far we have worked over only an algebraically closed field. However, in many questions we are interested in solutions of systems of algebraic equations over an algebraically nonclosed field $\Bbbk$. It is enough to recall the real geometry (when $\Bbbk = \mathbb{R}$), or Diophantine problems, and more generally the number theory (here $\Bbbk = \mathbb{Q}$, or $\Bbbk$ is a finite extension of $\mathbb{Q}$). At last, in coding theory the ground field is finite.

To work over nonclosed fields there exists a well-developed theory of algebraic varieties defined over an arbitrary field, but to discuss its details even in part would have taken several other books. Fortunately, for algebraic curves there is a purely algebraic approach to the theory that makes it possible to work with curves over a nonclosed field using a more elementary technique.

We present the basics of this alternative language in Sec. 2.5.1. In the second subsection we explain how to define in these terms a point on a curve, and in the third how to define a divisor. In the last subsection we shall see that in fact the language of fields $\Bbbk(X)$ is totally equivalent to that of algebraic curves, this being true over any ground field $\Bbbk$.

Note that for the case of an algebraically closed field $\Bbbk$ we have already encountered this approach. At the end of Sec. 2.1.2 it was explained (Theorem 2.1.46) that a smooth projective curve X is totally determined by its field of rational functions $\Bbbk(X)$. Moreover, Proposition 2.1.49 completely characterizes the fields $\Bbbk(X)$, Theorem 2.1.50 describes points of a curve X in terms of the field $\Bbbk(X)$, and in Sec. 2.1.3 we explained how to express the group $\text{Div}(X)$ of divisors on X.

For simplicity we always suppose the ground field $\Bbbk$ to be perfect. Recall that the field $\Bbbk$ is called *perfect* if its algebraic closure $\overline{\Bbbk}$ is separable over $\Bbbk$. This is true for any field of characteristic zero and for any finite field.

2.5.1. Function Fields

An *algebraic function field* $\mathbb{K}/\Bbbk$ *of one variable over* $\Bbbk$ (or, for short, simply a *function field* $\mathbb{K}/\Bbbk$) is any field $\mathbb{K} \supseteq \Bbbk$ which is a finite algebraic extension of $\Bbbk(x)$, where x is an element of $\mathbb{K}$ transcendental over $\Bbbk$ or, equivalently, a finite-generated extension of $\Bbbk$ of transcendence degree 1.

The set of elements of $\mathbb{K}$ that are algebraic over $\Bbbk$ is a subfield of $\mathbb{K}$, which is called the *field of constants* of $\mathbb{K}/\Bbbk$ and is denoted by $\widetilde{\Bbbk}$. We have $\Bbbk \subseteq \widetilde{\Bbbk} \subset \mathbb{K}$, and it is easily verified that $\mathbb{K}/\widetilde{\Bbbk}$ is also a function field over $\widetilde{\Bbbk}$. We say that $\Bbbk$ is *algebraically closed in* $\mathbb{K}$ (or is the *full constant field* of $\mathbb{K}$) if $\widetilde{\Bbbk} = \Bbbk$. (Often, one assumes the latter condition to be satisfied when saying that a function field $\mathbb{K}$ over $\Bbbk$ is given.)

Recall (Sec. 2.1.2) that a *valuation ring* of the function field $\mathbb{K}/\Bbbk$ is a proper subring $\mathcal{O} \subset \mathbb{K}$ such that $\Bbbk \subset \mathcal{O}$, and for any $f \in \mathbb{K}$ at least one of the elements f and f^{-1} belongs to $\mathcal{O}$. The set $\mathcal{O}^* = \{f \in \mathbb{K} \mid f \in \mathcal{O}, f^{-1} \in \mathcal{O}\}$ is called the *group of units* of $\mathcal{O}$.

Properties of valuation rings are given in the following statement.

PROPOSITION 2.5.1. *Let $\mathcal{O}$ be a valuation ring of a function field $\mathbb{K}/\Bbbk$. Then*

(a) *$\mathcal{O}$ is a local ring; i.e., $\mathcal{O}$ has a unique maximal ideal P, and $P = \mathcal{O} \setminus \mathcal{O}^*$.*

(b) *If $f \in \mathbb{K}$, $f \neq 0$, then $f \in P$ if and only if $f^{-1} \notin \mathcal{O}$.*

(c) $\widetilde{\Bbbk} \subseteq \mathscr{O}$ and $\widetilde{\Bbbk} \cap P = \{0\}$.

(d) P is a principal ideal.

(e) If $P = t\mathscr{O}$, then any nonzero element $f \in \mathbb{K}$ has a unique representation of the form $f = t^n u$, where $n \in \mathbb{Z}$ and $u \in \mathscr{O}^*$.

(f) $\mathscr{O}$ is a principal ideal domain; namely, if I is a nontrivial proper ideal in $\mathscr{O}$ and $P = t\mathscr{O}$, then $I = t^n \mathscr{O}$ for some $n \in \mathbb{N}$.

A ring having the above properties is called a *discrete valuation ring*.

PROOF. (a) Consider the set $P = \mathscr{O} \setminus \mathscr{O}^*$. Let us prove that it is an ideal of $\mathscr{O}$ (this immediately implies that P is a unique maximal ideal, since a proper ideal of $\mathscr{O}$ cannot contain a unit, i.e., an element of $\mathscr{O}^*$). First, if $f \in P$ and $z \in \mathscr{O}$, then $fz \notin \mathscr{O}^*$ (otherwise, f would be a unit); hence, $fz \in P$. It remains to prove that for any $f, g \in P$ their sum also belongs to P. Consider the elements f/g and g/f; at least one of them belongs to $\mathscr{O}$. Assume that it is f/g; then $f + g = g(f/g + 1) \in P$.

(b) is now obvious.

(c) Assume that $z \in \widetilde{\Bbbk}$ and $z \notin \mathscr{O}$. Then $z^{-1} \in \mathscr{O}$ by the definition of a valuation ring. Since z^{-1} is algebraic over $\Bbbk$, we have $a_n(z^{-1})^n + \ldots + a_1 z + 1 = 0$ for some constants $a_1, \ldots, a_n \in \Bbbk$. Then $z = -a_n(z^{-1})^{n-1} - \ldots - a_1 \in \mathscr{O}$. Now the statement $\widetilde{\Bbbk} \cap P = \{0\}$ is trivial.

(d) The proof of this statement is based on the following lemma.

LEMMA 2.5.2. *Let $\mathscr{O}$ be a valuation ring of an algebraic function field $\mathbb{K}/\Bbbk$, P be its maximal ideal, and x be a nonzero element of P. Let $x_1, \ldots, x_n \in P$ be such that $x_1 = x$ and $x_i \in x_{i+1} P$ for $i = 1, \ldots, n-1$. Then $n \leq [\mathbb{K} : \Bbbk(x)] < \infty$.*

Let us apply the lemma. If P is not a principal ideal, then for a nonzero $x_1 \in P$ we have $P \neq x_1 \mathscr{O}$, and therefore there is $x_2 \in P \setminus x_1 \mathscr{O}$. Then $x_2 x_1^{-1} \notin \mathscr{O}$, and therefore $x_2^{-1} x_1 \in P$ according to (b), whence $x_1 \in x_2 P$. By induction, we obtain an arbitrarily long sequence $x_1, x_2, \ldots$ such that $x_i \in x_{i+1} P$ for all i. This contradicts the lemma.

PROOF OF THE LEMMA. From the definition of a valuation ring and from (c) it follows immediately that $[\mathbb{K} : \Bbbk(x)] < \infty$, so it suffices to prove that $x_1, \ldots, x_n$ are linearly independent over $\Bbbk(x)$. Assume the contrary; i.e., let there be a nontrivial linear combination $\sum \lambda_i x_i = 0$ with $\lambda_i \in \Bbbk(x)$. We can assume that all λ_i are polynomials in x and that at least one of them is not divisible by x. Since $x \in P$, we have $\lambda_i \in \mathscr{O}$. Let a_i be the constant term of λ_i. Define $j = \max\{i \mid a_i \neq 0\}$. Then for $i < j$ we have $x_i \in x_j P$, and for $i > j$ we get $\lambda_i = x g_i$, where g_i is a polynomial in x. Thus,

$$-\lambda_j = \frac{1}{x_j} \sum_{i \neq j} \lambda_i x_i = \sum_{i<j} \lambda_i \frac{x_i}{x_j} + \sum_{i>j} \frac{x}{x_j} g_i x_i.$$

In the expression on the right-hand side, all summands belong to P, whence $\lambda_j \in P$. On the other hand, $\lambda_j = a_j + x g_j$, which implies $a_j \in P$ since $x \in P$ and $g_j \in \mathscr{O}$. Since $a_j \neq 0$, this contradicts (c). $\square$

We continue the proof of the theorem:

(e) The uniqueness of such a representation is obvious; let us show the existence. At least one of the elements f and f^{-1} belongs to $\mathscr{O}$; for definiteness, assume that it is f. The case $f \in \mathscr{O}^*$ is trivial; it remains to consider the case $f \in P$. It follows from Lemma 2.5.2 that there is a maximal $m \geq 1$ with $f \in t^m \mathscr{O}$ (the length of a sequence $f, t^{m-1}, \ldots, t \in P$ is bounded). Then $f = t^m u$, where $u \in \mathscr{O}$. Since m is maximal, we have $u \notin P$, i.e., $u \in \mathscr{O}^*$.

(f) For any $f \in I$ according to (e) we have $f = t^r u$, $f \in \mathscr{O}^*$, whence $t^r = f u^{-1} \in I$; therefore, there is $n = \min\{r \in \mathbb{N} \mid t^r \in I\}$. It is easily checked that $I = t^n \mathscr{O}$. □

2.5.2. Places of a Function Field

Now we are going to define the notion of a place, which replaces that of a point when we turn to the language of function fields.

A *place* P of a function field $\mathbb{K}/\mathbb{k}$ is the maximal ideal of some valuation ring $\mathscr{O}$ of $\mathbb{K}/\mathbb{k}$. Any element $t \in P$ such that $P = t\mathscr{O}$ is called a *local parameter at* P (also called a *prime element* or *uniformizing variable*). The set of all places of a function field $\mathbb{K}$ is denoted by $\mathbb{P}_\mathbb{K}$.

Since a valuation ring $\mathscr{O}$ is uniquely determined by its maximal ideal P, namely, $\mathscr{O} = \{f \in \mathbb{K} \mid f^{-1} \notin P\}$, $\mathscr{O} = \mathscr{O}_P$ is called the *valuation ring of the place* P.

REMARK 2.5.3. Theorem 2.1.50 states that for the case $\mathbb{K} = \mathbb{k}(X)$ and $\mathbb{k} = \overline{\mathbb{k}}$ there is a natural one-to-one correspondence between $\mathbb{P}_\mathbb{K}$ and $\{P \in X\}$.

Discrete valuation. In Sec. 2.1.2 we defined a discrete valuation for a field $\mathbb{k}(X)$ in the case of an algebraically closed $\mathbb{k}$. This notion is easily extended to an arbitrary field $\mathbb{k}$.

A *discrete valuation* of $\mathbb{K}/\mathbb{k}$ is a function $v \colon \mathbb{K}^* \to \mathbb{Z}$ with the following properties:

(1) $v(fg) = v(f) + v(g)$ for any $f, g \in \mathbb{K}$;
(2) $v(f+g) \geq \min\{v(f), v(g)\}$ for any $f, g \in \mathbb{K}$;
(3) there exists an element $f \in \mathbb{K}$ with $v(f) = 1$;
(4) $v(c) = 0$ for any $c \in \mathbb{k} \setminus \{0\}$.

We extend v to the whole $\mathbb{K}$, putting by definition $v(0) = \infty$.

By Proposition 2.5.1e, for each place $P \in \mathbb{P}_\mathbb{K}$, the choice of a local parameter t yields a unique representation of every nonzero element $f \in \mathbb{K}$ in the form $f = t^n u$, $u \in \mathscr{O}^*$.

EXERCISE 2.5.4. Prove that for any $f \neq 0$ the exponent n in this representation depends only on P but does not depend on the choice of local parameter.

Thus, the function $v_P \colon \mathbb{K} \setminus \{0\} \to \mathbb{Z}$ equal to the exponent n in this representation is well defined. We have the following theorem.

THEOREM 2.5.5. *Let $\mathbb{K}/\mathbb{k}$ be a function field.*

(a) *For any place $P \in \mathbb{P}_\mathbb{K}$, the function v_P is a discrete valuation. Moreover, we have*
$$\mathscr{O}_P = \{f \in \mathbb{K} \mid v_P(f) \geq 0\},$$
$$\mathscr{O}_P^* = \{f \in \mathbb{K} \mid v_P(f) = 0\},$$
$$P = \{f \in \mathbb{K} \mid v_P(f) > 0\},$$

and an element $t \in \mathbb{K}$ is a local parameter for P if and only if $v_P(t) = 1$.

(b) *Conversely, let v be a discrete valuation of the function field $\mathbb{K}/\mathbb{k}$. Then the set $P = \{f \in \mathbb{K} \mid v(f) > 0\}$ is a place of $\mathbb{K}$, and $\mathscr{O}_P = \{f \in \mathbb{K} \mid v(f) \geq 0\}$ is the corresponding valuation ring.*

(c) *Any valuation ring $\mathscr{O}$ of $\mathbb{K}/\mathbb{k}$ is a maximal proper subring of $\mathbb{K}$.*

PROOF. Obviously, v_P obeys properties (1), (3), and (4) of a discrete valuation. We need only to check the *triangle inequality* $v_P(f+g) \geq \min\{v_P(f), v_P(g)\}$. Let $v_P(f) = n$ and $v_P(g) = m$; we can assume that $n \leq m < \infty$. Then $f = t^n u$ and $g = t^m v$ with $u, v \in \mathscr{O}_P^*$; therefore, $f + g = t^n(u + t^{m-n}v) = t^n h$ with $h \in \mathscr{O}_P$. If $h = 0$, the triangle inequality is obviously satisfied; otherwise, we have $h = t^k u'$ with $u' \in \mathscr{O}_P^*$ and $k \geq 0$, whence the triangle inequality also follows. The remaining assertions of (a) are trivial, as is (b).

Let us prove (c). Let $\mathscr{O}$ be a valuation ring of $\mathbb{K}/\mathbb{k}$, P be its maximal ideal, and v_P be the discrete valuation associated to this ideal and $z \in \mathbb{K} \setminus \mathscr{O}$. We have to show that $\mathbb{K} = \mathscr{O}[z]$. Since $z \notin \mathscr{O}$, we have $v_P(z^{-1}) > 0$; therefore, for an arbitrary $f \in \mathbb{K}$ and for $k \geq 0$ large enough, we have $v_P(fz^{-k}) \geq 0$, whence $g = fz^{-k} \in \mathscr{O}$, i.e., $f = gz^k \in \mathscr{O}[z]$. $\square$

Value at a place. Since P is a maximal ideal, the residue class ring $\mathscr{O}_P/P$ is a field. This field $\mathbb{k}_P = \mathscr{O}_P/P$ is called the *residue class field* of P. For $f \in \mathscr{O}_P$ we denote by $f(P) = f + P \in \mathbb{k}_P$ the residue class of f modulo P. For $f \in \mathbb{K} \setminus \mathscr{O}_P$ we put $f(P) = \infty$ by definition. Thus, there is a map $f \mapsto f(P)$ from $\mathbb{K}$ to $\mathbb{k}_P \cup \{\infty\}$. The image $f(P)$ of f under this map is called its *value at* P; elements $f \in \mathscr{O}_P$ are called *regular at* P.

Now it becomes clear why the fields in question are called function fields and their elements are often called *functions*.

REMARK 2.5.6. Note that values of the same element at different places belong to different fields. We can easily avoid this, observing that $\mathbb{k}_P \subseteq \overline{\mathbb{k}}$ for any P.

Degree of a place. From Proposition 2.5.1c we know that $\mathbb{k} \subseteq \mathscr{O}_P$ and $\mathbb{k} \cap P = \{0\}$, so the canonical residue class map $\mathscr{O}_P \to \mathscr{O}_P/P$ induces a canonical embedding $\mathbb{k} \hookrightarrow \mathbb{k}_P$. Therefore, we shall hereafter always consider $\mathbb{k}$ as a subfield of $\mathbb{k}_P$ (this also applies to the constant field $\widetilde{\mathbb{k}}$). We call $\deg P = [\mathbb{k}_P : \mathbb{k}]$ the *degree* of a place P. The degree of a place is always finite; moreover, we have the following statement.

PROPOSITION 2.5.7. *Let P be a place of a function field $\mathbb{K}/\mathbb{k}$, and let x be a nonzero element of P. Then*
$$\deg P \leq [\mathbb{K} : \mathbb{k}(x)].$$

PROOF. We have to show that any elements $z_1, \ldots, z_n \in \mathscr{O}_P$ whose residue classes $z_1(P), \ldots, z_n(P) \in \mathbb{k}_P$ are linearly independent over $\mathbb{k}$ are linearly independent over $\mathbb{k}(x)$. Assume the contrary: let there be a nontrivial linear combination
$$0 = \sum_{i=1}^{n} \lambda_i z_i$$
with $\lambda_i \in \mathbb{k}(x)$. As in the proof of Lemma 2.5.2, we may assume that all λ_i are polynomials and not all of them are divisible by x; i.e., $\lambda_i = a_i + xg_i$ with $a_i \in \mathbb{k}$ and $g_i \in \mathbb{k}[x]$, and not all a_i are zero. Since $x \in P$ and $g_i \in \mathscr{O}_P$, we have $\lambda_i(P) = a_i(P) = a_i$. Applying the residue class map to $0 = \sum \lambda_i z_i$, we obtain
$$0 = 0(P) = \sum_{i=1}^{n} a_i z_i(P),$$
which contradicts the linear independence of the residues $z_i(P)$ over $\mathbb{k}$. $\square$

Ramification, splitting, inertia. Let $\mathbb{K} \subset \mathbb{L}$ be a finite extension of function fields, P be a place of the field $\mathbb{K}$. This place corresponds to the local ring $\mathscr{O}_P$ and the maximal ideal P in it. We say that the place Q of $\mathbb{L}$ *lies over* P if the ideal P lies in the ideal Q; in this case we write $Q \mid P$.

PROPOSITION 2.5.8. *The following conditions are equivalent*:
- *the place Q lies over the place P*;
- $\mathscr{O}_P \subseteq \mathscr{O}_Q$;
- *there exists a positive integer $e \geq 1$ such that $v_Q(f) = e \cdot v_P(f)$ for any $f \in \mathbb{K}$.*

These conditions being satisfied, $P = Q \cap \mathbb{K}$ and $\mathscr{O}_P = \mathscr{O}_Q \cap \mathbb{K}$. □

From this it follows, in particular, that any place of $\mathbb{L}$ lies over some place of $\mathbb{K}$. Thus, a field extension defines a map from the set of places of the larger field to that of the smaller.

Further on we suppose that the fields of constants of $\mathbb{K}$ and $\mathbb{L}$ coincide (this case corresponds to the coverings of curves). Then e is called the *ramification index* of the place Q over P. If $e = 1$, we say that Q is *unramified* over P, and if $e > 1$ that there is *ramification*. If all places over P are unramified, we say that P is *unramified* (*in the extension* $\mathbb{L}/\mathbb{K}$). If all places of $\mathbb{K}$ are unramified, then the extension is called *unramified*.

EXERCISE 2.5.9. Let the place Q lie over P. Prove the divisibility $\deg P \mid \deg Q$.

PROPOSITION 2.5.10. *Let Q run over all places lying over P. Then*
$$\sum_{Q \mid P} \deg Q \leq n \deg P,$$
where $n = [\mathbb{L} : \mathbb{K}]$ is the degree of the extension, and the equality takes place if and only if the extension is unramified at P. □

Thus, over P there are at most n places. If their number equals n, we say that P is *totally split*. (In this case P is unramified.) If $\deg Q > \deg P$, we say that there is *inertia*. If $\deg Q = n \deg P$, then we say that P is *inert* in $\mathbb{L}/\mathbb{K}$; in this case it is, of course, unramified. If over P there is a unique place Q and $\deg Q = \deg P$, then P is *totally ramified*.

EXERCISE 2.5.11. Let $\mathbb{K} = \Bbbk(t)$, $\mathbb{L} = \Bbbk(u)$, $t = u^2$. Describe completely the behaviour of places of $\mathbb{K}$ in this extension.

As we have seen before in Sec. 2.1.1 and as we shall see in the next subsection, the field extension $\mathbb{L}/\mathbb{K}$ corresponds to a map of curves $Y \to X$, where $\mathbb{K} = \Bbbk(X)$, $\mathbb{L} = \Bbbk(Y)$. The place Q lies over P if and only if in the language of curves the point Q is taken to P by this map.

EXERCISE 2.5.12. Reread the end of Sec. 2.1.3, and (for an algebraically closed field) establish the equivalence of corresponding notions.

Zeroes and poles. Let $f \in \mathbb{K}$ and $P \in \mathbb{P}_{\mathbb{K}}$. We say that P is a *zero* of f if $v_P(f) > 0$; in this case $v_P(f)$ is called the *zero order*. A place P is called a *pole* of f if $v_P(f) < 0$; in this case, $-v_P(f)$ is the *pole order*.

PROPOSITION 2.5.13. *Let $f \in \mathbb{K}/\Bbbk$ be transcendental over $\Bbbk$. Then f has at least one zero and one pole.* □

COROLLARY 2.5.14. *The set $\mathbb{P}_\mathbb{K}$ is nonempty.* □

In view of the latter result, Proposition 2.5.7 immediately implies the following fact.

COROLLARY 2.5.15. *The constant field $\widetilde{\mathbb{k}}$ of a function field $\mathbb{K}/\mathbb{k}$ is a finite-degree extension of $\mathbb{k}$.* □

An important property of discrete valuations of a function field is their independence in the following sense: for different discrete valuations $v_1, \ldots, v_n$ of $\mathbb{K}/\mathbb{k}$, if for some $f \in \mathbb{K}$ we know $v_i(f)$, $i = 1, \ldots, n-1$, we have no information on $v_n(f)$. More precisely, we have the following theorem.

THEOREM 2.5.16 (weak approximation theorem). *Let $\mathbb{K}/\mathbb{k}$ be a function field, $P_1, \ldots, P_n$ be pairwise distinct places of $\mathbb{K}$, $f_1, \ldots, f_n \in \mathbb{K}$, and $r_1, \ldots, r_n \in \mathbb{Z}$. Then there exists some $f \in \mathbb{K}$ such that*

$$v_{P_i}(f - f_i) = r_i, \quad i = 1, \ldots, n.$$
□

COROLLARY 2.5.17. *Any function field has infinitely many places.*

PROOF. Assume the contrary: let $P_1, \ldots, P_n$ be all places of the function field. By the weak approximation theorem we can find a nonzero element $f \in \mathbb{K}$ with $v_{P_i}(f) > 0$, $i = 1, \ldots, n$. Then f is transcendental over $\mathbb{k}$ since it has zeroes, but f has no pole, which contradicts Proposition 2.5.13. □

Using the weak approximation theorem, one can also prove the following result.

PROPOSITION 2.5.18. *Let $P_1, \ldots, P_r$ be zeroes of an element f of a function field $\mathbb{K}/\mathbb{k}$. Then*

$$\sum_{i-1}^{r} v_{P_i}(f) \deg P_i \leq [\mathbb{K} : \mathbb{k}(f)].$$
□

COROLLARY 2.5.19. *Any nonzero element of a functional field has finitely many zeroes and poles.*

PROOF. If f is a constant, it has neither zeroes nor poles. If f is transcendental over $\mathbb{k}$, the number of its zeroes is at most $[\mathbb{K} : \mathbb{k}(f)]$. Similarly, f^{-1} has only finitely many zeroes. □

2.5.3. Divisors

Just as for points on curves (cf. Sec. 2.1.3), for places of a function field we can introduce the notion of a divisor.

The free abelian group generated by places of a function field $\mathbb{K}/\mathbb{k}$ is called the *divisor group* of $\mathbb{K}/\mathbb{k}$ and is denoted by $\mathrm{Div}(\mathbb{K})$. Its elements—formal sums of the form

$$D = \sum_{P \in \mathbb{P}_\mathbb{K}} a_P P, \quad a_P \in \mathbb{Z},$$

where almost all a_P are zero—are called *divisors*. The *support* of a divisor D is defined by

$$\mathrm{Supp}\, D = \{P \in \mathbb{P}_\mathbb{K} \mid a_P \neq 0\}.$$

A divisor of the form $D = P$ with $P \in \mathbb{P}_\mathbb{K}$ is called a *prime divisor*.

For a divisor $D = \sum a_P P$ and a place $Q \in \mathbb{P}_\mathbb{K}$, put $v_Q(D) = a_Q$; in other words,
$$\operatorname{Supp} D = \{P \in \mathbb{P}_\mathbb{K} \mid v_P(D) \neq 0\}, \qquad D = \sum_{P \in \operatorname{Supp} D} v_P(D) P.$$

On the set of divisors there is a partial ordering: $D_1 \geq D_2$ if and only if $v_P(D_1) \geq v_P(D_2)$ for any $P \in \mathbb{P}_\mathbb{K}$. A divisor $D \geq 0$ is called *effective*.

The *degree* of a divisor is defined by
$$\deg D = \sum_{P \in \mathbb{P}_\mathbb{K}} v_P(D) \deg P;$$
thus, we have a homomorphism $\deg\colon \operatorname{Div}(\mathbb{K}) \to \mathbb{Z}$.

According to Corollary 2.5.19, any nonzero element $f \in \mathbb{K}$ has a finite set of zeroes Z and a finite set of poles N. Define for f its *divisor of zeroes*
$$(f)_0 = \sum_{P \in Z} v_P(f) P,$$

divisor of poles
$$(f)_\infty = \sum_{P \in N} (-v_P(f)) P,$$

and *principal divisor*
$$(f) = (f)_0 - (f)_\infty.$$

Nonzero constants have the following characterization: for $f \in \mathbb{K} \setminus \{0\}$,
$$f \in \widetilde{\mathbb{k}} \iff (f) = 0.$$

The following theorem describes the relation between zeroes and poles of an element of a function field. Also, it refines the result of Proposition 2.5.18.

THEOREM 2.5.20. *The degree of a principal divisor equals zero. Moreover, for any $f \in \mathbb{K} \setminus \widetilde{\mathbb{k}}$ we have*
$$\deg(f)_0 = \deg(f)_\infty = [\mathbb{K} : \mathbb{k}(f)]. \qquad \square$$

REMARK 2.5.21. It is significant that all we have said in Sec. 2.1.3 about local lattices, line bundles, and their connection with divisors is also valid for an arbitrary function field.

2.5.4. Function Fields and Algebraic Curves

As we have already seen, for an irreducible smooth projective curve X over an algebraically closed field $\mathbb{k}$ its field of rational functions $\mathbb{k}(X)$ is a function field in one variable (see Proposition 2.1.49). The inverse is also true: for any function field in one variable $K/\mathbb{k}$ there exists a (unique up to an isomorphism) smooth irreducible projective curve X, such that the field $\mathbb{k}(X)$ is isomorphic (as an extension of $\mathbb{k}$) to the field $\mathbb{K}$.

For the last statement to make sense we need to explain what is an algebraic curve over a nonclosed field.

Algebraic varieties. Let $\mathbb{k}$ be a field, $\overline{\mathbb{k}}$ be its algebraic closure, $\mathbb{k}^s$ be its separable closure (the union in $\overline{\mathbb{k}}$ of all separable extensions of $\mathbb{k}$), and $\operatorname{Gal}(\mathbb{k}) = \operatorname{Gal}(\overline{\mathbb{k}}/\mathbb{k}) = \operatorname{Gal}(\mathbb{k}^s/\mathbb{k})$ be the Galois group of $\mathbb{k}$.

EXERCISE 2.5.22. Let $\mathbb{k} = \mathbb{F}_q$ be a finite field. Prove that
$$\overline{\mathbb{F}}_q = \mathbb{F}_q^s = \bigcup_{m=1}^{\infty} \mathbb{F}_{q^m}.$$
Compute the group $\operatorname{Gal} \mathbb{F}_q$. (*Answer:* $\operatorname{Gal} \mathbb{F}_q = \widehat{\mathbb{Z}} = \varprojlim \mathbb{Z}/n\mathbb{Z}$, the profinite completion of the group $\mathbb{Z}$.)

We would like to preserve the fundamental for algebraic geometry correspondence between varieties given by systems of algebraic equations and ideals generated by the equations of this system (and also with corresponding rings and fields of functions). The banal definition of a variety as the set of solutions of the system of equations does not give the desired correspondence.

EXERCISE 2.5.23. Let $\mathbb{k} = \mathbb{R}$. Give an example of two polynomials $F(x,y)$ and $G(x,y)$ such that the ideals (F) and (G) in the ring $\mathbb{k}[x,y]$ are prime and different from one another and the sets of solutions of the equations $F = 0$ and $G = 0$ coincide.

A way out is the classical method of considering "complex points of real curves", which motivates the following definition.

A *projective* (respectively, *affine*) *variety* X *over the field* $\mathbb{k}$, or just a $\mathbb{k}$-*variety*, is a variety $\overline{X} \subseteq \mathbb{P}^m(\overline{\mathbb{k}})$ (respectively, $\overline{X} \subseteq \mathbb{A}^m(\overline{\mathbb{k}})$) that can be defined by a system of equations with coefficients in $\mathbb{k}$. The algebraic set and quasiprojective variety are defined in the same way.

EXERCISE 2.5.24. Reread Sec. 2.1 and find out which facts are true for the case of a nonclosed field and which are not.

Let $X/\mathbb{k}$ be a quasiprojective variety. For a field $\mathbb{k}'$ such that $\mathbb{k} \subset \mathbb{k}' \subset \overline{\mathbb{k}}$, we denote by $X(\mathbb{k}')$ the set of solutions of the corresponding system whose coordinates lie in $\mathbb{k}'$. In particular, $\overline{X} = X(\overline{\mathbb{k}})$. The Galois group $\operatorname{Gal}(\overline{\mathbb{k}}/\mathbb{k}')$ acts on $X(\overline{\mathbb{k}})$, and $X(\mathbb{k}') = X(\overline{\mathbb{k}})^{\operatorname{Gal}(\overline{\mathbb{k}}/\mathbb{k}')}$ is the set of elements of $X(\overline{\mathbb{k}})$ stable under this action; i.e.,
$$X(\mathbb{k}') = \{x \in X(\overline{\mathbb{k}}) \mid \sigma x = x \text{ for any } \sigma \in \operatorname{Gal}(\overline{\mathbb{k}}/\mathbb{k}')\}.$$

Therefore, the action of $\operatorname{Gal}(\mathbb{k})$ on $X(\overline{\mathbb{k}})$ is intrinsic for the variety X. The variety is defined over $\mathbb{k}$ if and only if $\sigma(\overline{X}) = \overline{X}$ for any $\sigma \in \operatorname{Gal}(\mathbb{k})$.

The variety X is called *absolutely irreducible* if $\overline{X}$ is irreducible. The *dimension* of X is defined as $\dim X = \dim \overline{X}$. The variety is called X smooth if $\overline{X}$ is smooth.

EXERCISE 2.5.25. Introduce the Zariski topology on X.

Functions and maps. Let X and Y be varieties over $\mathbb{k}$, and $f \colon X(\overline{\mathbb{k}}) \to Y(\overline{\mathbb{k}})$ be a map (regular or rational). The Galois group action of $\operatorname{Gal}(\mathbb{k})$ on the maps is defined as
$$(\sigma f)(x) = \sigma(f(\sigma^{-1}x)).$$

EXERCISE 2.5.26. Check that this is an action, i.e., that it respects the multiplication in the group $\operatorname{Gal}(\mathbb{k})$. Why are the definitions
$$(\sigma f)(x) = \sigma(f(x)) \quad \text{or} \quad (\sigma f)(x) = \sigma(f(\sigma x))$$
bad?

We say that a map f is defined over $\mathbb{k}$ if $\sigma f = f$ for any $\sigma \in \operatorname{Gal}(\mathbb{k})$.

EXERCISE 2.5.27. Check that a map f is defined over $\Bbbk$ if and only if it can be given by rational functions with coefficients in $\Bbbk$.

Since a rational function is a rational map $f\colon X \to \mathbb{P}^1$, we have also defined the action of $\mathrm{Gal}(\Bbbk)$ on the field of rational functions $\overline{\Bbbk}(\overline{X})$. We set $\mathbb{K} = \Bbbk(X) = \overline{\Bbbk}(\overline{X})^{\mathrm{Gal}(\Bbbk)}$ and call it the field of rational functions on the variety X. In dimension one this field uniquely determines the curve.

THEOREM 2.5.28. *There is a one-to-one correspondence between isomorphism classes of smooth absolutely irreducible curves $X/\Bbbk$ and isomorphism classes of fields $\mathbb{K}$ of transcendental degree 1 over $\Bbbk$ such that $\mathbb{K} \cap \overline{\Bbbk} = \Bbbk$.* □

This correspondence is given by mapping a curve X to its field of rational functions $\Bbbk(X)$.

Points (places). So, a variety X over $\Bbbk$ is a set $X(\overline{\Bbbk})$ of its $\overline{\Bbbk}$-points, equipped with an action of $\mathrm{Gal}(\Bbbk)$. This motivates the following definition.

A *point* P of a variety X is an orbit of the $\mathrm{Gal}(\Bbbk)$ action on $X(\overline{\Bbbk})$, i.e., a set $P = \{\sigma \overline{P}\}_{\sigma \in \mathrm{Gal}(\Bbbk)}$, where $\overline{P} \in X(\overline{\Bbbk})$ is a $\overline{\Bbbk}$-point.

The *decomposition field* of a point P is the field $\Bbbk'$ defined over $\Bbbk$ by the points $\sigma \overline{P}$, in other words, the smallest field over which the orbit is decomposed into separate points.

The *degree* $\deg P$ of a point P is defined as the order of the orbit.

EXERCISE 2.5.29. Show that the degree of a point is always finite.

This definition of a point of the curve X and that of a place of the function field are in one-to-one correspondence: $P \in X$ corresponds to a maximal ideal $\mathfrak{m}_P$ of the local ring $\mathscr{O}_{X,P}$. This is the ideal consisting of functions (elements of the field $\mathbb{K}$) vanishing at the point P, i.e., of $\mathrm{Gal}(\Bbbk)$-invariant functions on $\overline{X}$ vanishing at one element of the orbit that we call the point P (and therefore at all its elements). This correspondence permits the transfer of definitions and facts from the language of algebraic curves to that of function fields and vice versa. So, in terms of function fields we can define the notion of genus of a function field (which will coincide with the genus of the corresponding curve X, defined as the genus of the curve $\overline{X}$), of a divisor and of a principal divisor (as is done in Sec. 2.5.3), of the order of a rational function at a point, of the Jacobian (the quotient of the group of divisors of degree 0 over the group of principal divisors), etc.

Points and norm rings. Let us comment in more detail on the relation between the points of a curve X and corresponding norm rings in $\Bbbk(X)$.

Let $P \in \overline{X}$. There exists a finite normal extension $\Bbbk(P)$ of the field $\Bbbk$ such that $P \in X(\Bbbk(P))$ and $P \notin X(\Bbbk')$ for any normal extension $\Bbbk'$ of $\Bbbk$ that does not contain $\Bbbk(P)$ (the field $\Bbbk(P)$ is the normal extension generated over $\Bbbk$ by the coordinates of P). Let $s = \deg P$; one can show that $s = [\Bbbk(P):\Bbbk]$. Since the curve $\overline{X}$ can be defined by equations with coefficients in $\Bbbk \subseteq \Bbbk(P)$, for any element $\sigma \in \mathrm{Gal}(\overline{\Bbbk}/\Bbbk)$ the conjugate point $\sigma(P)$ also lies in $X(\Bbbk(P))$. Since $s = [\Bbbk(P):\Bbbk]$, the set $\{\sigma(P) \mid \sigma \in \mathrm{Gal}(\overline{\Bbbk}/\Bbbk)\}$ contains exactly s elements $P_1, \ldots, P_s$. Let $\overline{V}_i = \mathscr{O}_{P_i}$ be local rings of points P_i in the field $\mathbb{K}' = \overline{\Bbbk}(\overline{X})$.

PROPOSITION 2.5.30. *Let $V = \mathbb{K} \cap \overline{V}_1$. Then $\mathbb{K} \cap \overline{V}_i = V$ for any $i = 1, \ldots, s$. The ring V is a norm ring $\mathbb{K} = \mathbb{k}(X)$ defining a point of degree s, and*

$$V = \bigcap_{i=1}^{s} \overline{V}_i, \qquad \overline{\mathbb{k}} \cdot V = \bigcup_{i=1}^{s} \overline{V}_i.$$

The other way round, for any norm ring V such that $\mathbb{k} \subset V \subset \mathbb{K}$ defines a point of X of degree s, there is a point $P \in X(\mathbb{k}')$, where $\mathbb{k}'$ is an extension of $\mathbb{k}$ of degree s such that $V = \mathbb{K} \cap \mathscr{O}_P$.

PROOF. Note first that if $\sigma \in G = \operatorname{Gal}(\overline{\mathbb{k}}/\mathbb{k})$ is such that $\sigma(P_1) = P_r$, then $\sigma(\overline{V}_1) = \overline{V}_r$, $r = 1, \ldots, s$. It follows immediately that $\mathbb{K} \cap \overline{V}_1 = \mathbb{K} \cap \overline{V}_r$ for any $r = 1, \ldots, s$ since $\mathbb{K} = (\mathbb{K}')^G$. Then, $\overline{V}_1 \cap \mathbb{K}$ is a norm ring, since, if $f \in \mathbb{K}$ and $f \notin V = \overline{V}_1 \cap \mathbb{K}$, then $f^{-1} \in \overline{V}_1$, so $f^{-1} \in V$. It is clear that $\overline{V}_i \supseteq \overline{V}_i \cap \mathbb{K}$ for any $i = 1, \ldots, s$; therefore, $\bigcap_{i=1}^{s} \overline{V}_i \supseteq V$. On the other hand,

$$\left(\bigcap_{i=1}^{s} \overline{V}_i\right) \subset (\mathbb{K}')^G = \mathbb{K},$$

and thus

$$\bigcap_{i=1}^{s} V_i = \left(\bigcap_{i=1}^{s} \overline{V}_i\right) \cap \mathbb{K} = V.$$

The proof of the opposite implication is similar, though somewhat more difficult. □

EXERCISE 2.5.31. Prove the second part of Proposition 2.5.30.

Thus, just as we have mentioned, any place of degree s of the curve X defines a set of s points of $\overline{X}$ conjugate under the Galois action and, conversely, any set of s conjugate points of $\overline{X}$, i.e., a point of degree s of the curve X, defines a place of degree s of the field $\mathbb{K}$.

Rational divisors. A *divisor* D on a curve X ($\mathbb{k}$-*rational divisor*, or $\mathbb{k}$-*divisor*) is a formal sum $D = \sum a_P P$, where a_P are integers, and only a finite number of a_P is nonzero and P are points of the curve X (in the sense of norm rings). The *degree* of a divisor D is the number

$$\deg D = \sum a_P \deg P.$$

Since any point P of degree s of the curve X defines a set of s points $\{P_1, \ldots, P_s\}$ of the curve $\overline{X}$, the divisor D defines uniquely the divisor

$$\overline{D} = \sum a_P \left(\sum_{i=1}^{s} P_i\right)$$

of degree

$$\deg \overline{D} = \sum a_P \deg P = \deg D$$

on the curve $\overline{X}$. The converse is also clear: if $\overline{D} = \sum a_P P$ is a divisor on the curve $\overline{X}$ such that $a_P = a_{\sigma(P)}$ for any element $\sigma = \operatorname{Gal}(\overline{\mathbb{k}}/\mathbb{k})$, then there is a unique $\mathbb{k}$-divisor D on X such that $\overline{D}$ is obtained from it in the described way.

Thus, we have two points of view on the notion of a $\Bbbk$-divisor: it is either a divisor D on the curve X or a divisor $\overline{D}$ on the curve $\overline{X}$ invariant under the action of the Galois group $\mathrm{Gal}(\Bbbk)$. In what follows we shall identify the divisors D and $\overline{D}$, passing freely from one point of view to the other.

PROPOSITION 2.5.32. *Let $\overline{D} = \sum a_P P$ be a $\mathrm{Gal}(\Bbbk)$-invariant divisor on the curve $\overline{X}$ (i.e., $\overline{D}$ corresponds to a $\Bbbk$-divisor D on X). Then the space $L(\overline{D})$ has a basis consisting of functions in $L(\overline{D}) \cap \Bbbk(X)$.*

PROOF. Let $f \in L(\overline{D})$. It is clear that $f \in \Bbbk'(X)$ for some finite extension $\Bbbk'$ of the field $\Bbbk$, which can be supposed to be normal over $\Bbbk$. Let $\mathrm{Gal}(\Bbbk'/\Bbbk) = \{\sigma_1, \ldots \sigma_n\}$, and let $\omega_1, \ldots, \omega_n$ be a basis of the extension $\Bbbk'/\Bbbk$. Set

$$g_j = \sum_{i=1}^n \sigma_i(\omega_j f) = \sum_{i=1}^n \sigma_i(\omega_j)\sigma_i(f), \quad j = 1, \ldots, n.$$

By the Galois theory $\det(\sigma_i(\omega_j)) \neq 0$. Hence, if $f_1, \ldots, f_m$ generate $L(\overline{D})$ over $\overline{\Bbbk}$, then the corresponding functions $g_{11}, \ldots, g_{1n}, \ldots, g_{m1}, \ldots, g_{mn}$ also generate $L(\overline{D})$ over $\overline{\Bbbk}$; obviously, $g_{ij} \in \Bbbk(X)$. □

Define the space $L(D)$ for a $\Bbbk$-divisor D on the curve X as $L(\overline{D}) \cap \Bbbk(X)$. From Proposition 2.5.32 it follows that

$$L(D) = \{f \in \Bbbk(X)^* \mid (f) + \overline{D} \geq 0\} \cup \{0\}.$$

By $\ell(D)$ we denote the dimension $\dim_\Bbbk L(D)$.

COROLLARY 2.5.33. *Let D be a $\Bbbk$-divisor on the curve X. Then $L(\overline{D}) = \overline{\Bbbk} \otimes L(D)$. In particular, $\ell(\overline{D}) = \ell(D)$.* □

EXERCISE 2.5.34. Show that if divisors $\overline{D}$ and $\overline{D}'$ on $\overline{X}$ corresponding to the divisors D and D' on X are linearly equivalent, then there is a function $f \in \Bbbk(X)$ such that $\overline{D} = \overline{D}' + (f)$.

Line bundles. An *asymptotically standard family of lattices* $\mathscr{L} = (\mathscr{L}_P)_{P \in X}$ on a curve X is the set $(\mathscr{L}_P)$, $P \in X$, of one-dimensional free $\mathscr{O}_P$-submodules $\mathscr{L}_P \subset \Bbbk(X)$ such that $\mathscr{L}_P = \mathscr{O}_P$ for almost any $P \in X$ (except for a finite number). The fibre $\overline{\mathscr{L}}_P$ of the set $\mathscr{L}$ at $P \in X$ is the one-dimensional space $\overline{\mathscr{L}}_P = \mathscr{L}_P / \mathfrak{m}_P \mathscr{L}_P$ over $\Bbbk(P) = \mathscr{O}_P / \mathfrak{m}_P$. As before, by $H^0(\mathscr{L})$ we denote the intersection $\bigcap_{P \in X} \mathscr{L}_P \subset \Bbbk(X)$. The set $\mathscr{L} = (\mathscr{L}_P)_{P \in X}$ on X defines a set $\overline{\mathscr{L}} = ((\overline{\mathscr{L}})_Q)_{Q \in \overline{X}}$ on $\overline{X}$, where $(\overline{\mathscr{L}})_Q = \overline{\Bbbk} \cdot \mathscr{L}_P \subset \overline{\Bbbk}(\overline{X})$ for any point Q on $\overline{X}$ corresponding to the point $P \in X$.

EXERCISE 2.5.35. Show that the correspondence between asymptotically standard families of lattices and divisors is preserved on the curve X and is also bijective. Define the notion of a line bundle on X and show that line bundles on X can be identified with line bundles on $\overline{X}$ invariant under the Galois action.

EXERCISE 2.5.36. Deduce from Corollary 2.5.33 that if $\mathscr{L}$ is a line bundle on X and $\overline{\mathscr{L}}$ is the corresponding bundle on $\overline{X}$, then $h^0(\mathscr{L}) = h^0(\overline{\mathscr{L}})$.

These results imply that for divisors and line bundles on X the Riemann–Roch theorem and its corollaries are true. In particular, the genus of the curve X, coinciding with the genus of $\overline{X}$, is well defined.

Historical and Bibliographic Notes

The material of this chapter is quite classical. We are far from the idea of exploring in detail the rich history of algebraic geometry (this being absolutely impossible on the few pages we possess). For classical results we just outline the development of mathematics that led to them. For a reader interested in the history of algebraic geometry we recommend the book by J. Dieudonné [Die74] and the historical appendix to the book by I.R. Shafarevich [Sha72]. The history of elliptic functions is described in chapter VII of [Die78] written by C. Houzél. Much interesting information on the history of algebraic geometry can be found in A. Weil's book on number theory [Wei84].

The study of algebraic curves goes back at least to Diophantos, who, speaking in modern terms, studied in his *Arithmetica* the structure of $\mathbb{Q}$-rational points of various curves. All curves considered by Diophantos were of genus 0 or 1; one of his methods was to find a rational parametrization of curves of genus 0, i.e., to find an explicit isomorphism of this curve with $\mathbb{P}^1$. Note that all reasoning of Diophantos was in terms of equations and their solutions. Only after the discovery of coordinates, i.e., after R. Descartes, did I. Newton realize that this is in fact geometry. After a long break P. Fermat returned to the subject. While studying problems set by Diophantos and those similar to them, he managed to get a number of properties of elliptic curves, always in terms of equations. Thus, interest in algebraic curves was first generated by solving equations in rational numbers.

Further development continued in the frame of analysis and concerned the study of elliptic integrals, the name being due to the fact that the length of an arc of an ellipse is given by such integral. A number of curious properties of these integrals was discovered at the end of the 17th century by brothers Bernoulli. Then G.C. Di Fagnano continued these observations and constructed many examples of transforms of elliptic integrals in the first half of the 18th century. Being informed of the results of Di Fagnano, L. Euler started an intense study of the transform problem for elliptic integrals. He discovered the addition law for them, which was the main result of that period of development. As we understand it today, it is nothing but the group law on an elliptic curve. The theory of elliptic integrals was further developed by A.M. Legendre. At the same time when Legendre's book finalizing his research was published, the work of N.H. Abel and C.G.J. Jacobi was starting a new period of development of the theory, characterized by the study of functions inverse to elliptic integrals, i.e., elliptic functions. Abel and Jacobi obtained a series of fundamental results concerning these functions—first of all the addition theorem. Abel did not restrict himself to elliptic integrals; he turned to the study of arbitrary integrals of the form $\int f(x,y)\,dx$, where f is a rational function, x and y being connected by a polynomial equation $\chi(x,y) = 0$. For such integrals, which we now call abelian, Abel proved the addition theorem, implicitly including the notion of the genus of an algebraic curve and leading to the theory of Jacobians of algebraic curves.

The next important step was made by B. Riemann in the middle of the 19th century, still in the frame of complex analysis. Riemann discovered a totally new approach to the study of functions in one complex variable that led him to the notion of a Riemann surface. He got a part of the Riemann–Roch theorem (the Riemann inequality $\ell(D) \geq \deg D - g + 1$), still in analytic terms; later on, the theorem was completely proved by his student G. Roch.

The first algebraic geometry approach to Abel's and Riemann's results is due to R. Clebsch. It flourished in the work of his student M. Noether. The study of Clebsch and Noether exposed the algebraic geometry nature of Abel's and Riemann's results and created a rich theory of algebraic curves. The algebraic approach to algebraic curves over an algebraically closed field of zero characteristic is to consider the field of rational functions on the curve as the main object of study; it is due to R. Dedekind and H. Weber.

Curves over finite fields became a subject of interest in the 1920s in the work of E. Artin, F.K. Schmidt, and H. Hasse; for them the main motive lay in the analogy between these curves and number fields. Following this course, Hasse and his students created the theory of curves over arbitrary fields (the theory of function fields, speaking the algebraic language). The theory of algebraic varieties of higher dimension over an arbitrary field can be credited to B.L. van der Waerden and A. Weil. Weil managed to use the technique he developed (the use of multidimensional objects) to prove the Riemann hypothesis for algebraic curves. Further development of algebraic geometry in the work of J-P. Serre led to the definition of a variety based on the notion of a vector bundle. A. Grothendieck rebuilt the whole edifice of algebraic geometry on the foundation of his scheme theory.

Here are some basic sources for the topics we have seen in this chapter.

The theory of algebraic curves as seen by an algebraic geometer is exposed in the book of W. Fulton [Ful69]. The analytic aspect (the theory of Riemann surfaces) is contained in the book of J. Springer [Spr57]. A purely algebraic approach is taken in the books of C. Chévalley [Che51] and N.G. Chebotarev [Che48]. A perfect modern introduction to the function field theory is given in the book of H. Stichtenoth [Sti93], which we heavily used to write Sec. 2.5. The book of J.-P. Serre [Ser59] contains an easy proof of the Riemann–Roch theorem, the theory of singular curves and of Jacobians. The book of I.R. Shafarevich [Sha72], devoted not only to curves, contains many results on curves; in this chapter we often followed it. The theory of elliptic functions is the subject of a book of S. Lang [Lan73]. The standard textbook on schemes, containing in particular many results on algebraic curves, is that of R. Hartshorne [Har77]. A perfect textbook of algebraic geometry over the field of complex numbers is written by P.A. Griffiths and J. Harris [GH78]. It uses both algebraic geometry and complex analysis methods. The largest part of curve theory that it examines needs only algebraic geometry means. These books contain almost everything we were speaking about. Here are a few exceptions.

For Remark 2.2.14 see P. Roquette [Roq70]. Proposition 2.2.49 and Corollary 2.2.50 are due to B. Schoeneberg [Sch51]. The Cartier operator theory is exposed in the work of P. Cartier [Car57]; see also Appendix 2 in [Lan73].

CHAPTER 3

Curves over Finite Fields

In this chapter we discuss the theory of curves defined over a finite field. This theory plays the key role in applications of algebraic geometry to coding theory. One of the main points is the study of places of degree one ($\mathbb{F}_q$-points). We plan to discuss further theory in *Advanced Chapters*.

Finiteness of the ground field gives us auxiliary numeric invariants of the curve, namely, numbers of points of a given degree $m = 1, 2, \ldots$. These numbers are gathered into a generating function called the zeta function of the curve. Its properties are studied in Sec. 3.1. In Sec. 3.2 we study the asymptotic behaviour of the maximal possible number of points on a curve and on its Jacobian when the genus of the curve tends to infinity. In Sec. 3.3 we study in detail elliptic curves over a finite field; in Sec. 3.4 we give some examples of particularly interesting families of curves. Section 3.5 is devoted to links between the theory of curves and that of exponential sums.

Note that a finite field is always perfect (see Exercise 2.5.22).

3.1. Zeta Function

All mathematicians are familiar with the *Riemann zeta function*

$$\zeta_\mathbb{Q}(s) = \sum_{n=1}^\infty n^{-s} = \prod_p (1-p^{-s})^{-1},$$

a meromorphic complex-variable function which has an analytic continuation on the whole complex plane $\mathbb{C}$ with a unique pole at $s = 1$ and which satisfies the *functional equation*

$$\pi^{-s/2} \Gamma\left(\frac{s}{2}\right) \zeta(s) = \pi^{-\frac{1-s}{2}} \Gamma\left(\frac{1-s}{2}\right) \zeta(1-s).$$

The *Riemann hypothesis* claims that all nontrivial zeroes of $\zeta(s)$ lie on the critical line $\operatorname{Re} s = 1/2$.

The Riemann zeta function has natural generalizations, both to algebraic number fields (the Dedekind zeta function) and to algebraic curves over a finite field. It turns out that the zeta function of an algebraic curve over $\mathbb{F}_q$ has a natural representation as a function of the variable $t = q^{-s}$:

$$\zeta_X(s) = Z_X(q^{-s}).$$

In Sec. 3.1.1 we define the zeta function of a curve over a finite field and prove its rationality. In Sec. 3.1.2 we establish a functional equation for it. In Sec. 3.1.3 we state the Weil theorem (an analogue of the Riemann hypothesis for this zeta function) and obtain important consequences of this result. Then (Sec. 3.1.4) we prove the so-called explicit formula. Finally, Sec. 3.1.5 is devoted to another generating function, the Pellikaan zeta function, which is closely related to algebraic geometry codes from Chapter 4.

3.1.1. Definition and Rationality

Now let $\Bbbk = \mathbb{F}_q$ be a finite field of characteristic p with $q = p^a$ elements. For any r, the set of points of degree 1 in the projective space $\mathbb{P}^m(\mathbb{F}_{q^r})$ (where $\mathbb{F}_{q^r}$ is the unique extension of $\mathbb{F}_q$ of degree r)—and therefore the number of points of $X(\mathbb{F}_{q^r})$—is finite.

EXERCISE 3.1.1. Calculate $|\mathbb{A}^m(\mathbb{F}_q)|$ and $|\mathbb{P}^m(\mathbb{F}_q)|$.

Define $N_r = N_r(X) = |X(\mathbb{F}_{q^r})|$; in particular, $N = N_1$ is the number of $\mathbb{F}_q$-points of X. To study the numbers N_r, it is convenient to consider the generating function, a formal power series with rational coefficients

$$Z(t) = Z(X;t) = \exp\left(\sum_{r=1}^{\infty} \frac{N_r t^r}{r}\right),$$

called the *zeta function of the curve X over $\mathbb{F}_q$*.

In addition to N_r, it is sometimes important to consider B_r, the number of points of degree r of X. Since each point of X of degree r generates precisely r points of $X(\mathbb{F}_{q^r})$, we get the formula

$$N_r = \sum_{d|r} dB_d.$$

EXERCISE 3.1.2. Prove this formula. Derive the inverse formula for B_r in terms of N_r. (*Answer*: Formula (3.2.2), p. 148.)

Let us also introduce M_n, the number of effective divisors of degree n on X. Note that $M_0 = 1$ since the zero divisor is effective.

PROPOSITION 3.1.3. *For the zeta function of a curve X, we have the identities*

$$Z(t) = \prod_{d=1}^{\infty}(1-t^d)^{-B_d} = \sum_{n=0}^{\infty} M_n t^n.$$

PROOF. On the one hand,

$$\ln \prod_{d=1}^{\infty}(1-t^d)^{-B_d} = -\sum_{d=1}^{\infty} B_d \ln(1-t^d)$$

$$= \sum_{d=1}^{\infty} B_d \left(t^d + \frac{t^{2d}}{2} + \frac{t^{3d}}{3} + \ldots\right) = \sum_{r=1}^{\infty} t^r \left(\sum_{d|r} B_d \frac{d}{r}\right),$$

which proves the first equality. On the other hand,

$$\prod_{d=1}^{\infty}(1-t^d)^{-B_d} = \prod_{P \in X(\mathbb{F}_q)}(1-t^{\deg P})^{-1} = \prod_{P \in X(\mathbb{F}_q)} \sum_{n=0}^{\infty} t^{\deg(nP)}$$

$$= \sum_{D \in \mathrm{Div}^+(X(\mathbb{F}_q))} t^{\deg D} = \sum_{n=0}^{\infty} M_n t^n,$$

as required. □

EXERCISE 3.1.4. Express M_n in terms of B_m.

EXERCISE 3.1.5. Why is $\zeta_X(s) = Z(q^{-s})$ an analogue of the Riemann zeta function? (*Hint*: We return to this question later in Sec. 3.1.3.)

The definition of the zeta function immediately yields the following relation.

PROPOSITION 3.1.6. *We have*
$$N_r = \frac{1}{(r-1)!} \frac{d^r}{dt^r} (\log Z(t)) \bigg|_{t=0}.$$
□

All the above statements should be considered as statements about formal power series; we did not touch on the question of convergence. However, let us now demonstrate that $Z(t)$ converges for $|t| < 1/q$. To this end, we need some auxiliary arguments.

First, we introduce an important parameter h, called the *class number* of the function field $\mathbb{F}_q(X)$, which equals the order of the group $\mathrm{Pic}^0_{\mathbb{F}_q}(X)$ of classes of $\mathbb{F}_q$-divisors of degree zero on X or, equivalently, the number of $\mathbb{F}_q$-points of the Jacobian J_X. Throughout this chapter, we write $\mathrm{Pic}^0(X)$ instead of $\mathrm{Pic}^0_{\mathbb{F}_q}(X)$, but we always mean $\mathrm{Pic}^0_{\mathbb{F}_q}(X)$.

PROPOSITION 3.1.7. *The class number is finite.*

PROOF. Choose an arbitrary $\mathbb{F}_q$-divisor B of degree $n \geq g(X)$ and consider the set $\mathrm{Pic}^n(X) = \mathrm{Pic}^n_{\mathbb{F}_q}(X)$ of divisor classes of degree n. Since the map $A \mapsto A + B$, where $A \in \mathrm{Pic}^0(X)$, obviously is a bijection from $\mathrm{Pic}^0(X)$ to $\mathrm{Pic}^n(X)$, it suffices to show that $\mathrm{Pic}^n(X)$ is finite. For any $D \in \mathrm{Pic}^n(X)$ by the Riemann–Roch theorem we have
$$\ell(D) \geq n - g + 1 \geq 1,$$
which implies that such a class D necessarily contains an effective divisor, but the number M_n of effective divisors of degree n is finite. □

Now define the integer $\partial > 0$ by
$$\partial = \min\{\deg D \mid D \in \mathrm{Div}_{\mathbb{F}_q}(X),\ \deg D > 0\},$$
where $\mathrm{Div}_{\mathbb{F}_q}(X)$ is the group of $\mathbb{F}_q$-rational divisors on X. Throughout this chapter, we shall write $\mathrm{Div}(X)$ and $\mathrm{Div}^+(X)$ instead of $\mathrm{Div}_{\mathbb{F}_q}(X)$ and $\mathrm{Div}^+_{\mathbb{F}_q}(X)$.

Actually, as we shall soon see, ∂ is always equal to 1, but for technical reasons it is more convenient to use this parameter for a while. Obviously, the degree of any divisor on X is a multiple of ∂.

LEMMA 3.1.8. (a) *If $\partial \nmid n$, then $M_n = 0$.*
(b) *For any divisor class $D \in \mathrm{Pic}(X)$, we have*
$$|D \cap \mathrm{Div}^+(X)| = \frac{1}{q-1}\left(q^{\ell(D)} - 1\right).$$
(c) *For any integer $n > 2g - 2$ such that $\partial \mid n$, we have*
$$M_n = \frac{h}{q-1}\left(q^{n-g+1} - 1\right).$$

PROOF. (a) is trivial. Let us prove (b). The set $D \cap \mathrm{Div}^+(X)$ consists of divisors of the form $D_0 + (f)$, where D_0 is some fixed representative of the class and $f \in L(D) \setminus \{0\}$. Here, two elements of $L(D) \setminus \{0\}$ define the same principal divisor if and only if they differ by a constant factor $\alpha \in \mathbb{F}_q^*$. Assertion (c) follows from (b), Exercise 2.2.23c, and the definition of h. □

Rationality. Now it is time to study the convergence of the zeta function.

PROPOSITION 3.1.9. *The power series* $Z(t) = \sum_{n=0}^{\infty} M_n t^n$ *converges for* $|t| < q^{-1}$. *Moreover, for* $|t| < q^{-1}$ *we have the following inequalities:*

(a) *If the curve has genus* $g = 0$, *then*

$$Z(t) = \frac{1}{q-1}\left(\frac{q}{1-(qt)^\partial} - \frac{1}{1-t^\partial}\right).$$

(b) *If* $g \geq 1$, *then* $Z(t) = F(t) + G(t)$ *with*

$$F(t) = \frac{1}{q-1} \sum_{\substack{D \in \mathrm{Pic}(X) \\ 0 \leq \deg D \leq 2g-2}} q^{\ell(D)} t^{\deg D},$$

$$G(t) = \frac{h}{q-1}\left(q^{1-g}(qt)^{2g-2+\partial}\frac{1}{1-(qt)^\partial} - \frac{1}{1-t^\partial}\right).$$

PROOF. (a) For $g = 0$, the class number h is 1 (see Example 2.1.83a), and $2g - 2 < 0$. Therefore, Lemma 3.1.8 yields

$$\sum_{n=0}^{\infty} M_n t^n = \sum_{n=0}^{\infty} M_{\partial n} t^{\partial n} = \sum_{n=0}^{\infty} \frac{1}{q-1}\left(q^{\partial n+1} - 1\right) t^{\partial n}$$

$$= \frac{1}{q-1}\left(q\sum_{n=0}^{\infty}(qt)^{\partial n} - \sum_{n=0}^{\infty} t^{\partial n}\right)$$

$$= \frac{1}{q-1}\left(\frac{q}{1-(qt)^\partial} - \frac{1}{1-t^\partial}\right)$$

for $|qt| < 1$.

(b) For $g \geq 1$, we again apply Lemma 3.1.8 and similarly obtain

$$\sum_{n=0}^{\infty} M_n t^n = \sum_{\substack{D \in \mathrm{Pic}(X) \\ \deg D \geq 0}} |D \cap \mathrm{Div}^+(X)| t^{\deg D}$$

$$= \sum_{\deg D \geq 0} \frac{1}{q-1}\left(q^{\ell(D)} - 1\right) t^{\deg D}$$

$$= \frac{1}{q-1} \sum_{0 \leq \deg D \leq 2g-2} q^{\ell(D)} t^{\deg D}$$

$$+ \frac{1}{q-1} \sum_{\deg D > 2g-2} q^{\deg D + g - 1} t^{\deg D} - \frac{1}{q-1} \sum_{\deg D \geq 0} t^{\deg D}$$

$$= F(t) + G(t),$$

where $F(t)$ is as required, and

$$(q-1)G(t) = \sum_{n=\lfloor \frac{2g-2}{\partial} \rfloor + 1}^{\infty} hq^{n\partial - g + 1} t^{n\partial} - \sum_{n=0}^{\infty} h t^{n\partial}$$
$$= hq^{1-g}(qt)^{2g-2+\partial} \frac{1}{1-(qt)^{\partial}} - h \frac{1}{1-t^{\partial}}. \qquad \square$$

COROLLARY 3.1.10. *The function $Z(t)$ can be extended to a rational function on $\mathbb{C}$. It has a simple pole at $t = 1$.* $\qquad \square$

For further analysis, it is helpful to consider the functions $Z_r(t) = Z(X_r, t)$, i.e., zeta functions of curves $X_r = X \otimes_{\mathbb{F}_q} \mathbb{F}_{q^r}$ over $\mathbb{F}_{q^r}$. Here, $X \otimes_{\mathbb{F}_q} \mathbb{F}_{q^r}$ denotes the curve X considered over the ground field $\mathbb{F}_{q^r}$ instead of $\mathbb{F}_q$.

PROPOSITION 3.1.11. *We have*

$$Z_r(t^r) = \prod_{\zeta \in \mu_r} Z(\zeta t),$$

where $\mu_r = \{\zeta \in \mathbb{C} \mid \zeta^r = 1\}$ is the group of the r-th roots of unity.

PROOF. By definition, $Z(t) = \exp\left(\sum_j N_j t^j / j\right)$; therefore,

$$Z_r(t) = \exp\left(\sum_{j=1}^{\infty} N_{rj} \frac{t^j}{j}\right).$$

Hence, due to the property of roots of unity

$$\sum_{\zeta \in \mu_r} \zeta^m = \begin{cases} r & \text{if } r \mid m, \\ 0 & \text{otherwise,} \end{cases} \qquad (3.1.1)$$

we obtain

$$Z_r(t^r) = \exp\left(\sum_{j=1}^{\infty} N_{rj} \frac{t^{rj}}{rj} r\right) = \exp\left(\sum_{m=1}^{\infty} N_m \frac{\sum_{\zeta \in \mu_r} (\zeta t)^m}{m}\right) = \exp\left(\sum_{\zeta \in \mu_r} Z(\zeta r)\right),$$

as required. $\qquad \square$

COROLLARY 3.1.12 (Schmidt theorem). *We have $\partial = 1$.*

PROOF. Choose $\zeta \in \mu_\partial$. It follows immediately from Proposition 3.1.3 that $Z(\zeta t) = Z(t)$ since ∂ divides the degree of any point and any divisor on X. Then, by the proposition just proved, we have $Z_\partial(t^\partial) = (Z(t))^\partial$. Since the rational function $Z_\partial(t^\partial)$ has a simple pole at $t = 1$ by Corollary 3.1.10 and $(Z(t))^\partial$ has a pole of order ∂, we necessarily have $\partial = 1$. $\qquad \square$

COROLLARY 3.1.13. (a) *For a curve of genus 0 (i.e., for the projective line), we have*

$$Z(t) = \frac{1}{(1-t)(1-qt)}.$$

(b) *For a curve of genus $g \geq 1$, the zeta function is of the form $Z(t) = F(t) + G(t)$* with

$$F(t) = \frac{1}{q-1} \sum_{\substack{D \in \operatorname{Pic}(X) \\ 0 \leq \deg D \leq 2g-2}} q^{\ell(D)} t^{\deg D}$$

and

$$G(t) = \frac{h}{q-1}\left(q^g t^{2g-1} \frac{1}{1-qt} - \frac{1}{1-t} \right). \quad \square$$

EXERCISE 3.1.14. Derive from the explicit form of $G(t)$ the following formula for the number of effective divisors of degree $n \geq 2g - 1$:

$$M_n = \frac{h}{q-1}(q^{n-g+1} - 1).$$

3.1.2. Functional Equation

One of the key properties of the zeta function of a curve is its functional equation.

THEOREM 3.1.15. *The zeta function satisfies the functional equation*

$$Z(t) = q^{g-1} t^{2g-2} Z\left(\frac{1}{qt}\right).$$

PROOF. For $g = 0$ this follows immediately from Corollary 3.1.13a. For $g \geq 1$, by Corollary 3.1.13b, we have $Z(t) = F(t) + G(t)$, the functional equation for $G(t)$ being easily verified by direct substitution (EXERCISE!). To verify the functional equation for $F(t)$, apply the Riemann–Roch theorem:

$$\begin{aligned}
(q-1)F(t) &= \sum_{0 \leq \deg D \leq 2g-2} q^{\ell(D)} t^{\deg D} \\
&= \sum_{0 \leq \deg D \leq 2g-2} q^{\deg D - g + 1 + \ell(K-D)} t^{\deg D} \\
&= q^{g-1} t^{2g-2} \sum_{0 \leq \deg D \leq 2g-2} q^{\deg D - (2g-2) + \ell(K-D)} t^{\deg D - (2g-2)} \\
&= q^{g-1} t^{2g-2} \sum_{0 \leq \deg D \leq 2g-2} q^{\ell(K-D)} \left(\frac{1}{qt}\right)^{\deg(K-D)} \\
&= q^{g-1} t^{2g-2} (q-1) F\left(\frac{1}{qt}\right).
\end{aligned}$$

Here we used the fact that $\deg K = 2g - 2$ and that the condition $0 \leq \deg D \leq 2g - 2$ is equivalent to $0 \leq \deg(K - D) \leq 2g - 2$. $\square$

Corollary 3.1.13 in particular shows that the zeta function is of the form

$$Z(t) = \frac{P(t)}{(1-t)(1-qt)}, \tag{3.1.2}$$

$P(t)$ being a polynomial of degree at most $2g$ with integer coefficients. Obviously, it contains all the information contained in the zeta function and should also be closely investigated.

PROPOSITION 3.1.16. *The polynomial $P(t)$ has the following properties:*
(a) $P(0) = 1$, and $P(1) = h$ is the class number.
(b) (*functional equation*)
$$P(t) = q^g t^{2g} P\left(\frac{1}{qt}\right).$$
(c) Let $P(t) = a_0 + a_1 t + \ldots + a_{2g} t^{2g}$. Then
$$a_{2g-i} = q^g a_i, \quad 0 \le i \le g;$$
in particular, $a_{2g} = q^g$, whence $\deg P(t) = 2g$.
(d) $a_1 = N - (q+1)$, where $N = N_1$ is the number of $\mathbb{F}_q$-points of the curve. □

EXERCISE 3.1.17. Prove the proposition. (*Hint*: To prove (a), apply Corollary 3.1.13b; for (d), use the identity $P(t) = (1-t)(1-qt) \sum_{n=0}^{\infty} M_n t^n$.)

EXERCISE 3.1.18. (a) Consider the curve over $\mathbb{F}_2$ with equation
$$y^2 + xy = x^3 + x.$$
Check that this is a curve of genus $g = 1$ with the zeta function
$$Z(t) = \frac{2t^2 + t + 1}{(1-t)(1-2t)};$$
in particular, $B_1 = 4$, $B_2 = 2$, $B_3 = 0$, and $B_4 = 2$.
(b) Consider the curve over $\mathbb{F}_2$ with equation
$$y^2 + y = x^3.$$
Check that this is a curve of genus $g = 1$ with the zeta function
$$Z(t) = \frac{2t^2 + 1}{(1-t)(1-2t)};$$
in particular, $B_1 = 3$, $B_2 = 3$, and $B_3 = 2$.
(c) Consider the curve over $\mathbb{F}_3$ with equation
$$y^2 = 2x^4 + x^2 + 1.$$
Check that this is a curve of genus $g = 1$ with the zeta function
$$Z(t) = \frac{3t^2 + 2t + 1}{(1-t)(1-3t)};$$
in particular, $B_1 = 6$, $B_2 = 3$, and $B_3 = 4$.
(d) Consider the curve over $\mathbb{F}_3$ with equation
$$y^2 = x^3 + 2x^2 + 1.$$
Check that this is a curve of genus $g = 1$ with the zeta function
$$Z(t) = \frac{3t^2 + t + 1}{(1-t)(1-3t)};$$
in particular, $B_1 = 5$ and $B_2 = 5$.

We shall use the results of this exercise in Sec. 4.4.5.

PROPOSITION 3.1.19. (a) *In the expansion*

$$P(t) = \prod_{i=1}^{2g}(1-\omega_i t),$$

the complex coefficients ω_i are algebraic integers, and they can be arranged in such a way that $\omega_i \omega_{g+i} = q$ holds for $i = 1, \ldots, g$.

(b) *The polynomial $P_r(t)$ corresponding to the zeta function $Z_r(t)$ of the curve $X(\mathbb{F}_{q^r})$ can be represented as*

$$P_r(t) = \prod_{i=1}^{2g}(1-\omega_i^r t)$$

with the same ω_i.

PROOF. (a) The numbers ω_i are roots of the reciprocal polynomial

$$P^\perp(t) = a_0 t^{2g} + a_1 t^{2g-1} + \ldots + a_{2g} = t^{2g} P\left(\frac{1}{t}\right).$$

Since $a_0 = P(0) = 1$, they are algebraic integers. From the functional equation (Proposition 3.1.16b) it follows that if t_0 is a root of $P(t)$, then $1/(qt_0)$ is a root too. Since the roots of $P(t)$ are ω_i^{-1}, we conclude that ω_i can be arranged as

$$\omega_1, \frac{q}{\omega_1}, \ldots, \omega_k, \frac{q}{\omega_k}, \sqrt{q}, \ldots, \sqrt{q}, -\sqrt{q}, \ldots, -\sqrt{q},$$

where $\sqrt{q}$ occurs m times, and $-\sqrt{q}$ occurs n times. Since the product of all roots of the even-degree polynomial $P^\perp(t)$ equals its free term $a_{2g} = q^g$, this product is positive; i.e., the number n of negative roots is even. Since $2k + m + n = 2g$, we see that m is also even.

(b) Apply Proposition 3.1.11:

$$P_r(t^r) = (1-t^r)(1-q^r t^r)Z_r(t^r) = (1-t^r)(1-q^r t^r)\prod_{\zeta \in \mu_r} Z(\zeta t)$$

$$= (1-t^r)(1-q^r t^r)\prod_{\zeta \in \mu_r}\frac{P(\zeta t)}{(1-\zeta t)(1-q\zeta t)} = \prod_{\zeta \in \mu_r} P(\zeta t)$$

$$= \prod_{i=1}^{2g}\prod_{\zeta \in \mu_r}(1-\omega_i \zeta t) = \prod_{i=1}^{2g}(1-\omega_i^r t^r).$$

Here, we have used (3.1.1) several times. Hence, $P_r(t) = \prod_{i=1}^{2g}(1-\omega_i^r t)$. □

COROLLARY 3.1.20. *For any $r \geq 1$,*

$$N_r = q^r + 1 - \sum_{i=1}^{2g}\omega_i^r. \qquad (3.1.3)$$

In particular,

$$|X(\mathbb{F}_q)| = q + 1 - \sum_{i=1}^{2g}\omega_i. \qquad (3.1.4)$$

PROOF. Apply Proposition 3.1.16d to $P_r(t)$. □

The numbers ω_i are often called the *(inverse) Frobenius roots*, or *Frobenius eigenvalues*; we shall not discuss the reason here.

3.1.3. Weil Theorem and Its Corollaries

Now we come to the following fundamental result.

THEOREM 3.1.21 (Weil theorem). *We have*
$$|\omega_i| = \sqrt{q}, \quad i = 1, \ldots, 2g.$$
□

REMARK 3.1.22. This theorem is often referred to as the *Riemann hypothesis for algebraic function fields*. The explanation is as follows. Define the *absolute norm* of a divisor $D \in \mathrm{Div}(X)$ by $\mathcal{N}(D) = q^{\deg D}$. Consider the function
$$\zeta_X(s) = Z(X, q^{-s}).$$
It can be written as
$$\zeta_X(s) = \sum_{n=0}^{\infty} M_n q^{-sn} = \sum_{D \in \mathrm{Div}^+(X)} \mathcal{N}(D)^{-s};$$
therefore, $\zeta_X(s)$ is the analogue to the classical Riemann zeta function $\zeta(s) = \zeta_{\mathbb{Q}}(s) = \sum_{n=1}^{\infty} n^{-s}$ and the Dedekind zeta function $\zeta_{\mathbb{K}}(s)$ of a number field $\mathbb{K}$. The analogue of the classical Riemann hypothesis (which states that all nontrivial zeroes of $\zeta(s)$ lie on the line $\mathrm{Re}(s) = 1/2$) is precisely the Weil theorem, which states that
$$\zeta_X(s) = 0 \iff Z(q^{-s}) = 0 \implies |q^{-s}| = q^{-1/2} \iff \mathrm{Re}(s) = 1/2.$$

From the Weil theorem one can derive the following important conclusions.

PROPOSITION 3.1.23. *For the class number h (the number of $\mathbb{F}_q$-points of the Jacobian), we have the bound*
$$(\sqrt{q} - 1)^{2g} \le h \le (\sqrt{q} + 1)^{2g}.$$

PROOF. Indeed (see Proposition 3.1.16a),
$$h = P(1) = \prod_{i=1}^{2g} (\omega_i - 1), \tag{3.1.5}$$
and the proposition follows immediately. □

Formula (3.1.3) yields the following bound:
$$|N_r - (q^r + 1)| \le 2q^{r/2} g. \tag{3.1.6}$$

An important particular case is as follows:

THEOREM 3.1.24 (Weil bound). *For the number of points $N = N_1 = |X(\mathbb{F}_q)|$, we have*
$$|N - (q+1)| \le 2g\sqrt{q}.$$
□

In fact, this bound can be improved a little (the improvement is especially appreciable for large g).

THEOREM 3.1.25 (Serre bound). *For the number of points N, we have*
$$|N - (q+1)| \leq g\lfloor 2\sqrt{q}\rfloor.$$

PROOF. We have to show that $\left|\sum_{i=1}^{2g} \omega_i\right| \leq g\lfloor 2\sqrt{q}\rfloor$. As in Proposition 3.1.19a, arrange ω_i so that
$$\omega_{g+i} = \frac{q}{\omega_i} = \overline{\omega}_i, \quad 1 \leq i \leq g.$$

Set
$$\gamma_i = 2\operatorname{Re}(\omega_i) + \lfloor 2\sqrt{q}\rfloor + 1,$$
$$\delta_i = -2\operatorname{Re}(\omega_i) + \lfloor 2\sqrt{q}\rfloor + 1.$$

Since ω_i are algebraic integers, and $\omega_i + \omega_{g+i} = 2\operatorname{Re}(\omega_i)$, we conclude that γ_i and δ_i are real algebraic integers; moreover, they are positive.

Any field embedding $\mathbb{Q}(\omega_1, \ldots, \omega_{2g}) \hookrightarrow \mathbb{C}$ acts as a permutation σ on the set $\{\omega_i\}$. Moreover, if $\sigma(\omega_i) = \omega_j$, then
$$\sigma(\overline{\omega}_i) = \sigma(q/\omega_i) = q/\sigma(\omega_i) = \overline{\omega}_j;$$
i.e., σ acts as a permutation on the sets $\{\gamma_i\}$ and $\{\delta_i\}$. Thus, the algebraic integers $\gamma = \prod_{i=1}^g \gamma_i$ and $\delta = \prod_{i=1}^g \delta_i$ are invariant under all embeddings of $\mathbb{Q}(\omega_1, \ldots, \omega_{2g})$ in $\mathbb{C}$; hence, $\gamma, \delta \in \mathbb{Z}$. Therefore, $\prod_{i=1}^g \gamma_i \geq 1$.

Applying the inequality of arithmetic and geometric means, we obtain
$$\frac{1}{g}\sum_{i=1}^g \gamma_i \geq \left(\prod_{i=1}^{2g} \gamma_i\right)^{1/g} \geq 1.$$

Hence,
$$g \leq \left(\sum_{i=1}^g (\omega_i + \overline{\omega}_i)\right) + g\lfloor 2\sqrt{q}\rfloor + g;$$

thus,
$$\sum_{i=1}^{2g} \omega_i \geq -g\lfloor 2\sqrt{q}\rfloor.$$

Analogous computations for δ_i yield the other bound. □

Of interest is the question on the maximum possible value of N for given q and g. Denote $N_q(g) = \max_X |X(\mathbb{F}_q)|$, the maximum being taken over all curves of genus g over $\mathbb{F}_q$. The Serre bound states that
$$N_q(g) \leq q + 1 + g\lfloor 2\sqrt{q}\rfloor. \tag{3.1.7}$$

Curves with $N = q + 1 + 2g\sqrt{q}$ are called *maximal*. Obviously, they can exist only if q is a square. Moreover, it can be shown that the genus of a maximal curve cannot be too large as compared with q:

PROPOSITION 3.1.26. *For a maximal curve over $\mathbb{F}_q$, we have*
$$g \leq \frac{q - \sqrt{q}}{2}.$$

PROOF. Indeed, it follows immediately from (3.1.4) and the Weil theorem that all ω_i for a maximal curve are equal to $-\sqrt{q}$. Then $\sum_{i=1}^{2g} \omega_i^2 = 2gq$, and (3.1.3) implies
$$N_2 = q^2 + 1 - 2gq.$$
It remains to use the inequality $N_2 \geq N$. □

Another application of the Weil theorem is the so-called explicit formula, considered in the next subsection.

3.1.4. Explicit Formula

Since the Weil theorem reads that $|\omega_j| = \sqrt{q}$, it is sometimes convenient to represent ω_j as $\omega_j = \sqrt{q}\alpha_j$ with $\alpha_j = e^{i\theta_j}$. In these terms, the following formula for the number of points on a curve can be written.

THEOREM 3.1.27 (explicit formula). *Let $c_1, c_2, \ldots$ be arbitrary real numbers. Consider the functions*
$$f_m(\varphi) = 1 + 2\sum_{r=1}^{m} c_r \cos(r\varphi)$$
and
$$\psi_m(t) = \sum_{r=1}^{m} c_r t^r.$$
Then
$$N\psi_m(q^{-1/2}) = \psi_m(q^{1/2}) + \psi_m(q^{-1/2}) + g - \sum_{i=1}^{g} f_m(\theta_i) - \sum_{r=1}^{m}(N_r - N)c_r q^{-r/2},$$
where $\omega_1, \ldots, \omega_{2g}$ are indexed in such a way that $\theta_{g+i} = -\theta_i$.

PROOF. Divide (3.1.3) by $q^{r/2}$ to obtain
$$N_r q^{-r/2} = q^{r/2} + q^{-r/2} - \sum_{i=1}^{g}(\alpha_i^r + \alpha_i^{-r}).$$
Multiplying this by c_r, after obvious transformations we get
$$Nc_r q^{-r/2} = c_r q^{r/2} + c_r q^{-r/2} - 2\sum_{i=1}^{g} c_r \cos(r\theta_i) - (N_r - N)c_r q^{-r/2}.$$
Summing up these equations over all $r = 1, \ldots, m$ gives the desired formula. □

Since $N_r \geq N$, the explicit formula implies the following family of bounds on N:

PROPOSITION 3.1.28. *Let $c_r \geq 0$, $r = 1, \ldots, m$, and let not all c_r be zero. Let $f_m(\varphi) \geq 0$ for all φ. Then*
$$N \leq \frac{g}{\psi_m(q^{-1/2})} + \frac{\psi_m(q^{1/2})}{\psi_m(q^{-1/2})} + 1.$$
□

Choosing values of $c_1, \ldots, c_m$ in an appropriate way, one can obtain various bounds. We shall apply this method in the next section.

EXERCISE 3.1.29. Does choosing $m = 2$, $c_1 = 0$, and $c_2 = 1/2$ yield a bound (which one, if any)?

EXERCISE 3.1.30. Let $q=2$. Take $m=6$,
$$c_1 = \frac{184}{203}, \quad c_2 = \frac{20}{29}, \quad c_3 = \frac{90}{203}, \quad c_4 = \frac{89}{406}, \quad c_5 = \frac{2}{29}, \quad c_6 = \frac{2}{203}.$$
Check that the conditions of Proposition 3.1.28 are satisfied, and prove the bound
$$N_2(g) \leq 0.83g + 5.35. \tag{3.1.8}$$

We shall advert to the explicit formula many a time, in particular, when studying asymptotics in Sec. 3.2.

3.1.5. Pellikaan's Two-Variable Zeta Function

By analogy with the definition of the zeta function in the form $Z(t) = \sum_{n=0}^{\infty} M_n t^n$, where M_n is the number of effective divisors of degree n on a curve X, it is quite natural to try to consider the two-variable series
$$\sum_{n=0}^{\infty} \sum_{k=1}^{\infty} M_{n,k} t^n v^k,$$
where $M_{n,k}$ is the number of effective divisors of degree n and dimension k. (With $v=1$ we obtain the usual zeta function of the curve.) It turns out that this series defines a rational function. However, much more convenient is the following way to define a two-variable zeta function, which also involves the dimension of a linear system and, in addition, yields nice analogues of other properties of the zeta function—first of all, the functional equation.

Consider the series
$$Z(t,u) = Z(X, \mathbb{F}_q; t, u) = \sum_{n=0}^{\infty} \sum_{k=0}^{\infty} m_{n,k} \frac{u^k - 1}{u - 1} t^n,$$
where $m_{n,k}$ is the number of divisor classes of degree n and dimension k. It follows directly from Lemma 3.1.8b that $Z(t) = Z(t, q)$ and
$$M_{n,k} = \frac{q^k - 1}{q - 1} m_{n,k}.$$

EXERCISE 3.1.31. Prove that for $n > 2g - 2$ we have
$$m_{n,k} = \begin{cases} h & \text{if } k = n + 1 - g, \\ 0 & \text{otherwise,} \end{cases}$$
and if $0 \leq n \leq 2g - 2$, then $m_{n,k} = m_{2g-2-n, k-n-1+g}$. (*Hint*: Apply the Riemann–Roch theorem.)

The two-variable zeta function thus defined (the *Pellikaan zeta function*) possesses the following remarkable properties:

PROPOSITION 3.1.32. *The series $Z(t,u)$ defines a rational function. Furthermore*:

(a) *If $g = 0$, then*
$$Z(t,u) = \frac{1}{(1-t)(1-ut)}.$$

(b) If $g \geq 1$, then $Z(t,u) = F(t,u) + G(t,u)$ with
$$F(t,u) = \frac{1}{u-1} \sum_{\substack{D \in \text{Pic}(X) \\ 0 \leq \deg D \leq 2g-2}} u^{\ell(D)} t^{\deg D},$$
$$G(t,u) = \frac{h}{u-1}\left(\frac{u^g t^{2g-1}}{1-ut} - \frac{1}{1-t}\right).$$

(c) $Z(t,u)$ satisfies the functional equation
$$Z(t,u) = u^{g-1} t^{2g-2} Z\left(\frac{1}{ut}, u\right).$$

(d) The two-variable zeta function is of the form
$$Z(t,u) = \frac{P(t,u)}{(1-t)(1-ut)},$$
where $P(t,u)$ is a polynomial of degree $2g$ in t and degree g in u with integer coefficients. Moreover,
$$P(t,u) = \sum_{i=0}^{2g} P_i(u) t^i,$$
where $P_0(u) = 1$, $P_{2g}(u) = u^g$, $\deg P_i(u) \leq i/2$, and $P_{2g-i}(u) = u^{g-i} P_i(u)$ for $i = 0,\ldots,2g$. Also, $P(1,u) = h$. □

EXERCISE 3.1.33. Prove the proposition.

EXAMPLE 3.1.34. Let X be an elliptic curve with N rational points over $\mathbb{F}_q$. Then
$$Z(t,u) = \frac{1 + (N-1-u)t + ut^2}{(1-t)(1-ut)}.$$

EXERCISE 3.1.35. Check this assertion.

Recently, a number of very interesting studies appeared for which the Pellikaan zeta function was a starting point (see Historical and Bibliographic Notes to this chapter).

3.2. Asymptotics

In this section we are interested in curves of very large genus over a fixed finite field $\mathbb{F}_q$. To be precise, we investigate the ratio of the maximum possible number of points on a curve to its genus.

In Sec. 3.2.1 we prove the Drinfeld–Vlăduţ inequality, an upper bound for this ratio; in Sec. 3.2.2 we present some lower bounds. Section 3.2.3 shows that if there are many points of degree 1 on a curve, then a fortiori there are few points of larger degrees. In Sec. 3.2.4 we study the asymptotic behaviour of the class number (i.e., the number of $\mathbb{F}_q$-points on the Jacobian of a curve). Finally, Sec. 3.2.5 is an introduction to the theory of asymptotically exact families, which considerably strengthens the preceding results. This theory is of great importance for what follows; we hope to return to it in *Advanced Chapters*.

3.2.1. Drinfeld–Vlăduţ Theorem

Fix a finite field $\mathbb{F}_q$. Consider the question on the behaviour (as $g \to \infty$) of the parameter $N_q(g)$, the maximal number of points on a curve of genus g. To state results, it is convenient to introduce the parameter

$$A(q) = \limsup_{g \to \infty} \frac{N_q(g)}{g}.$$

The Serre bound (see formula (3.1.7)) yields

$$A(q) \leq \lfloor 2\sqrt{q} \rfloor.$$

In fact, there is a much tighter bound:

THEOREM 3.2.1 (Drinfeld–Vlăduţ bound). *We have*

$$A(q) \leq \sqrt{q} - 1.$$

PROOF. We apply Proposition 3.1.28. Fix $m \geq 1$, and set

$$c_r = 1 - \frac{r}{m}, \quad r = 1, \ldots, m.$$

It is easily checked that for this choice we get

$$\psi_m(t) = \sum_{r=1}^{m} \left(1 - \frac{r}{m}\right) t^r = \sum_{r=1}^{m} t^r - \frac{t}{m} \frac{d}{dt} \sum_{r=1}^{m} t^r = \frac{t}{m(t-1)} \left(\frac{t^m - 1}{t - 1} - m\right)$$

and

$$f_m(\varphi) = 1 + \psi_m(e^{i\varphi}) + \psi_m(e^{-i\varphi}) = \frac{1 - \cos(m\varphi)}{m(1 - \cos\varphi)} \geq 0.$$

Therefore,

$$\frac{N}{g} \leq \frac{1}{\psi_m(q^{-1/2})} + \frac{1}{g}\left(1 + \frac{\psi_m(q^{1/2})}{\psi_m(q^{-1/2})}\right).$$

Now it is easy to check that

$$\psi_m(q^{-1/2}) \to \frac{1}{\sqrt{q} - 1}$$

as $m \to \infty$, and the term

$$\frac{1}{g}\left(1 + \frac{\psi_m(q^{1/2})}{\psi_m(q^{-1/2})}\right)$$

tends to zero provided that q^m/g tends to zero. Hence, $N/g < \sqrt{q}-1+\varepsilon$ for any $\varepsilon > 0$. □

EXERCISE 3.2.2. Make explicit computations in the proof of Theorem 3.2.1.

3.2.2. Lower Asymptotic Bounds

The previous subsection gives an upper bound on $A(q)$. Remarkably, this bound is tight if q is a square. Indeed, towers from Sec. 3.4.3 yield the following result.

THEOREM 3.2.3. *Let q be an even power of a prime. Then*
$$A(q) = \sqrt{q}-1.$$
□

We do not know the answer for odd powers. The best known upper bound is always $\sqrt{q}-1$. Lower bounds are rather far from it. Here is a list of some of them.

THEOREM 3.2.4. *For any $\mathbb{F}_q$, $q = p^e$, we have*

(1) $A(q) \geq \dfrac{1}{96}\log_2 q$;

(2) *for $e > 1$,* $A(q) \geq \dfrac{2}{q-2}$;

(3) *for $e = 3$,* $A(q) \geq \dfrac{2(q^{2/3}-1)}{q^{1/3}+2}$.

(4) *If ℓ is a prime, $\ell \mid (q-1)$, and $q > 4\ell+1$, then*
$$A(q^\ell) \geq \frac{\sqrt{\ell(q-1)}-2\ell}{\ell-1}.$$

(5) *If q is odd and is not a square, then*
$$A(q^m) \geq \frac{2q+2}{\lceil 2\sqrt{2q+3}\rceil +1}.$$

(6) *If $q \geq 8$ is even and $m \geq 3$ is odd, then*
$$A(q^m) \geq \frac{q+1}{\lceil 2\sqrt{2q+2}\rceil +2}.$$

(7) $A(2) \geq 81/317 = 0.2555\ldots$;

(8) $A(3) \geq 8/17 = 0.4705\ldots$;

(9) $A(5) \geq 8/11 = 0.7272\ldots$. □

The spread between the upper bound and lower ones is so large that we can only formulate an interesting and important unsolved problem:

PROBLEM 3.2.5. Find $A(q)$ for odd powers of a prime. Is it also true that $A(q) = \sqrt{q}-1$?

3.2.3. Points of Higher Degrees

Here we are interested in the asymptotics of the numbers B_r, which are defined at the beginning of Sec. 3.1.1.

A family of curves $\{X\}$ with
$$\lim_{g \to \infty} \frac{N(X)}{g(X)} = \sqrt{q}-1$$

is called *asymptotically optimal*. One can derive from the proof of Theorem 3.2.1 (more precisely, from the explicit formula and Proposition 3.1.28, on which the proof of the theorem is based) that, for a curve X from an asymptotically optimal family and for any fixed r, we have

$$\lim_{g\to\infty} \frac{N_r(X) - N(X)}{g(X)} = 0.$$

Since $rB_r \leq N_r - N$, we also have

$$\lim_{g\to\infty} \frac{B_r(X)}{g(X)} = 0. \tag{3.2.1}$$

EXERCISE 3.2.6. Prove these equalities.

We also need the following refinement of the last relation.

PROPOSITION 3.2.7. *Let $q \geq 4$. Let X be a curve with $N(X) \geq (\sqrt{q}-1)g(X)$, and let $c = \lfloor 2\log_q g \rfloor$. Then, for any s, $2 \leq s \leq c$, we have*

$$B_s \leq \frac{8gq^{s/2}}{s(c+1-s)}.$$

SKETCH OF THE PROOF. While deriving Proposition 3.1.28 from the explicit formula, we used the fact that the differences $N_r - N$ are nonnegative and neglected the corresponding terms. Now, let us keep one of them (with $r = s$) and use the fact that $N_s - N \geq sB_s$. Then the refined inequality of Proposition 3.1.28 takes the form

$$N\psi_m(q^{-1/2}) \leq g + \psi_m(q^{1/2}) + \psi_m(q^{-1/2}) - sB_s c_s q^{-s/2}.$$

From this, choosing $c_r = 1 - r/m$, as in the proof of the Drinfeld–Vlăduţ theorem, and using the inequality $N \geq (\sqrt{q}-1)g$, we obtain

$$\frac{sB_s(m-s)}{mq^{s/2}} \leq g\left(1 - (\sqrt{q}-1)\psi_m(q^{-1/2})\right) + \psi_m(q^{1/2}) + \psi_m(q^{-1/2}),$$

where $\psi_m(t)$ are the same functions as in Theorem 3.2.1. Setting $m = c+1 > s$ and making necessary transformations, we obtain the proposition. $\square$

REMARK 3.2.8. In the conditions of Proposition 3.2.7, we assume that not only

$$\limsup_{g\to\infty}(N_1/g) = \sqrt{q}-1,$$

but also $N_1 \geq (\sqrt{q}-1)g$. In fact, for all presently known families of curves with $N_1/g \to \sqrt{q}-1$, the inequality $N_1 \geq (\sqrt{q}-1)g$ holds; thus, Proposition 3.2.7 is sufficient for us. Also, it is clear that a similar proposition as well holds for $q = 2$ and $q = 3$ with a somewhat larger numerical factor on the right-hand side, but we do not need this fact.

EXERCISE 3.2.9. Make detailed computations and prove Proposition 3.2.7.

Another asymptotic inequality can be obtained with the help of the *Möbius inversion formula*: since $N_r = \sum_{d|r} dB_d$, we have

$$rB_r = \sum_{d|r} \mu\left(\frac{r}{d}\right) N_d, \tag{3.2.2}$$

where μ is the *Möbius function*, defined by

$$\mu(n) = \begin{cases} 1 & \text{if } n = 1, \\ 0 & \text{if } n \text{ is divisible by a square of a prime,} \\ (-1)^\ell & \text{if } n \text{ is a product of } \ell \text{ distinct primes.} \end{cases}$$

Since the Möbius function possesses the property $\sum_{d|r} \mu(r/d) = 0$ for $r > 1$, the inversion formula together with (3.1.3) yields

$$B_r = \frac{1}{r} \sum_{d|r} \mu\left(\frac{r}{d}\right)\left(q^d + 1 - \sum_{i=1}^{2g} \omega_i^d\right) = \frac{1}{r} \sum_{d|r} \mu\left(\frac{r}{d}\right)\left(q^d - \sum_{i=1}^{2g} \omega_i^d\right) \quad (3.2.3)$$

for any $r \geq 2$.

From this formula, one can derive the following bound.

PROPOSITION 3.2.10. *For any $r \geq 1$, we have*

$$\left|B_r - \frac{q^r}{r}\right| \leq \left(\frac{q}{q-1} + 2g\frac{\sqrt{q}}{\sqrt{q}-1}\right)\frac{q^{r/2}-1}{r} < (2+7g)\frac{q^{r/2}}{r}.$$

PROOF. For $r = 1$, the statement easily follows from the Weil bound. For $r \geq 2$, apply representation (3.2.3):

$$B_r - \frac{q^r}{r} = \frac{1}{r}\sum_{\substack{d|r \\ d<r}} \mu\left(\frac{r}{d}\right)q^d - \frac{1}{r}\sum_{d|r}\sum_{i=1}^{2g} \omega_i^d.$$

Hence, using the fact that $|\omega_i^d| \leq q^{d/2}$, we get

$$\left|B_r - \frac{q^r}{r}\right| \leq \frac{1}{r}\sum_{d=1}^{\lfloor r/2 \rfloor} q^d + \frac{2g}{r}\sum_{d=1}^{r} q^{d/2} \leq \frac{q}{r}\frac{q^{r/2}-1}{q-1} + \frac{2g\sqrt{q}}{r}\frac{q^{r/2}-1}{\sqrt{q}-1},$$

and the proposition easily follows. $\square$

COROLLARY 3.2.11. (a) *If $g = 0$, then $B_r > 0$ for any $r \geq 1$.*

(b) *For any r such that $2g + 1 \leq q^{(r-1)/2}(\sqrt{q}-1)$, there exists at least one point of degree r. In particular, if $r \geq 4g + 3$, then $B_r \geq 1$.* $\square$

EXERCISE 3.2.12. Prove the corollary.

PROBLEM 3.2.13. Find the exact bound $\rho(q,g)$ such that $B_r \geq 1$ whenever $r \geq \rho(q,g)$ for any curve of genus g over $\mathbb{F}_q$.

3.2.4. Asymptotics for the Jacobian

Along with the asymptotics for the number of rational points on a curve of growing genus, of interest is the question on the asymptotics of the number $h = h(X) = |J_X(\mathbb{F}_q)|$ of rational points on the Jacobian J_X of a curve X (the class number). Proposition 3.1.23 states that

$$2\log_q(\sqrt{q}-1) \leq \frac{\log_q h}{g} \leq 2\log_q(\sqrt{q}+1).$$

This bound can also be refined. We consider a refinement in the following situation: Let X be a curve from a family of curves of growing genus such that

$$\lim_{g \to \infty} \frac{N(X)}{g(X)} = A > 0$$

(such families of curves are called *asymptotically good*). Then there is the following bound:

PROPOSITION 3.2.14. *We have*
$$\liminf_{g\to\infty} \frac{\log_q h}{g} \geq 1 + A\log_q\left(\frac{q}{q-1}\right).$$

PROOF. According to (3.1.5), we have
$$h = \prod_{i=1}^{2g}(\omega_i - 1) = q^g \prod_{i=1}^{2g}\left(1 - \alpha_i q^{-1/2}\right)$$
since $\prod_{i=1}^{2g}\omega_i = q^g$. Hence,
$$\ln h = g\ln q + \sum_{i=1}^{2g}\ln\left(1 - \alpha_i q^{-1/2}\right)$$
$$= g\ln q - \sum_{i=1}^{2g}\sum_{m=1}^{\infty}\frac{\alpha_i^m q^{-m/2}}{m}$$
$$= g\ln q - \sum_{m=1}^{\infty}\frac{q^{-m/2}}{m}\sum_{i=1}^{2g}\alpha_i^m.$$

Note that (3.1.3) can be rewritten as
$$-\sum_{i=1}^{2g}\alpha_i^m = N_m q^{-m/2} - \left(q^{m/2} + q^{-m/2}\right).$$

If we substitute the right-hand side of this equality for the left-hand one in all terms of the sum with $m \leq b = \lfloor \log_q g \rfloor$, we obtain
$$\ln h = g\ln q + \sum_{m=1}^{b}\frac{q^{-m/2}}{m}\left(N_m q^{-m/2} - q^{m/2} - q^{-m/2}\right) - \sum_{m=b+1}^{\infty}\frac{q^{-m/2}}{m}\sum_{i=1}^{2g}\alpha_i^m.$$

Note that
$$\sum_{m=1}^{b}\frac{q^{-m/2}}{m}\left(N_m q^{-m/2} - q^{m/2} - q^{-m/2}\right) = \sum_{m=1}^{b}\frac{N_m q^{-m}}{m} - \sum_{m=1}^{b}\frac{1}{m} - \sum_{m=1}^{b}\frac{q^{-m}}{m};$$
here,
$$\sum_{m=1}^{\infty}\frac{q^{-m}}{m} = \ln\frac{q}{q-1}.$$

Note also that
$$\sum_{m=1}^{b}\frac{N_m}{m}q^{-m} \geq N\sum_{m=1}^{b}\frac{q^{-m}}{m} = N\left(\sum_{m=1}^{\infty}\frac{q^{-m}}{m} - \sum_{m=b+1}^{\infty}\frac{q^{-m}}{m}\right);$$
furthermore,
$$\sum_{m=b+1}^{\infty}\frac{q^{-m}}{m} < q^{-(b+1)}\sum_{j=0}^{\infty}q^{-j} = \frac{q}{q-1}q^{-(b+1)} < \frac{1}{g}\frac{q}{q-1}$$

since $b+1 > \log_q g$. Therefore,

$$\sum_{m=1}^{b} \frac{q^{-m/2}}{m}\left(N_m q^{-m/2} - q^{m/2} - q^{-m/2}\right)$$
$$> N\ln\frac{q}{q-1} - \frac{Nq}{g(q-1)} - b - \ln\frac{q}{q-1} = N\ln\frac{q}{q-1} - o(g).$$

Also, we have

$$\left|\sum_{m=b+1}^{\infty} \frac{q^{-m/2}}{m} \sum_{i=1}^{2g} \alpha_i^m\right| \leq 2g \sum_{m=b+1}^{\infty} \frac{q^{-m/2}}{m} < 2gq^{-(b+1)/2} \sum_{j=0}^{\infty} q^{-j/2} < 2\sqrt{g}\frac{\sqrt{q}}{\sqrt{q}-1}.$$

Thus,

$$\ln h \geq g\ln q + N\ln\frac{q}{q-1} - o(g).$$

Dividing by $g\ln q$, we obtain the proposition. □

In the next subsection, we improve this result.

Note that we know from Proposition 3.1.16a,b that $h = P(1) = q^g P(q^{-1})$; i.e., in the proof of Proposition 3.2.14 we actually computed the asymptotics for $P(q^{-1})$. In fact, for an asymptotically optimal curve, one can similarly obtain an asymptotic expression for $P(q^{-s})$ for any $s \in \mathbb{N}$.

EXERCISE 3.2.15. Show that for a family of curves of growing genus with $\lim_{g\to\infty} N/g = \sqrt{q}-1$, we have the asymptotic equality

$$\lim_{g\to\infty} \frac{1}{g}\log_q P(q^{-s}) = -(\sqrt{q}-1)\log_q(1-q^{-s})$$

for any $s \in \mathbb{N}$. (*Hint*: Use an estimate for $B_s - B_1$ that follows from Proposition 3.2.7 and Remark 3.2.8.)

REMARK 3.2.16. In fact, one can show (see Exercise 3.2.39) that for an asymptotically optimal curve we have an even stronger result: we have

$$\frac{1}{g}\log_q P(q^{-s}) \to -(\sqrt{q}-1)\log_q(1-q^{-s})$$

for any $s \in \mathbb{C}$ in the region $\{\operatorname{Re} s > 1/2\}$, uniformly in any half-plane $\operatorname{Re} s > \sigma$, where $\sigma > 1/2$.

3.2.5. Asymptotically Exact Families

Now we want to refine Theorem 3.2.1 by introducing points of higher degrees into its statement. This leads to the following definition.

A family $\mathcal{X} = \{X_i\}$ of (smooth, projective, absolutely irreducible) curves over $\mathbb{F}_q$ is called *asymptotically exact* if $g(X_i) \to \infty$ and for any $m \in \mathbb{N}$ the limit

$$\beta_m = \beta_m(\mathcal{X}) = \lim_{i\to\infty} \frac{B_m(X_i)}{g(X_i)}$$

exists.

Of course, an arbitrarily taken family of curves by no means obeys this property. In fact, asymptotically exact families are exceptions rather than the rule. However, the study of any family reduces in a sense to the study of asymptotically exact families due to the following result.

PROPOSITION 3.2.17. *Any family of curves with $g(X_i) \to \infty$ contains an asymptotically exact subfamily.* □

EXERCISE 3.2.18. Prove the proposition. (*Hint*: First, choose a subfamily where β_1 exists; within it, take a subfamily with β_2, etc.; then take the diagonal subsequence.)

EXERCISE 3.2.19. Let
$$\ldots \to X_i \to \ldots \to X_2 \to X_1$$
be a tower of curves with $g(X_i) \to \infty$. Prove that $\{X_i\}$ is asymptotically exact. (*Hint*: Induction on m; use the Hurwitz formula for the genus of a cover.)

EXERCISE 3.2.20. Show that an asymptotically optimal family is asymptotically exact. (*Hint*: Use Theorem 3.2.1 and formula (3.2.1).)

Let us try to guess what conditions are satisfied by β_m.

LEMMA 3.2.21. *Let X be a curve over $\mathbb{F}_q$, $P_X(t) = \prod_{i=1}^{2g}(1 - \omega_i t)$. Then we have the following equality of formal power series in t:*
$$\frac{1}{t^{-1} - 1} + \frac{1}{q^{-1}t^{-1} - 1} - \sum_{i=1}^{2g} \frac{1}{\omega_i^{-1} t^{-1} - 1} = \sum_{i=1}^{\infty} mB_m \frac{1}{t^{-m} - 1}. \tag{3.2.4}$$

PROOF. We know that
$$Z_X(t) = \frac{P_X(t)}{(1-t)(1-qt)} = \prod_{m=1}^{\infty}(1-t^m)^{-B_m}.$$
It suffices now to take the logarithm of this equality, then take the derivative, and multiply the obtained equality by t. □

HOAX 3.2.22. Now put $t = q^{-1/2}$ in (3.2.4). Reordering ω_i so that $\omega_{i+g} = \bar{\omega}_i$, $i = 1, \ldots, g$, we obtain
$$\sum_{m=1}^{\infty} mB_m \frac{1}{q^{m/2} - 1}$$
$$= \left(\frac{1}{q^{1/2} - 1} + \frac{1}{q^{-1/2} - 1} \right) - \sum_{i=1}^{g} \left(\frac{1}{q^{-1/2}\omega_i - 1} + \frac{1}{q^{1/2}\omega_i^{-1} - 1} \right) = g - 1.$$

Now let $\mathcal{X} = \{X_i\}$ be an asymptotically exact family of curves. Dividing by g and tending i to infinity, in the limit we get
$$\sum_{m=1}^{\infty} \frac{m\beta_m}{q^{m/2} - 1} = 1. \tag{3.2.5}$$

EXERCISE 3.2.23. Where do we cheat? For which $t \in \mathbb{C}$ does the series on the right-hand side of (3.2.4) converge?

Unfortunately, inequality (3.2.5) not only cannot be considered to be proved but is not true at all. However, a slightly weaker statement is valid:

THEOREM 3.2.24. *For an asymptotically exact family of curves, we have*
$$\sum_{m=1}^{\infty} \frac{m\beta_m}{q^{m/2} - 1} \le 1. \qquad \square$$

EXERCISE 3.2.25. Deduce the Drinfeld–Vlăduţ inequality from this theorem.

REMARK 3.2.26. In fact, we know the "defect" in the inequality of the theorem, i.e., the difference
$$\delta = 1 - \sum_{m=1}^{\infty} \frac{m\beta_m}{q^{m/2}-1}.$$
Namely, one can show that δ is the "limit number of Frobenius roots lying very close to $\sqrt{q}$, divided by g," under an appropriate definition of this vague notion.

Theorem 3.2.24 is an easy consequence of the following statement:

THEOREM 3.2.27. *Let $\mathcal{X} = \{X_i\}$ be an asymptotically exact family, and let $f: \mathbb{N} \to \mathbb{N}$ be a function such that*
$$\lim_{i \to \infty} \frac{f(g(X_i))}{\log g(X_i)} = 0.$$
Then
$$\limsup \frac{1}{g} \sum_{m=1}^{f(g)} \frac{mB_m}{q^{m/2}-1} \leq 1.$$

To prove the theorem, we need the following lemma.

LEMMA 3.2.28. *For any $b \in \mathbb{N}$, we have*
$$1 \geq \sum_{j=1}^{b} \left(1 - \frac{j}{b+1}\right) \frac{N_j}{g} q^{-j/2} - \frac{1}{g} \sum_{j=1}^{b} \left(1 - \frac{j}{b+1}\right) \left(q^{j/2} + q^{-j/2}\right).$$

PROOF. As above, let $\alpha_i = \omega_i/\sqrt{q}$. Since for any α_i we have
$$0 \leq \left|\alpha_i^b + \alpha_i^{b-1} + \ldots + 1\right|^2 = (b+1) + \sum_{j=1}^{b}(b+1-j)\left(\alpha_i^j + \alpha_i^{-j}\right),$$
we find $b+1 \geq -\sum_{j=1}^{b}(b+1-j)\left(\alpha_i^j + \alpha_i^{-j}\right)$. Summing up these inequalities over all $i = 1, \ldots, 2g$ and using the properties $\sum_{i=1}^{2g} \alpha_i^j = \sum_{i=1}^{2g} \alpha_i^{-j}$ and $\sum_{i=1}^{2g} \alpha_i^j = N_j q^{-j/2} - q^{j/2} - q^{-j/2}$, we obtain
$$g(b+1) \geq \sum_{j=1}^{b}(b+1-j)\left(N_j q^{-j/2} - q^{j/2} - q^{-j/2}\right). \qquad \square$$

PROOF OF THEOREM 3.2.27. The second term on the right-hand side of the inequality in Lemma 3.2.28 is not greater than
$$\frac{2}{g} \sum_{j=1}^{b} q^{j/2} = \frac{2(q^{b/2}-1)\sqrt{q}}{g(\sqrt{q}-1)};$$
for $b = f(g)$, it tends to zero by the condition of the theorem. Furthermore, there always exists a function $\ell(g)$ such that for $b = \ell(g)$ this term also tends to zero but $f(g)/\ell(g) \to 0$.

Then Lemma 3.2.28 gives

$$1 + o(1) \geq \sum_{j=1}^{\ell(g)} \left(1 - \frac{j}{\ell(g)+1}\right) \frac{N_j}{g} q^{-j/2}$$

$$\geq \sum_{j=1}^{f(g)} \left(1 - \frac{j}{\ell(g)+1}\right) \frac{N_j}{g} q^{-j/2}$$

$$\geq \left(1 - \frac{f(g)}{\ell(g)+1}\right) \sum_{j=1}^{f(g)} \frac{N_j}{g} q^{-j/2}.$$

The last expression tends to

$$\sum_{j=1}^{f(g)} \frac{N_j}{g} q^{-j/2} = \sum_{j=1}^{f(g)} \frac{1}{g} q^{-j/2} \sum_{m|j} m B_m$$

$$= \sum_{m=1}^{f(g)} \frac{m B_m}{g} \sum_{i=1}^{\lfloor f(g)/m \rfloor} q^{-im/2}$$

$$= \sum_{m=1}^{f(g)} \frac{m B_m}{g} q^{-m/2} \frac{1 - q^{-m \lfloor f(g)/m \rfloor /2}}{1 - q^{-m/2}},$$

which tends to the required limit. □

To what extent is the condition $\lim \frac{f(g)}{\log g} = 0$ in Theorem 3.2.27 necessary? A partial answer is given by the following proposition.

PROPOSITION 3.2.29. *Let $f: \mathbb{N} \to \mathbb{N}$ be a function such that*

$$\lim \frac{f(g(X_i))}{\log g(X_i)} = \infty.$$

Then

$$\lim \frac{1}{g} \sum_{m=1}^{f(g)} \frac{m B_m}{q^{m/2} - 1} = \infty.$$
□

EXERCISE 3.2.30. Prove the proposition.

Limit formula for the class number. Now we are going to derive a formula for the limit behaviour of the class number

$$h = h(X) = |J_X(\mathbb{F}_q)| = P_X(1)$$

for an asymptotically exact family (Proposition 3.2.14 gives an estimate only).

LEMMA 3.2.31. *We have the equality of formal power series*

$$P(t) = t^{2g-2}(1-t)(1-qt)q^{g-1} \prod_{m=1}^{\infty} (1 - q^{-m} t^{-m})^{-B_m}. \quad (3.2.6)$$

PROOF. This time, in addition to the equality

$$\frac{P(t)}{(1-t)(1-qt)} = \prod_{m=1}^{\infty} (1 - t^m)^{-B_m},$$

we also apply the functional equation (Theorem 3.1.15). We obtain
$$P(t) = (1-t)(1-qt)Z(t)$$
$$= (1-t)(1-qt)Z(q^{-1}t^{-1})(\sqrt{q}t)^{2g-2}$$
$$= (1-t)(1-qt)q^{g-1}t^{2g-2} \prod_{m=1}^{\infty} (1-q^{-m}t^{-m})^{-B_m},$$

as required. □

HOAX 3.2.32. Substituting $t = 1$ into (3.2.6), we get
$$h = P(1) = (1-1)(1-q)q^{g-1} \prod_{m=1}^{\infty} (1-q^{-m})^{-B_m}.$$

If we take the logarithm, divide by g, and pass to the limit as $g \to \infty$, assuming here that $\lim_{g \to \infty} \frac{1}{g} \log(1-1) = \lim_{t \to \infty} \lim_{g \to \infty} \frac{1}{g} \log(1-t) = 0$, we "obtain" the following formula for an arbitrary asymptotically exact family:
$$\lim_{i \to \infty} \frac{1}{g(X_i)} \log_q h(X_i) = 1 + \sum_{m=1}^{\infty} \beta_m \log_q \frac{q^m}{q^m - 1}.$$

Strangely enough, despite the cheating so unconcealed, this formula turns out to be correct:

THEOREM 3.2.33. *For an asymptotically exact family of curves $\mathcal{X} = \{X_i\}$, the limit*
$$H = H(\mathcal{X}) = \lim_{i \to \infty} \frac{1}{g(X_i)} \log_q h(X_i)$$

exists and equals
$$H = 1 + \sum_{m=1}^{\infty} \beta_m \log_q \frac{q^m}{q^m - 1}. \qquad \Box$$

The theorem is directly implied by the following statement.

THEOREM 3.2.34. *Let $\mathcal{X} = \{X_i\}$ be an asymptotically exact family, and let $f: \mathbb{N} \to \mathbb{N}$ be a function such that*
$$\lim_{g \to \infty} f(g) = \infty$$

and
$$\lim_{g \to \infty} \frac{\log f(g)}{g} = 0.$$

Then
$$\lim_{i \to \infty} \left(\frac{1}{g(X_i)} \log_q h(X_i) - \frac{1}{g(X_i)} \sum_{m=1}^{f(g(X_i))} B_m \log_q \frac{q^m}{q^m - 1} \right) = 1.$$

EXERCISE 3.2.35. Deduce Theorem 3.2.33 from Theorem 3.2.34. For any (not necessarily asymptotically exact) family of curves with $g(X_i) \to \infty$, deduce from this the inequality
$$\liminf_{i \to \infty} \frac{1}{g(X_i)} \log_q h(X_i) \geq 1 + \sum_{i=1}^{k} \beta_{m,\inf} \log_q \frac{q^m}{q^m - 1}$$

for any $k \in \mathbb{N}$, where
$$\beta_{m,\inf} = \liminf_{i\to\infty} \frac{B_m(X_i)}{g(X_i)}.$$

To prove Theorem 3.2.34, we need the following lemma.

LEMMA 3.2.36. *For any $b \in \mathbb{N}$, we have*
$$\frac{1}{g}\ln h = \ln q + \frac{1}{g}\sum_{j=1}^{b}\frac{q^{-j}}{j}N_j - \frac{1}{g}\sum_{j=1}^{b}\frac{1+q^{-j}}{j} - \frac{1}{g}\sum_{j=b+1}^{\infty}\frac{q^{-j/2}}{j}\sum_{i=1}^{2g}\alpha_i^j.$$

PROOF. We have
$$\frac{1}{g}\ln h = \frac{1}{g}\ln\prod_{i=1}^{2g}(1-\omega_i) = \frac{1}{g}\ln\left(q^g\prod_{i=1}^{2g}(1-\alpha_i q^{-1/2})\right)$$
$$= \ln q + \frac{1}{g}\sum_{i=1}^{2g}\ln(1-\alpha_i q^{-1/2}) = \ln q - \frac{1}{g}\sum_{j=1}^{\infty}\frac{q^{-j/2}}{j}\sum_{i=1}^{2g}\alpha_i^j$$
$$= \ln q + \frac{1}{g}\sum_{j=1}^{b}\frac{q^{-j/2}}{j}\left(N_j q^{-j/2} - q^{j/2} - q^{-j/2}\right) - \frac{1}{g}\sum_{j=b+1}^{\infty}\frac{q^{-j/2}}{j}\sum_{i=1}^{2g}\alpha_i^j,$$
as required. □

PROOF OF THEOREM 3.2.34. The last term in the equality of Lemma 3.2.36 tends to zero for any b tending to infinity. Note that
$$\frac{1}{g}\sum_{j=1}^{f(g)}\frac{1+q^{-j}}{j} \le \frac{2}{g}(1+\ln f(g))$$
also tends to zero by the condition imposed on the function $f(g)$. Thus, by Lemma 3.2.36 we get
$$\frac{1}{g}\ln h - \ln q = \frac{1}{g}\sum_{j=1}^{f(g)}\frac{q^{-j}}{j}N_j = \frac{1}{g}\sum_{j=1}^{f(g)}\frac{q^{-j}}{j}\sum_{m|j}mB_m$$
$$= \frac{1}{g}\sum_{m=1}^{f(g)}B_m\sum_{i=1}^{\lfloor f(g)/m\rfloor}\frac{q^{-im}}{i}$$
$$= \frac{1}{g}\sum_{m=1}^{f(g)}B_m\ln\frac{q^m}{q^m-1} - \frac{1}{g}\sum_{m=1}^{f(g)}B_m\sum_{\lfloor f(g)/m\rfloor+1}^{\infty}\frac{q^{-im}}{i}.$$

The second term in the obtained expression is estimated as follows:
$$\frac{1}{g}\sum_{m=1}^{f(g)}B_m\sum_{\lfloor f(g)/m\rfloor+1}^{\infty}\frac{q^{-im}}{i} \le \frac{1}{g}\sum_{m=1}^{f(g)}\frac{N_m}{m}\frac{q^{-m(\lfloor f(g)/m\rfloor+1)}}{(\lfloor f(g)/m\rfloor+1)(1-q^{-m})};$$
the right-hand side tends to
$$\frac{1}{g}\sum_{m=1}^{f(g)}N_m\frac{q^{-f(g)}}{f(g)(1-q^{-m})} \le \frac{1}{g}\sum_{m=1}^{f(g)}(q^m+1+2gq^{m/2})\frac{q^{-f(g)}}{f(g)(1-q^{-m})},$$

which, in turn, is not greater than
$$\frac{1}{g}\left(q^{f(g)}+1+2gq^{f(g)/2}\right)\frac{q^{-f(g)}}{1-q^{-1}}.$$
The last expression tends to zero since both g and $f(g)$ tend to infinity. □

For an arbitrary family of curves with $g(X_i) \to \infty$, we can estimate the asymptotic behaviour of the class number from both sides.

THEOREM 3.2.37. *Consider the family $\{X_i\}$ of all curves over $\mathbb{F}_q$ (one curve from each isomorphism class). Then*
$$1 \le \liminf \frac{1}{g(X_i)} \log_q h(X_i) \le \limsup \frac{1}{g(X_i)} \log_q h(X_i)$$
$$\le 1 + (\sqrt{q}-1)\log_q \frac{q}{q-1}.$$
□

EXERCISE 3.2.38. Prove the theorem. (*Hint*: Estimate the maximum and minimum of
$$\sum_{m=1}^{\infty} \beta_m \log \frac{q^m}{q^m-1}$$
provided that
$$\sum_{m=1}^{\infty} \frac{m\beta_m}{q^{m/2}-1} \le 1,$$
and then use the fact that any accumulation point of $\frac{1}{g(X_i)}\log_q h(X_i)$ is realized by an asymptotically exact subfamily.)

EXERCISE 3.2.39. Show that for an asymptotically exact family we have
$$\lim_{i\to\infty} \frac{1}{g(X_i)} \log_q P_{X_i}(t) = -\sum_{m=1}^{\infty} \beta_m \log_q(1-t^m)$$
uniformly in the region $0 < |t| < q^\sigma$ for any σ with $\operatorname{Re}\sigma < -1/2$. (*Hint*: In this region, the product representing the zeta function of the curve uniformly converges; then it is necessary to pass to the limit accurately.)

3.3. Elliptic Curves over Finite Fields

Elliptic curves over a finite field are of special interest not only as the first nontrivial (after the projective line) example but also due to the property of their points over $\mathbb{F}_{q^r}$ to form a finite group; such groups find numerous applications, including those in cryptography.

In Sec. 3.3.1 we study elliptic curves over $\mathbb{F}_q$ up to isomorphism; in particular, we find the number of isomorphism classes of such curves. In Sec. 3.3.2 the same is done for isogeny classes. Then we analyze endomorphism rings of these curves and their zeta functions. In Sec. 3.3.4 we specify groups that occur as $E(\mathbb{F}_q)$.

3.3.1. Isomorphism Classes

The theory of elliptic curves presented in Sec. 2.4 concerns the case of an algebraically closed ground field; whereas we are mainly interested in elliptic curves E over a finite field $\Bbbk = \mathbb{F}_q$ of $q = p^a$ elements (i.e., $\operatorname{char} \Bbbk = p$). In this case we have to make a number of changes in the theory given above. Let us begin from the very definition. From the definition of an elliptic curve as a curve of genus 1, it is not in the least clear why such a curve has a $\Bbbk$-point, which is necessary for introducing the group law and developing further theory. Moreover, for example, for $\Bbbk = \mathbb{Q}$ this is not true at all. However, in the case of a finite ground field, existence of such a point follows from the Weil bound (Theorem 3.1.24):

$$|E(\mathbb{F}_q)| \geq q + 1 - 2\sqrt{q} = (\sqrt{q} - 1)^2 > 0;$$

hence, $|E(\mathbb{F}_q)| \geq 1$, so any curve of genus 1 over a finite field has an $\mathbb{F}_q$-point P_0. Taking it as a starting point, we can proceed with defining the group structure using $\mathbb{F}_q$-points of the curve only. Thus, we obtain a finite group $E(\mathbb{F}_q)$ consisting of $\mathbb{F}_q$-points of E. Furthermore, we can always represent the equation of E in a canonical form:

THEOREM 3.3.1. *Over any field $\Bbbk$, an elliptic curve (that has a point of degree 1) is isomorphic to a plane curve with equation*

$$y^2 + a_1 xy + a_3 y = x^3 + a_2 x^2 + a_4 x + a_6; \quad (3.3.1)$$

moreover, we may assume that, if $\operatorname{char} \Bbbk > 3$, then $a_1 = a_2 = a_3 = 0$; if $\operatorname{char} \Bbbk = 3$, then $a_1 = a_3 = 0$; and if $\operatorname{char} \Bbbk = 2$, then $a_2 = 0$. □

The proof of this fact (see Sec. 2.4.1) uses the Riemann–Roch theorem only, which is valid over any field. Equation (3.3.1) is known as the *(generalized) Weierstrass equation*.

On the other hand, for $\operatorname{char} \Bbbk \neq 2$, representation of E in the form

$$y^2 = x(x-1)(x-\lambda)$$

with $\lambda \in \mathbb{F}_q$ is not always possible since roots of a cubic polynomial with coefficients from $\mathbb{F}_q$ need not lie in the same field $\mathbb{F}_q$. Also, in the case of a nonclosed ground field, the j-invariant $j(E)$ does not provide a criterion for isomorphism of curves. One can only state the following:

PROPOSITION 3.3.2. *Let E and E' be elliptic curves over a perfect field $\Bbbk$ such that $j(E) = j(E')$. Then E and E' are isomorphic over some finite extension $\widehat{\Bbbk}$ of $\Bbbk$; the degree $[\widehat{\Bbbk} : \Bbbk]$ divides 24; for $p = \operatorname{char} \Bbbk > 3$, it divides either 4 or 6. If $j(E) \neq 0$ and $j(E) \neq 1728$, then $[\widehat{\Bbbk} : \Bbbk] \leq 2$.*

In fact, as we shall now show, for $j(E) \neq 0$ or 1728 over a finite field there always exist precisely two elliptic curves with a given value of $j(E)$. To prove this, let us find when two elliptic curves, E and E', with equations

$$y^2 + a_1 xy + a_3 y = x^3 + a_2 x^2 + a_4 x + a_6$$

and

$$y^2 + \bar{a}_1 xy + \bar{a}_3 y = x^3 + \bar{a}_2 x^2 + \bar{a}_4 x + \bar{a}_6$$

are isomorphic over $\mathbb{F}_q$.

THEOREM 3.3.3. *The curves E and E' are isomorphic if and only if there exist constants $u, r, s, t \in \mathbb{F}_q$, $u \neq 0$, such that*

$$u\bar{a}_1 = a_1 + 2s,$$
$$u^2 \bar{a}_2 = a_2 - s a_1 + 3r - s^2,$$
$$u^3 \bar{a}_3 = a_3 + r a_1 + 2t,$$
$$u^4 \bar{a}_4 = a_4 - s a_3 + 2 r a_2 - (t + sr) a_1 + 3 r^2 - 2 s t,$$
$$u^6 \bar{a}_6 = a_6 + r a_4 + r^2 a_2 + r^3 - t a_3 - t^2 - r t a_1.$$

REMARK 3.3.4. In fact, this theorem is valid for any perfect ground field $\Bbbk$ (not only for $\mathbb{F}_q$).

SKETCH OF THE PROOF. Direct computation using an *admissible change of variables* of the form

$$(x, y) \mapsto (u^2 x + r, u^3 y + u^2 s x + t),$$

the inverse being given by

$$(x, y) \mapsto (u^{-2}(x - r), u^{-3}(y - s x - t + r s)). \qquad \square$$

In the case where the characteristic p of the ground field $\mathbb{F}_q$ is strictly greater than 3, successively making the change of variables

$$(x, y) \mapsto \left(x, y - \frac{a_1}{2} x - \frac{a_3}{2}\right),$$

which reduces (3.3.1) to

$$y^2 = x^3 + b_2 x^2 + b_4 x + b_6,$$

and the second change

$$(x, y) \mapsto \left(x - \frac{b_2}{3}, y\right),$$

one can reduce (3.3.1) to the form

$$y^2 = x^3 + a x + b, \qquad (3.3.2)$$

which we already know (the Weierstrass equation; see Remark 2.4.8), i.e., to equation (3.3.1) with $a_1 = a_2 = a_3 = 0$. Then the assertion of Theorem 3.3.3 is essentially simplified.

COROLLARY 3.3.5. *Let $p \neq 2, 3$. Then curves E and E' with equations*

$$E: \quad y^2 = x^3 + a x + b, \quad a, b \in \mathbb{F}_q,$$
$$E': \quad y^2 = x^3 + \bar{a} x + \bar{b}, \quad \bar{a}, \bar{b} \in \mathbb{F}_q,$$

are isomorphic if and only if there exists $u \in \mathbb{F}_q^*$ such that
$$u^4 \bar{a} = a, \qquad u^6 \bar{b} = b.$$

COROLLARY 3.3.6. *The group of $\mathbb{F}_q$-automorphisms of E is*
$$\mathrm{Aut}_{\mathbb{F}_q}(E) = \begin{cases} \{\pm 1\} & \text{if } a \neq 0, \ b \neq 0, \\ \mu_4 \cap \mathbb{F}_q^* & \text{if } b = 0, \\ \mu_6 \cap \mathbb{F}_q^* & \text{if } a = 0, \end{cases}$$
where μ_n is the group of n-th roots of unity in $\overline{\mathbb{F}}_q$.

We can now give a proof of Proposition 3.3.2 for our case of a finite field $\mathbb{F}_q$ of characteristic $p \neq 2, 3$.

PROOF OF PROPOSITION 3.3.2. Assume that $j(E) = j(E')$, i.e.,
$$\frac{a^3}{a^3 - 27b^2} = \frac{\bar{a}^3}{\bar{a}^3 - 27\bar{b}^2}.$$
If $j(E) \neq 0, 1728$, then neither of the coefficients is zero; hence it follows that $a^3/\bar{a}^3 = b^2/\bar{b}^2 \in \mathbb{F}_q^*$. If we put $u = \dfrac{\bar{a}b}{a\bar{b}} \in \mathbb{F}_q$, then the conditions of Corollary 3.3.5 are satisfied: indeed,
$$u^4 \bar{a} = \frac{\bar{a}^2 b^2}{a^2 \bar{b}^2} \bar{a} = \frac{\bar{a}^3}{a^3} \frac{b^2}{\bar{b}^2} a = a,$$
and similarly $u^6 \bar{b} = b$.

If $b = 0$, i.e., $j = 1728$, then put $u^4 = a/\bar{a}$, $u \in \mathbb{F}_{q^4}$; for $a = j = 0$, take $u^6 = b/\bar{b}$, $u \in \mathbb{F}_{q^6}$. This completes the proof in these cases. In characteristics 2 and 3, the proof is considerably more complicated. □

THEOREM 3.3.7 (mass formula). *Let $p \neq 2, 3$. Then*
$$\sum_E \frac{1}{|\mathrm{Aut}_{\mathbb{F}_q}(E)|} = q, \qquad (3.3.3)$$
where the sum is taken over isomorphism classes of elliptic curves over $\mathbb{F}_q$.

In view of the description of $\mathrm{Aut}_{\mathbb{F}_q}(E)$ given in Corollary 3.3.6, the theorem implies the following result.

COROLLARY 3.3.8. *Let $p \neq 2, 3$. The number of isomorphism classes of elliptic curves over $\mathbb{F}_q$ equals*

$$\begin{array}{lll} 2q + 6 & \text{if} & q \equiv 1 \pmod{12}, \\ 2q + 2 & \text{if} & q \equiv 5 \pmod{12}, \\ 2q + 4 & \text{if} & q \equiv 7 \pmod{12}, \\ 2q & \text{if} & q \equiv 11 \pmod{12}. \end{array}$$
□

EXERCISE 3.3.9. Prove the corollary.

PROOF OF THEOREM 3.3.7. Consider the set $\mathcal{E}_q$ of all pairs $(a,b) \in \mathbb{F}_q^2$ such that $y^2 = x^3 + ax + b$ defines an elliptic curve. Obviously,
$$\mathcal{E}_q = \mathbb{F}_q^2 \setminus \{(a,b) \mid a^3 - 27b^2 = 0\} = \mathbb{F}_q^2 \setminus \mathcal{D}_q,$$

where the set $\mathcal{D}_q$ contains precisely q pairs. This is easily verified by making the substitution $t = a/b$, whereafter the equation which connects t and a becomes that of a parabola. Hence,
$$|\mathcal{E}_q| = q^2 - q.$$

The group $\mathbb{F}_q^*$ acts on $\mathcal{E}_q$ according to the rule
$$(a,b) \mapsto (u^4 a, u^6 b);$$
orbits of this action are precisely the isomorphism classes of elliptic curves over $\mathbb{F}_q$. The orbit that corresponds to E is of length $(q-1)/|\mathrm{Aut}_{\mathbb{F}_q}(E)|$ since $\mathrm{Aut}_{\mathbb{F}_q}(E)$ is a stationary subgroup of $\mathbb{F}_q^*$ under this action. Hence we conclude that
$$\sum_E \frac{q-1}{|\mathrm{Aut}_{\mathbb{F}_q}(E)|} = |\mathcal{E}_q| = q^2 - q,$$
which gives (3.3.3) after dividing by $q-1$. □

3.3.2. Isogeny Classes

As we have seen in the previous subsection, the number of isomorphism classes of elliptic curves over $\mathbb{F}_q$ with growing q equals $2q + O(1)$. Let us now consider the question on isogeny classes. Recall that E and E' are isogeneous over $\mathbb{F}_q$ if there is a nonconstant map $f \colon E \to E'$ defined over $\mathbb{F}_q$ (one can check that this is an equivalence relation). Obviously, curves isomorphic over $\mathbb{F}_q$ are isogeneous over $\mathbb{F}_q$.

The problem of isogeny of curves E and E' over $\mathbb{F}_q$ is closely connected to that of the number of $\mathbb{F}_q$-points on them:

THEOREM 3.3.10. *Elliptic curves E and E' defined over $\mathbb{F}_q$ are isogeneous over $\mathbb{F}_q$ if and only if they have the same number of $\mathbb{F}_q$-points.* □

EXERCISE 3.3.11. Consider the elliptic curves
$$E \colon y^2 + y = x^3 + x \quad \text{and} \quad E' \colon y^2 + y = x^3 + x + 1$$
over $\mathbb{F}_2$. Show that $j(E) = j(E')$. Compute $|E(\mathbb{F}_{2^r})|$ and $|E'(\mathbb{F}_{2^r})|$ for $r = 1, 2, \ldots$. Find the smallest field over which E and E' are isomorphic.

Theorem 3.3.10 shows that the number of isogeny classes of elliptic curves over $\mathbb{F}_q$ is for sure not greater than $2\lfloor 2\sqrt{q} \rfloor + 1$ since, by the Weil theorem, the number of $\mathbb{F}_q$-points of an elliptic curve lies in the interval $[q+1-2\sqrt{q}, q+1+2\sqrt{q}]$. The next theorem describes which values of $m = q+1-N \in [-2\sqrt{q}, 2\sqrt{q}]$ really occur.

THEOREM 3.3.12 (Deuring–Waterhouse theorem). *There is a natural bijection between the set of classes of isogeneous elliptic curves over $\mathbb{F}_q$ and the set of integers m with $|m| \leq 2\sqrt{q}$ that satisfy one of the following conditions:*
  (1) $(q, m) = 1$;
  (2) *q is a square and $m = \pm 2\sqrt{q}$;*
  (3) *q is a square, $p \not\equiv 1 \pmod{3}$, and $m = \pm\sqrt{q}$;*
  (4) *q is not a square, $p = 2$ or 3, and $m = \pm\sqrt{pq}$;*
  (5) *q is not a square and $m = 0$; or q is a square, $p \not\equiv 1 \pmod{4}$, and $m = 0$.*

Furthermore, $|E(\mathbb{F}_q)| = q + 1 - m$ for all curves E from the isogeny class corresponding to m. □

Hence, the number of isogeny classes of elliptic curves over $\mathbb{F}_q$ is $4\sqrt{q}(1-1/p) + O(1)$, which is interesting to compare with the number $2q + O(1)$ of isomorphism classes. Thus, on the average, each isomorphism class of elliptic curves over $\mathbb{F}_q$ contains $\frac{\sqrt{q}}{2}\left(1-\frac{1}{p}\right)^{-1} + o(1)$ isogeny classes.

3.3.3. Endomorphism Ring and the Zeta Function

Elliptic curves over a finite field have one more feature: the number of their endomorphisms is large enough. More precisely, let $\mathrm{End}_{\mathbb{F}_q}(E)$ be the ring formed by endomorphisms of a curve E that are defined over $\mathbb{F}_q$. Of course, $\mathrm{End}_{\mathbb{F}_q}(E) \supseteq \mathbb{Z}$.

PROPOSITION 3.3.13. $\mathrm{End}_{\mathbb{F}_q}(E) \neq \mathbb{Z}$.

SKETCH OF THE PROOF. We shall prove the proposition for ordinary (non-supersingular) curves, as well as for supersingular curves E with $j(E) \in \mathbb{F}_p$. Indeed, let $q = p^a$; consider the map $F^a \colon E \to E^{(p^a)}$, the composition of a Frobenius morphisms. Since E is defined over $\mathbb{F}_q$, we have $E^{(p^a)} = E$, and therefore $F^a \in \mathrm{End}_{\mathbb{F}_q}(E)$ (this endomorphism of E is known as the *Frobenius endomorphism* over $\mathbb{F}_q$). If E is an ordinary curve, then $F^a \notin \mathbb{Z}$, since multiplication by an element $f \in \mathbb{Z}$ is never purely inseparable. If E is defined over a prime field $\mathbb{F}_p$, then $a = 1$, and $F \notin \mathbb{Z}$ since its degree is p (the degree of an element $n \in \mathbb{Z}$ is n^2). The case of a supersingular curve E with $j(E) \notin \mathbb{F}_p$ requires subtler considerations. □

Hence, by Theorem 2.4.23, we get the following corollary.

COROLLARY 3.3.14. *The group* $\mathrm{End}_{\mathbb{F}_q}(E)$ *always contains an order in some complex quadratic field.* □

Zeta function of an elliptic curve. According to formula (3.1.2), p. 138, the zeta function of an elliptic curve E over $\mathbb{F}_q$ is of the form

$$Z_E(t) = \frac{1 - mt + qt^2}{(1-t)(1-qt)},$$

where m is an integer, and Proposition 3.1.16d implies that $|E(\mathbb{F}_q)| = N = q + 1 - m$. The Weil theorem (proved by Hasse for the case of elliptic curves) states that $|m| \leq 2\sqrt{q}$. Let us give a sketch of the proof of the Hasse theorem for a nonsupersingular curve E.

SKETCH OF THE PROOF. By virtue of Corollary 3.3.14, the algebra $\mathrm{End}^0(E) = \mathrm{End}_{\mathbb{F}_q}(E) \otimes \mathbb{Q}$ is a complex quadratic field. On the endomorphism ring $\mathrm{End}_{\mathbb{F}_q}(E)$, which is an order in the field $\mathrm{End}^0(E)$, the morphism degree

$$\deg \colon \mathrm{End}_{\mathbb{F}_q}(E) \longrightarrow \mathbb{Z}$$

is defined; it has the following properties (see Proposition 2.4.12):

1. $\deg \psi = 0 \Leftrightarrow \psi = 0$,
2. $\deg(\psi \circ \varphi) = \deg \psi \cdot \deg \varphi$,
3. $\deg \psi^* = \deg \psi$ for the dual isogeny ψ,
4. $\varphi \circ \varphi^* = \deg \varphi$.

These properties follow directly from the definitions of the morphism degree and of the dual isogeny (Sec. 2.4.3).

Note that
$$\deg(1-F) = (1-F)\circ(1-F^*) = 1-(F+F^*)+F\circ F^* = 1-\operatorname{Tr} F + q.$$
On the other hand,
$$\deg(1-F) = |E(\mathbb{F}_q)|$$
since
$$E(\mathbb{F}_q) = \{P \in E(\overline{\mathbb{F}}_q) \mid F(P) = P\}.$$
Thus, $m = \operatorname{Tr} F$, and we have to prove that $|m| \le 2\sqrt{q}$. To this end, consider the endomorphism $t+sF$ with $t,s \in \mathbb{Z}$. Its degree is
$$\deg(t+sF) = t^2 + s^2 q + tsm \ge 0$$
for all s and t. This implies that the quadratic polynomial $x^2 + mx + q$ has a nonpositive discriminant, i.e., $m^2 - 4q \le 0$. $\square$

The above sketch of the proof explains why m is often called the *Frobenius trace* of E.

3.3.4. Structure of $E(\mathbb{F}_q)$

We conclude our brief discussion of elliptic curves over a finite field with a complete description of possible types of the groups $E(\mathbb{F}_q)$. Possible orders of these groups are already described in the Deuring–Waterhouse theorem (Theorem 3.3.12).

Possible types of $E(\mathbb{F}_q)$ are listed in the following theorem.

THEOREM 3.3.15. *A group G of order $N = q+1-m$ is isomorphic to $E(\mathbb{F}_q)$ for some elliptic curve E over $\mathbb{F}_q$ if and only if one of the conditions below is satisfied:*
  (1) $(q,m) = 1$, $|m| \le 2\sqrt{q}$, *and* $G \cong \mathbb{Z}/A\mathbb{Z} \times \mathbb{Z}/B\mathbb{Z}$, *where* $B \mid A$ *and* $B \mid (m-2)$;
  (2) q *is a square*, $m = \pm 2\sqrt{q}$, *and* $G \cong (\mathbb{Z}/A\mathbb{Z})^2$, *where* $A = \sqrt{q} \mp 1$;
  (3) q *is a square*, $p \not\equiv 1 \pmod{3}$, $m = \pm\sqrt{q}$, *and G is cyclic*;
  (4) q *is not a square*, $p = 2$ *or 3*, $m = \pm\sqrt{pq}$, *and G is cyclic*;
  (5a) *either q is not a square and $p \not\equiv 3 \pmod{4}$, or q is a square and $p \not\equiv 1$* (mod 4), $m = 0$, *and G is cyclic*;
  (5b) q *is not a square*, $p \equiv 3 \pmod{4}$, $m = 0$, *and either G is cyclic or* $G = \mathbb{Z}/M\mathbb{Z} \times \mathbb{Z}/2\mathbb{Z}$, *where* $M = (q+1)/2$.

SKETCH OF THE PROOF. Since $E(\mathbb{F}_q)$ is annihilated by multiplication by N, we have $E(\mathbb{F}_q) \subseteq E_N = (\mathbb{Z}/N\mathbb{Z})^2$. Hence, $E(\mathbb{F}_q) \cong \mathbb{Z}/A\mathbb{Z} \times \mathbb{Z}/B\mathbb{Z}$, where $B \mid A$ and $AB = N$. Choose a basis in E_N such that $E(\mathbb{F}_q)$ is generated by $\begin{pmatrix} A \\ 0 \end{pmatrix}$ and $\begin{pmatrix} 0 \\ B \end{pmatrix}$. Let M_F be the matrix of the action of the Frobenius endomorphism in this basis. Then
$$M_F = \begin{pmatrix} a & b \\ c & d \end{pmatrix}, \quad a,b,c,d \in \mathbb{Z}/N\mathbb{Z},$$
where $\operatorname{Tr} M_F \equiv m \pmod{N}$ and $\det M_F \equiv q \pmod{N}$. The subgroup $E(\mathbb{F}_q)$ in E_N consists of fixed elements under the action of M_F; therefore, $aA \equiv A \pmod{N}$ and $dB \equiv B \pmod{N}$. Hence, $m-2 = (a-1) + (d-1)$ is divisible by B since $B \mid A$. On the other hand, if $B \mid (m-2)$, then the matrix
$$M'_F = \begin{pmatrix} m-1 & -A \\ B & 1 \end{pmatrix}$$

satisfies the conditions $\operatorname{Tr} M'_F \equiv m$, $\det M'_F \equiv q$, $aA \equiv A$, and $dB \equiv B \pmod N$.

Now let us examine the cases of Theorem 3.3.12 one by one.

In case (1), one can show that, for any matrix M'_F such that $\operatorname{Tr} M'_F \equiv m$ and $\det M'_F \equiv q \pmod N$, there exists an elliptic curve E over $\mathbb{F}_q$ for which M'_F is the matrix of the action of the Frobenius endomorphism (this follows from the fact that any endomorphism E can be lifted to characteristic 0; this fact is proved by the same methods as Theorem 2.4.23 and Proposition 3.3.13). Applying this result to the matrix $M'_F = \begin{pmatrix} m-1 & -A \\ B & 1 \end{pmatrix}$, we obtain part (1) of Theorem 3.3.15.

In case (2) of Theorem 3.3.12, it is easy to check that M_F is a scalar matrix (e.g., using the fact that any ℓ-adic lifting of M_F is semi-simple), whence part (2) of Theorem 3.3.15 follows.

In case (3), we have $q \equiv N - 1 + m \equiv 1 \pmod B$ since $B \mid (m-2)$ and $B \mid N$. Next, $q = m^2 \equiv 4 \pmod B$. Note that the case $B = 3$ is impossible, since in this case we would have $m \equiv 2, 5$, or $8 \pmod 9$, and $q \equiv 4, 7$, or $1 \pmod 9$, respectively, so that $N \equiv 3 \pmod 9$, which contradicts the condition $B^2 \mid N$. Hence, $B = 1$, and therefore $E(\mathbb{F}_q)$ is cyclic.

In case (4), we have either $p = 2$, $m \equiv 2 \pmod B$, N and B odd, $2q = m^2 \equiv 4$, $q \equiv 2$, and $N \equiv 1 \pmod B$, whence $B = 1$; or $p = 3$, $(B, 3) = 1$, $m \equiv 2$, $3q = m^2 \equiv 4$, and $3N \equiv 1 \pmod B$, which again yields $B = 1$. Hence, $E(\mathbb{F}_q)$ is cyclic.

In case (5), $m = 0$; hence, $B = 1$ or 2. If $B = 2$, then $4 \mid N$, and therefore $q \equiv 3 \pmod 4$.

Since in cases (2), (3), (4), and (5a) the group $E(\mathbb{F}_q)$ is uniquely determined by its order N, and since by Theorem 3.3.12 a curve E with $|E(\mathbb{F}_q)| = N$ does exist, in these cases everything is proved. It remains to examine case (5b) and to show that both possibilities can be realized: $B = 1$ and $B = 2$.

Let us first show that, in this case, for a curve E with a cyclic group $E(\mathbb{F}_q)$, there is an isogeny $E \to E'$ of degree 2 such that $E'(\mathbb{F}_q)$ is not cyclic. Indeed, let $E(\mathbb{F}_q)$ be cyclic; then the subgroup $E_4 \cap E(\mathbb{F}_q)$ is also cyclic. Denote its generator by v, and let $E_2 = \{0, 2v, e, f = e + 2v\}$. Since E_2 does not lie in $E(\mathbb{F}_q)$, the Frobenius endomorphism interchanges e and f. Let $\varphi \colon E \to E'$ be the isogeny with the kernel generated by $2v$. Obviously, the Frobenius endomorphism preserves $\varphi(v)$ and $\varphi(e) = \varphi(f)$, so that $E'_2(\mathbb{F}_q) = (\mathbb{Z}/2\mathbb{Z})^2$.

On the other hand, let $E(\mathbb{F}_q) \supseteq (\mathbb{Z}/2\mathbb{Z})^2$. Then $E_4(\mathbb{F}_q)$ is of the form either $(\mathbb{Z}/2\mathbb{Z})^2$ or $\mathbb{Z}/2\mathbb{Z} \times \mathbb{Z}/4\mathbb{Z}$ (since $E_4(\mathbb{F}_q) \subseteq E(\mathbb{F}_q) \subseteq \mathbb{Z}/2\mathbb{Z} \times \mathbb{Z}/N\mathbb{Z}$). In the first case, for any isogeny $\varphi \colon E \to E'$ of degree 2, the group $E'(\mathbb{F}_q)$ is cyclic. In the second case, one has to consider the isogeny $\varphi \colon E \to E'$ with the kernel generated by $u + 2v$, where u is the generator of the quotient group $\mathbb{Z}/2\mathbb{Z}$, and v is that of $\mathbb{Z}/4\mathbb{Z}$ in $E_4(\mathbb{F}_q)$. □

3.4. Remarkable Examples

In this section we give some examples of curves with many points, namely, those of "elementary" character. In *Advanced Chapters*, we shall consider other examples, based on a more elaborate technique.

In Sec. 3.4.1 we briefly discuss a remarkable example of maximal curves, namely, the Hermitian curves, as well as their images under covering maps. In Sec. 3.4.2 we examine Kummer and Artin–Schreier covers, which also provide a lot of interesting examples. In Sec. 3.4.3 we present several constructions of asymptotically good (in particular, meeting the asymptotic Drinfeld–Vlăduţ bound) towers of curves due to García and Stichtenoth. Their advantage is that equations defining the towers are simple and explicit. Sec. 3.4.2 is devoted to several cases where $N_q(g)$ is known precisely, namely, curves of genus 1 and 2 and curves with $g=3$ for some values of q.

In this section, we discuss a negligible part of known examples. Regularly updated tables of parameters of curves with many points over finite fields are maintained by G. van der Geer and M. van der Vlugt [vdGvdV].

3.4.1. Hermitian, Sub-Hermitian, and Maximal Curves

A Hermitian form over the complex field is given by $\sum x_i \bar{x}_i$. If $q = r^2$, a Hermitian form has an obvious analogue over $\mathbb{F}_q$: it suffices to put $\bar{x} = x^r$ (cf. Remark 1.1.16).

A *Hermitian curve* over $\mathbb{F}_q$, $q = r^2$, is a curve X_q with the projective equation

$$X^{r+1} + Y^{r+1} + Z^{r+1} = 0$$

or, equivalently, with the affine equation

$$x^{r+1} + y^{r+1} + 1 = 0.$$

Since raising to the power r in $\mathbb{F}_q$ is an analogue of the Hermitian conjugation, the projective equation can be rewritten as

$$X\overline{X} + Y\overline{Y} + Z\overline{Z} = 0,$$

which explains the name of the curves.

EXERCISE 3.4.1. Check that Hermitian curves are plane smooth curves and hence that the genus of a Hermitian curve X_q is $g_q = r(r-1)/2 = (q - \sqrt{q})/2$.

To describe $\mathbb{F}_q$-points of X_q, it is convenient to make the following change of variables:

$$u = b/(y - bx), \qquad v = xu - a,$$

where $a, b \in \mathbb{F}_q$ satisfy $a^r + a = b^{r+1} = -1$.

EXERCISE 3.4.2. Prove that such a and b do exist. (*Hint*: For b, note that $|\mathbb{F}_q^*| = r^2 - 1$. For a, note that the image of $F(a) = a^r + a + 1$ lies in $\mathbb{F}_r$ and the degree of $F(a)$ is r.)

EXERCISE 3.4.3. Show that the above change of variables defines an isomorphism over $\mathbb{F}_q$ from X_q onto a curve X'_q with the affine equation

$$X'_q: \quad v^r + v = u^{r+1}.$$

EXERCISE 3.4.4. Show that X'_q contains at least $r^3 + 1$ points over $\mathbb{F}_q$. (*Hint*: X'_q consists of a single point at infinity, which is the common pole of u and v, and points with coordinates (α, β), where α is an arbitrary element of $\mathbb{F}_q$ and β is a solution of $\beta^r + \beta = \alpha^{r+1}$, all solutions lying in $\mathbb{F}_q$.)

This immediately implies the following result.

THEOREM 3.4.5. *Hermitian curves are maximal over $\mathbb{F}_q$.*

PROOF. Indeed, since
$$r^3 + 1 = r^2 + 1 + 2rg_q,$$
the curve X'_q contains precisely $r^3 + 1$ points over $\mathbb{F}_q$; therefore, X'_q (as well as the isomorphic curve X_q) is maximal over $\mathbb{F}_q$. □

Recall that Proposition 3.1.26 states that a curve X maximal over $\mathbb{F}_q$ always has $g(X) \leq g_q$.

EXERCISE 3.4.6. Prove that for a maximal curve X, any nontrivial cover $Y \to X$ is ramified.

PROPOSITION 3.4.7. *Let X be a curve over $\mathbb{F}_q$. If X is covered by a curve X_q, i.e., there is a surjective $\mathbb{F}_q$-morphism $\varphi \colon X_q \to X$, then X is also maximal.* □

Such curves X are known as *sub-Hermitian*.

REMARK 3.4.8. The proposition follows from the fact that for any map $X \to Y$ of curves over $\mathbb{F}_q$, the Frobenius roots for Y are a subset of the Frobenius roots for X, which, in turn, is due to the fact that J_Y is embedded in J_X. In *Advanced Chapters*, we shall return to this fundamental fact.

Strangely enough, presently we do not know any example of a maximal curve that cannot be obtained from Proposition 3.4.7; i.e., for each known maximal curve X, a required morphism φ can be constructed. This phenomenon gives rise to a natural question:

PROBLEM 3.4.9. *Is it true that any maximal curve X is covered by a Hermitian curve?*

There are a large number of results indicating the possible positive answer to this question.

Note that if $g(X) = g_q$, then the positive answer is equivalent to the following result.

THEOREM 3.4.10. *If $g(X) = g_q$ and X is a maximal curve, then X is isomorphic to X_q over $\mathbb{F}_q$.* □

Let us present some other results in this direction.

THEOREM 3.4.11. *If X is a maximal curve of genus g, then*
$$g = g_q = \frac{r(r-1)}{2} \quad or \quad g \leq \frac{(r-1)^2}{4}.$$
□

Note that this theorem is necessary for the positive answer, since a curve with $g_q > g > (r-1)^2/4$ cannot be covered by X_q, which follows from the Hurwitz formula.

THEOREM 3.4.12. *Let $q = r^2$ be odd, and let X be a maximal curve over $\mathbb{F}_q$ of genus $g = (r-1)^2/4$. Then X is isomorphic to the curve*

$$y^r + y = x^{\frac{r+1}{2}}$$

(which is covered by X_q under the obvious map $(x,y) \mapsto (x^2, y)$). □

There are a number of existence theorems, for instance, the following ones:

PROPOSITION 3.4.13. *Let p be odd and $q = p^n$. Then for any pair (v,w), $0 \le v \le n$, $0 \le w \le n-1$, there exists a maximal curve over $\mathbb{F}_q$ of genus*

$$g = \frac{1}{2} p^{n-v}(p^{n-w} - 1).$$
□

PROPOSITION 3.4.14. *Let $q = 2^n$. If $m \mid (q^2 - q + 1)$, then there exists a maximal curve over $\mathbb{F}_q$ of genus*

$$g = \frac{m-1}{2}.$$
□

The equation of the Hermitian curve can be generalized as follows. The field $\mathbb{F}_q$, $q = r^{2k}$, always contains an element α such that $\alpha^{r^k - 1} = -1$ (EXERCISE!). Define the curve X by the equation

$$\sum_{j=0}^{k-1} y^{r^j} = \alpha x^{r^k + 1}.$$

EXERCISE 3.4.15. Prove that X is maximal over $\mathbb{F}_q$.

Here, we have only presented the most elementary part of the theory of Hermitian curves; more detailed and technically elaborated exposition is postponed to *Advanced Chapters*.

3.4.2. Kummer and Artin–Schreier Covers

The second class of interesting examples of curves, often having many points, is given by curves that cover $\mathbb{P}^1$.

The reader should remember that many of the examples are created in search of improvement in tables of curves with many points. Accordingly, they are compared with the tables (see [vdGvdV]). The tables contain, for a given q and a given genus, a *lower bound*, i.e., the largest known number of points on a curve of this genus, and an *upper bound*, i.e., a value that certainly cannot be exceeded.

Hyperelliptic curves. A curve X is called *hyperelliptic* if it has a map $f\colon X \to \mathbb{P}^1$ of degree 2. Such a curve always has a singular plane model with an affine equation

$$y^2 + A(x)y + B(x) = 0,$$

and the map f for this model is given by the projection $f(x,y) = x$. If $g \ge 2$, then f is unique up to an automorphism of $\mathbb{P}^1$ (i.e., for any $f'\colon X \to \mathbb{P}^1$ with $\deg f' = 2$ there exists an automorphism $\sigma \in \mathrm{Aut}(\mathbb{P}^1)$ such that $\sigma \circ f' = f$) and hence is defined over the ground field $\mathbb{F}_q$. Hyperelliptic curves exist for any genus g. Obviously, $|X(\mathbb{F}_q)| \le 2g + 2$. Therefore, these curves can be maximal only if $g \le (q+1)/\lfloor 2\sqrt{q} \rfloor$; in this respect, they are much worse than Hermitian curves, which have $g = (q - \sqrt{q})/2$.

Kummer covers. Let $f \in \mathbb{F}_q(x)$, and let n be a divisor of $q-1$. Consider the function field $\mathbb{F}_q(x,y)$, where
$$y^n = f(x), \quad f(x) \in \mathbb{F}_q(x).$$
We assume that $f \notin \overline{\mathbb{F}}_q(x)^d$ for $d \mid (q-1)$, $d > 1$. Then the curve $X_K(f)$ corresponding to this field is called the *Kummer curve* corresponding to f. The natural cover $X_K(f) \to \mathbb{P}^1$ of degree n (the *Kummer cover*) is given by the projection $(x,y) \mapsto x$.

We assume in addition that $f(x)$ satisfies the following condition: there exists a place (of the function field $\mathbb{F}_q(x)$) where the order of $f(x)$ is relatively prime to p. Then the Hurwitz formula yields the following result:

THEOREM 3.4.16. *The genus of a Kummer curve is*
$$g(X_K) = 1 - n + \frac{1}{2} \sum_Q (n - r_Q) \deg Q,$$
where the sum is taken over all places Q of the function field $\mathbb{F}_q(x)$, and
$$r_Q = \gcd(n, v_Q(f)). \qquad \square$$

Note that $\gcd(n,0) = n$, so the sum involves only places Q with $v_Q(f) \neq 0$, i.e., zeroes and poles of f.

In particular, if $f(x)$ is a polynomial, its degree m is relatively prime to n, and $f(x)$ has no multiple irreducible factors in $\mathbb{F}_q[x]$, then
$$g(X_K) = \frac{(n-1)(m-1)}{2}.$$

The question arises of how to choose $f(x)$ in such a way that the Kummer curve contains a large number of $\mathbb{F}_q$-points. One possible method is as follows.

Let us consider covers of degree $q-1$ only, and let us require $f(x)$ to satisfy the following conditions: First, $f(x) = 1$ for a large enough subset $\mathcal{P} \subset \mathbb{P}^1(\mathbb{F}_q)$ (then the curve contains at least $(q-1)|\mathcal{P}|$ rational points). Second, $f(x)$ has many multiple zeroes and poles; then we can bound the genus of the curve. Indeed, let the divisor of f over the algebraic closure of the ground field be $(f) = \sum_{i=1}^\ell d_i P_i$, where $P_i \in \mathbb{P}^1(\overline{\mathbb{F}}_q)$. Then, by Theorem 3.4.16, we have
$$g = 1 + \frac{1}{2}(\ell - 2)(q-1) - \frac{1}{2} \sum_{i=1}^\ell \gcd(q-1, d_i); \qquad (3.4.1)$$
i.e., we would like ℓ to be as small as possible and the common divisors to be large.

In view of this, let us choose functions f as follows. Let $q = p^m$; consider $\mathbb{F}_q$ as a vector space over $\mathbb{F}_p$. Let L be its r-subspace, $r \geq 2$.

EXERCISE 3.4.17. *Prove that the polynomial*
$$R(x) = \prod_{c \in L}(x - c)$$
is p-linearized, i.e., is of the form
$$R(x) = \sum_{i=0}^r a_i x^{p^i} \in \mathbb{F}_q[x],$$
and moreover $a_0 a_r \neq 0$; conversely, zeroes of any polynomial of this type form a linear space.

Let us represent $R(x)$ as a sum $R(x) = R_1(x) + R_2(x)$ of two polynomials of the form $R_1(x) = \sum_{i=s}^{r} b_i x^{p^i}$ and $R_2(x) = \sum_{i=0}^{t} c_i x^{p^i}$, where $t \leq s$, $b_s b_r \neq 0$, and $c_0 c_t \neq 0$. Denote by L_i the space of zeroes of R_i; in particular, $|L_1| = p^{r-s}$ and $|L_2| = p^t$. Furthermore, we require $L_1 \neq L_2$.

Now, let

$$f(x) = -\frac{R_1(x)}{R_2(x)} = -\frac{\left(\sum_{i=s}^{r} b_i^{p^{-s}} x^{p^{i-s}}\right)^{p^s}}{\sum_{i=0}^{t} c_i x^{p^i}}. \qquad (3.4.2)$$

Then $f(x) = 1$ for $x \in L \setminus (L_1 \cup L_2)$. The condition $R = R_1 + R_2$ implies that $L \cap (L_1 \cup L_2) = L_1 \cap L_2$, i.e., $L \setminus (L_1 \cup L_2) = L \setminus (L_1 \cap L_2)$. Moreover, the zeroes of R_1 and the pole at infinity are multiple. Finally, the condition $L_1 \neq L_2$ guarantees that $f \neq g^d$ for $g \in \overline{\mathbb{F}}_q(x)$, $d > 1$, and $d \mid (q-1)$. Thus, such a function $f(x)$ satisfies all our requirements.

From all this, one can deduce the following proposition.

PROPOSITION 3.4.18. *The Kummer curve X_K given by $y^{q-1} = f(x)$, where $f(x) = -R_1(x)/R_2(x)$ is of the form (3.4.2), is a curve of genus*

$$g(X_K) = \frac{1}{2}\Big((p^{r-s} + p^t - \delta - 1)(q-2) - \delta p^{\gcd(m,s)} - p^{\gcd(m,r-t)} + 2\delta + 2\Big),$$

and the number of $\mathbb{F}_q$-points of X satisfies

$$|X_K(\mathbb{F}_q)| \geq (p^r - \delta)(q-1),$$

where $\delta = |L_1 \cap L_2|$. □

REMARK 3.4.19. Additional rational points could also exist, namely, rational ramification points on the Kummer curve. The set of ramification points on $\mathbb{P}^1$ is $L_1 \cup L_2 \cup \infty$.

EXAMPLE 3.4.20. Let $q = 16$; take $L = \mathbb{F}_{16}$. Then $R(x) = x^{16} + x$, and we represent it as the sum of the polynomials $R_1(x) = x^{16} + x^2$ and $R_2(x) = x^2 + x$. In this case $r = 4$, $s = t = 1$, $L_1 = \mathbb{F}_8$, $L_2 = \mathbb{F}_2$, and $\delta = 2$. Then by Proposition 3.4.18 we see that the curve X_K defined over $\mathbb{F}_{16}$ by

$$y^{15} = \frac{x^{16} + x^2}{x^2 + x} = x^{14} + x^{13} + \ldots + x$$

is of genus 49, and $|X_K(\mathbb{F}_{16})| \geq 14 \cdot 15 = 210$. Actually, $|X_K(\mathbb{F}_{16})| = 213$ since there are also three rational ramification points over points in $\mathbb{F}_2 \cup \infty$. This example, as well as almost all examples in this section, gives an entry in the table of curves with the best parameters [vdGvdV], i.e., gives the best lower bound on $N_q(g)$ presently known.

The case $L = \mathbb{F}_q$. Note that for curves from Proposition 3.4.18, the ratio N/g is lower bounded by $2p^r/(p^{r-s} + p^t)$, which is optimal if $s = t = \lfloor r/2 \rfloor$. Therefore, the case $L = \mathbb{F}_{p^m}$, i.e., $R(x) = x^{p^m} - x$, might be especially interesting.

For an odd m, put

$$x^{p^m} - x = R_1 + R_2 = \left(x^{p^m} - ax^{p^{(m-1)/2}}\right) + \left(ax^{p^{(m-1)/2}} - x\right),$$

where $a \in \mathbb{F}_q^*$; i.e., consider the case $s = t = \lfloor m/2 \rfloor$. Since
$$\gcd\left(x^{p^m} - ax^{p^{(m-1)/2}}, ax^{p^{(m-1)/2}} - x\right) = \gcd\left(x^{p^m} - x, ax^{p^{(m-1)/2}} - x\right),$$
for $u \in \mathbb{F}_q^*$ we obtain
$$u \in L_1 \cap L_2 \quad \text{if and only if} \quad u^{p^{(m-1)/2}-1} = 1/a.$$
This equation has no solutions in $\mathbb{F}_q^*$ if a is not a $(p^{(m-1)/2} - 1)$st power in $\mathbb{F}_q^*$; otherwise, the number of solutions in $\mathbb{F}_q^*$ is $\gcd(p^{(m-1)/2} - 1, p^m - 1) = p - 1$ (the latter, in particular, always holds for $p = 2$).

The set of elements $a \in \mathbb{F}_q^*$ that can be represented as $a = b^d$, $b \in \mathbb{F}_q^*$, will be denoted by $(\mathbb{F}_q^*)^d$. First, let us consider the case where $a \in (\mathbb{F}_q^*)^{p^{(m-1)/2}-1}$.

PROPOSITION 3.4.21. *For an odd $m \geq 3$, the curve X_m defined over $\mathbb{F}_q = \mathbb{F}_{p^m}$ by*
$$y^{q-1} = -\frac{\left(x^{p^{(m+1)/2}} - a^{p^{(m+1)/2}} x\right)^{p^{(m-1)/2}}}{ax^{p^{(m-1)/2}} - x},$$
where $a \in (\mathbb{F}_q^*)^{p^{(m-1)/2}-1}$, is of genus
$$g(X_m) = \frac{1}{2}\left((p^{(m+1)/2} + p^{(m-1)/2} - p - 1)(q-2) - p^2 + p + 2\right)$$
and has the following number of rational points:
$$N = |X_m(\mathbb{F}_q)| = \begin{cases} (q-1)(q-p), & p \neq 2, \\ (q-1)(q-p) + 3, & p = 2. \end{cases} \qquad \square$$

EXAMPLE 3.4.22. As an example, take $p = 3$ and $m = 3$; we obtain a curve over $\mathbb{F}_{27}$ given by
$$-y^{26} = x^{24} + x^{22} + \ldots + x^2.$$
It has $g = 98$ and $N = 624$, which is not too far from the best known upper bound on the number of points of a curve of this genus, $N \leq 745$.

If we take $\mathbb{F}_q = \mathbb{F}_{32}$, then the curve with equation
$$y^{31} = (x^8 + x)^4/(x^4 + x)$$
has genus 135 and $N = 31 \cdot 30 + 3 = 933$; the upper bound in this case is $N \leq 1098$.

For $a \notin (\mathbb{F}_q^*)^{p^{(m-1)/2}-1}$ there is a similar statement:

PROPOSITION 3.4.23. *For an odd $m \geq 3$, the curve X_m defined over $\mathbb{F}_q = \mathbb{F}_{p^m}$ by*
$$y^{q-1} = -\frac{\left(x^{p^{(m+1)/2}} - a^{p^{(m+1)/2}} x\right)^{p^{(m-1)/2}}}{ax^{p^{(m-1)/2}} - x},$$
where $a \notin (\mathbb{F}_q^*)^{p^{(m-1)/2}-1}$, is of genus
$$g(X_m) = \frac{1}{2}\left((p^{(m+1)/2} + p^{(m-1)/2} - 2)(q-2) - 2p + 4\right)$$
and has the following number of rational points:
$$N = |X_m(\mathbb{F}_q)| = \begin{cases} (q-1)^2, & -a \notin (\mathbb{F}_q^*)^{p-1}, \\ (q-1)^2 + 2(p-1), & -a \in (\mathbb{F}_q^*)^{p-1}. \end{cases} \qquad \square$$

EXAMPLE 3.4.24. For an odd p, we can take $a = -1$. Over $\mathbb{F}_{27}$, the curve X_m has genus 124 and the number of points $N = 680$, while the upper bound is $N \leq 901$.

Now consider the case of an even m. In this case, even the decomposition
$$x^q - x = \left(x^q - ax^{\sqrt{q}}\right) + \left(ax^{\sqrt{q}} - x\right),$$
where $a \in \mathbb{F}_q^*$ and $a \notin (\mathbb{F}_q^*)^{\sqrt{q}-1}$, yields very good curves.

PROPOSITION 3.4.25. *For an even m, the curve X_m defined over $\mathbb{F}_q = \mathbb{F}_{p^m}$ by*
$$y^{q-1} = -\frac{x^q - ax^{\sqrt{q}}}{ax^{\sqrt{q}} - x},$$
where $a \in \mathbb{F}_q^*$, $a \notin (\mathbb{F}_q^*)^{\sqrt{q}-1}$, is of genus $g(X_m) = (\sqrt{q}-1)(q-2) - \sqrt{q} + 2$ and the number of points $N = |X_m(\mathbb{F}_q)| = (q-1)^2$. □

EXAMPLE 3.4.26. For $q = 9$, we obtain $g = 13$ and $N = 64$, which is very close to the upper bound $N \leq 66$ and might well be the actual value of $N_q(g)$. For $q = 16$, we find $g = 40$ and $N = 225$; the upper bound in this case is $N \leq 244$. For $q = 64$, we get $g = 428$ and $N = 3969$, the upper bound being $N \leq 4786$.

Subspaces of codimension 1. Now let L be the subspace
$$L = \{x \in \mathbb{F}_q : \operatorname{tr}_{\mathbb{F}_q/\mathbb{F}_p} x = 0\},$$
where
$$\operatorname{tr}_{\mathbb{F}_q/\mathbb{F}_p} x = x^{p^{m-1}} + \ldots + x^p + x;$$
put
$$R(x) = \sum_{i=0}^{m-1} x^{p^i} = \sum_{i=s}^{m-1} x^{p^i} + \sum_{i=0}^{s-1} x^{p^i}.$$
Applying Proposition 3.4.18, we obtain the following result:

PROPOSITION 3.4.27. *For $m \geq 3$ and $0 < s < m-1$, provided that $\gcd(m,s) = 1$, the curve X_m defined over $\mathbb{F}_q = \mathbb{F}_{p^m}$ by*
$$y^{q-1} = -\frac{\left(x^{p^{m-1-s}} + \ldots + x\right)^{p^s}}{x^{p^{s-1}} + \ldots + x}$$
is of genus
$$g(X_m) = \frac{1}{2}\left((p^{m-1-s} + p^{s-1} - 2)(q-2) - 2p + 4\right)$$
and has the following number of rational points:
$$N = |X_m(\mathbb{F}_q)| = \begin{cases} (p^{m-1}-1)(q-1) \\ \quad \text{if } pm \text{ is odd and } p \nmid s(m-s), \\ (p^{m-1}-1)(q-1) + p - 1 \\ \quad \text{if } pm \text{ is odd and } p \mid s(m-s), \\ (p^{m-1}-1)(q-1) + 2(p-1) \\ \quad \text{if } pm \text{ is even and } p \nmid s(m-s), \\ (p^{m-1}-1)(q-1) + 3(p-1) \\ \quad \text{if } pm \text{ is even and } p \mid s(m-s). \end{cases}$$
□

EXAMPLE 3.4.28. Take $\mathbb{F}_{27}$; then L is given by $R(x) = x^9 + x^3 + x$, and the corresponding curve is
$$y^{26} = -(x^8 + x^2).$$
It has $g = 24$ and $N = 208$, which gives an entry in the table of optimal curves.

For $\mathbb{F}_{32}$ we get the curve
$$y^{31} = (x^4 + x^2 + x)^4 / (x^2 + x)$$
of genus 60 with $N = 468$, the upper bound being $N \leq 542$.

Now let us consider an example with $\gcd(m, s) \neq 1$.

EXAMPLE 3.4.29. Take $\mathbb{F}_{64}$ and
$$f(x) = \frac{x^{32} + x^{16}}{x^8 + x^4 + x^2 + x} = \frac{(x^2 + x)^{15}}{(x^4 + x + 1)(x^2 + x + 1)}.$$
For the curve X with equation $y^{63} = f(x)$, formula (3.4.1) gives
$$g = 1 + \frac{1}{2}(7 \cdot 63 - 3 \cdot 3 - 6 \cdot 1) = 214.$$
Each of the ramification points 0, 1, and ∞ gives three more rational points on X; each zero of $x^2 + x + 1$ gives one more point; finally, ramification points that correspond to zeroes of $x^4 + x + 1$ are not rational. Therefore, the total number of rational points on X is $N = (32 - 2) \cdot 63 + 11 = 1901$; the upper bound is $N \leq 2553$.

Note that for an even m and $s = m/2$, the decomposition
$$\sum_{i=0}^{m-1} x^{p^i} = R_1(x) + R_2(x) = \sum_{i=m/2}^{m-1} x^{p^i} + \sum_{i=0}^{m/2-1} x^{p^i}$$
does not satisfy the condition $L_1 \neq L_2$. Therefore, the corresponding equation $y^{q-1} = -R_1/R_2 = -R_2^{\sqrt{q}-1}$ leads to the cover
$$y^{\sqrt{q}+1} = aR_2 = a\left(x^{p^{m/2}-1} + x^{p^{m/2}-2} + \ldots + x\right), \qquad (3.4.3)$$
where $a \in \mathbb{F}_q^*$ is a root of $a^{\sqrt{q}} + a = 0$.

PROPOSITION 3.4.30. *For curves of the form* (3.4.3) *over* $\mathbb{F}_q$, *the genus is* $(q - p\sqrt{q})/(2p)$, *and the number of points is* $q\sqrt{q}/p + 1$. *In particular, these curves are maximal.* □

REMARK 3.4.31. Making the substitution $x = z^p - z$ in (3.4.3), we obtain the Hermitian curve $y^{\sqrt{q}+1} = a(z^{\sqrt{q}} - z)$. Thus, a curve from the above proposition is sub-Hermitian.

Finally, note that we are not obliged to start with linearized polynomials only. Thus, the following example concerns the curve of the form
$$y^{q-1} = xf^p(x), \quad f(x) \in \mathbb{F}_q[x].$$

EXAMPLE 3.4.32. Take $\mathbb{F}_{16}$ and consider the irreducible complete nonsingular curve with equation
$$y^{15} = x(x^2 + x + 1)^2.$$
Note that $f(x) = x(x^2 + x + 1)^2 = x^5 + x^3 + x$ satisfies all the conditions we have imposed on $f(x)$. Thus, according to (3.4.1), this curve is of genus $g = 12$ and has

the number of points $N = 15 \cdot 5 + 8 = 83$, where the ramification point ∞ gives five rational points, and each of the ramification points in $\mathbb{F}_4 \setminus \{1\}$ gives one point more.

Artin–Schreier covers. Another interesting class of examples is produced by Artin–Schreier covers. Let $f \in \mathbb{F}_q(x)$, char $\mathbb{F}_q = p$, and let $f \neq g - g^p$ for any function $g \in \overline{\mathbb{F}}_q(x)$. Consider the function field $\mathbb{F}_q(x, y)$, where

$$y^p - y = f(x), \quad f(x) \in \mathbb{F}_q(x).$$

Then the curve $X_{AS}(f)$ corresponding to this field is called the *Artin–Schreier curve corresponding to* f. The natural cover $X_{AS}(f) \to \mathbb{P}^1$ of degree p (*Artin–Schreier cover*) is defined by the projection $(x, y) \mapsto x$.

In particular, for $f = x^{p+1}$, the curve $X_{AS}(f)$ is the Hermitian curve $y^p - y = x^{p+1}$.

We shall assume that $f(x)$ satisfies the following condition: $f(x)$ has a pole of order relatively prime to p. Then we have the following statement.

PROPOSITION 3.4.33. *Let the function $f(x)$ be represented as an irreducible fraction $g(x)/h(x)$, where g and h are polynomials. Let*

$$h(x) = h_1(x)^{m_1} \ldots h_t(x)^{m_t}$$

be a prime factorization in $\mathbb{F}_q[x]$. If all m_i are relatively prime to p, then all ramification places of the Artin–Schreier cover (possibly except for infinity) are precisely the places corresponding to the factors $h_i(x)$, and the ramification index at them is precisely p. The place at infinity is ramified if $\deg g(x) > \deg h(x)$ and $\deg g(x) - \deg h(x)$ is not divisible by p. □

In particular, if $f(x)$ is a polynomial whose degree is not a multiple of p, then the Artin–Schreier cover is ramified at infinity only.

REMARK 3.4.34. Note that this phenomenon—ramification at a single point—is possible in positive characteristic only; over a field of characteristic zero, any nontrivial cover of $\mathbb{P}^1$ is ramified at least at two points (cf. Exercise 2.2.39).

In view of Proposition 3.4.33, the Hurwitz formula yields the following expression for the genus of X_{AS}:

THEOREM 3.4.35. *Let all the m_i be relatively prime to p. Let $\deg g(x) > \deg h(x)$, and assume that p does not divide $\deg g(x) - \deg h(x)$. Then*

$$g(X_{AS}) = \frac{p-1}{2} \left(\sum_i \deg h_i(x) + \deg g(x) - 1 \right).$$
□

Choosing f in an appropriate way, one can find curves with many points. However, still better results are obtained using the following construction.

Fibre product of curves. Let $\varphi \colon Y \to X$ and $\psi \colon Z \to X$ be two nontrivial (i.e., nonconstant) $\mathbb{F}_q$-morphisms of curves. Then we can consider the following set (the *fibre product*):

$$Y \underset{X}{\times} Z = \{(y, z) \in Y \times Z \mid \varphi(y) = \psi(z)\}.$$

It can be shown that the set $Y \times_X Z$ is always closed in $Y \times Z$ but is not always irreducible. However, if $Y \times_X Z$ is irreducible, then it is an algebraic curve, which is smooth under some natural conditions on φ and ψ. We shall use this construction in the case where φ and ψ are Artin–Schreier covers.

To be precise, let X be a smooth complete irreducible curve over $\mathbb{F}_q$ with a sufficiently large number of $\mathbb{F}_q$-points, $q = p^m$. Consider the following family of Artin–Schreier covers. Let $\mathcal{P}$ be a (large enough) set of rational points on X, and let D be an effective divisor defined over $\mathbb{F}_q$ such that $\operatorname{Supp} D \cap \mathcal{P} = \varnothing$. Finally, consider the vector $\mathbb{F}_p$-subspace $F \subset L(D)$ consisting of functions f that satisfy the following conditions: first,

$$F \cap \{g^p - g \mid g \in \mathbb{F}_q(X)\} = \{0\},$$

and second,

$$\operatorname{tr}_{\mathbb{F}_q/\mathbb{F}_p}(f(P)) = 0 \quad \text{for any } P \in \mathcal{P}.$$

Consider an arbitrary basis $f_1, \ldots, f_r$ of such F. Let X_{f_i} be the Artin–Schreier cover of X given by the equation $z_i^p - z_i = f_i$. Finally, consider the normalization of the fibre product of these covers

$$X_F = \left(X_{f_1} \underset{X}{\times} \ldots \underset{X}{\times} X_{f_r} \right)^{\nu}.$$

By a curve over X, we mean a curve Y together with a morphism $f \colon Y \to X$. We say that Y is *isomorphic over X* (or *X-isomorphic*) to a curve Z, where Z is defined together with a morphism $g \colon Z \to X$, if there is an isomorphism $h \colon Y \to Z$ such that $f = g \circ h$.

PROPOSITION 3.4.36. *Up to an X-isomorphism, the curve X_F does not depend on the choice of basis in F. The genus of X_F is*

$$g(X_F) = g(X) + \sum_{f \in \mathbb{P}F} \bigl(g(X_f) - g(X) \bigr),$$

where $\mathbb{P}F$ denotes the projectivization of F as a vector space over $\mathbb{F}_p$, and the notation $f \in \mathbb{P}F$ means that we take one representative from each projective equivalence class. □

For an arbitrary divisor $D = \sum_Q a_Q Q$, denote by $\lceil D/p \rceil$ the divisor $\sum_Q \lceil a_Q/p \rceil Q$.

LEMMA 3.4.37. *The dimension of the space*

$$V = \{ f \in L(D) \mid f = g^p - g \text{ for some } g \in \overline{\mathbb{F}}_q(X) \}$$

equals $\dim_{\mathbb{F}_p} L(\lceil D/p \rceil)$. □

Now we can formulate a statement on the number of points on the curves C_f and C_F.

PROPOSITION 3.4.38. *Let D be an $\mathbb{F}_q$-divisor on a curve X, $\mathcal{P} = \{P_1, \ldots, P_n\}$ be a subset of $\mathbb{F}_q$-points on X, and*

$$\delta = |\operatorname{Supp} D \cap \mathcal{P}|.$$

Then, for any integer r from the interval

$$1 \le r \le \dim_{\mathbb{F}_p} L(D) - \dim_{\mathbb{F}_p} L(\lceil D/p \rceil) + 1 - n + \delta,$$

there exists an $\mathbb{F}_p$-subspace $F_r \in L(D)$ of dimension r such that for any $f \in F_r$ the curve X_f possesses the property

$$|C_f(\mathbb{F}_q)| = p(n - \delta) + \varepsilon_f,$$

where ε_f is the number of rational points on X_f lying in the inverse image of points of Supp D.

PROOF. Let V be a subspace defined in Lemma 3.4.37. Take a complementary subspace, i.e., a linear $\mathbb{F}_p$-subspace $W \in L(D)$ of dimension $\dim_{\mathbb{F}_p} L(D) - \dim_{\mathbb{F}_p} L(\lceil D/p \rceil)$ such that $W \cap V = \{0\}$. Take any element c of $\mathbb{F}_q$ such that $\mathrm{tr}_{\mathbb{F}_q/\mathbb{F}_p}(c) \neq 0$, put
$$\widetilde{W} = W \oplus (\mathbb{F}_p \cdot c),$$
and consider in $\widetilde{W}$ the system of linear equations
$$\mathrm{tr}_{\mathbb{F}_q/\mathbb{F}_p}(f(P_i)) = 0, \quad \text{for all } P_i \notin \mathrm{Supp}\, D. \tag{3.4.4}$$
The solution space of this homogeneous system of linear equations over $\mathbb{F}_p$ is of dimension at least $\dim_{\mathbb{F}_p} \widetilde{W} - (n - \delta)$. Equations of the system precisely express the condition that all the points P_i that do not belong to Supp D are completely split; i.e., each of them has the maximum possible number of distinct preimages defined over $\mathbb{F}_q$ (see p. 124). □

COROLLARY 3.4.39. *In the conditions of Proposition 3.4.38, for each subspace F_r we have*
$$|X_{F_r}(\mathbb{F}_q)| = p^r(n - \delta) + \delta + \sum_{f \in \mathbb{P} F_r} (\varepsilon_f - \delta).$$
□

EXAMPLE 3.4.40. Let X be the elliptic curve given over $\mathbb{F}_2$ by
$$y^2 + y = x^3 + x.$$
Rational points on this curve are P_∞, $P_1 = (0,0)$, $P_2 = (1,0)$, $P_3 = (1,1)$, and $P_4 = (0,1)$. Take the divisor $D = 3P_\infty + P_1 + P_3$ and consider the subspace $\widetilde{W}$ over $\mathbb{F}_2$ with the basis
$$1,\ x,\ y,\ \frac{x}{x+y},\ \frac{xy}{x+y}.$$
The solution space of system (3.4.4) for P_2 and P_4 is generated by the functions
$$f_1 = x + \frac{x}{x+y},\quad f_2 = 1 + x + y,\quad f_3 = \frac{xy}{x+y}.$$
The two-dimensional $\mathbb{F}_2$-subspace $F = \langle f_1, f_2 + f_3 \rangle$ gives a curve of genus 10 with 13 points.

For $D = 3P_\infty + 3P_4$, take $\widetilde{W} = \langle 1, y/x, x, y, y/(x(x+y+1)) \rangle$; then the subspace $F = \langle 1 + y/x + x + y, (xy + y^2)/(x(x+y+1)) \rangle$ gives a curve of genus 11 with 14 points.

For $D = 7P_\infty + P_4$, the solution space of system (3.4.4) for $\{P_1, P_2, P_3\}$ is $H = \langle 1 + x + y + y/x, y + xy, y + x^2 y \rangle$. The subspace $F = \langle 1 + x + y + y/x, xy + x^2 y \rangle$ gives a curve X_F of genus 13 with 15 points. Finally, the whole subspace H gives a curve C_H of genus 29 with 25 points.

The first three curves of this example are optimal; the last one gives the best known lower bound, the upper bound being $N \leq 27$.

EXAMPLE 3.4.41. Consider the elliptic curve X over $\mathbb{F}_3$ with equation
$$y^3 - y = x^2 - 1.$$

It has 7 rational points. Letting $P_\infty = (0:1:0)$, $P_1 = (1,0)$, and $D = 8P_\infty + 2P_1$, we may take the subspace $\widetilde{W}$ with the $\mathbb{F}_3$-basis

$$1, \frac{y+2}{2x+y+1}, \frac{x+y+1}{y}, y, x+2, y^2, (x+2)y, (x+2)y^2, (x+2)^2 y.$$

Solving in $\widetilde{W}$ the system

$$\mathrm{tr}_{\mathbb{F}_3/\mathbb{F}_3}(f(P_i)) = f(P_i) = 0, \quad i = 2, \ldots, 6,$$

for the remaining five rational points P_i, we obtain a four-dimensional solution space F, generated by

$$f_1 = \frac{y+2}{2x+y+1} + y + x + y^2,$$

$$f_2 = \frac{y+2}{2x+y+1} + \frac{x+y+1}{y} + x + 2xy,$$

$$f_3 = 2 + \frac{y+2}{2x+y+1} + y + 2y^2 + xy^2, \quad f_4 = 2y + x^2 y.$$

Taking $F = \langle f_1, f_2 \rangle$, we get a curve of genus 35 with 47 points (this is the best known lower bound, the upper bound being 51); taking $F = \langle f_1, f_2, f_3 \rangle$, we get a curve of genus 128 with 136 points (the upper bound is 149).

EXAMPLE 3.4.42. Consider the same elliptic curve X over $\mathbb{F}_3$ as in the previous example. Take a point Q of degree 4 given by the ideal $(x^2 + xy + 1, x^4 + x^3 + 2)$ and consider the divisor $D = 3Q$. Solutions of system (3.4.4) for all seven rational points form a two-dimensional linear space F with the basis

$$1 + f_2 + 2f_3 + 2f_1^2 + f_2^2 + f_3^2, \quad 2 + f_1^2 + f_2^2 + f_1 f_2 f_3,$$

where

$$f_1 = \frac{2x+y}{x^2+xy+1}, \quad f_2 = \frac{2x^2+1}{x^2+xy+1}, \quad f_3 = \frac{2x^2+y^2}{x^2+xy+1}.$$

Then X_F is a curve of genus 49 with $N = 63$ (the upper bound is $N \leq 67$, and the best known example has $N = 64$).

EXAMPLE 3.4.43. Let X be the elliptic curve given over $\mathbb{F}_4$ by

$$y^2 + y = x^3;$$

it has 9 rational points. Take $P_\infty = (0:1:0)$ and let $D = 11 P_\infty$. Elementary computations give $\dim_{\mathbb{F}_2} \widetilde{W} = 13$; as an $\mathbb{F}_2$-basis, we may take

$$\alpha, \; x, \; \alpha x, \; y, \; \alpha y, \; xy, \; \alpha xy, \; x^2 y, \; \alpha x^2 y, \; y^3, \; \alpha y^3, \; xy^3, \; \alpha xy^3,$$

where $\mathbb{F}_4 = \mathbb{F}_2(\alpha)$. Solving in $\widetilde{W}$ system (3.4.4) for the remaining eight points, we obtain a five-dimensional solution space, generated by

$$f_1 = x + xy + x^2 y, \quad f_2 = \alpha x + \alpha xy + \alpha^2 x^2 y, \quad f_3 = y^3,$$

$$f_4 = x + xy^3, \quad f_5 = \alpha x + \alpha xy^3.$$

With $F = \langle f_1, f_2 \rangle$, we obtain an optimal curve of genus 13 with 33 points; with $F = \langle f_1, f_3 \rangle$, we get a curve of genus 15 with 33 points (this is the lower bound, the upper bound being 37); with $F = \langle f_1, f_2, f_3 \rangle$, we get a curve of genus 33 with 65 points (the upper bound is 66); and with $F = \langle f_1, f_3, f_4 \rangle$, we obtain a curve of genus 39 with 65 points (this is also the best known lower bound).

3.4.3. García–Stichtenoth Towers

Let us give some explicit examples of families of function fields that asymptotically meet the Drinfeld–Vlăduţ bound. Let $q = r^2$, r being an integer.

Consider first the tower $(K_1, K_2, \ldots)$ of function fields over $\mathbb{F}_q$ defined as follows: $K_n = \mathbb{F}_q(x_1, \ldots, x_n)$, where

$$x_{i+1}^r + x_{i+1} = \frac{x_i^r}{x_i^{r-1} + 1}, \quad i = 1, \ldots, n-1.$$

Optimality of this tower is implied by the following result.

THEOREM 3.4.44. *The genus of $g(K_n)$ is given by*

$$g(K_n) = \begin{cases} (r^{n/2} - 1)^2 & \text{if } n \text{ is even,} \\ (r^{(n+1)/2} - 1)(r^{(n-1)/2} - 1) & \text{if } n \text{ is odd.} \end{cases}$$

The number of points $N(K_n)$ satisfies the inequality

$$N(K_n) \geq r^{n-1}(r^2 - r) + 1. \qquad \square$$

In fact, this tower is a subtower of the following tower $(E_1, E_2, \ldots)$ of function fields over $\mathbb{F}_q$. We again set $E_1 = \mathbb{F}_q(x_1)$, and for $n \geq 1$ we set $E_{n+1} = E_n(z_{n+1})$, where

$$z_{n+1}^r + z_{n+1} = x_n^{r+1}, \qquad x_{n+1} = \frac{z_{n+1}}{x_n}.$$

Note that for $n \geq 2$ we have

$$z_{n+1}^r + z_{n+1} = x_n^{r+1} = \frac{z_n^{r+1}}{x_{n-1}^{r+1}} = \frac{z_n^{r+1}}{z_n^r + z_n} = \frac{z_n^r}{z_n^{r-1} + 1},$$

whence it follows that the subfield $\mathbb{F}_q(z_2, \ldots, z_{n+1}) \subseteq E_{n+1}$ is isomorphic to K_n of the previous example. However, the tower $(E_1, E_2, \ldots)$ itself is also optimal:

THEOREM 3.4.45. *The genus $g(E_n)$ satisfies the inequality*

$$g(E_n) \leq r^n + r^{n-1} - r^{(n+1)/2} - 2r^{(n-1)/2} + 1,$$

and the number of points $N(E_n)$ satisfies

$$N(K_n) \geq (r^2 - 1)r^{n-1} + 1. \qquad \square$$

EXERCISE 3.4.46. What curve is E_2?

In the given examples, the extensions K_n/K_1 and E_n/E_1 have wildly ramified places. Let us give two examples of towers without wild ramification.

First, consider the following family of Kummer covers. Now let $q = p^e$, where p is a prime and $e > 1$ (q is not necessarily a square). Set $m = (q-1)/(p-1)$. Let $F_n = \mathbb{F}_q(x_1, \ldots, x_n)$, where

$$x_{i+1}^m + (x_i + 1)^m = 1, \quad i = 1, \ldots, n-1.$$

THEOREM 3.4.47. *The tower $(F_1, F_2, \ldots)$ has the following properties:*
- *For any $n \geq 1$, F_{n+1}/F_n is a cyclic extension of degree m.*
- *If $P \in F_1$ is a ramification place in F_n/F_1 for some $n > 1$, then P corresponds to an ideal $(x_1 - a)$, where $a \in \mathbb{F}_q$.*
- *The place at infinity is unramified in F_n/F_1 for any $n > 1$.* $\qquad \square$

These properties imply that
$$\frac{N(F_n)}{g(F_n)} \gtrsim \frac{2}{q-2}.$$
In particular, this tower is optimal for $q = 4$.

In the next example, q is a square again, $q = r^2$. Let $F_n = \mathbb{F}_q(x_1, \ldots, x_n)$, where
$$x_{i+1}^{r-1} + (x_i + 1)^{r-1} = 1, \quad i = 1, \ldots, n-1.$$

THEOREM 3.4.48. *For this tower,*
$$\frac{N(F_n)}{g(F_n)} \gtrsim \frac{2}{\sqrt{q}-2}.$$
□

In particular, this tower is optimal for $q = 9$.

To prove the theorems of this section one needs a subtle study of curve ramification and of how the points split. A wonderful fact is that all these curves happen to be modular; i.e., they parameterize families of elliptic curves with some additional structures. We hope to return to this topic in *Advanced Chapters*.

3.4.4. Curves of Small Genera

If $g = 0$, then a curve over a finite field is isomorphic to $\mathbb{P}^1$ and thus has $q + 1$ points. We would like if not to determine all possibilities for the number of points on a curve of small genus $g > 0$, then at least to find the maximal possible number $N_q(g)$ of points on a curve of genus g. However, we know not too much.

For $g = 1$ and 2, exact values of $N_q(g)$ are known.

THEOREM 3.4.49. *Let $q = p^e$, p prime. Then*
$$N_q(1) = q + 1 + \lfloor 2\sqrt{q} \rfloor,$$
except for the case where $p \mid \lfloor 2\sqrt{q} \rfloor$, $e \geq 3$, and e is odd, in which case
$$N_q(1) = q + \lfloor 2\sqrt{q} \rfloor.$$
□

THEOREM 3.4.50. *Let $q = p^e$, p prime. Then*
(a) *if e is even and $q \neq 4, 9$, we have*
$$N_q(2) = q + 1 + 4\sqrt{q}.$$
Furthermore,
$$N_4(2) = 10,$$
$$N_9(2) = 20.$$

(b) *Let e be odd; q is called special if either $p \mid \lfloor 2\sqrt{q} \rfloor$ or we have $q = m^2 + 1$, $q = m^2 + m + 1$, or $q = m^2 + m + 2$ for an integer m. If q is not special, then*
$$N_q(2) = q + 1 + 2\lfloor 2\sqrt{q} \rfloor;$$
if q is special, then
$$N_q(2) = \begin{cases} q + 2\lfloor 2\sqrt{q} \rfloor & \text{if } 2\sqrt{q} - \lfloor 2\sqrt{q} \rfloor > \frac{\sqrt{5}-1}{2}, \\ q + 2\lfloor 2\sqrt{q} \rfloor - 1 & \text{otherwise.} \end{cases}$$
□

For $g = 3$, the value of $N_q(g)$ is known for all $q \leq 100$ (and for some larger values of q).

3.4. REMARKABLE EXAMPLES

THEOREM 3.4.51. *We have the following table for $N_q(3)$, $q \le 27$:*

q	2	3	4	5	7	8	9	11	13	16	17	19	25	27
$N_q(3)$	7	10	14	16	20	24	28	28	32	38	40	44	56	56

□

Also, we have the following result.

THEOREM 3.4.52. *Let $q = p^{4m+2}$. Then $N_q(3)$ is maximal:*
$$N_q(3) = q + 1 + 6\sqrt{q}.$$
□

On the other hand, one may fix a small q and try to answer the question on the maximal number of points on a curve of small genus. For instance, for $q = 2$ we know the following.

THEOREM 3.4.53. *We have the following table for $N_2(g)$, $g \le 9$:*

g	0	1	2	3	4	5	6	7	8	9
$N_2(g)$	3	5	6	7	8	9	10	10	11	12

□

More detailed tables are presented in [vdGvdV].

The proof of Theorems 3.4.50, 3.4.51, and 3.4.53 is rather laborious and consists of two parts: the first is to obtain upper bounds for $N_q(g)$, based on the explicit formula for the number of points on a curve over a finite field (see Sec. 3.1.4), as well as on some arguments from the algebraic number theory, and the second part is to produce curves with many rational points, which is done using various methods of algebraic geometry.

Problems arising here are of high interest and afford a vast field of action. We shall return to these questions in *Advanced Chapters*.

3.5. Connection with Exponential Sums

In many problems of analytic number theory we come across sums of additive characters (exponents). These sums need to be calculated or estimated. It turns out that between exponential sums and curves over finite fields there is a close relation, useful for both domains.

In Sec. 3.5.1 we give an elementary example illustrating the connection between exponential sums (sums of roots of unity) with curves over finite fields; namely, we compute the number of rational points for some Fermat curves. Then, in Sec. 3.5.2, we introduce L-functions of characters, which are the main technical tool in the study of exponential sums. Section 3.5.3 shows how one can obtain estimates for exponential sums using the Riemann hypothesis for Kummer and Artin–Schreier curves.

3.5.1. Number of Points on Fermat Curves

We shall illustrate the connection between the number of points on a curve and exponential sums by the following simple example. Let us find the number of $\mathbb{F}_q$-points on the *Fermat curve*

$$x^r + y^r + z^r = 0$$

in the case where $q = p$ is a prime and r is relatively prime to $p-1$.

We shall identify elements of $\mathbb{F}_p$ with residues $\{0,1,\ldots,p-1\}$ modulo p. Let $\zeta \in \mathbb{C}$ be a primitive pth root of 1. Then, as is easily seen, for any $v \in \mathbb{F}_p$ we have

$$\sum_{u\in\mathbb{F}_p} \zeta^{uv} = \begin{cases} p & \text{if } v = 0, \\ 0 & \text{if } v \neq 0. \end{cases} \qquad (3.5.1)$$

EXERCISE 3.5.1. Prove this fact.

Let us use this relation. Let $F(x,y,z) \in \mathbb{F}_p[x,y,z]$ be an arbitrary (for a while) polynomial, and let N be the number of possible triples (x,y,z) for which $F(x,y,z) = 0$ (i.e., the number of $\mathbb{F}_p$-points on the corresponding affine surface). Then (3.5.1) implies that

$$\sum_{x,y,z\in\mathbb{F}_p} \sum_{u\in\mathbb{F}_p} \zeta^{uF(x,y,z)} = Np,$$

whence

$$N = \frac{1}{p} \sum_{u,x,y,z\in\mathbb{F}_p} \zeta^{uF(x,y,z)}.$$

Isolating the terms with $u = 0$, we obtain

$$N = p^2 + \frac{1}{p} \sum_{u\in\mathbb{F}_p^*} \sum_{x,y,z\in\mathbb{F}_p} \zeta^{uF(x,y,z)}.$$

In our case, for $F = x^r + y^r + z^r$, the last expression takes the form

$$N = p^2 + \frac{1}{p} \sum_{u\in\mathbb{F}_p^*} \sum_{x,y,z\in\mathbb{F}_p} \zeta^{ux^r} \zeta^{uy^r} \zeta^{uz^r} = p^2 + \frac{1}{p} \sum_{u\in\mathbb{F}_p^*} \left(\sum_{w\in\mathbb{F}_p} \zeta^{uw^r} \right)^3.$$

Thus, the problem is reduced to the analysis of sums of the form

$$\sum_{w \in \mathbb{F}_p} \zeta^{uw^r}, \quad u \neq 0. \tag{3.5.2}$$

It is clear that

$$\sum_{w \in \mathbb{F}_p} \zeta^{uw^r} = \sum_{t \in \mathbb{F}_p} m(t) \zeta^{ut},$$

where $m(t)$ is the number of solutions of the equation $w^r = t$ in the unknown w. In the general case, $m(t)$ is expressed in terms of multiplicative characters modulo p, and sums of the form (3.5.2) are expressed in terms of Gauss sums. However, under the assumption that r is relatively prime to $p-1$, one easily sees that $m(t)$ is identically equal to 1, and therefore the sum (3.5.2) is zero; i.e., $N = p^2$. Hence, the number of $\mathbb{F}_p$-points on a (projective) Fermat curve of degree relatively prime to $p-1$ equals

$$\frac{N-1}{p-1} = p+1.$$

3.5.2. L-functions of Characters

Recall that a *character* of a multiplicative group $\mathbb{F}_q^*$ is a homomorphism χ from $\mathbb{F}_q^*$ to the multiplicative group $\mathbb{C}^*$. Obviously, the image of such a homomorphism lies in the group μ_{q-1} of $(q-1)$st roots of unity.

EXERCISE 3.5.2. Show that the group of characters of $\mathbb{F}_q^*$ is isomorphic to $\mathbb{F}_q^*$; here, $1 \in \mathbb{F}_q^*$ corresponds to the *trivial character* $\chi_0(x)$ (where $\chi_0(x) = 1$ for any x).

The *order of a character* χ is the minimal positive integer d such that $\chi^d = \chi_0$. It is easily seen that the order of any character is a divisor of $q-1$. Each number $s \geq 1$ such that $\chi^s = \chi_0$ (i.e., a multiple of the order of χ) is called an *index* of χ; we write $s = \operatorname{ind} \chi$.

EXERCISE 3.5.3. Let $s > 0$ divide $q-1$. Show that there are precisely s characters of index s.

Let us extend characters of $\mathbb{F}_q^*$ onto the whole field $\mathbb{F}_q$, setting by definition

$$\chi(0) = \begin{cases} 1 & \text{if } \chi = \chi_0, \\ 0 & \text{if } \chi \neq \chi_0. \end{cases}$$

Such a character χ is called a *multiplicative character* of $\mathbb{F}_q$.

Denote by $(\mathbb{F}_q^*)^s$ the subgroup consisting of elements $x \in \mathbb{F}_q^*$ that can be represented as $x = y^s$, $y \in \mathbb{F}_q^*$. Here is a property of multiplicative characters of a field:

PROPOSITION 3.5.4. *Let $s > 0$ divide $q-1$. Then*

$$\sum_{\chi:\, s = \operatorname{ind} \chi} \chi(x) = \begin{cases} s & \text{if } x \in (\mathbb{F}_q^*)^s, \\ 0 & \text{if } x \notin (\mathbb{F}_q^*)^s \text{ and } x \neq 0, \\ 1 & \text{if } x = 0. \end{cases}$$

If η is a generator of $\mathbb{F}_q^*$, and $\chi \neq \chi_0$ is a multiplicative character of $\mathbb{F}_q$ of index s, then
$$\sum_{j=0}^{s-1} \chi(\eta^j) = 0. \qquad \square$$

EXERCISE 3.5.5. Prove the proposition. (*Hint*: To prove the second statement, show that χ is a nontrivial character of the quotient group $\mathbb{F}_q^*/(\mathbb{F}_q^*)^s$, which in turn consists of images of $1, \eta, \ldots, \eta^{s-1}$.)

Let $q = p^a$, p prime. Recall that the *trace* of $x \in \mathbb{F}_q$ is the element
$$\operatorname{tr} x = x + x^p + \ldots + x^{p^{a-1}}$$
of the field $\mathbb{F}_p$.

In Sec. 1.1.3 (p. 11), we considered additive characters ψ of the field $\mathbb{F}_q$ (i.e., characters of the additive group of $\mathbb{F}_q$). We have the following fact, formulated as an exercise:

EXERCISE 3.5.6. Show that any additive character ψ of $\mathbb{F}_q$ is of the form
$$\psi_\alpha(x) = e^{2\pi i \frac{\operatorname{tr}(\alpha x)}{p}}$$
for some $\alpha \in \mathbb{F}_q$. (*Hint*: Check that all the ψ_α are characters and they are pairwise distinct.).

Now, let $\mathbb{F}_{q^\nu}$ be an extension of $\mathbb{F}_q$ of degree ν, and let $\mathbb{F}_p$ be the prime subfield of $\mathbb{F}_q$. The Galois group $\operatorname{Gal}(\mathbb{F}_{q^\nu}/\mathbb{F}_q)$ is a cyclic group of order ν. Let σ be its generator; its action on $\mathbb{F}_{q^\nu}$ is given by the rule $\sigma(x) = x^q$. Then the element
$$\operatorname{tr}_\nu x = x + \sigma(x) + \ldots + \sigma^{\nu-1}(x) = x + x^q + \ldots + x^{q^{\nu-1}}$$
is called the *(relative) trace* of $x \in \mathbb{F}_{q^\nu}$, and
$$\operatorname{norm}_\nu x = x \cdot \sigma(x) \cdot \ldots \cdot \sigma^{\nu-1}(x) = x \cdot x^q \cdot \ldots \cdot x^{q^{\nu-1}}$$
is the *(relative) norm* of x in the extension $\mathbb{F}_{q^\nu}/\mathbb{F}_q$. The elements
$$\operatorname{Tr}_\nu x = \operatorname{tr}(\operatorname{tr}_\nu x) \qquad \text{and} \qquad \operatorname{Norm}_\nu x = \operatorname{norm}(\operatorname{norm}_\nu x),$$
where tr and norm are the trace and norm from $\mathbb{F}_q$ to $\mathbb{F}_p$, are called, respectively, the *absolute trace* and *absolute norm* of x.

EXERCISE 3.5.7. Let χ and ψ be a multiplicative and an additive character of $\mathbb{F}_q$ respectively. Check that
$$\chi_\nu(x) = \chi(\operatorname{norm}_\nu x)$$
is a multiplicative character and
$$\psi_\nu(x) = \psi(\operatorname{tr}_\nu x)$$
is an additive character of $\mathbb{F}_{q^\nu}$.

The characters χ_ν and ψ_ν are said to be *induced* by χ and ψ respectively.

Let $f(x)$ and $g(x)$ be nonconstant polynomials in $\mathbb{F}_q[x]$. Consider the sums
$$T_\nu = T_\nu(f, g) = \sum_{x \in \mathbb{F}_{q^\nu}} \chi_\nu(f(x)) \psi_\nu(g(x)).$$

In order to determine the dependence of T_ν on the integer parameter ν, introduce the complex-variable function

$$L(z) = L_{\chi,\psi}(z, f, g) = \exp\left(\sum_{\nu=1}^{\infty} \frac{T_\nu}{\nu} z^\nu\right),$$

called the *(Artin) L-function*. We have the following fundamental fact:

THEOREM 3.5.8. *Let $f, g \in \mathbb{F}_q[x]$, $\deg g = n$, and let $f(x) = f_1^{s_1}(x)\ldots f_r^{s_r}(x)$ be the irreducible factorization of f, where $\deg(f_1\ldots f_r) = m$. Let s be a positive integer divisor of $q - 1$. If at least one of the conditions*

(1) *χ is a nontrivial multiplicative character of $\mathbb{F}_q$ of index s, and the numbers $s, s_1, \ldots, s_r$ are coprime;*

(2) *ψ is a nontrivial additive character of $\mathbb{F}_q$, and n is relatively prime to q*

is satisfied, then the L-function is a polynomial:

$$L(z) = P(z) = \prod_{j=1}^{m+n-1} (1 - \omega_j z). \qquad \square$$

Now, let us define L-functions depending on a single character:
$$L_\chi(z) = L_\chi(z, f) = L_{\chi,\psi_0}(z, f, g),$$
$$L_\psi(z) = L_\psi(z, g) = L_{\chi_0,\psi}(z, f, g).$$

COROLLARY 3.5.9. *In the conditions of Theorem 3.5.8, we have $L_\chi(z, f) = \prod_{j=1}^{m-1}(1 - \omega_j z)$, so that*

$$\sum_{x \in \mathbb{F}_{q^\nu}} \chi_\nu(f(x)) = -\sum_{j=1}^{m-1} \omega_j^\nu(\chi);$$

also, $L_\psi(z, g) = \prod_{j=1}^{n-1}(1 - \omega_j z)$, so that

$$\sum_{x \in \mathbb{F}_{q^\nu}} \psi_\nu(g(x)) = -\sum_{j=1}^{n-1} \omega_j^\nu(\psi). \qquad \square$$

3.5.3. Estimates for Exponential Sums

Kummer and Artin–Schreier curves, introduced in Sec. 3.4.2 above, can be used to obtain estimates for exponential sums.

Kummer covers. Consider first the Kummer equation

$$y^s = f(x), \quad f \in \mathbb{F}_q[x]. \tag{3.5.3}$$

Let us present conditions for a Kummer curve to be absolutely irreducible:

PROPOSITION 3.5.10. *Let $f(x) = f_1^{s_1}(x)\ldots f_r^{s_r}(x)$ be the irreducible factorization of $f(x)$ in $\mathbb{F}_q[x]$. The polynomial $y^s - f(x)$ is absolutely irreducible if and only if $s, s_1, \ldots, s_r$ are coprime.*

PROOF. If $s, s_1, \ldots, s_r$ have a common divisor d, then $y^s - f(x)$ is factorized as a difference of dth powers.

Conversely, let a polynomial $y^s - f(x)$ be reducible in $\overline{\mathbb{F}}_q[x,y]$. Consider the field $\mathbb{K} = \overline{\mathbb{F}}_q(x)$. Over its algebraic closure $\overline{\mathbb{K}}$, we have
$$y^s - f(x) = (y - y_1(x)) \ldots (y - y_s(x));$$
here, elements $y_i(x)$ are of the form $y_i(x) = \zeta^i y(x)$, $1 \leq i \leq s$, where $y(x)$ is any root of $y^s - f(x)$ in the field $\overline{\mathbb{K}}$, and ζ is a primitive sth root of unity. Since the polynomial $y^s - f(x)$ is reducible in the ring $\overline{\mathbb{F}}_q[x,y]$, we conclude that for some $i_1 \ldots, i_t$, $t < s$,
$$\left(y - \zeta^{i_1} y(x)\right) \ldots \left(y - \zeta^{i_t} y(x)\right)$$
is an element of this ring. The free term of such a product,
$$(-1)^t \zeta^{i_1 + \ldots + i_t} y^t(x),$$
lies in $\overline{\mathbb{F}}_q[x]$; i.e., $y^t \in \overline{\mathbb{F}}_q[x]$. Denote by μ the minimal positive integer such that $y^\mu \in \overline{\mathbb{F}}_q[x]$. Then s is a multiple of μ; moreover, since $t < s$, we also have $\mu < s$. It remains to put $y^\mu = g(x)$; then $g^{s/\mu} = f(x)$, and therefore all the s_i are divisible by s/μ. $\square$

In particular, this immediately implies the following statement:

COROLLARY 3.5.11. *If* $\gcd(s, \deg f(x)) = 1$, *then* $y^s - f(x)$ *is absolutely irreducible.* $\square$

REMARK 3.5.12. The statement of Proposition 3.5.10 is valid for an arbitrary field; the proof does not use finiteness of the field.

It turns out that the number of rational points on a Kummer curve, i.e., the number of solutions $x, y \in \mathbb{F}_q$ of equation (3.5.3), can be expressed in terms of exponential sums with multiplicative characters. Indeed, let $s' = \gcd(s, q-1)$ and $s = s'r$. Since r is relatively prime to $q-1$, we see that $z = y^r$, together with y, runs over all elements of $\mathbb{F}_q$; hence, the number of solutions $x, y \in \mathbb{F}_q$ of (3.5.3) coincides with the number of solutions $x, z \in \mathbb{F}_q$ of the equation
$$z^{s'} = f(x),$$
where $s' \mid (q-1)$. Therefore, we may assume without loss of generality that the exponent s in equation (3.5.3) divides $q-1$. In the following, we assume this condition to be satisfied.

PROPOSITION 3.5.13. *The number N_ν of solutions of equation* (3.5.3) *in the field $\mathbb{F}_{q^\nu}$ satisfies*
$$N_\nu = \sum_{\chi: s = \mathrm{ind}\,\chi} \sum_{x \in \mathbb{F}_{q^\nu}} \chi_\nu(f(x)) = \sum_{\chi: s = \mathrm{ind}\,\chi} \sum_{x \in \mathbb{F}_{q^\nu}} \chi(\mathrm{norm}_\nu f(x)).$$

PROOF. Let t be an element of $\mathbb{F}_{q^\nu}$. To prove the proposition, it suffices to show that the number of solutions of $y^s = t$ in elements $y \in \mathbb{F}_{q^\nu}$ is equal to
$$\sum_{\chi: s = \mathrm{ind}\,\chi} \chi_\nu(t) = \sum_{\chi: s = \mathrm{ind}\,\chi} \chi(\mathrm{norm}_\nu t).$$

The map $t \mapsto \mathrm{norm}_\nu t$ is a homomorphism of groups $\mathbb{F}_{q^\nu}^* \to \mathbb{F}_q^*$. The restriction of this map onto the subgroup $(\mathbb{F}_{q^\nu}^*)^s$ defines a homomorphism from this subgroup to $(\mathbb{F}_q^*)^s$. Next, by Proposition 3.5.4 we have

$$\sum_{\chi:\, s=\mathrm{ind}\,\chi} \chi(\mathrm{norm}_\nu t) = \begin{cases} s & \text{if } \mathrm{norm}_\nu t \in (\mathbb{F}_q^*)^s, \\ 0 & \text{if } \mathrm{norm}_\nu t \notin (\mathbb{F}_q^*)^s \text{ and } \mathrm{norm}_\nu t \neq 0, \\ 1 & \text{if } \mathrm{norm}_\nu t = 0. \end{cases}$$

Let us compare these values of $\sum_{\chi:\, s=\mathrm{ind}\,\chi} \chi(\mathrm{norm}_\nu t)$ with the number of solutions of $y^s = t$. In the first case, we have $t \in (\mathbb{F}_{q^\nu}^*)^s$, and the equation $y^s = t$ has s solutions in elements $y \in \mathbb{F}_{q^\nu}$. In the second case, we have $t \notin (\mathbb{F}_{q^\nu}^*)^s$, $t \neq 0$, and $y^s = t$ has no solutions. In the third case, we have $t = 0$, and $y^s = t$ has the unique solution $y = 0$. □

If we compare this result with the definition of the L-function and with Theorem 3.5.8, we observe that it can be reformulated as follows:

THEOREM 3.5.14. *For an absolutely irreducible Kummer curve $y^s = f(x)$, we have*

$$P_K(z) = \prod_{\chi:\, s=\mathrm{ind}\,\chi} L_\chi(z),$$

where $P_K(z)$ is the numerator of the zeta function of the Kummer curve. □

This theorem yields the following estimate for exponential sums with multiplicative characters:

THEOREM 3.5.15. *Let $f(x)$ be a polynomial with coefficients from $\mathbb{F}_q$, $f = f_1^{s_1} \ldots f_r^{s_r}$ be its irreducible factorization in $\mathbb{F}_q[x]$, and $m = \deg(f_1 \ldots f_r)$. Let χ be a nontrivial multiplicative character of $\mathbb{F}_q$ of index s, and let $s, s_1, \ldots, s_r$ be coprime. Then for any $\nu \geq 1$ we have*

$$\left| \sum_{x \in \mathbb{F}_{q^\nu}} \chi_\nu(f(x)) \right| \leq (m-1) q^{\nu/2}.$$

PROOF. It suffices to apply Corollary 3.5.9 and then use Theorem 3.5.14, the latter implying that all ω_i in the formula of Corollary 3.5.9 are the inverse Frobenius roots (see Sec. 3.1.2). Thus, the result follows from the Weil theorem. □

Artin–Schreier covers. Similar results can be obtained for Artin–Schreier curves

$$y^q - y = g(x), \quad g \in \mathbb{F}_q[x]; \tag{3.5.4}$$

one should only replace multiplicative characters by additive ones.

A sufficient condition for an Artin–Schreier curve to be absolutely irreducible is as follows:

PROPOSITION 3.5.16. *If $\gcd(q, \deg g(x)) = 1$, then the polynomial $y^q - y - g(x)$ is absolutely irreducible.* □

Similarly to Proposition 3.5.13, we have the following result:

PROPOSITION 3.5.17. *For the number N_ν of solutions of equation (3.5.4) in $\mathbb{F}_{q^\nu}$, we have*
$$N_\nu = \sum_\psi \sum_{x \in \mathbb{F}_{q^\nu}} \psi_\nu(g(x)) = \sum_\psi \sum_{x \in \mathbb{F}_{q^\nu}} \psi(\operatorname{tr}_\nu g(x)).$$

SKETCH OF THE PROOF. Let t be an element of $\mathbb{F}_{q^\nu}$. To prove the proposition, it suffices to verify that the number of solutions of $y^q - y = t$ in elements $y \in \mathbb{F}_{q^\nu}$ is equal to
$$\sum_\psi \psi_\nu(t) = \sum_\psi \psi(\operatorname{tr}_\nu t).$$

The equation $y^q - y = t$ in $\mathbb{F}_{q^\nu}$ is solvable in y if and only if $\operatorname{tr}_\nu t = 0$; this equation, along with a solution y, has at least q solutions $y + z$, $z \in \mathbb{F}_q$, and the degree of the equation is q. On the other hand, for any $\operatorname{tr}_\nu t = u \in \mathbb{F}_q$ we have
$$\sum_\psi \psi(u) = \begin{cases} q, & u = 0, \\ 0, & u \neq 0. \end{cases} \qquad \square$$

EXERCISE 3.5.18. *Give a detailed proof.*

Similarly to Proposition 3.5.13, this result can be reformulated as follows:

THEOREM 3.5.19. *For an absolutely irreducible Artin–Schreier curve $y^q - y = g(x)$, we have*
$$P_{AS}(z) = \prod_\psi L_\psi(z),$$
where $P_{AS}(z)$ is the numerator of the zeta function. $\qquad \square$

By the same arguments as in the proof of Theorem 3.5.15, we obtain the following estimate for sums of additive characters:

THEOREM 3.5.20. *Let $g(x)$ be a polynomial from $\mathbb{F}_q[x]$ of degree n relatively prime to q, and let ψ be a nontrivial additive character of $\mathbb{F}_q$. Then for any $\nu \geq 1$ we have*
$$\left| \sum_{x \in \mathbb{F}_{q^\nu}} \psi_\nu(g(x)) \right| \leq (n-1) q^{\nu/2}. \qquad \square$$

Historical and Bibliographic Notes

The subject of this chapter is mostly classical, with an addition of several recent results, mostly motivated by the theory of algebraic geometry codes.

The material of Secs. 3.1.1, 3.1.2 and the first half of Sec. 3.1.3 (up to Theorem 3.1.24 inclusive) is classical: the zeta function of a curve over a finite field and its basic properties were studied by E. Artin, H. Hasse, and F.K. Schmidt in the 1920s and 1930s. The Weil theorem was proved in 1940 by A. Weil [Wei48a, Wei48b] using multi-dimensional varieties (Jacobians of curves). Its particular case for genus 1 was proved by H. Hasse in 1930.

The key role in the theory was and continues to be played by the wonderful parallelism between curves over a finite field and algebraic number fields (we discuss it at the beginning of Sec. 3.1). Let us make several remarks on this parallelism, which was also very important for our own results, which we shall partly explore in *Advanced Chapters*. This parallelism is in fact a strict and consistent theory of global fields.

A *global field* is either a finite extension of the field $\mathbb{Q}$ of rational numbers or a finite extension of the field $\mathbb{F}_r(t)$ of rational functions over a finite field. In the first case it is called a *number field*, and in the second a *function field*. A function field is always the field of functions on a (smooth, projective, absolutely irreducible) curve X.

Consider a ring $\mathcal{O}$ such that the global field $\mathbb{K}$ is its field of fractions. This ring is of dimension 1; i.e., any nonzero prime ideal $\mathfrak{p}$ in $\mathcal{O}$ is maximal. Moreover, the residue field $\mathcal{O}/\mathfrak{p} = \mathbb{F}_q$ is finite. These two properties and the property to be finitely generated over the prime field (i.e., over $\mathbb{Q}$ or $\mathbb{F}_r$) determine the class of global fields.

In the number field case, in K there is a most important subring $\mathcal{O}_\mathbb{K}$, the integral closure of the ring $\mathbb{Z}$. In the function field case it has an analogue, the integral closure of the ring $\mathbb{F}_r[t]$; however, this ring is not defined uniquely, since there is no unique imbedding $\mathbb{F}_r(t) \subset \mathbb{K}$ and no unique choice of the infinity.

The notion of a place of the field is good for both cases. Let us remark, however, that for a place of a number field the degree is not defined; it is replaced by the notion of the norm of a place. If the place P corresponds to the maximal ideal $\mathfrak{p}$ in a subring $\mathcal{O}$ (which is true if and only if the norm is non-Archimedean; i.e., the strong triangle inequality $v_P(f+g) \geq \min\{v_P(f), v_P(g)\}$ is satisfied for any $f, g \in \mathbb{K}$), then we can define the *norm* of the place as $\mathrm{Norm}\, P = \mathrm{Norm}\, \mathfrak{p} = |\mathcal{O}/\mathfrak{p}|$. In the function field case (over $\mathbb{F}_r$) all places are non-Archimedean and $\mathrm{Norm}\, P = r^{\deg P}$; in particular, the norm is always the power of the same prime number, and all residue fields are of the same characteristic. That is why the function field case is also called "the case of equal characteristics". In the number field case among the characteristics we see all prime numbers (the "case of different characteristics"). In the number field case there also is a nonzero finite number of Archimedean places; the corresponding norms correspond to embeddings of the field $\mathbb{K}$ into $\mathbb{R}$ or into $\mathbb{C}$. For them the norm of a place is not defined; they are just distinguished as being real or complex.

The number field case is almost always more difficult than the function field case. An illustrious example is the analogy between zeta functions discussed in Sec. 3.1.3 (Remark 3.1.22). Whereas the zeta function of a curve over a finite field (and also of a multi-dimensional variety over a finite field) is a rational function in $t = r^{-s}$, the zeta function of a number field is much more complicated, the Riemann hypothesis on its zeroes remains unproved, and so on.

For a map $X \to Y$ the divisor of ramification points on X corresponds to the different defined for a field extension $\mathbb{L} \supset \mathbb{K}$, and the ramification divisor on Y to the discriminant $D_{\mathbb{L}/\mathbb{K}}$ of the extension. There is a number field analogue of the Hurwitz formula and other results. The analogue of the genus of the curve X is $g_{\mathbb{K}} = \ln \sqrt{|D_{\mathbb{K}}|} + 1$, where $D_{\mathbb{K}}$ is the absolute discriminant of the field $\mathbb{K}$. (Sometimes it is easier to consider just $\ln \sqrt{|D_{\mathbb{K}}|}$ as the analogue of the genus.)

Note that the fields $\mathbb{Q}$ and $\mathbb{F}_r(t)$ have no unramified extensions.

The fractional ideal $\mathfrak{a} = \prod_\mathfrak{p} \mathfrak{p}^{n_\mathfrak{p}}$ of the field $\mathbb{K}$ is the same divisor D written multiplicatively. Then the space $L(D)$ corresponds to $\mathfrak{a}^{-1}$. The ideal class group $\mathrm{Cl}_{\mathbb{K}}$ is almost the group of $\mathbb{F}_r$-points of the Jacobian of the curve; its cardinality corresponds in fact to the product of the class number of the number field by its regulator.

There are lots of other common features and analogies in notions, results, and methods. Good examples are the analytic number theory in the function field case in the style of L. Carlitz and the theory of Drinfeld modules, analogues of elliptic curves. We shall return to this topic in *Advanced Chapters*.

A great part of this theory for the function fields case is exposed in the book by H. Stichtenoth [Sti93]; we followed his excellent exposition while writing Sec. 3.1.

Quite the contrary to Secs. 3.1.1, 3.1.2, and to the first half of Sec. 3.1.3, the rest of the results of Sec. 3.1 was obtained in the last twenty years as a result of the activity surge in the theory generated by the discovery of algebraic geometry codes. Theorem 3.1.25 was proved by J-P. Serre in 1983 [Ser83b]. Theorem 3.1.27 (and Proposition 3.1.28)—which is the function field analogue of "explicit formulae" of E. Landau and A. Weil of the (algebraic) number field theory—is also due to him.

The result of Exercise 3.1.30 is an example of bounds obtained by J. Oesterlé applying linear programming to Proposition 3.1.28.

The Pellikaan zeta function was defined in [Pel96a]; recently the number field analogue of this function was studied [vdGS00, LR03], its interpretation as a regularized determinant given [Den03], and its motivic analogue defined [BDN03].

Almost all results of Sec. 3.2 were obtained while studying families of curves, this study being motivated by algebraic geometry codes. The only exception is the value $A(q)$ introduced independently of algebraic geometry codes by Y. Ihara [Iha81], who proved the inequality

$$A(q) \leq \sqrt{2q + \frac{1}{4}} - \frac{1}{2}.$$

The proof of Theorem 3.2.1 [VD83] has Ihara's method of [Iha81] for its starting point. The formula (3.2.1) follows directly from the proof of Theorem 3.2.1. Propositions 3.2.7 and 3.2.14 were proved in [Vlă87]. Proposition 3.2.10 is taken from [Sti93]. Exercise 3.2.15 is taken from Elkies's paper [Elk01a]. The content of Sec. 3.2.5 is taken from [Tsf92a, TV97].

The lower bounds for $A(q)$ of Theorem 3.2.4 are due to the following authors: (1) to H. Niederreiter and C. Xing [NX99b]—in this paper the authors calculated explicitly the constant in the inequality $A(q) \geq c \ln q$ due to J-P. Serre [Ser85b];

(2) to A. García and H. Stichtenoth [GS96a]; (3) to T. Zink [Zin85]; (4) to M. Perret [Per91]; (5) and (6) to H. Niederreiter and C. Xing [NX99c]; (7) is also due to them [XN98]; (8) and (9) to A. Temkine [Tem01] and independently to B. Anglès and C. Maire [AM02].

All cases of Theorem 3.2.4 except for (3) are proved by constructing some class field towers. We hope to return to this method in *Advanced Chapters*.

The material of Sec. 3.3 is mostly classical, the main results being mostly obtained by H. Hasse and M. Deuring in the 1930s and 1940s. The modern exposition was given by J. Tate and P. Deligne [Del75]. Theorem 3.3.12 was proved by M. Deuring [Deu41] in 1941, and its more precise and much more general version by W.C. Waterhouse [Wat69]. Theorem 3.3.15 was proved by M.A. Tsfasman [Tsf85a, Tsf85b] in 1985 and then independently by H.G. Rück and J.F. Voloch [Rüc87, Vol88], who followed the paper of R. Schoof [Sch87].

All examples of Sec. 3.4, except for Hermitian curves, were constructed after the discovery of algebraic geometry codes, being mostly stimulated by this discovery. The study of sub-Hermitian curves—motivated by Problem 3.4.9—was started by H. Stichtenoth and C. Xing [XS95], who proved Theorem 3.4.10 and a weaker version of Theorem 3.4.11. Theorem 3.4.11 was finally proved in [FT96]. Theorem 3.4.12 was obtained in [FGT97]. Sec. 3.4.2 is based on the works of G. van der Geer and M. van der Vlugt [vdGvdV97, vdGvdV99, vdGvdV00a]. García–Stichtenoth towers were constructed in [GS95a, GS95b, GS96b]. Later on N.D. Elkies proved [Elk01b] that in fact these towers are those of Drinfeld modular curves. The towers of Theorems 3.4.47 and 3.4.48 were constructed in [GST97].

The study of $N_q(g)$ was initiated by J-P. Serre [Ser83a, Ser83b, Ser83c, Ser85a, Ser85b], who proved in particular Theorems 3.4.49, 3.4.50, 3.4.51, and 3.4.53; for $N_q(3)$ see also [Top03]. Theorem 3.4.52 was proved by T. Ibukiyama [Ibu93].

The content of Sec. 3.5 is classical and goes back to H. Hasse, E. Artin and A. Weil. While writing it we followed S.A. Stepanov [Ste91].

CHAPTER 4

Algebraic Geometry Codes

In this chapter we at last come to algebraic geometry constructions of codes. Besides the striking link between two disciplines a priori so far from each other, algebraic geometry codes have very good parameters if, of course, we choose the varieties they come from in a proper way. Their advantages become obvious when we consider asymptotic problems. Here, the proper algebraic geometry objects are asymptotically optimal families of curves, such as modular curves of different types. We have already seen some examples of such families in Sec. 3.4.3. We shall return to them in *Advanced Chapters*. Here we mostly collect the consequences they have for coding theory.

In Sec. 4.1 we give basic constructions and analyze properties of algebraic geometry codes. All these constructions are different versions of V.D. Goppa's original construction. The study of properties of algebraic geometry codes includes detailed analysis of their parameters, duality, automorphisms, spectra, and so on. In Sec. 4.2 we present several versions of main bounds and constructions, which sometimes help to improve the parameters of algebraic geometry codes. In Sec. 4.3 we try to find out which codes are algebraic geometry ones. The next section is devoted to examples of algebraic geometry codes; in particular, we analyze in detail codes from curves of small genera. The case of elliptic curves is especially well studied. In Sec. 4.5 we pick the ripe fruit, studying asymptotic problems. Algebraic geometry codes lead to improvement of almost all non-algebraic-geometry lower bounds for the main asymptotic problem, for linear and for nonlinear codes (provided that q is sufficiently large), for codes with polynomial construction complexity (for all q), for self-dual codes, for constant-weight codes, etc. Section 4.6 deals with essentially different new constructions of asymptotically good nonlinear codes. As usual, we end the chapter with historical and bibliographic notes.

4.1. Constructions and Properties

There are several (essentially equivalent) ways to construct linear codes starting from algebraic curves (and also from varieties of higher dimensions). For curves, the codes we obtain can be rather well described: we can bound their parameters and weight spectra and we understand the duality.

In Sec. 4.1.1 we discuss four types of constructions of algebraic geometry codes and estimate the parameters. Section 4.1.2 is devoted to duality and spectra. In Sec. 4.1.3 we begin to discuss the decoding problem.

All the curves considered in this section are projective, smooth, and absolutely irreducible over a finite field $\mathbb{F}_q$. Recall some notation: let X be such a curve; then $\mathbb{K} = \mathbb{F}_q(X)$ is the field of rational functions on X, $\Omega(X)$ is the space of rational differential 1-forms on X, $X(\mathbb{F}_{q^r})$ is the set of $\mathbb{F}_{q^r}$-points of X and $N_r = |X(\mathbb{F}_{q^r})|$

is its cardinality, $\mathrm{Div}(X)$ is the group of $\mathbb{F}_q$-divisors, $\mathrm{Div}^+(X)$ is the semi-group of effective divisors, $\mathrm{Pic}\, X$ is the divisor class group (or, equivalently, the group of isomorphism classes of line bundles on X), $J_X = \mathrm{Pic}^0(X)$ is the Jacobian of X (identified with the algebraic group of divisor classes of degree zero); and if $D \in \mathrm{Div}(X)$, then

$$L(D) = \{f \in \mathbb{F}_q(X)^* \mid (f) + D \geq 0\} \cup \{0\},$$
$$\Omega(D) = \{\omega \in \Omega(X)^* \mid (\omega) + D \geq 0\} \cup \{0\}.$$

4.1.1. Basic Algebraic Geometry Constructions and Their Parameters

L-construction. Let X be a curve such that $X(\mathbb{F}_q) \neq \varnothing$. Let $\mathcal{P} \subseteq X(\mathbb{F}_q)$, $|\mathcal{P}| = n$, and let $D \in \mathrm{Div}(X)$. Assume that $\mathrm{Supp}\, D \cap \mathcal{P} = \varnothing$.

Consider the *evaluation map*

$$\mathrm{Ev}_{\mathcal{P}} \colon L(D) \to \mathbb{F}_q^n, \quad f \mapsto (f(P_1), \ldots, f(P_n)),$$

where $\mathcal{P} = \{P_1, \ldots, P_n\}$. We get a code $C = \mathrm{Ev}_{\mathcal{P}}(L(D)) \subseteq \mathbb{F}_q^n$. For this code, we use the notation $C = (X, \mathcal{P}, D)_L$.

Assume that D is chosen in such a way that any function $f \in L(D)$ has at most b zeroes at $\mathbb{F}_q$-points of X. If $n > b$, then $\mathrm{Ev}_{\mathcal{P}}$ is an embedding, and we have

$$k = \ell(D), \quad d \geq n - b.$$

The Riemann–Roch theorem allows us to estimate the parameters of C.

THEOREM 4.1.1. *Let X be a curve of genus g, and let $0 \leq \deg D = a < n = |\mathcal{P}|$. Then $C = (X, \mathcal{P}, D)_L$ is an $[n, k, d]_q$ code with*

$$k \geq a - g + 1, \quad d \geq n - a.$$

PROOF. Let $D = D_1 - D_2$, $D_1 \geq 0$, $D_2 \geq 0$. A nonzero function $f \in L(D)$ has at most $a_1 = \deg D_1$ poles and at least $a_2 = \deg D_2$ zeroes in $\mathrm{Supp}\, D$ since $D + (f)_0 - (f)_\infty \geq 0$. Hence, the number of its zeroes outside $\mathrm{Supp}\, D$ is at most $a_1 - a_2 = a$. We have assumed that $a < n$; therefore, $\mathrm{Ev}_{\mathcal{P}}(f) \neq 0$ for any nonzero $f \in L(D)$; i.e., $\mathrm{Ev}_{\mathcal{P}}$ is an embedding. Moreover, $\mathrm{Ev}_{\mathcal{P}}$ has at least $n - a$ nonzero coordinates, and thus

$$d \geq n - a.$$

On the other hand, $C = \mathrm{Ev}_{\mathcal{P}}(L(D)) \cong L(D)$; i.e.,

$$k = \ell(D) \geq a - g + 1$$

according to the Riemann–Roch theorem (Theorem 2.2.17). $\square$

REMARK 4.1.2. The statement of Theorem 4.1.1 is valid for $0 \leq a < n$, but the first of the estimates is nontrivial only for $g - 1 < a < n$.

REMARK 4.1.3. The reasoning that leads to the estimate for d can be reformulated as follows: let $f \in L(D)$ be a nonzero function with $n - d$ zeroes in $\mathcal{P}$. Let $\mathcal{P}_1 \subseteq \mathcal{P}$ be the set of these zeroes; then $f \in L(D - \mathcal{P}_1)$, whence

$$0 \leq \deg L(D - \mathcal{P}_1) = a - (n - d).$$

REMARK 4.1.4. As will be seen below, the above inequalities do not always appear to be equalities. Let $d_c = n - a$ and $k_c = a - g + 1$; the parameters d_c and k_c are called, respectively, the *designed distance* and *designed dimension* of an algebraic geometry code C. By the Riemann–Roch theorem, if $n > a \geq 2g-1$, then $k = k_c$.

Using the Riemann–Roch theorem, we can also write out a precise formula for the dimension of the code for any a:
$$k = \ell(D) - \ell(D - \boldsymbol{P}) = a - g + 1 + \ell(K - D) - \ell(D - \boldsymbol{P}),$$
where $\boldsymbol{P} = \sum_{P_i \in \mathcal{P}} P_i$. In particular, we have $k \geq n - g$ for $a \geq n$, since $\ell(D - \boldsymbol{P}) \leq \deg(D - \boldsymbol{P}) + 1 = a - n + 1$ (see Exercise 2.1.55). (Note also that in some papers on this subject, the code $(X, \mathcal{P}, D)_L$ is denoted by $C(G, D)$, where G is our D and D is our $\boldsymbol{P}$.)

EXAMPLE 4.1.5. Let $X = \mathbb{P}^1$ and $D = a \cdot \infty$; then $L(D)$ is the space of polynomials of degree at most a. If we take for $\mathcal{P}$ all $\mathbb{F}_q$-points of $\mathbb{P}^1$ except for ∞, i.e., elements of the field $\mathbb{F}_q$, then the obtained $[q, a+1, q-a]_q$ code is a Reed–Solomon code (see Sec. 1.2.1).

EXERCISE 4.1.6. Let
$$D \sim D', \qquad \mathcal{P} \subseteq X(\mathbb{F}_q) \setminus (\operatorname{Supp} D \cup \operatorname{Supp} D').$$
How are the codes $C = (X, \mathcal{P}, D)_L$ and $C' = (X, \mathcal{P}, D')_L$ obtained from these two divisors related?

REMARK 4.1.7. Let X be a curve of genus g, $N = |X(\mathbb{F}_q)|$. If D_1 and D_2 are two divisors with $D_2 - D_1 \geq 0$, then $C_1 = (X, \mathcal{P}, D_1)_L \subseteq C_2 = (X, \mathcal{P}, D_2)_L$ for any $\mathcal{P} \subseteq X(\mathbb{F}_q) \setminus \operatorname{Supp} D_1$. Choose a point $P_0 \in X(\mathbb{F}_q)$, take $D_a = aP_0$, and let $\mathcal{P} \subseteq X(\mathbb{F}_q) \setminus \{P_0\}$. Varying a, we obtain a family of embedded algebraic geometry codes $C_a = (X, \mathcal{P}, D_a)_L$ with $C_a \subseteq C_{a+1}$, $k_a + d_a \geq n - g + 1$. Here, $n \leq N - 1$.

Note that the L-construction just described can be applied not only to a curve but also to any smooth projective variety X over $\mathbb{F}_q$, any set $\mathcal{P}$ of its $\mathbb{F}_q$-points, and any $\mathbb{F}_q$-divisor D on X. As distinct from curves, in the general case it is difficult to give a good estimate for the parameters of the obtained code $C = (X, \mathcal{P}, D)_L$. Nevertheless, there are some particular cases where such an estimate can be given. We postpone examination of such codes to *Advanced Chapters*.

Ω-construction. As before, we assume X to be a curve. Let $\mathcal{P} = \{P_1, \ldots, P_n\}$. Set
$$\boldsymbol{P} = P_1 + \ldots + P_n \in \operatorname{Div}(X).$$
Let $D \in \operatorname{Div} X$, $\operatorname{Supp} D \cap \mathcal{P} = \varnothing$. Consider the space of differential forms
$$\Omega(\boldsymbol{P} - D) = \{\omega \in \Omega(X)^* \mid (\omega) + \boldsymbol{P} - D \geq 0\} \cup \{0\}.$$
If $D \geq 0$, then this is the space of forms with zeroes of appropriate multiplicities in the support of D and with at most simple poles at the points P_i.

Recall that in Sec. 2.2.2 for any point P of X and for any nonzero form $\omega \in \Omega(X)$ we defined the residue $\operatorname{Res}_P(\omega)$. It is easily seen that if $P \in X(\mathbb{F}_q)$ and ω is defined over $\mathbb{F}_q$, we also have $\operatorname{Res}_P(\omega) \subset \mathbb{F}_q$.

The map
$$\operatorname{Res}_\mathcal{P} \colon \Omega(\boldsymbol{P} - D) \to \mathbb{F}_q, \quad \omega \mapsto (\operatorname{Res}_{P_1}(\omega), \ldots \operatorname{Res}_{P_n}(\omega)),$$
defines a code $C = \operatorname{Res}_\mathcal{P}(\Omega(\boldsymbol{P} - D))$. We write $C = (X, \mathcal{P}, D)_\Omega$. (In many papers, the code $(X, \mathcal{P}, D)_\Omega$ is denoted by $C^*(G, D)$, where G is our D, and D is our $\boldsymbol{P}$.)

Let us estimate the parameters.

THEOREM 4.1.8. *Let X be a curve of genus g, and let $2g-2 < a$. Assume that $\mathcal{P} \cap \operatorname{Supp} D \neq \emptyset$. Then the code $C = (X, \mathcal{P}, D)_\Omega$ is an $[n, k, d]_q$ code with*
$$k \geq n - a + g - 1, \qquad d \geq a - 2g + 2.$$

PROOF. Let K be a divisor from the canonical class, $\deg K = 2g - 2$. Then
$$\Omega(\boldsymbol{P} - D) \cong L(K + \boldsymbol{P} - D)$$
and
$$\dim \Omega(\boldsymbol{P} - D) = \ell(K + \boldsymbol{P} - D) \geq (2g - 2 + n - a) - g + 1 = n - a + g - 1$$
since
$$\deg(K + \boldsymbol{P} - D) = 2g - 2 + n - a.$$
For any differential form ω, we have
$$K \sim (\omega) = (\omega)_0 - (\omega)_\infty;$$
i.e.,
$$2g - 2 = \deg K = \deg(\omega)_0 - \deg(\omega)_\infty.$$
Let $D = D_1 - D_2$, $D_i \geq 0$, $\operatorname{Supp} D_1 \cap \operatorname{Supp} D_2 = \emptyset$. If $\omega \in \Omega(\boldsymbol{P} - D)$, then $(\omega)_0 \geq D_1$; hence,
$$\deg(\omega)_\infty = \deg(\omega)_0 - 2g + 2 \geq a - 2g + 2 + \deg D_2;$$
i.e., ω has at least $a - 2g + 2$ poles outside $\operatorname{Supp} D_2$. Since $\mathcal{P} \cap \operatorname{Supp} D = \emptyset$, all poles of ω are among the points $P_i \in \mathcal{P}$, the poles are of order 1, and $\operatorname{Res}_{P_i}(\omega) \neq 0$ if and only if $P_i \in \operatorname{Supp}(\omega)_\infty$. We have assumed that $a > 2g - 2$. Therefore, $\operatorname{Res}_\mathcal{P}(\omega) \neq 0$ for any form $\omega \neq 0$; i.e., $\operatorname{Res}_\mathcal{P}$ is an embedding. Moreover, the number of nonzero coordinates is not less than $a - 2g + 2$. The dimension $k = \dim \Omega(\boldsymbol{P} - D)$ of the code is already estimated above. □

REMARK 4.1.9. The estimate for k is nontrivial only for $a \leq n + g - 1$. As above, we call $d_c = a - 2g + 2$ and $k_c = n - a + g - 1$, respectively, the *designed distance* and *designed dimension* of the algebraic geometry code $C = (X, \mathcal{P}, D)_\Omega$. For $2g - 2 < a < n$ we have the equality $k = k_c$. The precise value of k for any a is
$$k = \ell(K + \boldsymbol{P} - D) - \ell(K - D) = n - a + g - 1 - \ell(K - D) + \ell(D - \boldsymbol{P}).$$
In particular, $k \geq n - g$ for $a \leq 2g - 2$ since
$$\ell(K - D) \leq \deg(K - D) + 1 = 2g - 1 - a.$$

REMARK 4.1.10. The designed parameters for both the L- and Ω-construction satisfy
$$d_c + k_c = n - g + 1.$$

Both constructions have a serious disadvantage: we have to choose points $P_i \notin \operatorname{Supp} D$. We have the following three ways to dispose of this drawback.

Extension to points of $\operatorname{Supp} D$. For codes of the form $C = (X, \mathcal{P}, D)_L$, there is an elementary lengthening construction that makes it possible to eliminate the condition $\operatorname{Supp} D \cap \mathcal{P} = \emptyset$. Let X be a curve, $\mathcal{P} \subseteq X(\mathbb{F}_q)$, $|\mathcal{P}| = n$, $D \in \operatorname{Div}(X)$; and let $D = D' + D''$, where $\operatorname{Supp} D' \cap \mathcal{P} = \emptyset$ and $\operatorname{Supp} D'' \subseteq \mathcal{P}$. For each point

$Q_i \in \operatorname{Supp} D''$ choose a local parameter t_i. If $D'' = \sum_{i=1}^{s} b_i Q_i$, then for any $f \in L(D)$ the function $t_i^{b_i} f$ is regular at Q_i. Consider the map

$$\operatorname{Ev}_{\mathcal{P}}^{\operatorname{ext}} \colon L(D) \to \mathbb{F}_q^n, \quad f \mapsto \Big(f(P_1), \ldots, f(P_r), t_1^{b_1} f(Q_1), \ldots, t_s^{b_s} f(Q_s)\Big),$$

where $\{P_1, \ldots, P_r\} = \mathcal{P} \setminus \operatorname{Supp} D''$. Thus, the code $C^{\operatorname{ext}} = \operatorname{Ev}_{\mathcal{P}}^{\operatorname{ext}}(L(D))$, which we denote by $(X, \mathcal{P}, D)'_L$, is a lengthening of $(X, \mathcal{P}', D)_L$ (where $\mathcal{P}' = \mathcal{P} \setminus \operatorname{Supp} D''$) by s positions corresponding to the points of $\operatorname{Supp} D'' = \operatorname{Supp} D \cap \mathcal{P}$.

EXERCISE 4.1.11. Prove that the parameters of C^{ext} satisfy all the estimates of Theorem 4.1.1.

EXERCISE 4.1.12. Construct $[q+1, a+1, q+1-a]_q$ codes. (*Hint*: Extend codes on $\mathbb{P}^1$ (Example 4.1.5); i.e., lengthen the Reed–Solomon codes.)

In fact, the described lengthening construction can be changed for the following (much more conceptual) construction, which is valid for an arbitrary variety X.

H-construction. Let X be a smooth projective variety over $\mathbb{F}_q$, $\mathscr{L}$ be a line bundle on X defined over $\mathbb{F}_q$, and $H^0(\mathscr{L})$ be the space of its sections. Let $\mathcal{P} = \{P_1, \ldots, P_n\} \subseteq X(\mathbb{F}_q)$. It is impossible to map $H^0(\mathscr{L})$ to $\mathbb{F}_q^n$ by evaluation at points (as was done in the L-construction), since the value of a section at a point is not well defined. But vanishing of a section at a point is well defined, which is quite near to what we want. There is a natural map (see Sec. 2.1.3)

$$H^0(\mathscr{L}) \to \bigoplus_{i=1}^n \overline{\mathscr{L}}_{P_i},$$

where $\overline{\mathscr{L}}_{P_i}$ is the fibre of $\mathscr{L}$ at P_i, i.e., a one-dimensional vector space over $\mathbb{F}_q$.

Fix an arbitrary *trivialization* of the fibres $\overline{\mathscr{L}}_{P_i}$, i.e., an isomorphism $\overline{\mathscr{L}}_{P_i} \cong \mathbb{F}_q$ (this is equivalent to choosing a basis vector in each fibre $\overline{\mathscr{L}}_{P_i}$). We obtain a map

$$\begin{array}{rcl} \operatorname{Germ}_{\mathcal{P}} \colon H^0(\mathscr{L}) & \longrightarrow & \mathbb{F}_q^n \\ & \searrow & \| \\ & & \bigoplus_{i=1}^n \overline{\mathscr{L}}_{P_i}. \end{array}$$

Consider the code $C = \operatorname{Germ}_{\mathcal{P}}(H^0(\mathscr{L}))$. We write $C = (X, \mathcal{P}, \mathscr{L})_H$.

As above, good estimates for the parameters can be given in the case of a curve.

THEOREM 4.1.13. *Let X be a curve of genus g, $n = |\mathcal{P}|$, and $a = \deg \mathscr{L}$, $0 \le a < n$. Then the code $C = (X, \mathcal{P}, \mathscr{L})_H$ is an $[n, k, d]_q$ code with*

$$k \ge a - g + 1, \qquad d \ge n - a.$$

PROOF. Any nonzero section $s \in H^0(\mathscr{L})$ has the property that the divisor D of its zeroes belongs to the divisor class that corresponds to the line bundle $\mathscr{L}$ (see Sec. 2.1.3). Therefore, the total number of zeroes of s (counted with proper multiplicities) is a, and thus the number of zeroes among points of $\mathcal{P}$ is at most a. This guarantees $d \ge n - a$. In particular, if $a < n$, the map $\operatorname{Germ}_{\mathcal{P}}$ is an embedding, and $k = h^0(\mathscr{L}) \ge a - g + 1$ by the Riemann–Roch theorem. $\square$

COROLLARY 4.1.14. *Let X be a curve of genus g over $\mathbb{F}_q$, and let $N = |X(\mathbb{F}_q)| > g - 1$. Then for any $n = g, g+1, \ldots, N$ and any $k = 1, \ldots, n - g$ there exists a linear $[n, k, d]_q$ code whose parameters satisfy*
$$k + d = n - g + 1.$$

PROOF. Choose a set $\mathcal{P} = \{P_1, \ldots P_n\} \subseteq X(\mathbb{F}_q)$ and a line bundle $\mathscr{L} \in \operatorname{Pic} X$ of degree $a = k + g - 1$ (on a curve X there exist bundles of any given degree; indeed, $X(\mathbb{F}_q)$ is nonempty, and we may take the divisor class which is a multiple of an $\mathbb{F}_q$-point); $g - 1 < a < n$. By Theorem 4.1.13, the code $C' = (X, \mathcal{P}, \mathscr{L})_H$ has parameters
$$k' \geq a - g + 1 = k, \qquad d' \geq n - a.$$

By the spoiling lemma (Lemma 1.1.39), given this code one can construct an $[n, k, d]_q$ code C with $d = n - a$, i.e.,
$$k + d = n - g + 1. \qquad \square$$

REMARK 4.1.15. Strictly speaking, the code $C = (X, \mathcal{P}, \mathscr{L})_H$ depends also on the choice of trivializations; moreover, this code, as well as $(X, \mathcal{P}, D)_L$ and $(X, \mathcal{P}, D)_\Omega$, depends on the ordering of elements of $\mathcal{P}$. However, any other choice of trivialization corresponds to multiplication of the ith basis vector of $\mathbb{F}_q^n$ by a nonzero constant $y_i \in \mathbb{F}_q^*$, and the choice of another ordering in $\mathcal{P}$ corresponds to a permutation of coordinates; i.e., these operations lead to equivalent codes (see Sec. 1.1.1). Thus, the equivalence class of $(X, \mathcal{P}, \mathscr{L})_H$ is uniquely defined.

REMARK 4.1.16. Let D be a divisor corresponding to the bundle $\mathscr{L}$; i.e., D is the divisor of zeroes of some section $s_0 \in H^0(\mathscr{L})$. The section s_0 uniquely defines the following trivialization of $\overline{\mathscr{L}}_{P_i}$ for any $P_i \notin \operatorname{Supp} D$: any $t \in \overline{\mathscr{L}}_{P_i}$ is mapped to an element $x \in \mathbb{F}_q$ such that $t = x s_0(P_i)$. Thus, D defines trivializations of all fibres $\overline{\mathscr{L}}_{P_i}$ up to multiplication by a common nonzero constant (an element of $\mathbb{F}_q^*$).

EXERCISE 4.1.17. Using the preceding remark, prove that if $\mathcal{P} \cap \operatorname{Supp} D = \varnothing$, then the code $(X, \mathcal{P}, \mathscr{L})_H$ obtained by this choice of trivialization coincides with $(X, \mathcal{P}, D)_L$. What is the relation between $(X, \mathcal{P}, \mathscr{L})_H$ and $(X, \mathcal{P}, D)_L^{\text{ext}}$ in the case $\operatorname{Supp} D \cap \mathcal{P} \neq \varnothing$?

Let us now describe the H-construction in somewhat different language.

P-construction. Let us make use of the fact (see Sec. 1.1.2) that the equivalence class of a nondegenerate linear $[n, k, d]_q$ code C is uniquely determined by a projective $[n, k, d]_q$ system $\mathcal{P}$.

If a variety X is given with a fixed embedding $X \subset \mathbb{P}^m$, then, choosing any $\mathcal{P} \subseteq X(\mathbb{F}_q)$, we obtain a projective $[n, k, d]_q$ system with $n = |\mathcal{P}|$, $k = m + 1$, and $d = n - \max_H \{|H \cap \mathcal{P}|\}$, where the maximum is taken over all hyperplanes $H \subset \mathbb{P}^m$ defined over $\mathbb{F}_q$.

If X is a curve (with a fixed embedding), then by definition its degree $\deg X$ is the number of $\overline{\mathbb{F}}_q$-points (counted with appropriate multiplicities) in the intersection of X with an arbitrary hyperplane (see Sec. 2.1.3). Anyway, $\max_H \{|H \cap \mathcal{P}|\} \leq \deg X$. We have proved the following fact.

PROPOSITION 4.1.18. *Let a curve $X \subset \mathbb{P}^m$ be given, and let $N = |X(\mathbb{F}_q)| > 0$. Then for any n in the interval $N \geq n > \max\{m, \deg X\}$ there exists a nondegenerate linear $[n, k, d]_q$ code with $k = m + 1$ and $d \geq n - \deg X$.* $\square$

Let X be a variety, and let $\mathscr{L}$ be a line bundle (divisor class) on it. To the bundle $\mathscr{L}$ there corresponds the map
$$\varphi_{\mathscr{L}} \colon X \to \mathbb{P}(H^0(\mathscr{L})) \cong \mathbb{P}^{h^0(\mathscr{L})-1}$$
(see Sec. 2.1.3). For a projective $[n,k,d]_q$ system, consider any subset $\mathcal{P} \subseteq X(\mathbb{F}_q)$ mapped into $\mathbb{P}(H^0(\mathscr{L}))$ (if some points of $\mathcal{P}$ glue together, then their image is counted with the appropriate multiplicity); here, $n = |\mathcal{P}|$ and $k = h^0(\mathscr{L})$. According to Sec. 2.1.3, inverse images of hyperplane sections of $\mathbb{P}(H^0(\mathscr{L}))$ are effective divisors D belonging to the class $\mathscr{L}$. Hence,
$$d = n - \max_D\{|D \cap \mathcal{P}|\}$$
(the definition of $|D \cap \mathcal{P}|$ is obvious).

Consider now the case of curves. If X is a curve of genus g and $a = \deg \mathscr{L}$, then $k = h^0(\mathscr{L}) \geq a - g + 1$. Furthermore, $|D(\mathbb{F}_q)| \leq \deg D = a$; i.e., $d \geq n - a$. Thus, we have constructed a linear $[n,k,d]_q$ code $C = (X, \mathcal{P}, \mathscr{L})_P$.

EXERCISE 4.1.19. Show that the projective systems $(X, \mathcal{P}, \mathscr{L})_H$ and $(X, \mathcal{P}, \mathscr{L})_P$ are isomorphic; i.e., the corresponding codes are equivalent.

REMARK 4.1.20. If $a \geq 2g+1$, then by Corollary 2.2.29 the map $\varphi_{\mathscr{L}}$ is an embedding, and we obtain a projective system without multiplicities.

REMARK 4.1.21. Of interest is the question of how wide the class of algebraic geometry codes is. Let there be given an arbitrary linear $[n,k,d]_q$ code. To this code there corresponds a projective system $\mathcal{P}$, i.e., a set of points with multiplicities in $\mathbb{P}^{k-1}$. If the projective system has no repetitions, then it can be shown that there exists a smooth curve X passing through these points and no other $\mathbb{F}_q$-points of $\mathbb{P}^{k-1}$. In this sense, any projective system without repetitions is an algebraic geometry system (in fact, the same holds for systems with repetitions as well). A more precise answer to this question is given in Sec. 4.3.

Automorphisms. Let $G \subseteq \mathrm{Aut}_{\mathbb{F}_q}(X)$ be a subgroup in the group of $\mathbb{F}_q$-automorphisms of a curve X. In general, there is no natural action of G on an algebraic geometry code $(X, \mathcal{P}, D)_L$. However, if $\mathcal{P}$ is a G-invariant set, as well as the divisor D, then G acts on the algebraic geometry code $(X, \mathcal{P}, D)_L$. Indeed, in this case for any $g \in G$ and any $f \in L(D)$, the function $g^*(f)$ also belongs to $L(D)$ since $g(D) = D$. Therefore, a codeword of the form $(f(P_1), \ldots, f(P_n))$ is taken to the codeword
$$\bigl((g^*(f))(P_1), \ldots, (q^*(f))(P_n)\bigr) = \bigl(f(g(P_1)), \ldots, f(g(P_n))\bigr).$$
Thus, there is a natural map from G to the group $\mathrm{Aut}(C) \cap S_n$, and according to what was said in Sec. 1.2.2 on group codes, we have $C \subseteq \mathbb{F}_q[G/H_1] \oplus \ldots \oplus \mathbb{F}_q[G/H_m]$, where H_i are the stabilizers of some points $Q_i \in \mathcal{P}$. We obtain the following result.

PROPOSITION 4.1.22. *Let $G \subseteq \mathrm{Aut}_{\mathbb{F}_q}(X)$. Let $\mathcal{P}$ be a G-invariant subset of $X(\mathbb{F}_q)$, D be a G-invariant $\mathbb{F}_q$-divisor on X, and $\mathrm{Supp}\, D \cap \mathcal{P} = \varnothing$. Then $C = (X, \mathcal{P}, D)_L$ is a group code: $C \subseteq \mathbb{F}_q[G/H_1] \oplus \ldots \oplus \mathbb{F}_q[G/H_m]$, where H_i is the stabilizer of a point $Q_i \in \mathcal{P}$; $\{Q_1, \ldots, Q_m\}$ is the set of representatives of orbits of the action of G on $\mathcal{P}$.* □

In *Advanced Chapters* we intend to show that some codes on modular curves can be represented in this form.

EXAMPLE 4.1.23. Let
$$X = \mathbb{P}^1, \qquad D = a \cdot \infty, \qquad \mathcal{P} = X(\mathbb{F}_q) \setminus \{\infty\}, \qquad \mathrm{Aut}(X) = \mathrm{PGL}(2, \mathbb{F}_q).$$
The subgroup G of affine automorphisms of the form $x \mapsto ax + b$, $a \in \mathbb{F}_q^*$, $b \in \mathbb{F}_q$, fixes $\mathcal{P}$ and D. Hence, G acts on the Reed–Solomon code $C = (X, \mathcal{P}, D)_L$; moreover, since the action of G on $\mathcal{P}$ is transitive, we have $C \subseteq \mathbb{F}_q[G/H]$, where H is the stabilizer of $P_1 = 0 \in X(\mathbb{F}_q)$, $H = \{x \mapsto ax \mid a \in \mathbb{F}_q^*\} \cong \mathbb{F}_q^*$. Note that the group G/H is isomorphic to the additive group of $\mathbb{F}_q$.

EXERCISE 4.1.24. Examine the following example: $X = \mathbb{P}^1$, $D = a \cdot \infty$, and $\mathcal{P} = X(\mathbb{F}_q) \setminus \{0, \infty\}$.

4.1.2. Duality and Spectra

L/Ω-duality. Recall that for any differential form $\omega \in \Omega_X$ we have the residue formula
$$\sum_{Q \in \mathrm{Supp}(\omega)_\infty} \mathrm{Res}_Q \, \omega = 0.$$

THEOREM 4.1.25. *The codes $C_L = (X, \mathcal{P}, D)_L$ and $C_\Omega = (X, \mathcal{P}, D)_\Omega$ are dual.*

PROOF. First consider the case $2g - 2 < a < n$, in which case the maps $\mathrm{Ev}_\mathcal{P}$ and $\mathrm{Res}_\mathcal{P}$ are embeddings. Let $f \in L(D)$ and $\omega \in \Omega(\boldsymbol{P} - D)$, where $\boldsymbol{P} = \sum_{P_i \in \mathcal{P}} P_i$; then by the residue formula we have
$$\sum_{i=1}^n f(P_i) \mathrm{Res}_{P_i}(\omega) = \sum_{i=1}^n \mathrm{Res}_{P_i}(f\omega) = 0,$$
since all poles of the form $f\omega$ lie in $\boldsymbol{P}$ (possible poles of f are suppressed by zeroes of ω). Thus, each vector of C_L is orthogonal to any vector of C_Ω; i.e., $C_\Omega \subseteq C_L^\perp$. On the other hand,
$$\dim C_L^\perp = n - \dim C_L \leq n - (a - g + 1) \leq \dim C_\Omega,$$
and therefore $C_\Omega = C_L^\perp$. Here, the inequalities for the dimensions of C_Ω and C_L turn into equalities.

Now consider arbitrary a. In this case, $\mathrm{Ker}(\mathrm{Ev}_\mathcal{P}) = L(D - \boldsymbol{P})$ and $\mathrm{Ker}(\mathrm{Res}_\mathcal{P}) = \Omega(-D)$. The Riemann–Roch theorem shows that C_L and C_Ω have complementary dimensions. Since the proof of orthogonality does not use conditions on a, the theorem is proved. $\square$

H-duality. The line bundle $\Omega(\boldsymbol{P})$, where $\boldsymbol{P} = \sum_{P_i \in \mathcal{P}} P_i$, possesses the canonical trivialization
$$\mathrm{Res}_{P_i} : \overline{\Omega(\boldsymbol{P})}_{P_i} \xrightarrow{\sim} \mathbb{F}_q$$
at points $P_i \in \mathcal{P}$ (recall that elements of the fibre $\overline{\Omega(\boldsymbol{P})}_{P_i}$ are differential forms with poles of order at most 1 at P_i modulo forms that are regular at P_i).

Let $\mathscr{L}$ be a line bundle on X, and let $\mathscr{M} = \Omega(\boldsymbol{P}) \otimes \mathscr{L}^{-1}$. Fix a trivialization $t_i : \overline{\mathscr{L}}_{P_i} \xrightarrow{\sim} \mathbb{F}_q$, $\ell_0 = t_i^{-1}(1)$. Define a trivialization $t_i' : \overline{\mathscr{M}}_{P_i} \xrightarrow{\sim} \mathbb{F}_q$ by the condition $t_i'(m) = \mathrm{Res}_{P_i}(\ell_0 m)$. We say that the trivialization t_i' is *consistent* with t_i.

THEOREM 4.1.26. *Assume that there are chosen consistent trivializations on line bundles $\mathscr{L}$ and $\Omega(\boldsymbol{P}) \otimes \mathscr{L}^{-1}$. Then the codes $C = (X, \mathcal{P}, \mathscr{L})_H$ and $C' = (X, \mathcal{P}, \Omega(\boldsymbol{P}) \otimes \mathscr{L}^{-1})_H$ are dual.*

PROOF. By the Riemann–Roch theorem, the codes C and C' have complementary dimensions. Let vectors $v \in C$ and $v' \in C'$ correspond to sections $s \in H^0(\mathscr{L})$ and $s' \in H^0(\mathscr{M})$, and let $s_i \in \overline{\mathscr{L}}_{P_i}$ and $s'_i \in \overline{\mathscr{M}}_{P_i}$ be the images of s and s'. Then

$$(v, v') = \sum_i t_i(s_i) t'_i(s'_i) = \sum_i \mathrm{Res}_{P_i}(s_i s'_i) = \sum_i \mathrm{Res}_{P_i}(s \otimes s') = 0$$

by the residue formula (since all poles of sections of the line bundle $\Omega(\boldsymbol{P})$ are concentrated in $\mathcal{P}$). □

EXERCISE 4.1.27. Show that, for any a, $0 \le a \le n - g + 1$, the parameters of the code dual to C satisfy the condition

$$k^\perp + d^\perp \ge n - g + 1.$$

Self-dual algebraic geometry codes. Self-dual codes are of special interest as analogues of unimodular lattices, which are important in number theory. They are also useful for construction of quantum codes. Let us see what can be said concerning self-duality in the case of algebraic geometry codes.

Let $y = (y_1, \ldots, y_n) \in \mathbb{F}_q^n$, $y_i \ne 0$ for all i. Recall that a code C is called *quasi-self-dual with respect to y* if $n = 2k$ and for any $u, v \in C$ we have

$$\sum_{i=1}^n u_i v_i y_i = 0$$

(i.e., the vectors u and v are *y-orthogonal*). A code is called *quasi-self-dual* if there exists any y with this property and *self-dual* if $y = (1, \ldots, 1)$.

THEOREM 4.1.28. *Let $n > 2g - 2$, n being even, $a = n/2 + g - 1$. If $K + \boldsymbol{P} \sim 2D$, where $\boldsymbol{P} = \sum_{P_i \in \mathcal{P}} P_i$, then the $\left[n, \frac{n}{2}, \ge \frac{n}{2} - g + 1\right]_q$ code $C = (X, \mathcal{P}, D)_L$ is quasi-self-dual. Furthermore, there exists a unique (up to a factor) form $\omega_0 \in \Omega(\boldsymbol{P} - 2D)$ such that C is quasi-self-dual with respect to $y = (y_1, \ldots, y_n)$, where $y_i = \mathrm{Res}_{P_i}(\omega_0) \ne 0$. In particular, if $\mathrm{Res}_{P_1}(\omega_0) = \ldots = \mathrm{Res}_{P_n}(\omega_0)$, then C is self-dual.*

PROOF. Since $n > 2g - 2$, we have $n > a > 2g - 2$; hence, $k = a - g + 1 = n/2$ and $d \ge n - a = n/2 - g + 1$. Let $2D \sim K + \boldsymbol{P}$; then $\dim \Omega(\boldsymbol{P} - 2D) = \dim \Omega(-K) = 1$. Let $\omega_0 \in \Omega(\boldsymbol{P} - 2D)$, $\omega_0 \ne 0$. If we assume that $\mathrm{Res}_{P_i}(\omega_0) = 0$ for some i, then $\omega_0 \in \Omega(\boldsymbol{P} - 2D - P_i)$, which is impossible since $\deg(\boldsymbol{P} - 2D - P_i) < 2 - 2g$. Therefore, all residues $y_i = \mathrm{Res}_{P_i}(\omega_0)$ are nonzero. For any two elements $f, g \in L(D)$, the form $\omega = fg\omega_0$ belongs to the space $\Omega(\boldsymbol{P})$, i.e., has no poles outside $\mathcal{P}$. Therefore,

$$\sum_{i=1}^n f(P_i) g(P_i) y_i = \sum_{i=1}^n \mathrm{Res}_{P_i} \omega = 0;$$

i.e., any two elements of the code are y-orthogonal. □

COROLLARY 4.1.29. *If the image of the divisor $\boldsymbol{P} = \sum_{P_i \in \mathcal{P}} P_i$ is divisible by 2 in the group $(\mathrm{Pic}\, X)(\mathbb{F}_q)$, then there exists a quasi-self-dual algebraic geometry code with parameters $\left[n, \frac{n}{2}, \ge \frac{n}{2} - g + 1\right]_q$. If, moreover, q is even, then there exists a self-dual code with these parameters.*

To prove this fact, we need one special property of curves over $\mathbb{F}_q$, which concerns their canonical classes.

PROPOSITION 4.1.30. *Let X be a complete projective curve over $\mathbb{F}_q$, and let K_X be its canonical class. Then $K_X \sim 2D$ for some $\mathbb{F}_q$-divisor D on X.* □

In the general case, the proof of this proposition is based on the class field theory. However, in the case $p = \operatorname{char} \mathbb{F}_q = 2$ it can be proved by elementary methods. Indeed, let $f \in \mathbb{F}_q(X) \setminus \mathbb{F}_q$ and $\omega = df$. Let $P \in \operatorname{Supp}(\omega)$, and let $t = t_P$ be a local parameter at P. Then $df = \sum_{i=i_0}^{\infty} a_{2i+1} t^{2i} dt$, where $\tau_P(f) = \sum_{i=m}^{\infty} a_i t^i$ is a power series expansion at P and i_0 is the smallest integer i for which $a_{2i+1} \neq 0$. Hence, $(\omega) = (df) = 2i_0 P + (\omega')$, where $P \notin \operatorname{Supp}(\omega')$, and by induction we conclude that $(\omega) = 2D$ for some $\mathbb{F}_q$-divisor D.

PROOF OF COROLLARY 4.1.29. By Proposition 4.1.30, the canonical divisor K is always divisible by two in the group $(\operatorname{Pic} X)(\mathbb{F}_q)$. Put $D = (K + \boldsymbol{P})/2$, and apply Theorem 4.1.28. For q even, one should also apply Exercise 1.1.35. □

One would like to be able to construct sets $\mathcal{P} \subseteq X(\mathbb{F}_q)$ with even sums $\boldsymbol{P}$ and with even $n = |\mathcal{P}|$ as large as possible.

THEOREM 4.1.31. *Let $N = |X(\mathbb{F}_q)| > 2g$. Then there exists $\mathcal{P} \subseteq X(\mathbb{F}_q)$ with an even $n = |\mathcal{P}| \geq N - 2g - 1$ and an even sum.*

PROOF. Put $J = (\operatorname{Pic} X)(\mathbb{F}_q)$, and let $J_0 \subset J$ be the kernel of the degree map $\deg: J \to \mathbb{Z}$. Since the group $\mathbb{Z}$ is torsion-free, $J/2J = \mathbb{Z}/2\mathbb{Z} \times J_0/2J_0$. On the other hand, the group J_0 is finite, and therefore the order of $J_0/2J_0$ equals the order of the kernel of multiplication by 2. The latter group is embedded in $(\mathbb{Z}/2\mathbb{Z})^{2g}$; hence, $J/2J \subseteq (\mathbb{Z}/2\mathbb{Z})^{2g+1}$.

Let us order $\mathbb{F}_q$-points of X so that the images $P_{r+1}, \ldots, P_N$ form a basis of $J/2J$ over $\mathbb{F}_2$; as we have proved above, $r \geq N - 2g - 1$. We decompose the image of the sum $P_1 + \ldots + P_r$ in $J/2J$ in this basis:

$$P_1 + \ldots + P_r = \sum_{i=r+1}^{N} \varepsilon_i P_i \pmod{2},$$

where $\varepsilon_i = 0$ or 1. Then reorder $P_{r+1}, \ldots, P_N$ so that $\varepsilon_i = 1$ for $r+1 \leq i \leq s$ (if all $\varepsilon_i = 0$, then put $s = r$) and $\varepsilon_i = 0$ for $i > s$. Then $P_1 + \ldots + P_s \equiv 0 \pmod{2}$, $s \geq r \geq N - 2g - 1$, and we may put $\mathcal{P} = \{P_1, \ldots, P_s\}$. Here, s is always even since a divisor of an odd degree cannot be zero mod 2. □

COROLLARY 4.1.32. *In the notation of Theorem 4.1.31, there exists a quasi-self-dual $[n, n/2, \geq n/2 - g + 1]_q$ code (self-dual if q is even) with an even $n \geq N - 2g - 1$.* □

REMARK 4.1.33. In general, the sufficient condition of Theorem 4.1.28 is not necessary. Consider the elliptic curve $y^2 + y = x^3$. Over $\mathbb{F}_2$, this curve has 3 points; i.e., $|X(\mathbb{F}_2)| = 2 + 1 - (\omega + \overline{\omega}) = 3$, whence $\omega = \pm\sqrt{2}i$. Therefore, $|X(\mathbb{F}_4)| = 4 + 1 - (\omega^2 + \overline{\omega}^2) = 9$. By Theorem 3.3.15 we have $X(\mathbb{F}_4) \cong (\mathbb{Z}/3\mathbb{Z})^2$. Let P_0 be the point at infinity, $P_1 = (0,0)$, $P_2 = (\varepsilon, 1)$, $P_3 = (\varepsilon, \varepsilon)$, and $P_4 = (\varepsilon, \varepsilon^2)$, where ε is a generator of $\mathbb{F}_4^*$. The code $C_L = (X, \{P_1, \ldots, P_4\}, 2P_0)_L$ constructed from the divisor $2P_0$ and the system of points $\{P_1, \ldots, P_4\}$ is a $[4, 2, 3]_4$ code; its dual code C_Ω has the same parameters (the space $L(2P_0)$ is generated by the functions 1 and x, and the space $\Omega(P_1 + P_2 + P_3 + P_4 - 2P_0)$ by the forms $\omega_1 = \dfrac{x^2 \, dx}{y(y - \varepsilon)}$ and $\omega_2 = \dfrac{dx}{y - \varepsilon}$).

Computation of residues shows that $C_L = C_\Omega$. On the other hand, the divisor $P_1 + P_2 + P_3 + P_4 \sim P_1 + 3P_0$ is not equivalent to $4P_0$.

EXERCISE 4.1.34. Check all facts claimed in Remark 4.1.33.

However, there are situations where the condition of Theorem 4.1.28 is not only sufficient but also necessary.

THEOREM 4.1.35. *Let $g \geq 1$. Let $D = \sum a_j Q_j$ be an effective divisor of degree $a = n/2 + g - 1 \geq 4g - 1$ with $a_j \leq a - (2g-1)$ for any j, and let $\mathcal{P} \subseteq X(\mathbb{F}_q) \setminus \operatorname{Supp} D$. The code $C = (X, \mathcal{P}, D)_L$ is self-dual if and only if there exists a form $\omega \in \Omega(\boldsymbol{P} - 2D)$ with $\operatorname{Res}_{P_i}(\omega) = 1$ for all $i = 1, \ldots, n$.*

PROOF. Let $C = C^\perp = (X, \mathcal{P}, D)_\Omega$. Since $1 \in L(D)$, there exists a form $\eta \in \Omega(\boldsymbol{P} - D)$ with $\operatorname{Res}_{P_i}(\eta) = 1$. Let us demonstrate that $\eta \in \Omega(\boldsymbol{P} - 2D)$. This is a local condition; i.e., it suffices to prove that $(\eta) \geq 2 a_j Q_j$ for any j. Let Q be one of the points Q_j, and let $D = bQ + D_0$, $Q \notin \operatorname{Supp} D_0$; by the condition of the theorem, $\deg D_0 \geq 2g - 1$. We may assume that Q is defined over $\mathbb{F}_q$ (otherwise, we may extend the ground field; the condition $(\eta) \geq 2D$ will be preserved). By the condition of the theorem, there is a divisor $D' = bQ + D'_0$, $0 \leq D' \leq D$, such that $\deg(D - D') = 2g - 1$; here, $\deg D' \geq 2g$. Then by the Riemann–Roch theorem we have $L(D') \neq L(D' - Q)$. Let $f \in L(D') \setminus L(D' - Q)$, i.e., f has a pole of order precisely b at Q. By the self-duality, there exists a form $\omega \in \Omega(\boldsymbol{P} - D)$ with $\operatorname{Res}_{P_i}(\omega) = f(P_i)$ (since $f \in L(D') \subseteq L(D)$). Here, $f\eta \in \Omega(\boldsymbol{P} - (D - D'))$ and $\operatorname{Res}_{P_i}(f\eta) = \operatorname{Res}_{P_i}(\omega)$, i.e., $\omega - f\eta \in \Omega(-(D - D'))$; but $\deg(D - D') = 2g - 1$, and a nonzero regular differential form cannot have $2g - 1$ zeroes. Hence, $\omega = f\eta$. Therefore, the zero order of η at Q satisfies
$$v_Q(\eta) = v_Q(\omega) - v_Q(f) = v_Q(\omega) + b \geq 2b,$$
whence we get $(\eta) \geq 2bQ$. This holds at any point $Q \in \operatorname{Supp} D$; hence, $(\eta) \geq 2D$, i.e., $\eta \in \Omega(\boldsymbol{P} - 2D)$. $\square$

REMARK 4.1.36. One can show that for $g = 0$ the criterion of Theorem 4.1.35 is valid for *any* effective divisor D, and for $g = 1$ for any effective divisor D of degree $a \geq 3$.

Spectra. What can be said about the weight spectrum of an algebraic geometry code? Let
$$W_C(x:y) = \sum_{v \in C} x^{n-w(v)} y^{w(v)} = x^n + \sum_{r=d}^{n} A_r x^{n-r} y^r$$
be the enumerator of a code C, and $W_C(x)$ its inhomogeneous form. In Sec. 1.1.3 we expressed it in terms of the numbers B_ℓ: Let $d \geq n - a$ for some integer $a \geq 0$; then
$$W_C(x) = x^n + \sum_{\ell=0}^{a} B_\ell (x-1)^\ell, \quad B_\ell = \sum_{j=n-a}^{n-\ell} \binom{n-j}{\ell} A_j,$$
where we assume that $A_1 = \ldots = A_{d-1} = 0$.

In Exercise 1.1.29, B_ℓ is interpreted as the sum of the numbers $(q^{\ell(i_1, \ldots, i_\ell)} - 1)$ over all sets $\{i_1, \ldots, i_\ell\} \subseteq \{1, \ldots, n\}$ of cardinality ℓ, where $\ell(i_1, \ldots, i_\ell)$ is the dimension of the subcode consisting of the vectors of C that have zeroes at positions $i_1, \ldots, i_\ell$.

THEOREM 4.1.37. *Let X be a curve of genus g. Let $\mathscr{L} \in \operatorname{Pic} X$, $\deg \mathscr{L} = a$, $\mathcal{P} \subseteq X(\mathbb{F}_q)$, $|\mathcal{P}| = n$, and $C = (X, \mathcal{P}, \mathscr{L})_H$. Then*

$$W_C(x) = x^n + \sum_{\ell=0}^{a} B_\ell (x-1)^\ell,$$

where for $0 \le \ell \le a - 2g + 1$ we have

$$B_\ell = \binom{n}{\ell} (q^{a-\ell-g+1} - 1),$$

and for $a \ge \ell \ge a - 2g + 2$ we have

$$\binom{n}{\ell} (q^{\lfloor (a-\ell)/2 \rfloor + 1} - 1) \ge B_\ell \ge \max \left\{ 0, \binom{n}{\ell} (q^{a-\ell-g+1} - 1) \right\}.$$

PROOF. By Theorem 4.1.13 and Exercise 4.1.27, parameters of C satisfy

$$k + d \ge n - g + 1, \qquad (n-k) + d^\perp \ge n - g + 1;$$

i.e., C is a code of genus at most g (in the sense of the definition in Sec. 1.1.3).

For such a code, our theorem (except for the left-hand inequality) is always valid (see Theorem 1.1.18). Example 1.1.21 shows that for an arbitrary code of genus at most g, the left-hand inequality can be invalid, but its weakened version with the substitution of $a - \ell + 1$ for $\lfloor (a-\ell)/2 \rfloor + 1$ always holds. Let us show that for an algebraic geometry code this stronger inequality is always true. Indeed, the subcode consisting of vectors with zeroes at the positions $i_1, \ldots, i_\ell$ in C is the image in $\mathbb{F}_q^n$ of the subspace

$$H^0(\mathscr{L} \otimes \mathscr{O}(-P_{i_1} - \ldots - P_{i_\ell})) \subseteq H^0(\mathscr{L}).$$

If the line bundle $\mathscr{L}' = \mathscr{L} \otimes \mathscr{O}(-P_{i_1} - \ldots - P_{i_\ell})$ is not special, i.e., if we have $h^0(\Omega \otimes (\mathscr{L}')^{-1}) = 0$, then the Riemann–Roch theorem yields

$$h^0(\mathscr{L}') = \deg \mathscr{L}' - g + 1 = a - \ell - g + 1,$$

but $a - \ell \le 2g - 2$, and therefore

$$h^0(\mathscr{L}') = a - \ell - g + 1 \le (a-\ell)/2 + 1.$$

If $\mathscr{L}'$ is special, then one may apply the Clifford theorem (Theorem 2.2.43), which shows that $h^0(\mathscr{L}') \le (a-\ell)/2 + 1$. □

EXERCISE 4.1.38. Prove Theorem 4.1.37 not using Theorem 1.1.18. (*Hint*: Apply the Riemann–Roch theorem and the Clifford theorem.)

Thus, in the spectrum of an algebraic geometry code constructed from a curve of genus g, there are $2g - 1$ unknown parameters B_ℓ, $a - 2g + 2 \le \ell \le a$. For $g = 0$, Theorem 4.1.37 completely determines the spectrum of a Reed–Solomon code. Below, in Sec. 4.4.2, we shall compute the spectrum for the simplest nontrivial case $g = 1$ (elliptic codes).

4.1.3. Decoding Problem

In this subsection we describe the setting of the decoding problem for algebraic geometry codes and present the simplest (basic) algorithm for its partial solution. We hope to devote a special section of *Advanced Chapters* to a complete solution of the decoding problem for algebraic geometry codes and to its numerous modifications.

Syndromes and the basic system. Consider an algebraic geometry code $C = (X, \mathcal{P}, D)_\Omega$ over $\mathbb{F}_q$. If $\deg D = a$, $|\mathcal{P}| = n$, and $2g - 2 < a \leq n + g - 1$, then its designed parameters are

$$k_c = n - a + g - 1, \qquad d_c = a - 2g + 2.$$

Theorem 4.1.25 shows that a parity-check matrix of C is of the form $(f_i(P_j))$, where $f_1, \ldots, f_m$ is a basis of $L(D)$.

For a vector $v \in \mathbb{F}_q^n$ and a function $f \in L(D)$, define the syndrome

$$s(v, f) = \sum_{P_i \in \mathcal{P}} v_i f(P_i).$$

Note that if $v = u + e$, where $u \in C$ and e is an error vector, then

$$s(v, f) = \sum_{i \in I} e_i f(P_i),$$

where $I = \{i \mid e_i \neq 0\}$ is the set of "error coordinates" (or *error locators*). Set $t = |I|$.

Consider now an auxiliary divisor D' of degree b such that $\operatorname{Supp} D' \cap \mathcal{P} = \varnothing$. By $g_1, \ldots, g_\ell$ denote a basis in $L(D')$, and by $h_1, \ldots, h_r$ a basis in $L(D - D')$. Here, of course, $g_i h_j \in L(D)$. Let

$$s_{ij} = s_{ij}(v) = s(v, g_i h_j).$$

A crucial role for the decoding of C is played by the system of equations

$$\sum_{i=1}^{\ell} s_{ij} x_i = 0, \quad j = 1, \ldots, r. \tag{4.1.1}$$

PROPOSITION 4.1.39. *If*

$$\ell(D') > t,$$

then system (4.1.1) *has a nontrivial solution. If*

$$a - b > t + 2g - 2,$$

then, for any solution $y = (y_1, \ldots, y_\ell)$ *of the system, the function*

$$g_y = \sum_{i=1}^{\ell} y_i g_i$$

vanishes at P_i *for any* $i \in I$.

PROOF. If $\ell(D') > t$, then

$$\ell\left(D' - \sum_{i \in I} P_i\right) \geq \ell(D') - t > 0.$$

Let $g \in L\left(D' - \sum_{i \in I} P_i\right)$, $g \neq 0$. Expand g in the basis of $L(D')$: $g = \sum y_i g_i$. The vector $y = (y_1, \ldots, y_\ell)$ is a solution to (4.1.1), since $g(P_k) = 0$ for $k \in I$ and therefore

$$\sum_i s_{ij}(v) y_i = \sum_i s_{ij}(e) y_i = \sum_i \sum_{k \in I} e_k g_i(P_k) h_j(P_k) y_i = \sum_{k \in I} e_k g(P_k) h_j(P_k) = 0.$$

If $a - b > t + 2g - 2$, then by the Riemann–Roch theorem we have

$$\ell\left(D - D' - \sum_{i \in I} P_i\right) = a - b - t - g + 1$$

since
$$\deg\left(D - D' - \sum_{i \in I} P_i\right) \geq 2g - 1.$$

By the same reason,
$$\ell(D - D') = a - b - g + 1.$$

Let $\mathcal{P}_I = \{P_i\}_{i \in I}$. The kernel of the map
$$\mathrm{Ev}_{\mathcal{P}_I} : L(D - D') \longrightarrow \mathbb{F}_q^t$$
is $L\left(D - D' - \sum_{i \in I} P_i\right)$. However,
$$\ell(D - D') - \ell\left(D - D' - \sum_{i \in I} P_i\right) = y,$$

hence the map $\mathrm{Ev}_{\mathcal{P}_I}$ is surjective, and therefore for any $i \in I$ there exists a function $F_i \in L(D - D')$ such that $F_i(P_i) = 1$ and $F_i(P_j) = 0$ for $j \in I \setminus \{i\}$.

Since the functions $F_i \in L(D - D')$ are linear combinations of the functions h_j, the equations $\sum_i s(v, g_i F_j) x_i = 0$ are implied by system (4.1.1). For any solution $y = (y_1, \ldots, y_\ell)$, we have
$$0 = \sum_i s(v, g_i F_j) y_i = \sum_i \sum_{k \in I} e_k g_i(P_k) F_j(P_k) y_i = \sum_{k \in I} e_k g_y(P_k) F_j(P_k) = e_j g_y(P_j);$$
i.e., $g_y(P_j) = 0$ for any $j \in I$. $\square$

Thus, the above conditions on the divisor D' allow us to find a function g_y vanishing at all points P_i that correspond to error locators (and possibly also vanishing at some other "extra" points). Denote the set of points $P_i \in \mathcal{P}$ such that $g_y(P_i) = 0$ by I_y; we have proved that $I_y \supseteq I$.

Consider the system of equations
$$\sum_{i \in I_y} f_j(P_i) z_i = s(v, f_j), \tag{4.1.2}$$

where $f_1, \ldots, f_m$ is a basis of $L(D)$. The error vector e is a solution to this system since $s(v, f_i) = s(e, f_i)$.

PROPOSITION 4.1.40. *If*
$$a - b > 2g - 2,$$
then system (4.1.2) has at most one solution.

PROOF. Let z and z' be two distinct solutions of the system. Then $z - z'$ is a solution to the system
$$\sum_{i \in I_y} f_j(P_i) x_i = 0;$$

i.e., the vector $x = (x_1, \ldots, x_n)$, where $x_i = z_i - z_i'$ for $i \in I_y$ and $x_i = 0$ for $i \notin I_y$, is a nonzero code vector: $x \in C \setminus \{0\}$. Moreover, $\|x\| \leq |I_y| \leq \deg D' = b$ since $g_y \in L(D')$. On the other hand, by the condition we have $b < a - 2g + 2 = d_c \leq d$. But the weight of a nonzero vector cannot be less than the minimum distance of a code; this contradiction proves the proposition. $\square$

Algorithm. Thus, a choice of an auxiliary divisor D' yields the following decoding algorithm for the code $C = (X, \mathcal{P}, D)_\Omega$, called the *basic algorithm*:
1. Compute a basis $\{f_i\}$ of $L(D)$, a basis $\{g_j\}$ of $L(D')$, and a basis $\{h_k\}$ of $L(D - D')$.
2. For a given vector $v \in \mathbb{F}_q^n$ (the input of the algorithm), compute the syndromes $s(v, g_j h_k)$ and $s(v, f_i)$.
3. Find a solution y of linear system (4.1.1).
4. Find (by searching over the points P_i) those i for which $g_y(P_i) = 0$;
5. Solve linear system (4.1.2).

Propositions 4.1.39 and 4.1.40 immediately imply the following result.

THEOREM 4.1.41. *Let $C = (X, \mathcal{P}, D)_\Omega$, where $2g - 2 < a = \deg D \leq n - g + 1$. If for a positive integer t there is a divisor D' of degree b with $\operatorname{Supp} D' \cap \mathcal{P} = \varnothing$ and*
$$\ell(D') > t,$$
$$a - b > t + 2g - 2,$$
then the basic algorithm corrects t errors. □

COROLLARY 4.1.42. *The basic algorithm corrects any*
$$t \leq t_0 = \max_{D'} \{\min\{\ell(D'), a - \deg D' - 2g + 2\}\} - 1$$
errors, where the maximum is taken over all D' such that $\operatorname{Supp} D' \cap \mathcal{P} = \varnothing$. □

Note that to use the basic algorithm, we must know the corresponding divisor D' explicitly. Using the H-construction or choosing D in the form aP_0, where $P_0 \in X(\mathbb{F}_q)$, allows us to easily solve the problem of finding D'.

EXERCISE 4.1.43. Extend results of this subsection to codes $C = (X, \mathcal{P}, \mathscr{L})_H$, thus disposing of the conditions $\operatorname{Supp} D \cap \mathcal{P} = \varnothing$ and $\operatorname{Supp} D' \cap \mathcal{P} = \varnothing$.

Let us estimate t_0. Due to the result of Exercise 4.1.43, we may forget the condition $\operatorname{Supp} D' \cap \mathcal{P} = \varnothing$.

PROPOSITION 4.1.44. *We have*
$$t_0 \geq \left\lfloor \frac{d_c - 1 - g}{2} \right\rfloor.$$

PROOF. Choose a point P_0 and put $D' = bP_0$, where $b = t_0 + g$. By the Riemann–Roch theorem, $\ell(D') \geq t_0 + 1$. Hence,
$$t_0 \geq a - b - 2g + 1 = a - t_0 - 3g + 1;$$
i.e.,
$$2t_0 \geq a - 3g + 1 = d_c - 1 - g.$$
□

Thus, we are able to decode an algebraic geometry code provided that the number of errors is at most $\left\lfloor \dfrac{d_c - g - 1}{2} \right\rfloor$, which is by $g/2$ smaller than what we would like to have. Actually, there are numerous modifications of the basic algorithm (as well as other approaches to the decoding problem), which, in particular, yield the following result.

THEOREM 4.1.45. *Let $C = (X, \mathcal{P}, D)_L$ or $C = (X, \mathcal{P}, D)_\Omega$. There exists a polynomial-time algorithm which corrects any $t \leq \left\lfloor \dfrac{d_c - 1}{2} \right\rfloor$ errors.* □

We hope to return to this subject in *Advanced Chapters*.

4.2. Additional Bounds and Constructions

In Sec. 4.2.1 we give bounds on parameters of algebraic geometry codes, which in certain special cases make it possible to improve the parameters. Section 4.2.2 is devoted to some versions of the main algebraic geometry construction, which also in a number of cases yield codes with better parameters. In Sec. 4.2.3 we present a construction that allows us to deal with algebraic geometry codes if we have only partial information on the curve.

4.2.1. Extra Bounds

Parameters. It is worth mentioning that parameters of algebraic geometry codes were not found exactly but only estimated. Some of the algebraic geometry codes have better parameters. More precisely, if $\deg D = a < n$, then the L-construction gives $k = a - g + 1 + \ell(K - D)$, where $\ell(K - D) = 0$ for $a \geq 2g - 1$ and $0 \leq \ell(K - D) \leq 2g - 1 - a$ for $a < 2g - 1$. The problem of computing $\ell(K - D)$ is rather complicated.

On the other hand, a function $f \in L(D)$ has no more than a zeroes, but it can occur that for any function from $L(D)$ these zeroes are mainly outside the $\mathbb{F}_q$-points. The proof of the main result of Sec. 4.5.2 is based on this observation.

Thus, it may happen that $k > k_c$ or $d > d_c$, and there are many cases where one (or both) of these inequalities can be proved. First, we consider the simplest case of improving the estimate for k in the L- or H-construction. As we have seen above, such an improvement is only possible if $a = \deg G < 2g - 1$, and it does take place if $\ell(K - D) > 0$, i.e., if D is a special divisor. Unfortunately, our knowledge of special $\mathbb{F}_q$-divisors is quite poor. Nevertheless, some results in this direction can be obtained. First, let us find limits for possible improvements.

PROPOSITION 4.2.1. *Let D be a special divisor of degree a on a curve X, $\mathscr{L} = \mathscr{O}(D)$, and $C = (X, \mathcal{P}, \mathscr{L})_H$ be the corresponding $[n,k,d]_q$ code. Then $2k + d_c \leq n + 2$.*

PROOF. By the Clifford theorem (Theorem 2.2.43) we have $k \leq a/2 + 1$. Since $d_c = n - a$, the proposition follows. $\square$

Let us right away note that there exists a case where this bound is attained. Indeed, if X is a hyperelliptic curve and $D = mF$, where F is a *hyperelliptic divisor* (i.e., $F = \pi^*(P)$, where $\pi \colon X \to \mathbb{P}^1$ is a projection of degree 2 and P is an arbitrary $\mathbb{F}_q$-point of $\mathbb{P}^1$), then $a = \deg D = 2m$, $\ell(D) = m + 1$, and for the code $C = (X, \mathcal{P}, \mathscr{O}(D))_H$ with an arbitrary $\mathcal{P} \subseteq X(\mathbb{F}_q)$ we have $2k + d_c = n + 2$. Unfortunately, codes on hyperelliptic curves cannot be longer than $2(q+1)$.

REMARK 4.2.2. It follows from the second part of the Clifford theorem (Remark 2.2.44) that for an $[n,k,d]_q$ code $C = (X, \mathcal{P}, \mathscr{L})_H$ with $1 \leq \deg \mathscr{L} \leq 2g - 3$, the equality $2k + d_c = n + 2$ can occur for a hyperelliptic curve only.

The obtained estimate (Proposition 4.2.1) does not improve estimates derived from bounds of Sec. 1.1.4. Thus, bounds on parameters of a code result in some estimates for parameters of special $\mathbb{F}_q$-divisors on curves which we cannot derive from purely algebraic geometry considerations. Here is a simple example: the Plotkin bound applied to algebraic geometry codes can be rewritten as

$$n - a = d_c \leq d \leq \frac{nq^{k-1}(q-1)}{q^k - 1}.$$

Thus,

$$n \leq a \frac{q^k - 1}{q^{k-1} - 1}$$

for any $\mathbb{F}_q$-divisor D of degree a on a curve X, where $k = \ell(D)$ and $n = X(\mathbb{F}_q)$. For $k = 2$, we get an elementary estimate $n \leq a(q+1)$, which can also be obtained by purely geometric arguments.

EXERCISE 4.2.3. Write out the relations between $n = |X(\mathbb{F}_q)|$, $\deg D$, and $\ell(D)$ that correspond to other upper bounds for codes.

An important feature of the obtained upper bound $d_c + 2k \leq n + 2$ is that it does not prohibit algebraic geometry codes from being very good. In particular, there is no prohibition for binary algebraic geometry codes to lie above the Gilbert-Varshamov bound.

EXERCISE 4.2.4. Write out the relation between $|X(\mathbb{F}_q)|$, $\deg D$, and $\ell(D)$ which is sufficient for parameters of $C = (X, X(\mathbb{F}_q), \mathcal{O}(D))_H$ to be above the Gilbert-Varshamov bound.

Unfortunately, based on presently known results on special divisors, it is absolutely unclear how to construct special divisors satisfying this relation (and whether they do exist at all).

Employing Weierstrass points. One of the few classes of special divisors which we know a little better are multiples of Weierstrass points (see Sec. 2.2.3). Let $Q \in X(\mathbb{F}_q)$ be a Weierstrass point on a curve of genus $g \geq 2$, and let $(a_1, \ldots, a_{g-1})$ be its nongap sequence, $2 \leq a_1 < a_2 < \ldots < a_{g-1} \leq 2g - 1$; i.e., $\ell(a_i Q) > \ell((a_i - 1)Q)$ for all $i = 1, \ldots, g - 1$. Then $\ell(a_i) = i + 1$; choosing for D one of the divisors $a_i Q$, for the code $C = (X, \mathcal{P}, \mathcal{O}(D))_H$ we get $k = \ell(D) = i + 1$ and $d_c = n - a_i$. The obtained value of k is better than k_c for all i such that $i + 1 > a_i - g + 1$. Since Q is a Weierstrass point, this inequality holds at least for one i. If Q is a hyperelliptic point, we are again in the situation of a hyperelliptic divisor, described just before Remark 4.2.2.

Employing Weierstrass points and more subtle considerations with the use of the Cartier operator leads in some cases to sharper estimates for parameters of algebraic geometry codes obtained by the Ω-construction. Let $C = (X, \mathcal{P}, D)_\Omega$, and let $D = \sum_{i=1}^m a_i Q_i$, where Q_i are prime divisors, i.e., points of X that are rational over some finite extension of the ground field $\mathbb{F}_q$ (see Sec. 2.5.3). Let $a_i = \beta_i q + r_i$, where $\beta_i, r_i \in \mathbb{Z}$, $0 \leq r_i \leq q - 1$, and let $(s_{i,1} \leq \ldots \leq s_{i,g})$ be the gap sequence at Q_i. Define

$$w_i = \begin{cases} \sup\{h \mid s_{i,h} \leq \beta_i\} & \text{if } \beta_i > 0, \\ 0 & \text{if } \beta_i = 0; \end{cases}$$

$$\varepsilon_i = \begin{cases} 1 & \text{if } a_i \equiv -1 \pmod{q} \text{ and } (a_i + 1)/q \text{ is a nongap at } Q_i, \\ 0 & \text{otherwise.} \end{cases}$$

As usual,
$$a = \deg D = \sum a_i \deg Q_i, \qquad n = |\mathcal{P}|.$$

PROPOSITION 4.2.5. *For $n \leq a \leq n+2g-2$, the parameters of the code $C = (X, \mathcal{P}, D)_\Omega$ satisfy the inequalities*
$$k \geq n - \sum_{i=1}^{m}(a_i - \beta_i + w_i) \deg Q_i + g - 1$$

and
$$d \geq a + \sum_{i=1}^{m} \varepsilon_i \deg Q_i - 2g + 2. \qquad \square$$

We give the proof of the proposition in the form of a series of exercises which follow.

EXERCISE 4.2.6. Prove the estimate
$$k \geq n + g - 1 - a - \beta - w$$
in the case where $D = aQ$, $Q \in X(\mathbb{F}_q)$. (*Hint*: Consider the filtration
$$\Omega(\boldsymbol{P}) \supseteq \Omega(\boldsymbol{P} - Q) \supseteq \ldots \supseteq \Omega(\boldsymbol{P} - aQ).$$
Apply the result of Exercise 2.2.53 and the fact that $\Omega(-aQ) = \Omega(-(a+1)Q)$ whenever $a+1$ is a nongap at Q to prove that there are at least $\beta - w$ equalities in the filtration.)

EXERCISE 4.2.7. Prove the estimate
$$d \geq a + 1 - 2g + 2$$
in the same situation. (*Hint*: If $(a+1)/q$ is a nongap at Q, then $\Omega(\boldsymbol{P} - aQ) = \Omega(\boldsymbol{P} - (a+1)Q)$.)

EXERCISE 4.2.8. Examine the case $D = a + \sum_{i=1}^{m} \sigma^i(Q)$, where Q is a point of degree m on X and σ is a generator of the Galois group $\mathrm{Gal}(\mathbb{F}_{q^m}/\mathbb{F}_q)$. (*Hint*: Consider the filtration
$$\Omega(\boldsymbol{P}) \supseteq \Omega(\boldsymbol{P} - R) \supseteq \Omega(\boldsymbol{P} - 2R) \supseteq \ldots \supseteq \Omega(\boldsymbol{P} - aR),$$
where $R = \sum_{i=1}^{m} \sigma^i(Q)$.)

EXERCISE 4.2.9. Prove the proposition in the general case, considering the filtration
$$\Omega(\boldsymbol{P}) \supseteq \Omega(\boldsymbol{P} - R_1) \supseteq \ldots \supseteq \Omega(\boldsymbol{P} - aR_1)$$
$$\supseteq \Omega(\boldsymbol{P} - a_1 R_1 - R_2) \supseteq \ldots \supseteq \Omega(\boldsymbol{P} - D + R_m) \supseteq \Omega(\boldsymbol{P} - D),$$
where $R_i = \sum_{j=1}^{b_i} \sigma_i^j(Q_i)$ if Q_i is a point of degree b_i and σ_i is a generator of $\mathrm{Gal}(\mathbb{F}_{q^{b_i}}/\mathbb{F}_q)$.

The improvement of parameters given by Proposition 4.2.5 is also more significant in the case of hyperelliptic points.

EXERCISE 4.2.10. Let X be a hyperelliptic curve such that none of its hyperelliptic points $Q_1,\ldots,Q_{2g+2}$ lie in $X(\mathbb{F}_q)$, and let $D = \sum_{i=1}^{2g+2} a_i Q_i$ be an $\mathbb{F}_q$-divisor. Write out the estimates for k and d obtained from Proposition 4.2.5 in this case.

One-point codes. If the divisor D is of the simplest form, i.e., is a multiplicity of an $\mathbb{F}_q$-point Q, the corresponding algebraic geometry codes are called *one-point* codes. In a number of cases, estimates for the minimum distance of such codes (more precisely, of those obtained by the Ω-construction) can be improved by considering the gap sequence at Q. Let us present the simplest results.

As usual, $\mathcal{P} = \{P_1,\ldots,P_n\}$ is a set of n distinct $\mathbb{F}_q$-points, and $Q \notin \mathcal{P}$.

THEOREM 4.2.11. *Let α and β be gaps at Q. Then for the minimum distance d of the code $(X,\mathcal{P},D)_\Omega$, where $D = (\alpha+\beta-1)Q$, provided that $k > 0$, we have*

$$d \geq \deg D - (2g-2) + 1.$$
$\square$

This theorem is a particular case of a more general fact (Theorem 4.2.14), which we shall prove a bit later. Note that the estimate of the theorem is better by 1 than the designed distance obtained from the Riemann–Roch theorem.

If there are some consecutive (going in series) gaps at Q, this estimate can be improved still more:

THEOREM 4.2.12. *Let $\alpha, \alpha+1, \ldots, \alpha+t$ and $\beta-(t-1),\ldots,\beta-1,\beta$ be two series of consecutive gaps at Q such that $\alpha+t \leq \beta$ and $t \geq 1$. Then for the minimum distance d of the code $(X,\mathcal{P},D)_\Omega$, where $D = (\alpha+\beta-1)Q$, provided that $k > 0$, we have*

$$d \geq \deg D - (2g-2) + t + 1.$$
$\square$

Now let us consider a more general situation. Let B be a divisor defined over $\mathbb{F}_q$. We call a natural number α a *B-gap* at Q if there is no rational function f on X such that

$$((f)+B)_\infty = \alpha Q.$$

Note that usual gaps at Q are 0-gaps in these terms.

We say that a natural number γ is the *order at Q of a divisor D* if there exists an effective divisor E such that $Q \notin \operatorname{Supp} E$ and

$$D \sim \gamma Q + E.$$

EXERCISE 4.2.13. Prove that α is a B-gap at Q if and only if $\alpha - 1$ equals the order of $K - B$ at Q, where K is the canonical divisor. (*Hint*: Apply the Riemann–Roch theorem.)

Let α and β be B-gaps at Q. Put

$$D = (\alpha+\beta-1)Q + 2B,$$

and consider the code $C = (X,\mathcal{P},D)_\Omega$, where $\mathcal{P}$ consists of n distinct $\mathbb{F}_q$-points, each not belonging to $\operatorname{Supp} D$.

THEOREM 4.2.14. *If the dimension of such a code is positive, then its minimum distance satisfies*

$$d \geq \deg D - (2g-2) + 1.$$

PROOF. Put $w = \deg D - (2g-2)$. If there exists a codeword of weight w in C, then there exists $\eta \in \Omega(D - \boldsymbol{P})$ with exactly w simple poles $P_1, \ldots, P_w \in \mathcal{P}$. Then
$$(\eta)_0 \geq D_0 \quad \text{and} \quad (\eta)_\infty \leq D_\infty + P_1 + \ldots + P_w.$$
Hence,
$$2g - 2 = \deg(\eta) \geq \deg D_0 - \deg D_\infty - w = 2g - 2.$$
Therefore, we have $(\eta)_0 = D_0$ and $(\eta)_\infty = D_\infty + P_1 + \ldots + P_w$. Thus there exists a canonical divisor K of the form
$$K = D - (P_1 + \ldots + P_w).$$
Since β is a B-gap at Q, we have
$$K - B \sim (\beta - 1)Q + E,$$
where $E \geq 0$ and $Q \notin \operatorname{Supp} E$. Thus,
$$E + (P_1 + \ldots + P_w) - (B + \alpha Q) \sim 0.$$
Hence, there exists a rational function f such that $((f) + B)_\infty = \alpha Q$, which contradicts the definition of a B-gap. □

Obviously, for $B = 0$ we get the assertion of Theorem 4.2.11.

Weierstrass pairs. Similar results can also be obtained for *two-point codes*, i.e., codes constructed from divisors of the form $D = aQ_1 + bQ_2$, where Q_1 and Q_2 are two distinct $\mathbb{F}_q$-points that do not belong to $\mathcal{P}$.

A pair of nonnegative integers (α_1, α_2) is said to be a *nongap at the pair* (Q_1, Q_2) if there exists a function $f \in \mathbb{F}_q(X)$ such that $(f)_\infty = \alpha_1 Q_1 + \alpha_2 Q_2$; otherwise, we call it a *gap at* (Q_1, Q_2). Consider the set $G(Q_1, Q_2)$ of gaps at (Q_1, Q_2).

THEOREM 4.2.15. *Let Q_1 and Q_2 be two distinct rational points, each not belonging to $\mathcal{P}$, and let $(\alpha_1, \alpha_2) \in G(Q_1, Q_2)$, where $\alpha_1 \geq 1$ and*
$$\ell(\alpha_1 Q_1 + \alpha_2 Q_2) = \ell((\alpha_1 - 1)Q_1 + \alpha_2 Q_2).$$
Assume that for some β_1, β_2 we have $(\beta_1, \beta_2 - t - 1) \in G(Q_1, Q_2)$ for all t such that
$$0 \leq t \leq \min\{\beta_2 - 1, 2g - 1 - (\alpha_1 + \alpha_2)\}.$$
Put
$$D = (\alpha_1 + \beta_1 - 1)Q_1 + (\alpha_2 + \beta_2 - 1)Q_2;$$
then the minimum distance of the code $(X, \mathcal{P}, D)_\Omega$ satisfies
$$d \geq \deg D - (2g - 2) + 1. \quad \square$$

Developing the idea of the theorem, we call a pair (α_1, α_2) a *pure gap at a pair* (Q_1, Q_2) if
$$\ell(\alpha_1 Q_1 + \alpha_2 Q_2) = \ell((\alpha_1 - 1)Q_1 + \alpha_2 Q_2) = \ell(\alpha_1 Q_1 + (\alpha_2 - 1)Q_2).$$
Denote by $G_0(Q_1, Q_2)$ the set of all pure gaps at (Q_1, Q_2); obviously, we have $G_0(Q_1, Q_2) \subseteq G(Q_1, Q_2)$.

THEOREM 4.2.16. *Let Q_1 and Q_2 be two distinct rational points, each not belonging to $\mathcal{P}$. Let there exist $(\alpha_1, \alpha_2), (\beta_1, \beta_2) \in \mathbb{N} \times \mathbb{N}$ with $t_i = \beta_i - \alpha_i \geq 0$, $i = 1, 2$, such that the set of pairs*
$$\{(\gamma_1, \gamma_2) \mid \alpha_1 \leq \gamma_1 \leq \beta_1, \ \alpha_2 \leq \gamma_2 \leq \beta_2\}$$

is entirely contained in $G_0(Q_1, Q_2)$. Put
$$D = (\alpha_1 + \beta_1 - 1)Q_1 + (\alpha_2 + \beta_2 - 1)Q_2;$$
then the minimum distance of the code $(X, \mathcal{P}, D)_\Omega$ satisfies
$$d \geq \deg D - (2g-2) + t_1 + t_2 + 2. \qquad \square$$

Gonality and abundant codes. The *gonality* $\gamma(X)$ of a curve X over $\Bbbk$ is the minimal degree of a nonconstant map (defined over $\Bbbk$) from X to the projective line. As a rule, we simply write γ for $\gamma(X)$.

LEMMA 4.2.17. *If D is a divisor of degree $\deg D < \gamma(X)$, then $\ell(D) \leq 1$.*

PROOF. If $\ell(D) > 1$, then there exists a nonconstant rational function f such that $(f) \geq -D$, whence we have $(f)_\infty \leq D$. One can consider f as a nonconstant map, defined over the field of constants, from X to the projective line. The degree of this map is equal to $\deg(f)_\infty \leq \deg D < \gamma$, which contradicts the definition of gonality. $\square$

COROLLARY 4.2.18. *For a curve over a finite field, we have $\gamma \leq g+1$.*

PROOF. Here we use the fact that over a finite field there always exists a divisor of degree $g+1$ (since $N_{g+1} > 0$). But by the Riemann–Roch theorem, the dimension of such a divisor is at least 2. $\square$

COROLLARY 4.2.19. *If a curve X has a rational point and $\deg D \geq \gamma - 1$, then*
$$\ell(D) \leq \deg D + 2 - \gamma.$$

PROOF. Let P be a rational point on X. Let $b = \deg D - \gamma + 1$. Then $b \geq 0$, and we put $E = D - bP$. If $\ell(D) > \deg D + 2 - \gamma$, then $\ell(E) \geq \ell(D) - b > 1$. But $\deg E = \gamma - 1$, which contradicts Lemma 4.2.17. $\square$

EXERCISE 4.2.20. Prove that the gonality of a curve is 1 if and only if the curve is isomorphic to the projective line, and the gonality is 2 if and only if the curve is either elliptic or hyperelliptic.

REMARK 4.2.21. If a (smooth) plane curve of degree m has a rational point P, one can project the curve with center P to a line outside P. In this way we get a map from the curve to the projective line of degree at most $m-1$; i.e., the gonality of the curve is at most $m-1$. In fact, it can be proved that the gonality of such a curve is precisely $m-1$.

If the field of constants is algebraically closed, then $\gamma \leq \lfloor (g+3)/2 \rfloor$ (this follows from the Brill–Noether theory), and the equality holds for the general curve. For a finite field, however, this is not true, as the following example shows. Consider a smooth plane curve of degree 4 over a finite field without rational points. Such curves exist; take for example (*check!*) the curve with equation $x^4 + y^4 + z^4 = 0$ over $\mathbb{F}_5$ or the curve with equation
$$x^4 + y^4 + z^4 + y^2z^2 + x^2z^2 + x^2y^2 + x^2yz + xy^2z + xyz^2 = 0$$
over $\mathbb{F}_2$. A smooth plane curve of degree 4 has genus 3 (see Corollary 2.2.8). Let us show that if such a curve has no rational points, then its gonality is 4. Indeed, otherwise there exists an effective divisor of degree 3 and dimension 2, so this divisor is special. But by the Riemann–Roch theorem, if on a curve of genus g there exists

an effective special divisor of degree a and dimension k, then there also exists an effective special divisor of degree $2g-2-a$ and dimension $k-a-1+g$. Hence, in our case, there is an effective special divisor of degree 1, i.e., a rational point. Thus, the gonality of such a curve is 4, which is greater than $\lfloor (g+3)/2 \rfloor$.

This example shows that the gonality can change if one extends the field of constants.

The example also shows that the gonality of a curve of genus greater than 2 over a finite field may be equal to $g+1$. The following proposition shows that such cases are few.

PROPOSITION 4.2.22. *If a curve over $\mathbb{F}_q$ has gonality $\gamma = g+1 > 3$, then $g \leq 10$ and $q \leq 31$.*

PROOF. First let us show that such a curve has no effective divisors of degree $g-2$. Indeed, if such a divisor A exists, take a canonical divisor K such that $\operatorname{Supp} A \cap \operatorname{Supp} K = \varnothing$. Then $\ell(K-A) \geq \ell(K) - \deg(A) = 2$, and $\deg(K-A) = g$, which contradicts Lemma 4.2.17. Since the curve has no effective divisors of degree $g-2$, the curve over an extension of degree $g-2$ has no rational points. Apply the Weil bound (Theorem 3.1.24):

$$0 \leq q^{g-2} + 1 - 2gq^{\frac{g-2}{2}},$$

whence $g < 2\log_q(2g) + 1$; this holds with the indicated values of g and q only. □

EXERCISE 4.2.23. Check the last claim in the proof.

PROBLEM 4.2.24. Find a tighter estimate for γ for large g and q.

For a curve over a finite field, there is the following relation between the gonality and the number of rational points:

PROPOSITION 4.2.25. *Let X be a curve defined over $\mathbb{F}_q$, and let N be the number of $\mathbb{F}_q$-points on X. Then $\gamma(X) \geq N/(q+1)$.*

PROOF. Under a nonconstant map of degree γ from a curve to the projective line, each of the N rational points of the curve is mapped to one of the $q+1$ rational points of the projective line, and the inverse image of a point on the projective line consists of at most γ points. □

As above, let $\mathcal{P} = \{P_1, \ldots, P_n\}$ be a system of points, $\boldsymbol{P} = \sum_{i=1}^{n} P_i$. A divisor D is called $\boldsymbol{P}$-*abundant* if $\operatorname{Supp} D \cap \mathcal{P} = \varnothing$ and D is linearly equivalent to $\boldsymbol{P} + A$ for some effective divisor A. In this case $\deg A$ is called the $\boldsymbol{P}$-*abundance* of D. If it is clear from the context which divisor $\boldsymbol{P}$ is concerned, we omit its notation and simply speak about an abundant divisor and its abundance. If D is a $\boldsymbol{P}$-abundant divisor, $(X, \mathcal{P}, D)_L$ is called an *abundant* code.

LEMMA 4.2.26. *Let $D \sim \boldsymbol{P} + A$ for some effective divisor A. If an abundant code $(X, \mathcal{P}, D)_L$ has a codeword of weight $d > 0$, then*

$$\ell(P_{i_1} + \ldots + P_{i_d} + A) > 1$$

for some $i_1 < \ldots < i_d$.

PROOF. Let $c = (c_1, \ldots, c_n)$ be a codeword of weight d. Rearranging the points $P_1, \ldots, P_n$ if necessary, we may assume without loss of generality that the first d coordinates of c are nonzero. It follows from the construction of the code that there exists a function $f \in L(D)$ such that $f(P_i) = c_i$. Thus, $(f) \geq -D$ and $f(P_i) = 0$ for all $i > d$. Hence, $(f) \geq -D + P_{d+1} + \ldots + P_n$ since D and $\boldsymbol{P}$ have disjoint supports. Put $E = (f) + D - (P_{d+1} + \ldots + P_n)$. Then E is an effective divisor, and

$$E \sim D - (P_{d+1} + \ldots + P_n) \sim \boldsymbol{P} + A - (P_{d+1} + \ldots + P_n) \sim P_1 + \ldots + P_d + A.$$

If $P_i \in \mathrm{Supp}\, E$ for some $i \leq d$, then $c_i = f(P_i) = 0$ for $i \leq d$, which is a contradiction. Therefore, the divisors E and $P_1 + \ldots + P_d + A$ are equivalent but not equal; i.e., $\ell(P_{i_1} + \ldots + P_{i_d} + A) > 1$. □

THEOREM 4.2.27. *Let X be a curve over $\mathbb{F}_q$ of gonality γ. If D is an abundant divisor of abundance $a < \gamma$, then $(X, \mathcal{P}, D)_L$ is an $[n, k, d]$ code with*

$$d \geq \gamma - a, \qquad k \geq n + a - g.$$

If, moreover, $n + a > 2g - 2$, then $k = n + a - g$.

PROOF. The statement about the minimum distance follows directly from Lemmas 4.2.26 and 4.2.17. Next, the code $(X, \mathcal{P}, D)_L$ is the image of the vector space $L(D)$ under the evaluation map $\mathrm{Ev}_\mathcal{P}$; by the Riemann–Roch theorem, $\ell(D) \geq n + a + 1 - g$, and the equality holds if $n + a > 2g - 2$. The kernel of this map is equal to $L(D - \boldsymbol{P})$, which is isomorphic to $L(A)$ since $D \sim \boldsymbol{P} + A$. But $\ell(A) = 1$ by Lemma 4.2.17 since $a < \gamma$. Thus, the dimension of the code is at least $n + a - g$, and in the case $n + a > 2g - 2$ the equality holds. □

REMARK 4.2.28. If D is linearly equivalent to $\boldsymbol{P} - A$ for some effective divisor A, then the code $(X, \mathcal{P}, D)_L$ has minimum distance γ since $(X, \mathcal{P}, D)_L$ is contained in $(X, \mathcal{P}, D + A)_L$, the latter being an abundant code of abundance 0.

Feng–Rao bound. We shall use the notation $\mathbb{N}_0 = \mathbb{N} \cup \{0\}$. Let R be an $\mathbb{F}_q$-algebra. A *weight function* on R is a map $\rho \colon R \to \mathbb{N}_0 \cup \{-\infty\}$ that has the following properties: for all $f, g \in R$,

(0) $\rho(f) = -\infty$ if and only if $f = 0$;
(1) $\rho(\lambda f) = \rho(f)$ for any $\lambda \in \mathbb{F}_q^*$;
(2) $\rho(f + g) \leq \max\{\rho(f), \rho(g)\}$, and the equality holds whenever $\rho(f) \neq \rho(g)$;
(3) $\rho(fg) = \rho(f) + \rho(g)$;
(4) if $\rho(f) = \rho(g)$, then there exists $\lambda \in \mathbb{F}_q^*$ such that $\rho(f - \lambda g) < \rho(f)$.

An elementary example of a weight function is the degree of a polynomial, $\rho(f) = \deg f$, in $R = \mathbb{F}_q[x]$. Another example: let R be the affine coordinate ring of an affine absolutely irreducible curve over $\mathbb{F}_q$ that has exactly one rational point Q at infinity. Let $\mathbb{K}$ be the field of fractions of R. Then $\mathbb{K}$ is a function field over $\mathbb{F}_q$; let v_Q be the discrete valuation of $\mathbb{K}$ at Q. Then $\rho = v_Q$ is a weight function. In fact, it can be shown that any $\mathbb{F}_q$-algebra with a weight function is of this type.

To each pair (R, ρ) there corresponds its semigroup $H \subseteq \mathbb{N}_0$, which is the image $\rho(R \setminus \{0\})$. If moreover the affine curve is nonsingular, then we have an equivalent description of this semigroup in the language of function fields. Let $\mathbb{K}$ be a function field with $\mathbb{F}_q$ as the field of constants. For an $\mathbb{F}_q$-rational place $Q \in \mathbb{P}_\mathbb{K}$, consider its *Weierstrass semigroup* $H(Q)$ consisting of all nongaps at Q. In other words, $m \in H(Q)$ if and only if there is a function $f \in \mathbb{K}$ with no poles outside Q and with a pole of order m at Q. Let $K_\infty(Q)$ be the ring of all functions $f \in \mathbb{K}$ that have no poles outside Q. Then $(K_\infty(Q), -v_Q)$ is an $\mathbb{F}_q$-algebra with a weight function, with $H = H(Q)$ as its semigroup.

REMARK 4.2.29. Any numerical semigroup H, i.e., a subsemigroup of $\mathbb{N}_0$, is the semigroup of some $\mathbb{F}_q$-algebra with a weight function; it suffices to consider the linear space R generated by monomials $\{x^i \mid i \in H\}$ with the weight function $\rho = \deg$. It can be shown however that not every numerical semigroup is realizable as the Weierstrass semigroup of a point on a nonsingular curve.

Now, let R be the affine coordinate ring of an absolutely irreducible affine curve X over $\mathbb{F}_q$ with exactly one rational point Q at infinity. The *conductor* of $H(Q)$ is the largest element $m \in H(Q)$ such that $m - 1 \notin H(Q)$. It is known that either $c = g = 0$ or $g + 1 \leq c \leq 2g$ (cf. Proposition 2.2.45).

Let us enumerate elements of $H(Q)$ in ascending order: $H(Q) = \{\rho_\ell, \ell \in \mathbb{N}\}$, and $\rho_\ell < \rho_{\ell+1}$ for all $\ell \in \mathbb{N}$. Consider the code

$$E_\ell = (X, \mathcal{P}, \rho_\ell Q)_L$$

and its dual code

$$C_\ell = (X, \mathcal{P}, \rho_\ell Q)_\Omega,$$

where $\mathcal{P}$ is a set of n distinct $\mathbb{F}_q$-points of X.

EXERCISE 4.2.30. Show that the minimum distance of C_ℓ is at least $\ell + 1 - g$.

Define

$$\nu_\ell = \left|\{(i,j) \in \mathbb{N}^2 \mid \rho_i + \rho_j = \rho_{\ell+1}\}\right|,$$

and let

$$d_{FR}(\ell) = \min\{\nu_{\ell'} \mid \ell' \geq \ell\}.$$

Then we have the following theorem.

THEOREM 4.2.31. *For an $[n,k,d]_q$ code C_ℓ, we have*

$$d \geq d_{FR}(\ell).$$

Furthermore, $d_{FR}(\ell) \geq \ell + 1 - g$, and the equality holds whenever $\ell > 2c - g - 2$, where c is the conductor of $H(Q)$. □

The number $d_{FR}(\ell)$ is known as the *Feng–Rao designed distance* of C_ℓ.

Concatenation. Let us apply concatenation (see Sec. 1.2.3) to algebraic geometry codes. Recall that $N_q(g) = \max |X(\mathbb{F}_q)|$, where the maximum is taken over all curves over $\mathbb{F}_q$ of genus g. It is known that $N_q(g) \leq q + 1 + \lfloor 2\sqrt{q} \rfloor g$ (see Sec. 3.1.3).

Corollary 4.1.14 states that there exist $[N,K,D]_q$ codes with $N \leq N_q(g)$ and $K + D = N - g + 1$ for any "reasonable" K (i.e., $K \geq 0$ and $D = N - g + 1 - K \geq 1$).

PROPOSITION 4.2.32. *For any $[n,k,d]_q$ code C, any $g \geq 0$, and any K in the range $0 \leq K \leq N_{q^k}(g) - g$, there exists a linear code with parameters*

$$\left[nN_{q^k}(g), kK, d\left(N_{q^k}(g) - K - g + 1\right)\right]_q.$$
□

EXERCISE 4.2.33. Prove this proposition, using Corollary 4.1.14 and the concatenated construction.

In particular, the field descent (i.e., the $[n,n,1]_q$ code) yields the following result.

COROLLARY 4.2.34. *For any $g \geq 0$, $n \geq 1$, and any K in the range $0 \leq K \leq N_{q^n}(g) - g$, there exists a linear code with parameters*
$$[nN_{q^k}(g), nK, N_{q^k}(g) - K - g + 1]_q.$$
□

REMARK 4.2.35. Of course, to use Proposition 4.2.32 and Corollary 4.2.34, we need some lower bounds on $N_q(g)$; see tables in [vdGvdV].

EXERCISE 4.2.36. What codes can be obtained from algebraic geometry codes using other constructions of Sec. 1.2.3?

Subfield restriction. Restriction of algebraic geometry codes also gives many interesting codes. Note that restricting codes of genus zero, we obtain Goppa codes and, in particular, BCH codes (see Sec. 1.2.2).

PROPOSITION 4.2.37. *For any $g \geq 0$, $m \geq 1$, and any K in the range*
$$\left\lfloor \frac{(m-1)N_{q^m}(g)}{m} \right\rfloor \leq K \leq N_{q^m}(g) - g,$$
there exists a linear code with parameters
$$[N_{q^m}(g), mK - (m-1)N_{q^m}(g), N_{q^m}(g) - K - g + 1]_q.$$
□

EXERCISE 4.2.38. Prove the proposition.

REMARK 4.2.39. Let $n = N_{q^m}(g)$ and $K = n - a + g - 1$ (see the Ω-construction in Sec. 4.1.1). Then Proposition 4.2.37 yields an $[n, n - m(a - g + 1), a - 2g + 2]_q$ code; the same parameters, of course, are obtained for any $n \leq N_{q^m}(g)$. In the BCH case ($g = 0$ and $n = q^m - 1$), we get a $[q^m - 1, q^m - m(a+1), a+2]_q$ code.

Improvement of parameters. It is known (see Sec. 1.2.2) that sometimes parameters of the BCH codes can be improved (the codimension is approximately $(q-1)/q$ as large). It turns out that this phenomenon takes place in the general case as well.

THEOREM 4.2.40. *Let $m \geq 1$ and $2g - 2 < a < n$. Consider an $[n, n-a+g-1, \geq a - 2g + 2]_{q^m}$ code $C_0 = (X, \mathcal{P}, D)_\Omega$, where $\mathcal{P} \subseteq X(\mathbb{F}_{q^m})$, $|\mathcal{P}| = n$, $D \geq 0$, and $\deg D = a$. If $D = qD_1$ with $D_1 = \sum a_i Q_i$, $0 \leq a_1 \leq q - 1$, then the subfield restriction gives the code $C = C_0 \cap \mathbb{F}_q^n$ with parameters*
$$\left[n, \geq n - ma\frac{q-1}{q} - 1, \geq a - 2g + 2 \right]_q.$$

PROOF. As is shown in Sec. 4.1.2, the code C_0 is dual to $C_1 = (X, \mathcal{P}, D)_L$. Let $b = a/q = \deg D_1$, and assume that $b > g$. By the Riemann–Roch theorem, $\ell(D_1) \geq b - g + 1$. Let $1, f_1, \ldots, f_{b-g}$ be linearly independent vectors in $L(D_1)$. Let us show that in this case the vectors $1, f_1, \ldots, f_{b-g}, f_1^q, \ldots, f_{b-g}^q$ are also linearly independent. Indeed,
$$x_0 + \sum x_i f_i \neq \sum y_i f_i^q$$
since the left-hand side cannot have a pole of order greater than $q-1$ (by the condition of the theorem, $a_i \leq q - 1$), and the order of any pole of the right-hand side is divisible by q. (If the right-hand side has no poles, then $\sum y_i f_i^q = y_0$.

Let $z_i^q = y_i$; then $(\sum z_i f_i)^q = z_0^q$, i.e., $z_0 = \sum z_i f_i$, which contradicts the independence of $1, f_1, \ldots, f_{b-g}$.)

Let us complete the set $M = \{1, f_1, \ldots, f_{b-g}, f_1^q, \ldots, f_{b-g}^q\}$ to a basis of $L(D)$. Since $C = C_1^\perp \cap \mathbb{F}_q^n$, an element $v \in \mathbb{F}_q^n$ belongs to C if and only if $(v, f) = 0$ for any element f of this basis. Here, $(v, f) \in \mathbb{F}_{q^m}$; i.e., in general, each equation $(v, f) = 0$ gives m linear conditions over $\mathbb{F}_q$. However, for $f = 1$ this is only one linear condition, and the condition $(v, f_i) = 0$ is equivalent to $(v, f_i^q) = 0$. Thus, we see that the number of $\mathbb{F}_q$-conditions is at least by $(b-g)m + (m-1)$ less than the a priori estimate $m\ell(D) = m(a-g+1)$; i.e.,

$$\dim C \geq n - m(q-1)b - 1.$$

In the case $b \leq g$, take $M = \{1\}$ and proceed as above. We improve the a priori estimate $m(a-g+1)$ by $m-1$ and obtain

$$\dim C \geq n - m(a-g) - 1,$$

which, for $b \leq g$, is better than the estimate of the theorem. □

Below, in Sec. 4.5.3, we shall see that for large m this improvement is considerable. However, even for $m = 1$ it also makes sense.

COROLLARY 4.2.41. *Let $qg \leq a < n$. If $D = qD_1$ with $D_1 = \sum a_1 Q_i$, $0 \leq a_i \leq q-1$, then the code $C = (X, \mathcal{P}, D)_\Omega$ has parameters*

$$\left[n, \geq n - a\frac{q-1}{q} - 1, \geq a - 2g + 2\right]_q,$$

where $n = |\mathcal{P}|$ and $a = \deg D$. □

REMARK 4.2.42. Using a subtler linear algebra technique, one can show that the assertion of Theorem 4.2.40 holds in the following, more general, assumptions: Let D be an arbitrary divisor, and let D_1 be such that $D \geq qD_1$. Then

$$k \geq \begin{cases} n - 1 - m(\ell(D) - \ell(D_1)) & \text{if } D \text{ is effective,} \\ n - m(\ell(D) - \ell(D_1)) & \text{if } D \text{ is not effective.} \end{cases}$$

The same technique also leads to estimates for the dimension of an arbitrary linear code obtained by subfield restriction in terms of the filtration corresponding to the endomorphism of raising to the qth power.

4.2.2. Variants of the Basic Construction

Now let us consider some extra constructions of codes from curves. These constructions often yield better parameters of codes thus obtained.

Generalized algebraic geometry codes. Let $P_1, \ldots, P_s$ be distinct points of arbitrary degrees $\deg P_i = k_i$ on a curve X over $\mathbb{F}_q$. Let D be a divisor such that $\operatorname{Supp} D \cap \{P_1, \ldots, P_s\} = \varnothing$, and let C_i, $i = 1, \ldots, s$, be linear $[n_i, k_i]_q$ codes with fixed $\mathbb{F}_q$-isomorphisms $\pi_i : \Bbbk_{P_i} \to C_i$, where $\Bbbk_{P_i}$ is the residue field of P_i. Set $n = \sum_{i=1}^{s} n_i$. Consider the linear map

$$\operatorname{Ev}_{\mathcal{P}, \mathcal{C}} : L(D) \to \mathbb{F}_q^n, \quad f \mapsto \bigl(\pi_1(f(P_1)), \ldots, \pi_s(f(P_s))\bigr);$$

here, as above, $\mathcal{P} = \{P_1, \ldots, P_s\}$ and $\mathcal{C} = \{C_1, \ldots, C_s\}$. We obtain a code $C = \operatorname{Ev}_{\mathcal{P}, \mathcal{C}}(L(D)) \subseteq \mathbb{F}_q^n$, called a *generalized algebraic geometry code*; we use the notation $C = (X, \mathcal{P}, D, \mathcal{C})_L$.

Consider the set of tuples of indices

$$I = \left\{ S \subseteq \{1,\ldots,s\} \;\Big|\; \sum_{i \in S} k_i \le \deg D \right\},$$

and call

$$d_c = \min\left\{ \sum_{i \notin S} d_i \;\Big|\; S \in I \right\}$$

the *designed distance* of the generalized algebraic geometry code C. This term is justified by the following result.

PROPOSITION 4.2.43. *Let* $\deg D < \sum_{i=1}^{s} k_i$. *Then* $(X,\mathcal{P},D,\mathcal{C})_L$ *is an* $[n,k,d]_q$ *code with parameters*

$$k = \ell(D) \ge \deg D + 1 - g, \qquad d \ge d_c.$$

PROOF. Let us first show that $\mathrm{Ev}_{\mathcal{P},\mathcal{C}}$ is an injective map. Let $h \in L(D)$ and $\mathrm{Ev}_{\mathcal{P},\mathcal{C}}(h) = 0$; then $h \in L\left(D - \sum_{i=1}^{s} P_i\right)$. Since

$$\deg D < \sum_{i=1}^{s} k_i = \deg\left(\sum_{i=1}^{s} P_i\right),$$

we see that $h = 0$; thus, $\dim C = \dim L(D) = \ell(D)$.

Now consider an element $f \in L(D)$ such that the codeword $\mathrm{Ev}_{\mathcal{P},\mathcal{C}}(f)$ is of weight d (the minimum distance of C). If the code is not trivial, then $f \ne 0$. Let

$$S = \{i \in \{1,\ldots,s\} \mid f(P_i) = 0\}.$$

Since $0 \ne f \in L\left(D - \sum_{i \in S} P_i\right)$, we obtain

$$\deg D \ge \deg\left(\sum_{i \in S} P_i\right) = \sum_{i \in S} k_i.$$

On the other hand,

$$d = \|\mathrm{Ev}_{\mathcal{P},\mathcal{C}}(f)\| = \sum_{i=1}^{s} \|\pi_i(f(P_i))\| = \sum_{i \in S} \|\pi_i(f(P_i))\| \ge \sum_{i \notin S} d_i;$$

i.e., by the definition of the designed distance, we have $d \ge d_c$. □

REMARK 4.2.44. It is easily seen that any linear $[n,k,d]_q$ code C can trivially be obtained by the generalized algebraic geometry construction. Indeed, consider the function field $\mathbb{K} = \mathbb{F}_q(x)$, and let P_∞ be the pole of x. Choose an arbitrary place $P \in \mathbb{P}_{\mathbb{K}}$ of degree k and an arbitrary isomorphism $\pi\colon \Bbbk_P \to C$. Then, obviously,

$$C = (X, \{P\}, (k-1)P_\infty, \{C\})_L.$$

In Sec. 4.4.5 we consider nontrivial examples of using this construction.

Improved algebraic geometry codes. Recall that in Sec. 4.2.1, when we discussed the Feng–Rao bound, we considered one-point codes of the form

$$E_\ell = (X, \mathcal{P}, \rho_\ell Q)_L,$$

where $(\rho_\ell, \ell \in \mathbb{N})$ is the strictly increasing sequence of elements of the Weierstrass semigroup $H(Q)$ at $Q \notin \mathcal{P}$, as well as their dual codes C_ℓ. Let us show that the parameters of C_ℓ can in some cases be improved.

Recall that by ν_ℓ we denote the number of pairs $(i,j) \in \mathbb{N} \times \mathbb{N}$ such that

$$\rho_i + \rho_j = \rho_{\ell+1}.$$

Let $\{f_i\}$ be a basis in the space $L(\rho_\ell Q)$, and let $h_i = \mathrm{Ev}_\mathcal{P}(f_i)$ be the images of elements of this basis under the evaluation map $\mathrm{Ev}_\mathcal{P}$ (i.e., parity checks for the code C_ℓ). Since for large values of ℓ the vectors h_ℓ might be linearly dependent, the parameters of C_ℓ can be improved as compared to the Feng–Rao bound d_{FR} in the sense that, deleting some parity-checks, we can obtain a code of larger dimension with the same minimum distance.

Let d be a positive integer. Define the code

$$\widetilde{C}(d) = \left\{ c \in \mathbb{F}_q^n \mid (c, h_i) = 0 \text{ for all } i \text{ such that } \nu_{i-1} < d \right\}.$$

It has the following property.

PROPOSITION 4.2.45. *The minimum distance of $\widetilde{C}(d)$ is at least d.* □

The code $\widetilde{C}(d)$ has the *super code* property in the following sense: if $d = d_{FR}(\ell)$, then $C_\ell \subseteq \widetilde{C}(d)$. In other words, if $d = d_{FR}(\ell)$, then the lower bound on the minimum distance of C_ℓ and $\widetilde{C}(d)$ is d for both codes, but $\widetilde{C}(d)$ might be strictly larger. That is why such codes are called *improved algebraic geometry codes*. It is clear that from the algebraic geometry point of view they are defined with the help of incomplete linear systems, i.e., linear subspaces in $L(D)$.

For a positive integer d, define

$$r_d = |\{i : \nu_{i-1} < d\}|.$$

It follows from the definition of the $[n, k, \geq d]_q$ code $\widetilde{C}(\ell)$ that $n - k \leq r_d$, i.e., $k \geq n - r_d$. In some cases, the numbers r_d can be described in a more explicit way.

Let $\{\mathbb{K}_m, m \in \mathbb{N}\}$ be a tower of function fields over $\mathbb{F}_q$; i.e., $\mathbb{K}_{m-1} \subset \mathbb{K}_m$ for all $m > 1$. Let $a_m = [\mathbb{K}_m : \mathbb{K}_{m-1}]$ be the degree of extension. Let Q_m be a rational place of $\mathbb{K}_m$. Assume that Q_m is totally ramified over Q_{m-1} for any $m > 1$. Let $H_m = H(Q_m)$ be the Weierstrass semigroup of Q_m and c_m be the conductor of H_m. Then, as is easily seen,

$$a_m H_{m-1} \cup \{n \in \mathbb{N}_0 \mid n \geq c_m\} \subseteq H_m.$$

A sequence of semigroups $(H_m, m \in \mathbb{N})$ is called an *inductive tower of semigroups* if $H_1 = \mathbb{N}_0$ and there exist sequences of positive integers $(a_m, m \in \mathbb{N})$ and $(b_m, m \in \mathbb{N})$ such that

$$H_m = a_m H_{m-1} \cup \{n \in \mathbb{N}_0 \mid n \geq a_m b_{m-1}\} \quad \text{and} \quad a_m b_{m-1} \leq b_m$$

for any $m > 1$. We define *inductive semigroups* by induction: $H = \mathbb{N}_0$ is inductive; if H_0 is an inductive semigroup with conductor c_0 and there are positive integers a and b such that

$$H = aH_0 \cup \{n \in \mathbb{N}_0 \mid n \geq ab\} \quad \text{and} \quad b \geq c_0,$$

then H is inductive.

EXERCISE 4.2.46. Let H_0 be a semigroup with conductor c_0 and the number of gaps g_0. Let
$$H = aH_0 \cup \{n \in \mathbb{N}_0 \mid n \geq ab\}.$$
Prove that if $b \geq c_0$, then H has conductor $c = ab$ and the number of gaps $g = (a-1)b + g_0$. In particular, let (H_m) be an inductive tower of semigroups. Let c_m be the conductor of H_m and g_m be the number of gaps for H_m. Then for $m > 1$ we have
$$c_m = a_m b_{m-1} \qquad \text{and} \qquad g_m = (a_m - 1)b_{m-1} + g_{m-1}.$$

Finally, let us state a theorem on the dimension of improved algebraic geometry codes.

THEOREM 4.2.47. *Let d be a positive integer, and let $\widetilde{C}(d)$ be an improved algebraic geometry $[n,k]$ code corresponding to a function field with an inductive Weierstrass semigroup. Then $n - k \leq r_d$, and*
$$r_d = \rho_{\lceil d/2 \rceil} + \lfloor d/2 \rfloor. \qquad \square$$

In Sec. 4.4.3 we give some examples of improved algebraic geometry codes.

4.2.3. Partial Algebraic Geometry Codes

It should be noted that often precise computation of the genus of a curve with many rational points encounters certain difficulties, while estimating the genus is much easier. Furthermore, even if a curve is given explicitly, computation of a complete basis in $L(D)$ can be very complicated, whereas finding a basis in an incomplete (though large enough) linear system (i.e., an $\mathbb{F}_q$-subspace) in $L(D)$ might be considerably easier. This gives reasons for the following construction.

Let X be an algebraic curve over a finite field $\Bbbk$, and let $\mathbb{K}$ be its function field. Let Q be a rational point on X. Instead of assuming that we know the genus g of the curve, we assume that we can construct a (reasonably large) set A_Q of functions from $\mathbb{K}$ that have no poles outside Q. Obviously, the sum and product of two elements of A_Q also have no poles outside Q, so we may assume that A_Q is a subring in $\mathbb{K}$ containing $\Bbbk$. Such a ring A_Q is called the *partial affine ring* of Q.

Let V_Q be a subset of functions from A_Q for which we can explicitly compute $-v_Q$, the pole order at Q. Since for any $\varphi, \psi \in \mathbb{K}$ we have $v_Q(\varphi\psi) = v_Q(\varphi) + v_Q(\psi)$, we may assume that the set of pole orders of functions from V_Q is an additive sub-semigroup M in $\mathbb{N}_0$. We call elements of $\mathbb{N}_0 \setminus M$ gaps and elements of M nongaps. The only condition we impose on V_Q is that there are finitely many gaps. This condition is equivalent to saying that there is a set of functions in V_Q whose pole orders are coprime, which could be verified straightforwardly. Thus, we denote the number of gaps of V_Q by $\widetilde{g}$ and call it the gap number. It follows from the Riemann–Roch theorem that $\widetilde{g} \geq g$, but we shall make no use of this fact; moreover, even if the equality holds, it may not be easy to prove.

It follows from purely combinatorial reasons that the gap number of V_Q possesses some properties of the genus of a curve.

PROPOSITION 4.2.48. *The largest gap is at most $2\widetilde{g} - 1$. Furthermore, if for some natural h in the interval $[0, 2h+1]$ there are exactly h gaps, then there is no other gap, and $h = \widetilde{g}$.*

PROOF. Let $n > 2\widetilde{g} - 1$ be a gap. Then each of the pairs $(i, n-i)$ for $i = 0, \ldots, \widetilde{g}$ contains at least one gap. Since no two pairs have a value in common, this gives at least $\widetilde{g} + 1$ gaps, which contradicts the definition of $\widetilde{g}$.

Let us prove the second claim. Let $n > 2h + 1$, and assume that all gaps that are strictly less than n are not greater than $2h + 1$; we shall show that n is a nongap. Consider the pairs $(i, n-i)$ for $i = 1, \ldots, h+1$. Since there are only h gaps below n, at least one of the pairs must consist of two nongaps; the sum of these nongaps equals n. □

PROPOSITION 4.2.49. *Let φ and ψ be two functions with the same order n at Q. Then there exist nonzero constants a and b such that*
$$v_Q(a\varphi + b\psi) > n.$$

PROOF. Let t be a local parameter at Q (i.e., a generator of the maximal ideal $\mathfrak{m}_Q$ of the local ring $\mathcal{O}_Q$). Since Q is a rational point, there exist nonzero constants a and b such that $v_Q(\varphi - bt^n) > n$ and $v_Q(\psi + at^n) > n$. It remains to note that
$$a\varphi + b\psi = a(\varphi - bt^n) + b(\psi + at^n).$$
□

COROLLARY 4.2.50. *Let L be a $\Bbbk$-subspace of $\mathbb{K}$, and let T be a subset of L such that all of its elements have distinct orders at Q. Then elements of T are linearly independent.*

Let G be the set of orders at Q of all elements of L. If L is finite-dimensional, then G is finite. If G is bounded from above, then a set T of elements of L, one of each order in G, is a basis of L.

PROOF. Since the order of a linear combination of elements of T equals the order of one of its nonzero terms, linear independence is evident. The second claim is also obvious. Let us prove the third.

Let m be the maximal value in G, and assume that T contains exactly one element of each order. Let us show that any element ψ is a linear combination of elements of T using induction on the pole order n of ψ at Q. Let $\varphi_n \in T$ have the same pole order as ψ. Then by Proposition 4.2.49 there exist nonzero constants a and b such that $v_Q(a\varphi_n + b\psi) > -n$. On the other hand, $a\varphi_n + b\psi \in L$; hence either $v(a\varphi_n + b\psi) \leq m$ or $a\varphi_n + b\psi = 0$. If $-n = m$, the first case is impossible, and therefore $\psi = a\varphi_n/b$. This is the basis of induction. If $-n < m$, then by the induction hypothesis $a\varphi_n + b\psi$ is a linear combination of elements of T, and then
$$\psi = ((a\varphi_n + b\psi) - a\varphi_n)/b.$$
□

Now, for a $\Bbbk$-rational divisor D, let us define the associated *partial L-space*
$$L'(D) = L(D) \cap V_Q$$
and put
$$\ell'(D) = \dim_{\Bbbk} L'(D).$$

The next lemma describes properties of $\ell'(D)$, which are similar to standard properties of $\ell(D)$ (throughout this subsection, a "rational divisor" means a $\Bbbk$-rational one).

LEMMA 4.2.51. *Let B be an effective rational divisor, and let D and E be arbitrary rational divisors.*

(a) *If $D \leq E$, then $\ell'(D) \leq \ell'(E)$ and*
$$\ell'(D) - \deg D \geq \ell'(E) - \deg E.$$
In particular, if $D \leq mQ$ for some m, then $\ell'(D) < \infty$.

(b) *For divisors of the form $mQ - B$, we have*
$$\ell'(mQ - B) \geq m + 1 - \widetilde{g} - \deg B.$$

(c) *For fixed B, the sequence*
$$k_m = \ell'(mQ - B) - m + \deg B$$
is nonincreasing and becomes constant for $m > \deg B + 2\widetilde{g} - 2$. In particular,
$$\ell'(mQ) = m + 1 - \widetilde{g} \quad \text{for} \quad m > 2\widetilde{g} - 2.$$

PROOF. (a) The first part of the statement follows directly from the definition. The second is equivalent to the assertion that
$$\ell'(E) - \ell'(D) \leq \deg E - \deg D.$$
As in the standard case, it suffices to prove this for $E = D + P$. In our case, P is a rational point, and so we need only to prove that $\ell'(E) - \ell'(D) \leq 1$. This follows from Proposition 4.2.49. Indeed, let φ and ψ be two functions from $L'(E) \setminus L'(D)$; then
$$v_P(\varphi) = v_P(\psi) = -a_P,$$
where a_P is the coefficient at P in the divisor E. Hence, $a\varphi + b\psi \in L(D)$ for suitable a and b.

Finally, from Proposition 4.2.49 we see that $\ell'(mQ)$ equals $m + 1 - k$, where k is the number of gaps that are less than $m + 1$. Hence, if $D \leq mQ$ for some m, then $\ell'(D) \leq m + 1$.

(b) Apply part (a) with $D = mq - B$ and $E = mQ$:
$$\ell'(mQ - B) \geq \ell'(mQ) - m + \deg(mQ - B) \geq m + 1 - \widetilde{g} - m + m - \deg B.$$

(c) The statement that the sequence k_m is nonincreasing is a particular case of part (a). Next, we call a number $m \geq \deg B$ a *B-gap* if
$$\ell'(mQ - B) = \ell'((m-1)Q - B).$$
These are precisely the values of m for which $k_m < k_{m-1}$. Since we know that $\ell'((\deg B - 1)Q - B) = 0$, part (b) implies that the total number of B-gaps is at most $\widetilde{g}$. Furthermore, m is a B-nongap if and only if there exists a function $\varphi \in L'(mQ - B)$ such that $v_q(\varphi) = -m + B(Q)$. Therefore, the sum of an ordinary nongap n and a B-nongap m is a B-nongap.

Similarly to Proposition 4.2.48, one can easily prove that the largest possible B-gap is $\deg B + 2\widetilde{g} - 1$. Indeed, if $m > \deg B + 2\widetilde{g} - 1$ is a B-gap, then in each of the $2\widetilde{g} + 1$ pairs of the form $(m - n, n)$, $n = 0, \ldots, 2\widetilde{g}$, the left-hand element is at least $\deg B$, and each pair must have a B-gap on the left or an ordinary gap on the right. But for each of them there are at most $\widetilde{g}$ candidates; i.e., the condition cannot be satisfied. Hence, for $m > 2\widetilde{g} - 2 + \deg B$ the sequence k_m is constant.

Finally, the value for $\ell'(mQ)$ is established in the proof of part (a). $\square$

The notion of a B-gap, introduced in the proof, will be of use in the following. We denote the total number of B-gaps by $\widetilde{g}(B)$.

Now, let us define partial algebraic geometry codes. Let $\mathcal{P} = \{P_1, \ldots, P_n\}$ consist of distinct simple rational points different from Q. Consider the evaluation map
$$\mathrm{Ev}'_{\mathcal{P}}: L'(mQ) \to \Bbbk^n, \quad f \mapsto (f(P_1), \ldots, f(P_n)).$$
The image of this map, the code $C = \mathrm{Ev}'_{\mathcal{P}}(L'(mQ)) \subseteq \Bbbk^n$, is denoted by $(X, \mathcal{P}, mQ)'_L$. By definition, a *partial algebraic geometry code* is its dual code, which is denoted by $(X, \mathcal{P}, mQ)'_\Omega$.

As above, $\boldsymbol{P}$ denotes the divisor $\sum_{P_i \in \mathcal{P}} P_i$. Obviously, the kernel of the evaluation map $\mathrm{Ev}'_{\mathcal{P}}$ is the space $L'(mQ - \boldsymbol{P})$, so the dimension of $(X, \mathcal{P}, mQ)'_L$ is $\ell'(mQ) - \ell'(mQ - \boldsymbol{P})$, and the dimension of $(X, \mathcal{P}, mQ)'_\Omega$ is $n - \ell'(mQ) + \ell'(mQ - \boldsymbol{P})$.

As we step down from a divisor $mQ - B$ to a divisor $mQ - B_1$ with $\deg B_1 = \deg B + 1$, the value $\ell'(mQ - B) - m + \deg B$ either remains the same or increases by 1. The first stage at which an increase occurs is an important parameter for the construction of partial algebraic geometry codes. Thus, we define the *threshold* $\delta(mQ) = \delta(mQ, \mathcal{P})$ to be the smallest value n_1 such that there exists $\mathcal{P}_1 \subseteq \mathcal{P}$ with $\deg \boldsymbol{P}_1 = n_1$ and
$$\ell'(mQ - \boldsymbol{P}_1) > \ell'(mQ) - n_1.$$

EXERCISE 4.2.52. Show that the condition
$$\ell'(mQ - \boldsymbol{P}_1) = \ell'(mQ) - \deg \boldsymbol{P}_1$$
is equivalent to the requirement that $(X, \mathcal{P}_1, mQ)'_L$ coincides with the entire $\Bbbk^{n_1}$, and thus $\delta(mQ)$ can be defined as the smallest value n_1 for which there exists a system $\mathcal{P}_1 \subseteq \mathcal{P}$ with $\deg \boldsymbol{P}_1 = n_1$ such that $(X, \mathcal{P}_1, mQ)'_L \neq \Bbbk^{n_1}$.

Let us estimate the parameters of partial algebraic geometry codes:

THEOREM 4.2.53. *Let $\mathcal{P} = \{P_1, \ldots, P_n\}$ consist of distinct simple rational points different from Q. Then $(X, \mathcal{P}, mQ)'_\Omega$ is an $[n, k, d]_q$ code of dimension*
$$k \geq n - \ell'(mQ)$$
with equality if $m < n$; in particular,
$$k = n - m - 1 + \widetilde{g} \quad \text{if} \quad 2\widetilde{g} - 2 < m < n.$$
Furthermore, if the code $(X, \mathcal{P}, mQ)'_\Omega$ is nontrivial (nonzero), then
$$d = \delta(mQ).$$

PROOF. As is shown above, $k = n - \ell'(mQ) + \ell'(mQ - \boldsymbol{P})$, which immediately implies the desired inequality for k. If $m < n$, then $\ell'(mQ - \boldsymbol{P}) = 0$; and if $m > 2\widetilde{g} - 2$, then $\ell'(mQ) = m + 1 - \widetilde{g}$. Let us proceed with the minimum distance.

Let c be a nonzero codeword of $(X, \mathcal{P}, mQ)'_\Omega$, and let all of its nonzero coordinates correspond to a subset $\mathcal{P}_1 \subseteq \mathcal{P}$. Then by deleting all zero positions in c (those corresponding to points outside $\mathcal{P}_1$) we obtain a codeword of the code $(X, \mathcal{P}_1, mQ)'_\Omega$, and all codewords of $(X, \mathcal{P}_1, mQ)'_\Omega$ can be obtained in this way. Hence, the minimum distance of $(X, \mathcal{P}, mQ)'_\Omega$ is the minimum cardinality of a system $\mathcal{P}_1$ for which $(X, \mathcal{P}_1, mQ)'_\Omega$ has positive dimension. Now it remains only to apply Exercise 4.2.52. □

The main difficulty is to obtain estimates for the threshold $\delta(mQ)$. For full algebraic geometry codes (those with $\widetilde{g} = g$), the Riemann–Roch theorem gives

$\delta(mQ) \geq m+2-g$, but there is no straightforward generalization for partial codes. Nevertheless, we can establish the following properties:

PROPOSITION 4.2.54. 1. *For any $m > 0$,*
$$\delta(mQ) < \delta((m+1)Q).$$

2. *If for some m we have*
$$\delta(mQ) < m+2-2\widetilde{g},$$
then $\widetilde{g}(\boldsymbol{P}_1) < \widetilde{g}$ for some system $\mathcal{P}_1 \subseteq \mathcal{P}$ with $\deg \boldsymbol{P}_1 = \delta(mQ)$, and also $\delta(mQ) = \delta((m+1)Q)$.

PROOF. 1. If $(X, \mathcal{P}_1, mQ)'_L = \Bbbk^{n_1}$, we obviously have $(X, \mathcal{P}_1, (m+1)Q)'_L = \Bbbk^{n_1}$.
2. By the definition of $\delta(mQ)$, the condition $\delta(mQ) < m+2-2\widetilde{g}$ implies that there exists a system $\mathcal{P}_1 \subseteq \mathcal{P}$ with $\deg \boldsymbol{P}_1 = \delta(mQ)$ such that
$$\ell'(mQ - \boldsymbol{P}_1) > \ell'(mQ) - \deg \boldsymbol{P}_1 = m + 1 - \widetilde{g} - \deg \boldsymbol{P}.$$
Then by Lemma 4.2.51c we obtain
$$\ell'(m_1 Q - \boldsymbol{P}_1) = m_1 - \deg \boldsymbol{P}_1 + 1 - \widetilde{g}(\boldsymbol{P}_1) > m_1 - \deg \boldsymbol{P}_1 + 1 - \widetilde{g}$$
for all $m_1 \geq m$. Hence, $\widetilde{g}(\boldsymbol{P}_1) < \widetilde{g}$.

Finally, taking $m_1 = m+1$, we get
$$\delta((m+1)Q) \leq \deg \boldsymbol{P}_1 = \delta(mQ). \qquad \square$$

Nevertheless in order to obtain an analogue of the Riemann–Roch theorem, we have to impose on V_Q an additional (though simple) condition. We say that V_Q *separates* a system $\mathcal{P}$ if for each point $P \in \mathcal{P}$ there exists a function $\varphi \in V_Q$ such that $v_P(\varphi) = 0$ and $v_{P'}(\varphi) > 0$ for the other points $P' \in \mathcal{P} \setminus \{P\}$. Now the analogue of the Riemann–Roch theorem is as follows:

THEOREM 4.2.55. *Assume that a set of functions V_Q separates $\mathcal{P}$. Then for all $m > 2\widetilde{g} - 2$ we have*
$$\delta(mQ) \geq m + 2 - 2\widetilde{g}.$$
Furthermore, the number of $\boldsymbol{P}_1$-gaps $\widetilde{g}(\boldsymbol{P}_1)$ is equal to $\widetilde{g}$ for any system $\mathcal{P}_1 \subseteq \mathcal{P}$. In other words, if $m - \deg \boldsymbol{P}_1 > 2\widetilde{g} - 2$, then
$$\ell'(mQ - \boldsymbol{P}_1) = m - \deg \boldsymbol{P}_1 + 1 - \widetilde{g}.$$

PROOF. Assume that $\delta(mQ) < m+2-2\widetilde{g}$. Then there exists a system $\mathcal{P}_1 \subseteq \mathcal{P}$ for which $\deg \boldsymbol{P}_1 = \delta(mQ)$ and $\ell'(mQ - \boldsymbol{P}_1) > \ell'(mQ) - \deg \boldsymbol{P}_1$. At the same time, for any proper subset $\mathcal{P}_2 \subset \mathcal{P}_1$ we have
$$\ell'(mQ - \boldsymbol{P}_2) = \ell'(mQ) - \deg \boldsymbol{P}_2.$$
Choose $P \in \mathcal{P}_1$ and let $\mathcal{P}_2 = \mathcal{P}_1 \setminus \{P\}$. Then
$$\ell'(mQ - \boldsymbol{P}_2) = \ell'(mQ) - \deg \boldsymbol{P}_1 + 1 \leq \ell'(mQ - \boldsymbol{P}_1) \leq \ell'(mQ - \boldsymbol{P}_2),$$
and therefore $\ell'(mQ - \boldsymbol{P}_1) = \ell'(mQ - \boldsymbol{P}_2)$. By Lemma 4.2.51c it follows that $\ell'(m_1 Q - \boldsymbol{P}_1) = \ell'(m_1 Q - \boldsymbol{P}_2)$ for all $m_1 > m$. On the other hand, by the condition of the theorem, there exists a function $\varphi \in V_Q$ that is nonzero at P and vanishes at all the other points of $\mathcal{P}_2$. If $m_1 \geq -v_Q(\varphi)$, then
$$\varphi \in L'(m_1 Q - \boldsymbol{P}_2) \setminus L'(m_1 Q - \boldsymbol{P}_1),$$

whence $\ell'(m_1Q - \boldsymbol{P}_1) < \ell'(m_1Q - \boldsymbol{P}_2)$. This contradiction disproves the assumption that $\delta(mQ) < m + 2 - 2\widetilde{g}$.

Thus, for all subsets $\mathcal{P}_1 \subseteq \mathcal{P}$ such that $\deg \boldsymbol{P}_1 < m + 2 - 2\widetilde{g}$, we have
$$\ell'(mQ - \boldsymbol{P}_1) = \ell'(mQ) - \deg \boldsymbol{P}_1 = m + 1 - \widetilde{g} - \deg \boldsymbol{P}_1.$$
Hence, $\widetilde{g}(\boldsymbol{P}_1) = \widetilde{g}$. Taking $m = n + 2\widetilde{g} - 1$, we obtain the result for all subsets $\mathcal{P}_1 \subseteq \mathcal{P}$. $\square$

COROLLARY 4.2.56. *Let V_Q separate $\mathcal{P}$. Then the partial algebraic geometry code $(X, \mathcal{P}_1, mQ)'_\Omega$ is trivial for $m > n + 2\widetilde{g} - 2$. If $(X, \mathcal{P}_1, mQ)'_\Omega$ is nontrivial, then its minimum distance d satisfies*
$$d \geq m + 2 - 2\widetilde{g}.$$
$\square$

4.3. Characterization of Algebraic Geometry Codes

This section studies the question of which linear codes are algebraic-geometric. It turns out that, in a certain "weak" sense, all linear codes are. However, if we impose some natural conditions on the divisor involved in the construction of the code, the answer becomes more interesting.

In Sec. 4.3.1 we introduce three main definitions—weakly algebraic-geometric codes, algebraic-geometric codes, and strongly algebraic-geometric codes—and discuss their basic properties. Section 4.3.2 shows that any linear code is weakly algebraic-geometric. In Sec. 4.3.3 we give necessary conditions for a code to be algebraic-geometric.

4.3.1. Three AG Levels

Throughout all this section, we distinguish the following cases. We call a linear $[n,k]_q$ code that can be represented as $C = (X, \mathcal{P}, D)_L$ for some curve X of genus g a *weakly algebraic-geometric* code (*WAG code*). If a code has a representation $(X, \mathcal{P}, D)_L$ with $\deg D < n$, we call it simply an *AG code*. Finally, by a *strongly algebraic-geometric code* (*SAG code*) we call a code that has a representation $(X, \mathcal{P}, D)_L$ with $2g - 2 < \deg D < n$.

These definitions can also be given in terms of the Ω-construction. It can be shown that there exists a differential form ω with a simple pole at each point of $\mathcal{P}$ and such that $\operatorname{Res}_{P_i}(\omega) = 1$ for all i; this implies (*check!*) that

$$(X, \mathcal{P}, D)_\Omega = (X, \mathcal{P}, (\omega) + \boldsymbol{P} - D)_L.$$

Therefore, a WAG code is any code that can be represented as $(X, \mathcal{P}, D)_\Omega$; it is an AG code if $2g - 2 < \deg D$, and it is an SAG code if $2g - 2 < \deg D < n$.

In view of Theorem 4.1.25, this immediately yields the following result.

PROPOSITION 4.3.1. *If C is a WAG code (SAG code), then $C^\perp$ is also a WAG code (SAG code).* □

It can be shown that this is not the case for AG codes (i.e., a code dual to an AG code need not be an AG code).

Let C_i be the code obtained from C by puncturing at the ith coordinate (concerning the puncturing construction, see Sec. 1.1.6, p. 30).

LEMMA 4.3.2. *If C is a WAG code, then C_i is also WAG.*

PROOF. Let $C = (X, \mathcal{P}, D)_L$. In this case we obviously have $C_i = (X, \mathcal{P}_i, D)_L$, where $\mathcal{P}_i = (P_1, \ldots, P_{i-1}, P_{i+1}, \ldots, P_n)$. □

The analogous statement for AG and SAG codes is not true (see the end of Sec. 4.3.3).

LEMMA 4.3.3. *A code equivalent to a WAG code (AG code, SAG code) is also a WAG code (AG code, SAG code).*

PROOF. Permutation of coordinates corresponds to reordering of P_i.

Let a code C' be obtained from C by multiplying coordinates by nonzero constants $\lambda_1, \ldots, \lambda_n \in \mathbb{F}_q$. There exists a rational function f such that $f(P_i) = \lambda_i$ (this is easily deduced from the Riemann–Roch theorem). Consider the divisor $D' = D - (f)$; we have $\operatorname{Supp} D' \cap \mathcal{P} = \varnothing$ since all λ_i are nonzero. Then $C' = (X, \mathcal{P}, D')_L$, and $\deg D = \deg D'$. □

Let $C = (X, \mathcal{P}, D)_L$ be a nondegenerate linear code over $\mathbb{F}_q$ of dimension $k \geq 2$. Let $L(D) = L(D-P)$ for some point $P \in X$. Then we necessarily have $P \notin \mathcal{P}$ since otherwise the code would be degenerate (if $P = P_i$, then the ith coordinate of any codeword is zero). Thus, $\mathrm{Supp}(D-P) \cap \mathcal{P} = \varnothing$, and $C = (X, \mathcal{P}, D-P)_L$. Repeating this procedure if necessary, we may assume without loss of generality that

$$L(D) \neq L(D-P) \quad \text{for any } P \in X.$$

In this case, D is said to be a *base point free* divisor. Let $\ell = \ell(D)$, and let $f_0, \ldots, f_{\ell-1}$ be a basis in $L(D)$. Consider the embedding

$$\varphi_D \colon X \longrightarrow \mathbb{P}^{\ell-1}$$

by the complete linear system $|D|$ (see (2.1.2), p. 81); for points $P \in X \setminus \mathrm{Supp}\, D$ (in particular, for P_i), we have

$$\varphi_D(P) = (f_0(P) : \ldots : f_{\ell-1}(P)).$$

Let X_0 be the image of X under φ_D. The curve X_0 is not contained in any hyperplane since $f_0, \ldots, f_{\ell-1}$ are linearly independent. Therefore, φ_D is a dominant morphism of projective curves. Since X is absolutely irreducible, X_0 is absolutely irreducible too. Since D is base point free, we have (cf. Proposition 2.1.74)

$$\deg D = (\deg \varphi_D)(\deg X_0). \tag{4.3.1}$$

We call a representation $(X, \mathcal{P}, D)_L$ of a nondegenerate WAG code (AG code, SAG code) C with $k \geq 2$ a *minimal* representation if D is base point free (i.e., $\ell(D) \neq \ell(D-P)$ for any $P \in \mathrm{Supp}\, D$) and $\deg \varphi_D = 1$.

PROPOSITION 4.3.4. *Let C be a nondegenerate WAG code with no identical coordinates (i.e., all points in the corresponding projective system are distinct) of dimension $k \geq 2$. If C has a representation $(X, \mathcal{P}, D)_L$ with D base point free, then it has a minimal representation $(\widetilde{X}_0, \widetilde{\mathcal{P}}, \widetilde{D})_L$, and there exists a surjective morphism $\varphi \colon X \to \widetilde{X}_0$ with the following properties:*

(a) $\widetilde{\mathcal{P}} = \{\varphi(P_1), \ldots, \varphi(P_n)\}$;
(b) $\varphi^*(\widetilde{D}) \sim D$;
(c) $\deg \varphi = \deg \varphi_D$;
(d) $\deg \widetilde{D} = \deg D / \deg \varphi \leq \deg D$;
(e) $g(\widetilde{X}_0) \leq g(X)$, *with equality if and only if* $\deg \varphi = 1$;
(f) *if* $(X, \mathcal{P}, D)_L$ *is an AG-representation, then* $(\widetilde{X}_0, \widetilde{\mathcal{P}}, \widetilde{D})_L$ *is also an AG-representation;*
(g) *if* $(X, \mathcal{P}, D)_L$ *is an SAG-representation, then* $(\widetilde{X}_0, \widetilde{\mathcal{P}}, \widetilde{D})_L$ *is also an SAG-representation.* $\square$

4.3.2. All Linear Codes Are Weakly AG

Let q be a power of a prime p. Consider the following system of homogeneous equations in $\mathbb{P}^\ell = \mathbb{P}(\mathbb{F}_q^{\ell+1})$:

$$x_i^{q+1} - x_i^2 x_0^{q-1} + x_{i+1} x_0^q - x_{i+1}^q x_0 = 0, \quad i = 1, \ldots, \ell-1.$$

PROPOSITION 4.3.5. (a) *These equations define a projective absolutely irreducible curve $X(\ell, q)$.*

(b) *The curve $X(\ell,q)$ has a unique point P_∞ in the hyperplane H with equation $x_0 = 0$, is nonsingular outside P_∞, and passes through all the q^ℓ points of $\mathbb{P}^\ell$ that lie outside H.* □

EXERCISE 4.3.6. Check statement (b) of Proposition 4.3.5. (*Hint*: Consider the corresponding equations in the affine coordinates $y_i = x_i/x_0$ in $\mathbb{P}^\ell \setminus H$.)

Let $X^\nu(\ell,q)$ be the normalization of $X(\ell,q)$ (on normalization, see Sec. 2.3.1). Let $\widetilde{P}_\infty$ be an arbitrary point in $\nu^{-1}(P_\infty)$, and let v_∞ be the discrete valuation at $\widetilde{P}_\infty$. Consider the rational functions $z_i = y_i \circ \nu = (x_i/x_0) \circ \nu$ on $X^\nu(\ell,q)$; they have no poles outside $\nu^{-1}(P_\infty)$. The proof of Proposition 4.3.5, which we do not present here, is based on the following fact:

PROPOSITION 4.3.7. *The inverse image $\nu^{-1}(P_\infty)$ consists of a single point $\widetilde{P}_\infty$, and*
$$v_\infty(z_i) = -q^{\ell-i}(q+1)^{i-1}.$$
□

Furthermore, there is the following result:

PROPOSITION 4.3.8. *The genus $g(\ell,q)$ of $X^\nu(\ell,q)$ equals*
$$\frac{1}{2}\left(\sum_{i=1}^{\ell-1} q^{\ell+1-i}(q+1)^{i-1} - (q+1)^{\ell-1} + 1\right).$$
□

One can also describe the gap sequence (see p. 103) at the point $\widetilde{P}_\infty$ of the curve $X^\nu(\ell,q)$:

PROPOSITION 4.3.9. *The set of all nongaps at $\widetilde{P}_\infty$ is of the form*
$$\left\{\sum_{i=1}^\ell k_i q^{\ell-i}(q+1)^{i-1} \,\Big|\, k_i \in \mathbb{N} \cup \{0\}\right\}.$$
□

Hence, taking into account that
$$\ell(mP) = |\{n \in \mathbb{N} \mid n \leq m,\ n \text{ is a nongap at } P\}|,$$
one immediately gets the following statement.

PROPOSITION 4.3.10. *The space $L(m\widetilde{P}_\infty)$ is generated by functions of the form $z_1^{k_1}\ldots z_\ell^{k_\ell}$, where $k_i \in \mathbb{N} \cup \{0\}$ satisfy the condition*
$$\sum_{i=1}^\ell k_i q^{\ell-i}(q+1)^{i-1} \leq m.$$
□

COROLLARY 4.3.11. *If*
$$2q^{\ell-1} > q^{\ell-i}(q+1)^{i-1},$$
then $1, z_1, \ldots, z_i$ form the basis of
$$L\big(q^{\ell-i}(q+1)^{i-1}\widetilde{P}_\infty\big).$$

PROOF. Linear independence of the functions $1, z_1, \ldots, z_i$ follows from Proposition 4.3.7. □

PROPOSITION 4.3.12. *If a linear $[n,k]_q$ code C contains a codeword of weight n, then C is WAG.*

PROOF. The condition that a code contains a word of weight n is equivalent to the condition that all points of the projective system $\mathcal{Q} = \{Q_1,\ldots,Q_n\} \subseteq \mathbb{P}^{k-1}$ corresponding to C lie outside some hyperplane H. By Lemma 4.3.3 we may assume that the equation of H is $x_0 = 0$. Some of the points Q_i can be identical; let s be the maximal multiplicity of a point in $\mathcal{Q}$. Set $\ell = \lceil k + \log_k s \rceil$; then $s \le q^{\ell-k+1}$. Consider the map
$$\pi \colon \mathbb{P}^\ell \setminus H' \longrightarrow \mathbb{P}^{k-1}$$
defined by
$$\pi(x_0 : \ldots : x_\ell) = (x_0 : \ldots : x_{k-1}),$$
where H' is the hyperplane in $\mathbb{P}^\ell$ with equation $x_0 = 0$. Then all the Q_i lie in the image of π, and fibres of π are isomorphic to $\mathbb{A}^{\ell-k+1}$. Therefore, there are n distinct $\mathbb{F}_q$-points $P_i \in \mathbb{P}^\ell$ such that $\pi(P_i) = Q_i$. It remains to choose a power q_0 of q such that
$$2q_0^{\ell-1} > q_0^{\ell-k}(q_0+1)^{k-1}$$
and to put
$$X = X(\ell, q_0), \qquad D = q_0^{\ell-k}(q_0+1)^{k-1}\widetilde{P}_\infty, \qquad \mathcal{P} = \{P_1, \ldots, P_n\}.$$
Then by Corollary 4.3.11 we obtain $C = (X, \mathcal{P}, D)_L$. $\square$

THEOREM 4.3.13. *Any linear code is WAG.*

PROOF. If we extend C by the overall parity check position, we obtain a code C' whose dual code $(C')^\perp$ contains the word $(1,\ldots,1)$ and thus is a WAG code by Proposition 4.3.12. Then C' is also a WAG code by Proposition 4.3.1, and C is obtained from C' by puncturing at the last coordinate and therefore is also WAG by Lemma 4.3.2. $\square$

4.3.3. Criteria

Let us now see which codes can have an AG-representation, i.e., a representation $(X, \mathcal{P}, D)_L$ with $\deg D < n$ (in what follows, we denote $\deg D = a < n$).

Denote by $g_q(n)$ the minimal genus of a nonsingular absolutely irreducible curve over $\mathbb{F}_q$ with at least n rational points. For a fixed q, this is the "inverse function" for $N_q(g)$, the maximal number of $\mathbb{F}_q$-points on a curve of genus g (see Sec. 3.1.3).

PROPOSITION 4.3.14. *Let $(X, \mathcal{P}, D)_L$ be an AG-representation of an $[n,k]_q$ code.*

(a) *If $a \le 2g - 2$, then*
$$k \le \left\lfloor \frac{n+1}{2} \right\rfloor.$$

(b) *If $a > 2g - 2$, then*
$$q_g(n) \le g \le n - k.$$

PROOF. (a) The case $k = 0$ is trivial. Let $k > 0$; i.e., $\ell(D) = k > 0$. If $g \le n - k$, then the Riemann–Roch theorem gives
$$k = \ell(D) = a + 1 - g + \ell(K - D) \le g - 1 + \ell(K - D) \le n - k - 1 + \ell(K - D),$$
whence $2k \le n - 1 + \ell(K - D)$. Thus, we have either $k \le (n-1)/2$ (this is sufficient) or $\ell(K - D) > 0$.

If $g > n-k$, then
$$k = \ell(D) = a+1-g+\ell(K-D) < a+1-n+k+\ell(K-D) \leq k+\ell(K-D),$$
whence we again obtain $\ell(K-D) > 0$.

In the case $\ell(K-D) > 0$, by the Clifford theorem (Theorem 2.2.43) we have
$$k = \ell(D) \leq a/2 + 1 \leq (n+1)/2.$$

(b) If $a > 2g-2$, then $\ell(K-D) = 0$. Then the Riemann–Roch theorem gives $k = a+1-g \leq n-g$. □

COROLLARY 4.3.15. *There exists an SAG code with parameters $[n,k]_q$ if and only if*
$$g_q(n) \leq \min\{k, n-k\}.$$

PROOF. Proposition 4.3.14b immediately implies that for an SAG code we have $g_q(n) \leq n-k$. Furthermore, the dual $[n, n-k]_q$ code is also an SAG code (Proposition 4.3.1); hence, $g_q(n) \leq k$. Conversely, assume that the condition on the parameters is satisfied. By the definition of $g_q(n)$, there is a X of genus $g = g_q(n)$ over $\mathbb{F}_q$ with at least n distinct rational points P_i. Among divisors of degree $k+g-1$ there is a divisor D with $\operatorname{Supp} D \cap \mathcal{P} = \varnothing$. Note that the condition $g \leq n-k$ yields $\deg D < n$ (hence, in particular, $\ell(D) = k$), and the condition $g \leq k$ gives $2g - 2 < \deg D$. Thus, $(X, \mathcal{P}, D)_L$ is an SAG code. □

COROLLARY 4.3.16. *For an AG code with parameters $[n,k]_q$, we have*
$$k \leq \left\lfloor \frac{n+1}{2} \right\rfloor \qquad \text{if } g_q(n) > n-k,$$
$$k \leq \left\lfloor \frac{(\lfloor 2\sqrt{q}\rfloor - 1)n + q + 1}{\lfloor 2\sqrt{q} \rfloor} \right\rfloor \qquad \text{if } g_q(n) \leq n-k.$$

PROOF. Let $C = (X, \mathcal{P}, D)_L$, $a = \deg D$. If $g_q(n) > n-k$, then $a \leq 2g-2$ by Proposition 4.3.14b; thus, by Proposition 4.3.14a we get $k \leq \lfloor (n+1)/2 \rfloor$. If $g_q(n) \leq n-k$, apply the Serre bound (Theorem 3.1.25):
$$n \leq q + 1 + g_q(n)\lfloor 2\sqrt{q}\rfloor \leq q + 1 + (n-k)\lfloor 2\sqrt{q}\rfloor,$$
which immediately implies the desired inequality. □

For the sequel, we need the following result on embedding of curves in projective spaces.

THEOREM 4.3.17 (Castelnuovo bound). *Let X be an absolutely irreducible curve of degree m over $\mathbb{F}_q$ embedded in $\mathbb{P}^\ell$ but not contained in any hyperplane. Then the genus g of X is at most $\pi(m,\ell)$, which is defined as follows:*
$$\pi(m,1) = 0,$$
and for $\ell > 1$
$$\pi(m,\ell) = \frac{t(t-1)}{2}(\ell - 1) + t\varepsilon,$$
where t and ε are, respectively, the incomplete quotient and the remainder of division of $m-1$ by $\ell - 1$. □

PROPOSITION 4.3.18. *Let $(X,\mathcal{P},D)_L$ be a minimal representation of a nondegenerate $[n,k]_q$ code with no identical coordinates and with $k \geq 2$. Then*

$$g_q(n) \leq g(X) \leq \pi(\deg D, \ell-1),$$

where $\ell = \ell(D)$. In particular, for an AG code we have

$$g_q(n) \leq g(X) \leq \pi(\deg D, k-1).$$

PROOF. By the condition, D is base point free, and $\varphi_D \colon X \to \mathbb{P}^{\ell-1}$ is a morphism of degree 1. Therefore, by equation (4.3.1), p. 226, we have $\deg X_0 = \deg D$. Since X has (at least) n rational points, we have $g_q(n) \leq g(X)$. Furthermore, $g(X) = g(X_0)$ since $\deg \varphi_D = 1$. It remains to apply the Castelnuovo bound to the curve X_0, since it is absolutely irreducible and is not contained in any hyperplane. The second part of the proposition follows from the fact that $\ell = k$ for $\deg D < n$. □

COROLLARY 4.3.19. *Let $k \geq 2$. If there exists a nondegenerate AG $[n,k]_q$ code with no identical coordinates, then*

$$g_q(n) \leq \pi(n-1, k-1).$$

PROOF. It is easily checked by direct computation that for any fixed ℓ the function $\pi(m,\ell)$ is nondecreasing in m (EXERCISE!). Let $(\widetilde{X},\widetilde{\mathcal{P}},\widetilde{D})_L$ be a minimal representation of the given AG code. Apply Proposition 4.3.18 to this representation, taking into account that $\deg \widetilde{D} \leq n-1$. □

PROPOSITION 4.3.20. *If there exists a binary nondegenerate AG $[n,k]_q$ code with no identical coordinates, then*
   (a) *if $n = 12$ or $n \geq 14$, we have $k < \lfloor n/2 \rfloor$;*
   (b) *if $n = 11$ or $n = 13$, we have $k < n/2$.*

PROOF. We have to prove that if $k \geq \lfloor n/2 \rfloor$, then either $n \leq 10$ or case (b) takes place. So let $k \geq \lfloor n/2 \rfloor$. If here $g_2(n) \leq n-k$, then (3.1.8) (Exercise 3.1.30) implies

$$n \leq 0.83\left(n - \left\lfloor \frac{n}{2} \right\rfloor\right) + 5.35,$$

and hence $n < 10$. If $g_2(n) > n - k$, then Corollary 4.3.16 gives

$$\left\lfloor \frac{n}{2} \right\rfloor \leq k \leq \left\lfloor \frac{n+1}{2} \right\rfloor.$$

This means that for $n \geq 10$ there are the following possibilities:
   (1) $n = 2k$, $k \geq 5$;
   (2) $n = 2k+1$, $k \geq 5$;
   (3) $n = 2k-1$, $k \geq 6$.

In the first case, we have $\pi(n-1, k-1) = k+2$, whence $g_2(n) \leq k+2$ by Corollary 4.3.19. Then inequality (3.1.8) (Exercise 3.1.30) gives $2k \leq 0.83(k+2) + 5.35$, whence $k \leq 5$; i.e., $n \leq 10$. In the second case, we similarly obtain $k \leq 6$ and $n \leq 13$, but now $k < n/2$. Finally, in the third case we get $k \leq 5$, which is impossible. □

In particular, this proposition immediately implies the following result.

COROLLARY 4.3.21. *The binary Golay code and the extended binary Golay code are not AG.* □

REMARK 4.3.22. The answer to whether the ternary Golay and extended Golay codes are AG does not follow directly from the above results. This question is still open.

EXERCISE 4.3.23. Let $r \geq 2$ and $r < m \leq 2r+1$. Show that the binary rth-order Reed–Muller code of length 2^m is not AG.

PROPOSITION 4.3.24. *There exists a binary SAG code of length n if and only if $n \leq 8$.*

PROOF. By Proposition 4.3.20, SAG codes of length $n \geq 11$ do not exist, since both the SAG code and its dual (see Proposition 4.3.1) are AG codes, and one of them must be of dimension at least $n/2$. For $n \leq 10$, all values of $g_2(n)$ are known (see Theorem 3.4.53). It is easily seen that the condition of Corollary 4.3.15 is satisfied only for all $n \leq 8$. □

Thus, for instance, by Corollary 4.3.15 there exists a binary SAG $[5,4]$ code since $g_2(5) = 1$ according to Theorem 3.4.53. Puncturing this code gives a $[4,4]$ code, which is not SAG (and not even AG).

PROBLEM 4.3.25. Transfer the results of this section to the case of the H-construction.

4.4. Examples

In this section we give particular examples where parameters of algebraic geometry codes can be computed explicitly, and compare them with codes obtained by other constructions.

In Sec. 4.4.1 we consider algebraic geometry codes on curves of small genera and compute the parameters of some binary codes obtained from them. Section 4.4.2 is devoted to a detailed analysis of the first nontrivial case, that of curves of genus 1; we compute there spectra, answer the question of whether they can be MDS codes, and find when they are self-dual. In Sec. 4.4.3 we consider codes on Hermitian curves, in particular those obtained by variants of the main construction. In Sec. 4.4.4 we give some examples of codes on maximal curves. Section 4.4.5 contains several examples of generalized algebraic geometry codes with good parameters.

4.4.1. Codes of Small Genera

As a rule, the less the genus, the simpler the analysis of curves and the corresponding codes.

Codes of genus zero. Any smooth curve of genus zero over $\mathbb{F}_q$ has at least one $\mathbb{F}_q$-point (this follows from the Weil bound), so the curve is $\mathbb{F}_q$-isomorphic to the projective line $\mathbb{P}^1$; the number of $\mathbb{F}_q$-points on it equals $q+1$. Therefore we get a family of $[n,k,d]_q$ codes with parameters

$$n \leq q+1, \qquad d = 1, \ldots, n, \qquad k = n+1-d.$$

EXERCISE 4.4.1. Check that for $n \leq q$ it is possible to construct in this way a family of embedded codes with $k = 1, \ldots, n$. (*Hint*: Use the L-construction with $D = kP_0$.)

REMARK 4.4.2. For $n = q+1$ it is impossible to construct a family of embedded codes. The reason is that if such a family existed, it would be possible to construct codes with parameters $[q+2, k, q+3-k]_q$ from them (using the construction of a code from an embedded pair, Sec. 1.2.3), and for $k = 4$ such codes do not exist (see the properties of n-sets in Sec. 1.1.2).

As has been explained in Proposition 1.2.4, for a code C of genus zero the spectrum is uniquely determined:

$$W_C(x) = x^n + \sum_{i=0}^{k-1} \binom{n}{i} (q^{k-i} - 1)(x-1)^i.$$

The spectrum of the dual code is of the same form. For $n = 2k$ we have the equality $W_{C^\perp}(x) = W_C(x)$; i.e., the code is formally self-dual. Moreover, there is the following fact:

EXERCISE 4.4.3. Prove that any algebraic geometry $[n, n/2, n/2+1]_q$ code C of genus zero is quasi-self-dual. Prove that for an even q there are self-dual codes among them. (*Hint*: If $g = 0$, then any divisor of an even degree is divisible by 2, and all divisors of a given degree are equivalent.)

In fact, codes on curves of genus zero (i.e., on $X = \mathbb{P}^1$) have already been described in Sec. 1.2.1: for $\mathcal{P} = \mathbb{P}^1(\mathbb{F}_q) \setminus \{0, \infty\}$ they are exactly the classical Reed–Solomon codes, and for $\mathcal{P} = \mathbb{A}^1(\mathbb{F}_q)$ and $\mathcal{P} = \mathbb{P}^1(\mathbb{F}_q)$ their natural extensions.

Codes of genus one. *Elliptic codes* are codes obtained from elliptic curves (i.e., curves of genus $g = 1$). They are codes of genus at most 1 in the sense of the definition of Sec. 1.1.3. Elliptic codes exist for any

$$n \leq N_q(1), \quad k = 1, \ldots, n-1,$$

where $N_q(1)$ is the maximum possible number of $\mathbb{F}_q$-points on a curve of genus 1. By Theorem 3.4.49, for $q = p^m$ we have

$$N_q(1) = \begin{cases} q + \lfloor 2\sqrt{q} \rfloor & \text{for } p \mid \lfloor 2\sqrt{q} \rfloor \text{ and odd } m \geq 3, \\ q + \lfloor 2\sqrt{q} \rfloor + 1 & \text{otherwise.} \end{cases}$$

The spectrum of an elliptic code C expressed in terms of B_i (see Theorem 4.1.37) has only one unknown coefficient, B_k. Below, in Sec. 4.4.2, we shall show how to compute it; for different curves and divisors, B_k may be different.

Codes of genus two and three. Now let us consider codes obtained from curves of genus $g = 2$. Such codes exist for any

$$n \leq N_q(2), \quad k = 1, \ldots n - 2.$$

Moreover, by Theorem 3.4.50 for $q = p^m$, $m \equiv 0 \pmod{2}$, we have

$$N_q(2) = \begin{cases} q + 4\sqrt{q} + 1 & \text{for } q \neq 4, 9, \\ 20 & \text{for } q = 9, \\ 10 & \text{for } q = 4, \end{cases}$$

and for $m \equiv 1 \pmod{2}$,

$$N_q(2) = \begin{cases} q + 2\lfloor 2\sqrt{q} \rfloor + 1 & \text{if } q \text{ is nonspecial,} \\ q + 2\lfloor 2\sqrt{q} \rfloor & \text{if } q \text{ is special} \\ & \text{and } 2\sqrt{q} - \lfloor 2\sqrt{q} \rfloor > (\sqrt{5}-1)/2, \\ q + 2\lfloor 2\sqrt{q} \rfloor - 1 & \text{if } q \text{ is special} \\ & \text{and } 2\sqrt{q} - \lfloor 2\sqrt{q} \rfloor < (\sqrt{5}-1)/2. \end{cases}$$

Recall that q is special if either $p \mid \lfloor 2\sqrt{q} \rfloor$ or q is of the form $q = \ell^2 + 1$, $q = \ell^2 + \ell + 1$, or $q = \ell^2 + \ell + 2$ for some integer ℓ.

Codes on curves of genus 3 have the parameters

$$k + d \geq n - 2, \quad n \leq N_q(3), \quad k = 1, \ldots, n - 3.$$

For $N_q(3)$, the following values are known (see Theorem 3.4.51):

q	2	3	4	5	7	8	9	11	13	16	17	19	25	27
$N_q(3)$	7	10	14	16	20	24	28	28	32	38	40	44	56	56

furthermore (Theorem 3.4.52), if $q = p^{4m+2}$, then $N_q(3)$ is maximal: $N_q(3) = q + 1 + 6\sqrt{q}$.

Binary codes. Let us consider in brief some binary codes obtained from the above-mentioned algebraic geometry codes of small genera using constructions of Sec. 1.2.3.

First of all note that applying subfield restriction to codes of genus zero over large fields, we get BCH codes. These codes have many interesting properties; we have discussed them in Sec. 1.2.2.

One can also construct binary analogues of BCH codes from curves of genus $g \geq 1$ over $\mathbb{F}_q$, $q = 2^m$.

EXAMPLE 4.4.4. Consider an elliptic curve X over $\mathbb{F}_{16}$ with the maximum possible number of points. As explained in Sec. 3.3, on X there are 25 $\mathbb{F}_{16}$-points; later we shall give an explicit example of such a curve (Example 4.4.46). Let us construct an algebraic geometry code C from this curve X, a divisor D of degree 8, and any 21 of these $\mathbb{F}_q$-points. The parameters of C are not worse than $[21, 8, 13]_{16}$. Concatenated construction with the outer code C and inner $[5, 4, 2]_2$ parity-check code yields a $[105, 32, 26]_2$ code. Its shortening by one symbol results in a $[104, 32, 25]_2$ code with rather good parameters (the best $[104, k, 25]_2$ code that was known before algebraic geometry constructions was nonlinear and had $k = 31.585$).

EXAMPLE 4.4.5. Consider a curve X of genus 3 over $\mathbb{F}_8$ with 24 $\mathbb{F}_8$-points (see Example 4.4.51) and a line bundle of degree 11. We obtain a $[24, 9, 13]_8$ code C. Concatenating it with the parity-check $[4, 3, 2]_2$ code, we obtain a $[96, 27, 26]_2$ code. Shortening yields a $[95, 27, 25]_2$ code. Before algebraic geometry codes, the best known code had parameters $[95, 24, 25]_2$.

Considerations like these applied to curves of genus 1, 2, or 3 and combined with various linear and nonlinear combinatorial constructions (concatenation, generalized concatenation, shortening, etc.) yield many binary codes with good parameters. Tables at the end of the book give an insight into the potential of this approach.

4.4.2. Elliptic Codes

An elliptic curve X (curve of genus 1) is an algebraic group (see Sec. 2.4.1). This in many respects determines properties of the corresponding algebraic geometry codes.

Spectra of elliptic codes. An elliptic code is a code of genus at most 1. By Theorem 1.1.18, its weight enumerator is

$$W_C(x) = x^n + \sum_{i=0}^{k-1} \binom{n}{i}(q^{k-i} - 1)(x-1)^i + B_k(x-1)^k, \quad k + d \geq n.$$

According to Remark 1.1.19, we have $B'_{n-k} = B_k$, and for $n = 2k$ these codes are formally self-dual; i.e., $W_C(x) = W_{C^\perp}(x)$.

Two cases are possible: either

$$B_k = B'_{n-k} = 0, \quad k + d = n + 1$$

(i.e., the code is a code of genus zero; below, in Theorems 4.4.19 and 4.4.21, we shall see that this case is quite rare) or

$$0 < B_k = B'_{n-k} \leq \binom{n}{k}(q-1), \quad k + d = n.$$

Let us compute the precise value of B_k. If $B_k > 0$, then $B_k = A_d$ is the number of vectors of the minimum weight.

Let X be an elliptic curve. Recall (see Sec. 2.4.1) that the set $X(\mathbb{F}_q)$ has a structure of an abelian group, which depends on the choice of the zero element

$P_0 \in X(\mathbb{F}_q)$; this group is a subgroup of $X(\overline{\mathbb{F}}_q)$. Let us assign to any divisor $D = \sum a_i Q_i$, where $Q_i \in X(\overline{\mathbb{F}}_q)$, the point $P_D = \sum a_i Q_i$, the sum of all Q_i with multiplicities a_i, taken in the sense of the group law. If D is defined over $\mathbb{F}_q$, then $P_D \in X(\mathbb{F}_q)$; for a principal divisor (f), the sum is zero: $P_{(f)} = P_0$. Hence, P_D depends on the divisor class only. Thus, for a line bundle $\mathscr{L}$, the point $P_{\mathscr{L}}$ is well defined.

THEOREM 4.4.6. *Let $C = (X, \mathcal{P}, \mathscr{L})_H$ be an elliptic $[n,k,d]_q$ code. Then*
$$B_k = (q-1) M(X(\mathbb{F}_q), \mathcal{P}, P_{\mathscr{L}}, k),$$
where for any finite abelian group H, its element $h \in H$, a subset $\mathfrak{h} \subseteq H$, and an integer $a > 0$, by $M(H, \mathfrak{h}, h, a)$ we denote the number of representations of h as a sum of a distinct elements of $\mathfrak{h}$.

PROOF. $B_k = A_{n-k}$ is the number of nonzero sections $s \in H^0(\mathscr{L})$ having precisely k zeroes. According to Exercise 1.1.29,
$$B_k = \sum_{\substack{\mathcal{Q} \subset \mathcal{P} \\ |\mathcal{Q}| = k}} \left(q^{\ell(\mathcal{Q})} - 1\right).$$
It is easily seen that
$$\ell(\mathcal{Q}) = h^0(\mathscr{L}_{\mathcal{Q}}),$$
where $\mathscr{L}_{\mathcal{Q}} = \mathscr{L} \otimes \left(\bigotimes_{P \in \mathcal{Q}} \mathscr{O}(-P_\ell) \right)$ is a bundle of degree zero.

If $\mathscr{L}_{\mathcal{Q}} \cong \mathscr{O}$, then we have $h^0(\mathscr{L}_{\mathcal{Q}}) = 1$; otherwise, $h^0(\mathscr{L}_{\mathcal{Q}}) = 0$ (since any section $s \in H^0(\mathscr{L}_{\mathcal{Q}})$ defines a homomorphism $\mathscr{O} \to \mathscr{L}_{\mathcal{Q}}$, which is an isomorphism because $\deg \mathscr{L}_{\mathcal{Q}} = 0$; see Sec. 2.1.3). Therefore,
$$B_k = (q-1) M,$$
where M is the number of sets $\mathcal{Q}$ such that $\mathscr{L}_{\mathcal{Q}} \cong \mathscr{O}$; i.e., $\mathscr{L} \cong \bigotimes_{P \in \mathcal{Q}} \mathscr{O}(P)$. But we know that the group law on X is consistent with the group law on $\operatorname{Pic} X$, the group of line bundles; i.e.,
$$\bigotimes_{P \in \mathcal{Q}} \mathscr{O}(P) \cong \mathscr{O}(P_{\mathcal{Q}}) \otimes (\mathscr{O}(P_0))^{k-1},$$
where $P_{\mathcal{Q}} = \sum_{P \in \mathcal{Q}} P \in X(\mathbb{F}_q)$. On the other hand,
$$\mathscr{L} \cong \mathscr{O}(P_{\mathscr{L}}) \otimes (\mathscr{O}(P_0))^{k-1}.$$
Thus, M is the number of sets $\mathcal{Q}$ such that $P_{\mathcal{Q}} = P_{\mathscr{L}}$, and the theorem is proved. $\square$

In the following, we confine ourselves with the case of codes of maximum length, i.e., $\mathcal{P} = X(\mathbb{F}_q)$. For an arbitrary abelian group H, define
$$M(H, h, a) = M(H, H, h, a).$$

PROPOSITION 4.4.7. *The number $M(H, h, a)$ depends only on the class of h in the quotient group H/aH.*

PROOF. Let $h = \sum_{i=1}^{a} h_i$ be a representation of h as a sum of distinct elements. Then for any $h_0 \in H$ there is a representation
$$h + ah_0 = \sum_{i=1}^{a}(h_i + h_0),$$
and all elements $h_i + h_0$ are also distinct. Therefore, the number of representations of h and of $h + ah_0$ is the same. □

COROLLARY 4.4.8. *If a is coprime to the order N of H, then for any $h \in H$ we have*
$$M(H, h, a) = \frac{1}{N}\binom{N}{a}.$$

PROOF. In this case $aH = H$. □

COROLLARY 4.4.9. *Let the degree k of a bundle $\mathscr{L}$ be coprime to the order n of $X(\mathbb{F}_q)$. Then $C = (X, X(\mathbb{F}_q), \mathscr{L})_H$ is an $[n, k, n-k]_q$ code, and its enumerator is*
$$W_C(x) = x^n + \sum_{i=0}^{k-1}\binom{n}{i}(q^{k-i} - 1)(x-1)^i + \binom{n}{k}\frac{q-1}{n}(x-1)^k. \quad \square$$

If a and N are not coprime, then $M(H, h, a)$ can also be computed, but the computation is much more complicated. We just present the answer:

Let H be an abelian group of order N, and let $H = \sum_{j=1}^{G} H_j$, $H_j \cong \mathbb{Z}/p_j^{\alpha_j}\mathbb{Z}$, be its primary decomposition. Any element $h \in H$ is represented in the form $h = (h_1, \ldots, h_G)$, where $h_j \in \mathbb{Z}$, $0 \leq h_j \leq p_j^{\alpha_j} - 1$.

Define $\varepsilon_j(h) \in \mathbb{Z} \cup \{\infty\}$ as the maximum power of p_j that divides h_j (and put $\varepsilon_j(h) = \infty$ if $h_j = 0$). For any prime $p \mid N$, let $G_p = \{j \mid p_j = p\}$ and
$$e_p(h) = \begin{cases} \min_{j \in G_p} \varepsilon_j(h) & \text{if there exists } j \in G_p \text{ such that } \varepsilon_j(h) \neq \infty, \\ \sum_{j \in G_p} \alpha_j & \text{if } \varepsilon_j(h) = \infty \text{ for all } j \in G_p. \end{cases}$$

Define
$$\Delta(h) = \prod_{p \mid N} p^{e_p(h)}.$$

Let $t \mid N$ and $t = \prod_p p^{\delta_p}$. Define an integer
$$R_H(t) = \prod_{p \mid N} p^{\sum_{j \in G_p} \min\{\delta_p, \alpha_j\}}.$$

In particular, $R_H(1) = 1$. Let $\mu(m)$ be the Möbius function.

THEOREM 4.4.10. *The number $M(H, h, a)$ of representations of an element h of an abelian group H as a sum of a distinct summands is given by*
$$M(H, h, a) = \frac{1}{N}\sum_{\ell \mid (N, a)}\binom{N/\ell}{a/\ell}(-1)^{a - a/\ell}\sum_{t \mid (\ell, \Delta(h))}\mu(\ell/t)R_H(t). \quad \square$$

EXERCISE 4.4.11. Prove the theorem.

Let us consider another question: What is the minimum possible number $B_{k,\min}(q, N, a)$ of minimum-weight vectors in an elliptic code for fixed parameters of the code (i.e., for fixed q, N, and a)? To answer the question, we have to compute the number
$$M(H, a) = \min_{h \in H} M(H, h, a).$$
The formula for $M(H, a)$ appears to be simpler than that for $M(H, h, a)$:

For an abelian group H, define r_H to be the maximum integer such that $(\mathbb{Z}/2\mathbb{Z})^{r_H} \subseteq H$.

THEOREM 4.4.12. (a) *If* $N \equiv 0 \pmod 4$ *and* $a \equiv 2 \pmod 4$, *then*
$$M(H, a) = \frac{1}{N}\left[\sum_{m\mid(a,N)} \binom{N/m}{a/m}\mu(m)(-1)^{a/m} - 2^{r_H}\sum_{m\mid(a/2, N/2)} \binom{N/2m}{a/2m}\mu(m)\right].$$

(b) *For any other pair* (N, a), *we have*
$$M(H, a) = \frac{1}{N}\sum_{m\mid(a,N)} \binom{N/m}{a/m}\mu(m)(-1)^{a-a/m}. \qquad \square$$

EXERCISE 4.4.13. Prove the theorem.

To find the minimum possible number of minimum-weight vectors, one should also minimize $M(H, a)$ over all $H = X(\mathbb{F}_q)$ of order N. This is not difficult, since we know the possible structure of the group $X(\mathbb{F}_q)$ (Theorem 3.3.15). In particular, $r_H = 1$ or 2.

THEOREM 4.4.14. *Let N be such that there exists an elliptic curve X over $\mathbb{F}_q$, $q = p^e$, with exactly N points.*

(a) *If $N \equiv 0 \pmod 4$, $a \equiv 2 \pmod 4$, q is odd, and moreover either $N \not\equiv 1 \pmod p$ or q is a square and $N = (\sqrt{q} \pm 1)^2$, or q is not a square, $p \equiv 3 \pmod 4$, and $N = q + 1$, then*
$$B_{k,\min}(N, a) = \frac{q-1}{N}\left[\sum_{m\mid(a,N)} \binom{N/m}{a/m}\mu(m)(-1)^{a-a/m}\right.$$
$$\left. - 4\sum_{m\mid(a/2, N/2)} \binom{N/2m}{a/2m}\mu(m)\right].$$

(b) *If $N \equiv 0 \pmod 4$, $a \equiv 2 \pmod 4$, and either q is even or q is odd and none of the additional conditions of part (a) are satisfied, then*
$$B_{k,\min}(N, a) = \frac{q-1}{N}\left[\sum_{m\mid(a,N)} \binom{N/m}{a/m}\mu(m)(-1)^{a-a/m}\right.$$
$$\left. - 2\sum_{m\mid(a/2, N/2)} \binom{N/2m}{a/2m}\mu(m)\right].$$

(c) If $N \not\equiv 0 \pmod 4$ or $a \not\equiv 2 \pmod 4$, then
$$B_{k,\min}(N,a) = \frac{q-1}{N} \sum_{m | (a,N)} \binom{N/m}{a/m} \mu(m).$$
□

EXERCISE 4.4.15. Prove the theorem.

REMARK 4.4.16. For given q and N, it is easy to answer the question of whether there exists an elliptic curve over $\mathbb{F}_q$ having exactly N points (see Theorem 3.3.12).

The mere formulae of Theorems 4.4.12 and 4.4.14 are not too interesting. Of interest is the very fact that $B_{k,\min}$ can be computed explicitly.

REMARK 4.4.17. In the case of curves of genus $g \geq 2$, computation of the numbers $B_k, B_{k-1}, \ldots, B_{k-2g+2}$ is much more complicated. For instance, for $g = 2$ the enumerator equals
$$W_C(x) = x^n + \sum_{i=0}^{k-1} \binom{n}{i}(q^{k-i}-1)(x-1)^i + B_k(x-1)^k$$
$$+ B_{k-1}(x-1)^{k-1} + B_{k-2}(x-1)^{k-2}, \quad k+d \geq n-1.$$
It is related to the enumerator of the dual code as follows:
$$B'_{n-k+1} = \frac{B_{k-1} - \binom{n}{k-1}(q-1)}{q},$$
$$B'_{n-k} = B_k,$$
$$B'_{n-k-1} = qB_{k+1} + \binom{n}{k+1}(q-1).$$

These codes, unlike codes of genus zero and elliptic codes, are not necessarily formally self-dual.

PROBLEM 4.4.18. Compute B_{k-1}, B_k, B_{k+1} in terms of geometric properties of a curve of genus 2 and a divisor. Compute $B_{k-g+1}, \ldots, B_{k+1}$ for a curve of an arbitrary genus.

Elliptic MDS codes. Recall that an $[n,k,d]_q$ code is called an MDS code (or a code of genus zero) if $k+d = n+1$. Such are algebraic geometry codes on the projective line; their length is $n \leq q+1$. A code of genus at most 1 is an MDS code if and only if $B_k = 0$. A question arises: Is it possible to construct an elliptic MDS code of length greater than $q+1$?

Let us start with elliptic codes constructed from all points of a curve X.

THEOREM 4.4.19. Let $C = (X, X(\mathbb{F}_q), \mathscr{L})_H$ be an $[n,k,d]_q$ code constructed from all points of an elliptic curve. If $k > 0$ and $B_k = 0$, then either $X(\mathbb{F}_q) = \mathbb{Z}/2\mathbb{Z}$ or q is odd and $X(\mathbb{F}_q) \cong (\mathbb{Z}/2\mathbb{Z})^2$.
□

EXERCISE 4.4.20. Prove the theorem. (*Hint*: Either derive it from Theorem 4.4.12 or prove directly, using Theorem 4.4.6 only.)

For codes constructed from proper subsets $\mathcal{P} \subset X(\mathbb{F}_q)$, we cannot obtain such a precise result. However, it can be proved that the answer to the above question is negative.

4.4. EXAMPLES

First of all, note that for $q \leq 11$ there are no nontrivial MDS codes of length greater than $q+1$ (see Sec. 1.2.1, the remark after Problem 1.1.12).

For the remaining values of q, there is the following result.

THEOREM 4.4.21. *For $q \geq 13$ there are no elliptic codes of length $n > q+1$ such that $B_k = 0$.*

PROOF. Assume that $B_k = 0$ for some elliptic code $C = (X, \mathcal{P}, \mathscr{L})_H$. This means that $M(H, \mathcal{P}, h_0, a) = 0$, where $H = X(\mathbb{F}_q)$, $|H| = N$, $|\mathcal{P}| = n$, $h_0 = \mathcal{P}_{\mathscr{L}} \in X(\mathbb{F}_q)$ is the sum of the divisor points, and $a = \deg \mathscr{L}$.

Let $a = s + 2t$; define an $(a, s, \mathcal{P})$-representation of h_0 as the decomposition
$$h_0 = (h_1 + \ldots + h_s) + (h_{s+1} + (-h_{s+1}) + \ldots + h_{s+t} + (-h_{s+t})),$$
where all the elements $h_1, \ldots, h_s, h_{s+1}, \ldots, h_{s+t}; -h_{s+1}, \ldots, -h_{s+t}$ are distinct and belong to $\mathcal{P}$.

Let us prove several auxiliary lemmas.

LEMMA 4.4.22. *If h_0 has an $(a, s, \mathcal{P})$-representation and $s + a \leq 2n - 5 - N$, then h_0 also has an $(a+2, s, \mathcal{P})$-representation.*

PROOF. Define
$$\mathcal{P}^* = \{h \in \mathcal{P} \mid -h \in \mathcal{P}, \ h \neq -h\}.$$
Since $H = X(\mathbb{F}_q) \subseteq (\mathbb{Z}/N\mathbb{Z})^2$ (Theorem 3.3.15), there are at most four elements $h \in H$ such that $h = -h$; denote the set of these elements by H_0. An element h of $\mathcal{P}$ does not belong to $\mathcal{P}^*$ if either $-h \notin \mathcal{P}$ (the number of such elements h is at most $|H \setminus \mathcal{P}| = N - n$) or $h \in H_0$. Therefore,
$$|\mathcal{P}^*| \geq n - 4 - (N - n) = 2n - 4 - N.$$

For a given $(a, s, \mathcal{P})$-representation h_0, put
$$R = \mathcal{P}^* \setminus \{\pm h_i\}_{i=1,\ldots,s+t},$$
$$|R| \geq 2n - 4 - N - 2s - 2t = 2n - 4 - s - a.$$
If $h^* \in R$, then $-h^* \in R$ and $\pm h^*$ do not enter the given $(a, s, \mathcal{P})$-representation of h_0; adding $h^* + (-h^*)$ to it, we get an $(a+2, s, \mathcal{P})$-representation. The set R is certainly nonempty for $s + a \leq 2n - N - 5$. □

REMARK 4.4.23. It is easily seen that the assertion of Lemma 4.4.20 is valid as well in the cases where either H does not contain $(\mathbb{Z}/2\mathbb{Z})^2$ and $s + a \leq 2n - 3 - N$, or H does not contain $\mathbb{Z}/2\mathbb{Z}$ and $s + a \leq 2n - 2 - N$.

LEMMA 4.4.24. *If*
$$N - n < \left\lceil \frac{N-2}{3} - \frac{2}{3(N-1)} \right\rceil,$$
then for any $h_0 \in H$ there exists a $(3, 3, \mathcal{P})$-representation.

PROOF. By Theorem 4.4.12, the number of representations of h_0 as a sum of three distinct elements of H is
$$M(H, h_0, 3) \geq M(H, 3) \geq \frac{1}{N}\left(\binom{N}{3} - \frac{N}{3}\right).$$

Let us estimate the number of representations $h_0 = h + h_1 + h_2$ such that one of the elements does not belong to $\mathcal{P}$ (for definiteness, let us assume that $h \notin \mathcal{P}$).

For a fixed h, for a pair $\{h_1,h_2\}$ there are at most $(N-1)/2$ possibilities (of course, $\{h_1,h_2\} = \{h_2,h_1\}$). Therefore, for $h_0 = h+h_1+h_2$, $h \notin \mathcal{P}$, there are at most $(N-n)(N-1)/2$ possibilities, and the lemma easily follows. □

LEMMA 4.4.25. *If $N-n < \left\lceil \dfrac{N-4}{2} \right\rceil$, then for any h_0 there is a $(2,2,\mathcal{P})$-representation.* □

LEMMA 4.4.26. *If $a \leq 2n-6-N$ and*
$$N-n < A(N) = \min\left\{\left\lceil \frac{N-4}{2} \right\rceil, \left\lceil \frac{N-2}{3} - \frac{2}{3(N-1)} \right\rceil\right\},$$
then for any $h_0 \in H$ we have $M(H,\mathcal{P},h_0,a) > 0$. □

EXERCISE 4.4.27. Prove Lemma 4.4.25. (*Hint*: The proof is similar to that of Lemma 4.4.24.) Prove Lemma 4.4.26. (*Hint*: Use induction on a and apply Lemma 4.4.22.)

We continue the proof of Theorem 4.4.21. Recall that we are interested in the case $N > n > q+1$ with $q \geq 13$ (for $q \leq 11$ we know that there are no nontrivial MDS codes with $n > q+1$, and the case $N = n$ is considered in Theorem 4.4.19). In particular, $n \geq 15$ and $N \geq q+3 \geq 16$.

Let us find out when the conditions of Lemma 4.4.26 are violated. Start with the inequality for $N-n$. It is easily seen that for $N \geq 16$ we have
$$A(N) = \left\lceil \frac{N-2}{3} - \frac{2}{3(N-1)} \right\rceil \geq \left\lceil \frac{N-2}{3} \right\rceil.$$
Since $q + \lfloor 2\sqrt{q} \rfloor + 1 \geq N > n \geq q+2$, we have
$$N - n \leq \lfloor 2\sqrt{q} \rfloor - 1.$$
The inequality for $N-n$ can be violated only if
$$2\sqrt{q} - 1 \geq \lfloor 2\sqrt{q} - 1 \rfloor \geq N - n \geq A(N) \geq \left\lceil \frac{N-2}{3} \right\rceil \geq \frac{N-2}{3} \geq \frac{q+1}{3};$$
i.e., $q - 6\sqrt{q} + 4 \leq 0$, which is possible for $q \leq 27$ only. For $q = 27$ we have $\lfloor 2\sqrt{q} \rfloor - 1 = 9 \leq \lceil (q+1)/3 \rceil$. The remaining values of q are $q = 13, 16, 17, 19, 23, 25$; they should be considered individually.

EXERCISE 4.4.28. Check that in each of these cases the inequalities
$$n \geq q+2, \qquad N \leq q + \lfloor 2\sqrt{q} \rfloor + 1, \qquad N-n \geq \left\lceil \frac{N-2}{3} \right\rceil$$
are inconsistent, so that the condition of Lemma 4.4.26 for $N-n$ is always satisfied (for $q \geq 13$).

Let us pass to the next condition.

For $q \geq 13$, the other condition of the lemma, $a \leq 2n-6-N$, can in general be violated. Let us use the following trick: if an elliptic code constructed from a bundle of degree a has $B_k = 0$, then the dual code has $B'_{k'} = 0$ for $a' = n-a$. Therefore, Lemma 4.4.26 cannot be applied only if $a \geq 2n-N-5$ and $n-a \geq 2n-N-5$. Adding these inequalities together, we obtain $2(N-n)+10 \geq n$; i.e.,
$$2\lfloor 2\sqrt{q} \rfloor + 8 \geq 2(N-n) + 10 \geq n \geq q+2.$$

Hence,
$$q - 2\lfloor 2\sqrt{q} \rfloor - 6 \leq 0,$$
which is impossible for $q \geq 27$.

EXERCISE 4.4.29. Make a detailed analysis for the remaining cases $q = 13, 16, 17, 19, 23, 25$. (*Hint*: Use the structure of the group $X(\mathbb{F}_q)$, which is given by Theorem 3.3.15.) □

Self-duality. The formula for the spectrum immediately implies (*check!*) that all elliptic codes with $n = 2k$ are formally self-dual.

The question of the longest possible quasi-self-dual elliptic codes is almost completely analyzable. Let us start with a simple auxiliary statement.

PROPOSITION 4.4.30. *Let H be an abelian group of order N. If N is odd, then the equality $\sum_{h \in \mathcal{P}} h = 0$ holds for $\mathcal{P} = \mathcal{P}_0 = H$ and $\mathcal{P} = \mathcal{P}_1 = H \setminus \{0\}$. If N is even, then either this equality holds for $\mathcal{P} = \mathcal{P}_0$ and $\mathcal{P} = \mathcal{P}_1$ or there exists an element $h_0 \in H$ such that it holds for $\mathcal{P} = \mathcal{P}_2 = H \setminus \{h_0\}$ and $\mathcal{P} = \mathcal{P}_3 = H \setminus \{0, h_0\}$.*

PROOF. Let H_0 denote the subgroup of elements of order two. If $h \notin H_0$, then $-h \neq h$, and thus $\sum_{h \notin H_0} h = 0$. In a group of odd order, $H_0 = \{0\}$. In a group of even order, $H_0 \cong (\mathbb{Z}/2\mathbb{Z})^r$. If $r \geq 2$, then $\sum_{h \in H_0} h = 0$ (this is easily proved by induction on r). In the case $r = 1$, we have to exclude the unique nonzero element of order two. □

THEOREM 4.4.31. *Let N be the number of $\mathbb{F}_q$-points on an elliptic curve. If N is odd, then there exists a quasi-self-dual $\left[N-1, \frac{N-1}{2}, \frac{N-1}{2}\right]_q$ code; if N is even, then there exists either a quasi-self-dual $\left[N, \frac{N}{2}, \frac{N}{2}\right]_q$ code or a quasi-self-dual $\left[N-2, \frac{N-2}{2}, \frac{N-2}{2}\right]_q$ code. If q is even, there exist corresponding self-dual codes.* □

EXERCISE 4.4.32. Prove the theorem.

REMARK 4.4.33. Since (see Theorem 3.4.49) the maximum number of $\mathbb{F}_q$-points on an elliptic curve is at least $q + \lfloor 2\sqrt{q} \rfloor$, it follows from Theorem 4.4.31 that for some $n \geq q + \lfloor 2\sqrt{q} \rfloor - 2$ there exists a quasi-self-dual $[n, n/2, n/2]_q$ code (self-dual if $q = 2^m$).

REMARK 4.4.34. Note that Theorem 4.4.31 is a little more precise than Theorem 4.1.28. A similar result can be obtained for a hyperelliptic curve X of an arbitrary genus g. Such a curve, being embedded in its Jacobian J_X, possesses the same property: if $P \in X$, then $-P \in X$.

Thus, concerning elliptic codes we know almost everything. As compared with MDS codes, they have parameters worse by one, $k + d = n$; on the other hand, they are longer by approximately $2\sqrt{q}$.

4.4.3. Hermitian Codes

Here we give the most elementary facts from the vast theory of codes on Hermitian curves, hoping to return to this subject in *Advanced Chapters*.

Parameters of an algebraic geometry code (more precisely, the sum of the designed parameters k_c and d_c) are determined by the genus of the curve: $k_c + d_c = n - g + 1$. Hence, the parameters are better the more $\mathbb{F}_q$-points can be found on a curve of a given genus. There exists a class of maximal curves for which the properties of the corresponding codes are rather well studied and very interesting. These are Hermitian curves.

It was shown in Sec. 4.5.1 that any Hermitian curve over $\mathbb{F}_q$, $q = r^2$, is isomorphic to the curve X'_r with the affine equation

$$X'_r: \quad v^r + v = u^{r+1}.$$

It has $r^3 + 1$ points over $\mathbb{F}_q$, namely, the unique point at infinity (the common pole of the functions u and v) and points with coordinates (α, β), where α runs over all elements of $\mathbb{F}_q$ and β runs over all solutions of $\beta^r + \beta = \alpha^{r+1}$.

We denote the common pole of u and v by Q, and a point with coordinates (α, β) by $P_{\alpha, \beta}$.

EXERCISE 4.4.35. Check the following expressions for the principal divisors of the functions $u - \alpha$ and $v - \beta$:

$$(u - \alpha) = \sum_{\beta^r + \beta = \alpha^{r+1}} P_{\alpha, \beta} - rQ;$$

$$(v - \beta) = \begin{cases} (r+1)P_{0, \beta} - (r+1)Q & \text{if } \beta^r + \beta = 0, \\ \sum_{\alpha^{r+1} = \beta^r + \beta} P_{\alpha, \beta} - (r+1)Q & \text{if } \beta^r + \beta \neq 0. \end{cases}$$

EXERCISE 4.4.36. Check that the principal divisor of $z = u^q - u$ is $(z) = \boldsymbol{P} - r^3 Q$, where $\boldsymbol{P} = \sum_{\alpha, \beta} P_{\alpha, \beta}$. Deduce from this that the residues of the differential $\omega = dz/z$ at any point $P_{\alpha, \beta}$ equals 1 and that the divisor (ω) is

$$(\omega) = (r^3 + r^2 - r - 2)Q - \boldsymbol{P}.$$

For any $\gamma, \delta \in \mathbb{F}_q$ such that $\delta^r + \delta = \gamma^{r+1}$, there is an $\mathbb{F}_q$-automorphism $\sigma = \sigma(\gamma, \delta)$ of X'_r given by

$$\sigma u = u + \gamma, \qquad \sigma v = v + \gamma^r u + \delta.$$

EXERCISE 4.4.37. Show that automorphisms of the form $\sigma(\gamma, \delta)$ constitute a subgroup G_r of order r^3 in $\mathrm{Aut}_{\mathbb{F}_q}(X'_r)$, which acts transitively on the set $\{P_{\alpha, \beta}\}$ and stabilizes the point Q.

Now, let $n = r^3$, and let m be an arbitrary positive integer. Denote by C_m the following q-ary code:

$$C_m = (X'_r, \mathcal{P}, mQ)_L,$$

where $\mathcal{P} = \mathrm{Supp}\, \boldsymbol{P} = \{P_{\alpha, \beta}\} = X'_r(\mathbb{F}_q) \setminus \{Q\}$ is the set of affine $\mathbb{F}_q$-points of X'_r.

EXERCISE 4.4.38. Prove that for any $m \leq r^3 + r^2 - r - 2$, the codes C_m and $C_{r^3 + r^2 - r - 2 - m}$ are dual. (*Hint*: Apply Theorem 4.1.26 and the results of Exercise 4.4.36.)

In particular, if q is even and $m = (r^3 + r^2 - r - 2)/2$, the code C_m is self-dual. To compute $k_m = \dim C_m$, we need the following exercise.

EXERCISE 4.4.39. Check that, for any $m \geq 0$, the space $L(mQ)$ has a basis of the following form:
$$\{u^i v^j \mid 0 \leq i,\ 0 \leq j \leq r-1,\ ir + j(r+1) \leq m\}.$$

For $m \in \mathbb{N}$, define an integer $\nu(m)$ as follows: let $\tilde{m}$ be the largest integer of the form $ir + j(r+1)$, $i \geq 0$, $0 \leq j \leq r-1$, not greater than m, and let
$$\tilde{m} = tr + s, \quad 0 \leq s \leq r-1.$$

Define
$$\nu(m) = 1 + \frac{t(t+1)}{2} + \min\{t, s\}.$$

EXERCISE 4.4.40. Prove that $k_m = \dim C_m$ is given by
$$k_m = \begin{cases} \nu(m) & \text{if } m \leq r^2 - r - 2, \\ m + 1 - \dfrac{r^2 - r}{2} & \text{if } r^2 - r - 2 < m < r^3, \\ r^3 - \nu(r^3 + r^2 - r - 2 - m) & \text{if } m \geq r^3. \end{cases}$$

(*Hint*: Apply the results of Exercises 4.4.39 and 4.4.38.)

Moreover, we are able to explicitly write out a generator (or parity-check) matrix of the Hermitian code C_m. Indeed, denote by M_m the $\nu(m) \times n$ matrix whose rows are the following vectors $u(\alpha, \beta)$ of length $\nu(m)$:
$$u(\alpha, \beta) = (\alpha^i \beta^j), \quad ir + j(r+1) \leq m, \quad 0 \leq i, \quad 0 \leq j \leq r-1,$$
for all $\alpha, \beta \in \mathbb{F}_q$ such that $\alpha^{r+1} = \beta^r + \beta$.

PROPOSITION 4.4.41. (a) *For* $0 \leq m < n = r^3$, M_m *is a generator matrix of* C_m.
(b) *For* $r^2 - r - 2 < m \leq r^3 + r^2 - r - 2$, $M_{r^3 + r^2 - r - 2 - m}$ *is a parity-check matrix of* C_m. □

For the minimum distance d_m of C_m, we as usual have the estimate $d_m \geq r^3 - m$. In some cases, one can show this estimate to be tight.

PROPOSITION 4.4.42. *Let* $m = ir + j(r+1) \leq r^3 - 1$ *for some* $i \geq 0$ *and* $0 \leq j \leq r-1$. *If either* $j = 0$ *or* $i \leq r^2 - r - 1$ (*in particular, if* $m \leq r^3 - r^2$), *then* $d_m = r^3 - m$.

PROOF. Let $j = 0$, and let $\alpha_1, \ldots, \alpha_i$ be distinct elements of $\mathbb{F}_q$. Then the function $f = \prod_{\ell=1}^{i}(u - \alpha_\ell)$ has exactly ir different zeroes in $\mathcal{P}$. Now let $i \leq r^2 - r - 1$. Consider the set
$$A = \{\alpha \in \mathbb{F}_q \mid \alpha^{r-1} \neq 1\}.$$
Its cardinality is $r^2 - r - 1$, so we can choose i distinct elements $\alpha_1, \ldots, \alpha_i \in A$. Put $t_1 = \prod_{\ell=1}^{i}(u - \alpha_\ell)$. Choose j distinct elements $\varepsilon_1, \ldots, \varepsilon_j$ such that $\varepsilon_\ell^r + \varepsilon = 1$ and put $t_2 = \prod_{\ell=1}^{j}(v - \varepsilon_\ell)$. It is easily checked that the function $t = t_1 t_2 \in L(mQ)$ has exactly m different zeroes in $\mathcal{P}$.

Finally, if $m \leq r^3 - r^2$, then either $j = 0$ or $j \geq 1$ and $i \leq r^2 - r - 1$. □

Since the group G_r acts transitively on $\mathcal{P}$ and stabilizes Q, the code C_m can be represented as an ideal I_m in the group algebra $\mathbb{F}_q[G_r]$.

EXERCISE 4.4.43. Write out the ideal I_m in the group algebra $\mathbb{F}_q[G_r]$ explicitly, and estimate the parameters of C_m not using the Riemann–Roch theorem. (*Hint*: Use the above representation of elements of G_r by pairs (γ, δ) of elements of $\mathbb{F}_q$, and consider the following elements of the group algebra:

$$u\colon G_r \to \mathbb{F}_q, \quad u\colon (\gamma, \delta) \mapsto \gamma,$$
$$v\colon G_r \to \mathbb{F}_q, \quad v\colon (\gamma, \delta) \mapsto \delta.$$

Show that I_m is generated by elements $u^i v^j$ for $0 \leq i, 0 \leq j \leq r-1$, $ir + j(r+1) \leq m$; this yields a formula for k. To estimate d, check that a function $f\colon G_r \to \mathbb{F}_q$ that belongs to I_m and has more than m zeroes is identically zero. In doing this, use properties of the Vandermonde determinant.)

Using gonality and abundant divisors. Consider the following example.

EXAMPLE 4.4.44. The plain Hermitian curve with equation $x^{r+1} + y^{r+1} + z^{r+1} = 0$ over $\mathbb{F}_q$, where $q = r^2$, has $r^3 + 1$ rational points and genus $r(r-1)/2$. Thus, its gonality is at least $(r^3 + 1)/(r^2 + 1)$ by Proposition 4.2.25 and at most r by Remark 4.2.21. Thus, the gonality is precisely r. Fix one of the rational points and call it Q; let $\mathcal{P}$ consist of the remaining r^3 rational points. Put $n = r^3$ and $D_m = mQ$ and consider the code $C_m = (X, \mathcal{P}, D_m)_L$.

If $m < n$, the designed distance of this code is $n - m$. If $n < m$, and m is a multiple of r or $m \leq n - q$, then the true minimum distance of C_m is equal to the designed distance. Furthermore, for $m < n$ the minimum distance of C_m is not greater than $n - m + t$, where t is the remainder of division of m by r.

Next, it can be shown that the divisors $r^3 Q$ and $\boldsymbol{P}$ are equivalent. Therefore, if $m \geq n$, then D_m is an abundant divisor of abundance $m - n$. Thus, for any $a = 0, \ldots, r - 1$ by Theorem 4.2.27 the code C_{n+a} has parameters

$$\left[r^3, r^3 + a - r(r-1)/2, \geq r - a\right].$$

Every line over $\mathbb{F}_q$ either intersects the curve in exactly $r+1$ distinct rational points or is tangent to it at a rational point with multiplicity $r+1$. If ℓ_1 is the tangent line at Q and ℓ_2 is a line passing through Q and intersecting the curve at the points $P_1, \ldots, P_r$, then the divisor of the rational function ℓ_1/ℓ_2 is $P_1 + \ldots + P_r - rQ$; i.e., $\ell(P_1 + \ldots + P_r) > 1$. Thus, by Lemma 4.2.26 the minimum distance of C_n is precisely r. Similarly, there exists a rational function with divisor $P_1 + \ldots + P_{r-1} + Q - rP_r$, so the minimum distance of C_{n+1} is exactly $r - 1$. Moreover, it follows from Remark 4.2.28 that for any $m = n - r, \ldots, n - 1$ the minimum distance of C_m is at least r. One can prove that in fact the equality holds.

Improved codes. Here we give an example of using the construction of improved algebraic geometry codes, described in Sec. 4.2.2.

EXAMPLE 4.4.45. Consider the Hermitian curve over $\mathbb{F}_{16}$ given by

$$x^5 = y^4 + y.$$

This is a curve of genus 6; it has 65 rational points and is optimal (cf. Example 4.4.57). The Weierstrass semigroup of the point Q at infinity is generated by 4 and 5, i.e.,

$$H = \{0, 4, 5, 8, 9, 10\} \cup \{n \in \mathbb{N}_0 \mid n \geq 12\};$$

in particular, the conductor is 12. Hence, we can easily write down the sequence (ν_i):
$$1,2,2,3,4,3,4,6,6,4,5,8,9,8,9,10,12,12,13,14,15,\ldots,$$
as well as the sequence (r_d):
$$0,1,3,5,8,9,11,11,13,15,16,16,18,19,20,21,22,\ldots.$$
Thus, we obtain a set of improved $[64, 64 - r_d, \geq d]_{16}$ codes; for $d = 4$, 5, 6, and 9 they indeed improve the Feng–Rao bound, i.e., have a larger dimension than that determined by $d = d_{FR}(64-k)$.

4.4.4. Other Examples

Now let us discuss several examples of algebraic geometry codes, not pretending to be systematical (all these codes are codes on maximal curves).

EXAMPLE 4.4.46. Consider the plane curve E over $\mathbb{F}_2$ with the affine equation
$$y^2 + y = x^3 + x + 1,$$
i.e., with the homogeneous equation
$$y^2 z + y z^2 = x^3 + x z^2 + z^3.$$

EXERCISE 4.4.47. Check that the projective curve $E \subset \mathbb{P}^2$ is absolutely irreducible and nonsingular and hence is an elliptic curve. Check that $|E(\mathbb{F}_2)| = 1$, $|E(\mathbb{F}_4)| = 5$, and $|E(\mathbb{F}_{16})| = 25$. (*Hint*: The only $\mathbb{F}_2$-point is the point at infinity; $|X(\mathbb{F}_{2^r})| = 2^r + 1 - \omega^r - \overline{\omega}^r$, where $|\omega| = \sqrt{2}$. Since $|X(\mathbb{F}_2)| = 1$, $\omega = 1 + i$.) Write out the coordinates of $\mathbb{F}_4$- and $\mathbb{F}_{16}$-points of E.

Note that $25 = 16 + 1 + 2 \cdot 4$, and therefore E is maximal over $\mathbb{F}_{16}$.

EXERCISE 4.4.48. Let $Q = (0 : 1 : 0)$ be the only infinite point of E. Show that for any $m \geq 0$ the space $L(mQ)$ has a basis of the form $\{x^i y^j \mid 2i + 3j \leq m\}$. Write out generator matrices of codes $(E, \mathcal{P}, mQ)_L$ and $(E, \mathcal{P}', mQ)_L$ over $\mathbb{F}_4$ and $\mathbb{F}_{16}$, respectively, where $\mathcal{P} = E(\mathbb{F}_4) \setminus \{Q\}$ and $\mathcal{P}' = E(\mathbb{F}_{16}) \setminus \{Q\}$, for $m = 2, 8, 12$, and 16. Compute spectra of these codes.

EXAMPLE 4.4.49. Consider the curve E_1 over $\mathbb{F}_{25}$ with affine equation
$$y^2 + y = x^3.$$

EXERCISE 4.4.50. Show that $E_1 \subset \mathbb{P}^2$ is a maximal elliptic curve over $\mathbb{F}_{25}$; thus, $|E_1(\mathbb{F}_{25})| = 36$ (one point, Q, at infinity and 35 points in $\mathbb{A}^2$). Write out the coordinates of $\mathbb{F}_{25}$-points of E_1 and generator matrices of codes $(E_1, \mathcal{P}_1, mQ)_L$ for $m = 2, 8$, and 16, where $\mathcal{P}_1 = E(\mathbb{F}_{25}) \setminus \{Q\}$. (*Hint*: Similarly to Exercise 4.4.48.)

EXAMPLE 4.4.51 (Klein quartic). The *Klein quartic* is the curve $X \subset \mathbb{P}^2$ with the homogeneous equation
$$x^3 y + y^3 z + z^3 x = 0.$$

EXERCISE 4.4.52. Show that X is a smooth curve of genus 3 and that it is maximal over $\mathbb{F}_8$. (*Hint*: $24 = 8 + 1 + 3 \cdot \lfloor 2\sqrt{8} \rfloor$.) Write out the coordinates of these $\mathbb{F}_8$-points.

EXERCISE 4.4.53. Prove that the divisor of the linear form z on X is $(z) = Q_1 + 3Q_2$, where $Q_1 = (1 : 0 : 0)$ and $Q_2 = (0 : 1 : 0)$. Show that the space $L(m(z))$ is generated by elements of the form $\{x^i y^j \mid 2i + 3j \leq 3m, \ i - 2j \leq m\}$. From this generator system, select a basis of $L(m(z))$ and write out generator matrices of codes $(X, \mathcal{P}, m(z))_L$ for $m = 2, 3$, and 5, where $\mathcal{P} = X(\mathbb{F}_8) \setminus \{Q_1, Q_2\}$.

EXAMPLE 4.4.54. Consider a smooth model Y of the curve over $\mathbb{F}_{25}$ with equation
$$y^2 = x^6 + 1.$$
The projection $(x, y) \mapsto x$ defines a map $f \colon Y \to \mathbb{P}^1$ of degree 2. Thus, Y is a hyperelliptic curve.

EXERCISE 4.4.55. Show that the genus of Y equals 2. (*Hint*: Apply the Hurwitz formula.)

EXERCISE 4.4.56. Show that $|Y(\mathbb{F}_{25})| = 46$ and that Y is a maximal curve. Find the coordinates of $\mathbb{F}_{25}$-points. (*Attention*: There are two points $Q_1, Q_2 \in Y(\mathbb{F}_{25})$ corresponding to $x = y = \infty$; i.e., the functions x and y have two common poles.)

EXAMPLE 4.4.57. Consider the Hermitian curve
$$X_5 \colon x^5 + y^5 = 1$$
over $\mathbb{F}_{16}$ and the code $C = (X_5, \mathcal{P}, 37Q)_L$, where $\mathcal{P} = X_5(\mathbb{F}_{16}) \setminus \{Q\}$ (in the notation of the preceding subsection).

EXERCISE 4.4.58. Show that C is a self-dual code of length 64 over $\mathbb{F}_{16}$, and its minimum distance equals 27. (*Hint*: See Proposition 4.4.42.) Write out a generator matrix for C. Show that in the spectrum of C we have $A_r > 0$ for any $r \geq 27$. (*Hint*: For all values of r, except for $r = 63, 62, 61, 58, 57$, and 53, this follows from the arguments used in the proof of Proposition 4.4.42. Similar functions t can as well be constructed for other values of r. For instance, for $r = 53$ one can take
$$t = (u - \alpha_1)(u - \alpha_2)(v - \gamma_1)(v - \gamma_2)(v - \gamma_3),$$
where $\alpha_1, \alpha_2 \neq 0$; $\gamma_i^3 + 1 = 0$ for $i = 1, 2$, and 3; and $\gamma_i \neq \gamma_j$ for $i \neq j$.)

All the above codes were constructed from plane curves. The following example deals with a space curve.

EXAMPLE 4.4.59. Consider the curve Z in $\mathbb{P}^3$ given by the following system of equations over $\mathbb{F}_3$:
$$\begin{cases} xt - z^2 = 0, \\ y^3 - yt^2 - x^2 z = 0. \end{cases}$$

EXERCISE 4.4.60. Show that Z is a smooth absolutely irreducible curve of genus 4 canonically embedded in $\mathbb{P}^3$. (*Hint*: Consider the divisor of the differential form dx and show that $(dx) = 6Q$, where Q is the point with coordinates $x = 1$ and $t = y = z = 0$.)

EXERCISE 4.4.61. Show that Z is a maximal curve over the field $\mathbb{F}_{81}$; thus, $|Z(\mathbb{F}_{81})| = 154$. Find the coordinates of its $\mathbb{F}_{81}$-points.

EXERCISE 4.4.62. Show that the space $L(mQ)$, $m \geq 0$, has a basis of the form $\{y^i z^j \mid 5i + 3j \leq m,\ 0 \leq j \leq 2,\ i \geq 0\}$. Write out a generator matrix of the code $(Z, \mathcal{P}, mQ)_L$ for $m = 3, 6, 20$, and 50, where $\mathcal{P} = Z(\mathbb{F}_{81}) \setminus \{Q\}$.

4.4.5. Generalized Algebraic Geometry Codes

In this subsection we apply the construction of generalized algebraic geometry codes $(X, \mathcal{P}, D, \mathcal{C})_L$, described in Sec. 4.2.2.

In all examples, points of a curve are assumed to be ordered according to nondecreasing degrees; unless otherwise specified, we take $\mathcal{P} = \{P_1, \ldots, P_s\}$ to be the first s points from this list. Furthermore, for sufficiently large r there always exist points P and Q of degrees r and $r+1$ respectively; therefore, choosing the divisor $D = m(Q - P)$, we may always assume that $\deg D = m$ and $\operatorname{Supp} D \cap \mathcal{P} = \varnothing$. All results of the examples given below follow from Proposition 4.2.43.

EXAMPLE 4.4.63. Consider the curve X from Exercise 3.1.18a over $\mathbb{F}_2$.

(a) Take $s = 6$, $[n_i, k_i, d_i] = [1, 1, 1]$ for $1 \leq i \leq 4$, and $[n_i, k_i, d_i] = [3, 2, 2]$ for $i = 5, 6$. We obtain binary codes with parameters

$$[10, m, 8 - m], \quad 1 \leq m \leq 7.$$

(b) Add one more point; i.e., set $s = 7$, take the same codes for $1 \leq i \leq 6$, and add $[n_7, k_7, d_7] = [8, 4, 4]$. We obtain binary codes with parameters

$$[18, m, 12 - m], \quad 1 \leq m \leq 11.$$

(c) Add one more point; i.e., set $s = 8$, take the same codes for $1 \leq i \leq 7$, and add $[n_8, k_8, d_8] = [8, 4, 4]$. We obtain binary codes with parameters

$$[26, m, 16 - m], \quad 1 \leq m \leq 15.$$

EXAMPLE 4.4.64. Consider the curve X from Exercise 3.1.18b over $\mathbb{F}_2$.

(a) Take $s = 5$, $[n_i, k_i, d_i] = [1, 1, 1]$ for $1 \leq i \leq 3$ and $[n_i, k_i, d_i] = [3, 2, 2]$ for $i = 4, 5$. We obtain binary codes with parameters

$$[9, m, 7 - m], \quad 1 \leq m \leq 6.$$

(b) Add one more point; i.e., set $s = 6$, take the same codes for $1 \leq i \leq 5$, and add $[n_6, k_6, d_6] = [3, 2, 2]$. We obtain binary codes with parameters

$$[12, m, 9 - m], \quad 1 \leq m \leq 8.$$

(c) Add one more point; i.e., set $s = 7$, take the same codes for $1 \leq i \leq 6$, and add $[n_7, k_7, d_7] = [4, 3, 2]$. We obtain binary codes with parameters

$$[16, m, 11 - m], \quad 1 \leq m \leq 10.$$

(d) Add one more point; i.e., set $s = 8$, take the same codes for $1 \leq i \leq 7$, and add $[n_8, k_8, d_8] = [4, 3, 2]$. We obtain binary codes with parameters

$$[20, m, 13 - m], \quad 1 \leq m \leq 12.$$

(e) Let $s = 8$, for $1 \leq i \leq 7$ take the same codes as in (c), but now add $[n_8, k_8, d_8] = [6, 3, 3]$. Then we obtain binary codes with parameters

$$[22, m, 14 - m], \quad 1 \leq m \leq 13.$$

(f) Let $s = 8$, for $1 \leq i \leq 6$ take the same codes as in (b), and add $[n_i, k_i, d_i] = [6, 3, 3]$ for $i = 7, 8$. We obtain binary codes with parameters

$$[24, m, 15 - m], \quad 1 \leq m \leq 14.$$

EXAMPLE 4.4.65. Consider the curve X from Exercise 3.1.18c over $\mathbb{F}_3$.

(a) Take $s = 7$, $[n_i, k_i, d_i] = [1,1,1]$ for $1 \le i \le 6$ and $[n_7, k_7, d_7] = [3,2,2]$. We obtain ternary codes with parameters
$$[9, m, 8-m], \quad 1 \le m \le 7.$$

(b) Add one more point; i.e., set $s = 8$, take the same codes for $1 \le i \le 7$, and add $[n_8, k_8, d_8] = [3,2,2]$. We obtain ternary codes with parameters
$$[12, m, 10-m], \quad 1 \le m \le 9.$$

(c) Add one more point; i.e., set $s = 9$, take the same codes for $1 \le i \le 8$, and add $[n_9, k_9, d_9] = [3,2,2]$. We obtain ternary codes with parameters
$$[15, m, 12-m], \quad 1 \le m \le 11.$$

(d) This time we take only three points of degree one, set $s = 8$, $[n_i, k_i, d_i] = [1,1,1]$ for $1 \le i \le 3$, $[n_i, k_i, d_i] = [3,2,2]$ for $4 \le i \le 6$, and $[n_i, k_i, d_i] = [4,3,2]$ for $i = 7, 8$. Then we obtain ternary codes with parameters
$$[20, m, 13-m], \quad 1 \le m \le 12.$$

(e) Take the same $s = 8$ points; for $1 \le i \le 7$ take the same codes as in the previous case, but now add $[n_8, k_8, d_8] = [6,3,3]$. We obtain ternary codes with parameters
$$[22, m, 14-m], \quad 1 \le m \le 13.$$

EXAMPLE 4.4.66. Consider the curve X from Exercise 3.1.18d over $\mathbb{F}_3$.

(a) Take $s = 7$, $[n_i, k_i, d_i] = [1,1,1]$ for $1 \le i \le 5$ and $[n_i, k_i, d_i] = [3,2,2]$ for $i = 6, 7$. We obtain ternary codes with parameters
$$[11, m, 9-m], \quad 1 \le m \le 8.$$

(b) Add one more point; i.e., set $s = 8$, take the same codes for $1 \le i \le 7$, and add $[n_8, k_8, d_8] = [3,2,2]$. We obtain ternary codes with parameters
$$[14, m, 11-m], \quad 1 \le m \le 10.$$

(c) Add one more point and one more $[3,2,2]$ code. We obtain ternary codes with parameters
$$[17, m, 13-m], \quad 1 \le m \le 12.$$

(d) Again add a point and one more $[3,2,2]$ code. We obtain ternary codes with parameters
$$[20, m, 15-m], \quad 1 \le m \le 14.$$

EXAMPLE 4.4.67. Consider the smooth model of a curve defined over $\mathbb{F}_5$ defined by the equation
$$y^4 = 2x(x-1)(x-2)(x-3) + 1.$$
This is a curve of genus $g = 3$ with $B_1 = 16$ and $B_2 \ge 2$.

(a) Take $s = 17$, $[n_i, k_i, d_i] = [1,1,1]$ for $1 \le i \le 16$, and $[n_{17}, k_{17}, d_{17}] = [3,2,2]$. We obtain codes over $\mathbb{F}_5$ with parameters
$$[19, m-2, 18-m], \quad 5 \le m \le 17.$$

(b) Add one more point and one more $[3,2,2]$ code. We obtain codes over $\mathbb{F}_5$ with parameters
$$[22, m-2, 20-m], \quad 5 \leq m \leq 19.$$

The above examples produce many linear codes whose parameters are best possible or nearly best possible. In the table below we list those of them which are the best known at present. For $q=2$ and $q=3$ these codes are actually optimal; i.e., their parameters meet an upper bound for linear codes of given length and dimension.

q	$[n,k,d]$	Example	q	$[n,k,d]$	Example
2	$[9,3,4]$	4.4.64a	2	$[18,4,8]$	4.4.63b
2	$[10,2,6]$	4.4.63a	3	$[9,2,6]$	4.4.65a
2	$[10,3,5]$	4.4.63a	3	$[15,3,9]$	4.4.65c
2	$[10,4,4]$	4.4.63a	5	$[19,6,10]$	4.4.67a
2	$[12,3,6]$	4.4.64b	5	$[19,7,9]$	4.4.67a
2	$[12,5,4]$	4.4.64b	5	$[19,8,8]$	4.4.67a
2	$[16,3,8]$	4.4.64c			

Some generalized algebraic geometry codes

EXERCISE 4.4.68. Prove all statements in this subsection.

4.5. Asymptotic Results

The advantages of algebraic geometry codes become most prominent in the study of asymptotic problems. It turns out that parameters of codes constructed from families of algebraic curves become better and better when the asymptotic ratio of the number of $\mathbb{F}_q$-points on the curve to its genus grows. In Sec. 4.5.1 we establish the basic algebraic geometry asymptotic bound; it is a straight line, and when q is large enough it intersects the Gilbert–Varshamov bound, improving it on a certain segment. This result shows that highly non-random codes obtained by subtle algebraic geometry constructions (using modular curves or García–Stichtenoth towers) may be asymptotically better than randomly chosen codes (recall that with probability 1 the parameters of a linear code lie on the Gilbert–Varshamov bound).

In Sec. 4.5.2 we show that for algebraic geometry codes we can also exploit randomization. Expurgation method allows us, varying the divisor, to get the expurgation bound. For small q this bound coincides with the Gilbert–Varshamov bound, and for q large enough it slightly improves the bound defined as the maximum of the Gilbert–Varshamov and the basic algebraic geometry bound, smoothing the angles between the two. On the way we prove a result on existence of (binary) codes with exponential number of codewords of minimal weight, which is of importance for coding theory.

The basic algebraic geometry bound is constructive; in Sec. 4.5.3 we apply it to various constructions of Sec. 1.2.3 to improve the polynomial Bloch–Zyablov bound (for any q and every δ).

The progress due to algebraic geometry codes leads to improvement of other bounds (Sec. 4.5.4). For nonlinear codes (for q large enough) it can be shown that the expurgation bound is not the best possible; it can be shown that the Gilbert bound can be improved for arbitrary (large enough) alphabets. We also manage to improve corresponding bounds for self-dual and constant-weight codes.

4.5.1. The Basic Algebraic Geometry Bound and Its Variants

Let us apply Corollary 4.1.14 to families of curves of growing genus.

THEOREM 4.5.1. *If over $\mathbb{F}_q$ there is a family of curves X_i of genus $g_i \to \infty$ such that*
$$\gamma = \liminf \frac{g_i}{N_i} < 1,$$
where $N_i = |X_i(\mathbb{F}_q)|$, then
$$\alpha_q^{\lin}(\delta) \geq 1 - \gamma - \delta.$$

PROOF. According to Corollary 4.1.14, a code constructed from the curve X_i has parameters $[n_i, k_i, d_i]_q$, where $n_i = 1, 2, \ldots, N_i$, $k_i = 0, 1, \ldots, n_i - g_i$, and
$$k_i + d_i = n_i - g_i + 1;$$
i.e., $R_i = 1 - \dfrac{g_i - 1}{N_i} - \delta_i$. Passing to the limit in i, we see that the points (δ_i, R_i) are everywhere dense on the line segment
$$R = 1 - \gamma - \delta, \quad 0 \leq \delta \leq 1 - \gamma. \qquad \square$$

Introduce the quantity
$$A(q) = \limsup \frac{N}{g},$$

where the limit is taken over all curves defined over $\mathbb{F}_q$; put

$$\gamma_q = \frac{1}{A(q)}.$$

COROLLARY 4.5.2. *We have*

$$\alpha_q^{\text{lin}} \geq 1 - \gamma_q - \delta. \qquad \square$$

Recall (see Theorem 3.2.1) that $\gamma_q \geq (\sqrt{q}-1)^{-1}$ and that this bound is tight if q is a square (Theorem 3.2.3).

COROLLARY 4.5.3. *Let q be an even power of a prime. Then*

$$\alpha_q^{\text{lin}}(\delta) \geq R_{TVZ}(\delta) = 1 - (\sqrt{q}-1)^{-1} - \delta. \qquad \square$$

The obtained bound $R_{TVZ}(\delta)$ is known as the *basic algebraic geometry bound*, or *the Tsfasman–Vlăduţ–Zink bound*.

Now we are going to compare the obtained bound (here we restrict ourselves to the case $q = p^{2m}$) with the Gilbert–Varshamov bound

$$R_{GV}(\delta) = 1 - H_q(\delta)$$

(see Sec. 1.3.2). The Gilbert–Varshamov bound is ∪-convex; its endpoints are $(0,1)$ and $((q-1)/q, 0)$. Therefore, the basic algebraic geometry bound, which is a line segment connecting the points $(0, 1-(\sqrt{q}-1)^{-1})$ and $(1-(\sqrt{q}-1)^{-1}, 0))$, either totally lies below it or intersects it in two points or is tangent to it.

THEOREM 4.5.4. *The basic algebraic geometry bound R_{TVZ} totally lies below the Gilbert–Varshamov bound R_{GV} for $q = p^{2m} < 49$. For $q = p^{2m} \geq 49$ these bounds intersect, and the basic algebraic geometry bound improves the Gilbert–Varshamov bound on the segment (δ_1, δ_2), where δ_1 and δ_2 are the roots of the equation*

$$H_q(\delta) - \delta = (\sqrt{q}-1)^{-1}.$$

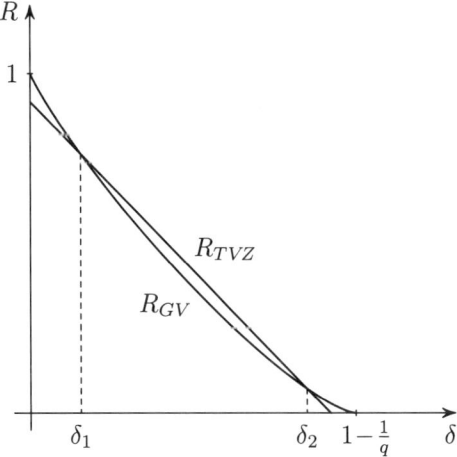

FIGURE 4.1. Lower bounds for $q = p^{2m} \geq 49$.

PROOF. Consider the tangent to the Gilbert–Varshamov bound parallel to the basic algebraic geometry bound. If it is above the basic algebraic geometry bound, then the curves do not intersect; otherwise, they do intersect, and the intersection equation is as above.

The tangent is given (see Exercise 1.3.19) by
$$R = 1 - (\log_q(2q-1) - 1) - \delta.$$
The equation $H_q(\delta) - \delta = \gamma_q$ has two roots if and only if
$$\gamma_q < \log_q(2q-1) - 1. \tag{4.5.1}$$

Theorem 3.2.1 states that $\gamma_q \geq (\sqrt{q}-1)^{-1}$; if $q = p^{2m}$, then by Theorem 3.2.3 we have the equality. It is easily seen that for q that are even powers of primes, inequality (4.5.1) holds for $q \geq 49$ and does not hold for smaller q. □

EXERCISE 4.5.5. Check the last claim.

EXERCISE 4.5.6. Check that for q that are odd powers of primes and are not greater than 41, the basic algebraic geometry bound lies below the Gilbert–Varshamov bound.

REMARK 4.5.7. The basic algebraic geometry bound does not depend on the choice of a construction of algebraic geometry codes. If we use the L-construction (or Ω-construction), we may take $D = aP_0$, where $P_0 \in X(\mathbb{F}_q)$, and construct codes of length $n = N - 1$ evaluating at the other $\mathbb{F}_q$-points of X. Losing 1 in length does not influence asymptotic properties. However, in some asymptotic situations it is more convenient to use the H-construction (cf. Sec. 4.5.2 below).

Consideration of abundant codes in some cases allows us to improve the basic algebraic geometry bound.

Asymptotically good abundant codes. Define the following function (the *Pellikaan bound*):
$$R_{Pel}(\delta) = \begin{cases} 1 + \dfrac{1}{q+1} - \gamma_q - \delta & \text{if } 0 \leq \delta \leq \dfrac{1}{q+1}, \\ 1 - (q+1)\gamma_q \delta & \text{if } \dfrac{1}{q+1} \leq \delta \leq \dfrac{\sqrt{q}-1}{q+1}. \end{cases}$$

PROPOSITION 4.5.8. *If q is an even power of a prime, then for any δ in the interval*
$$0 \leq \delta \leq \frac{\sqrt{q}-1}{q+1}$$
there exists a sequence of abundant codes of growing length with relative minimum distance δ and rate at least $R_{Pel}(\delta)$.

PROOF. According to Theorem 4.2.27 and Proposition 4.2.25, if there exists a curve with N rational points over $\mathbb{F}_q$, then for all n and a such that $0 \leq a < N/(q+1)$ there exist abundant codes with
$$k \geq n + a + \frac{N}{q+1} - g, \qquad d \geq \frac{N}{q+1} - a.$$
Since $\gamma_q = 1/(\sqrt{q}-1)$, if we take $n = N$ and $0 \leq a < N/(q+1)$, we get the result for δ in the first interval; taking $a = 0$ and $0 < n < N$, we get the result for δ in the second interval. □

4.5. ASYMPTOTIC RESULTS

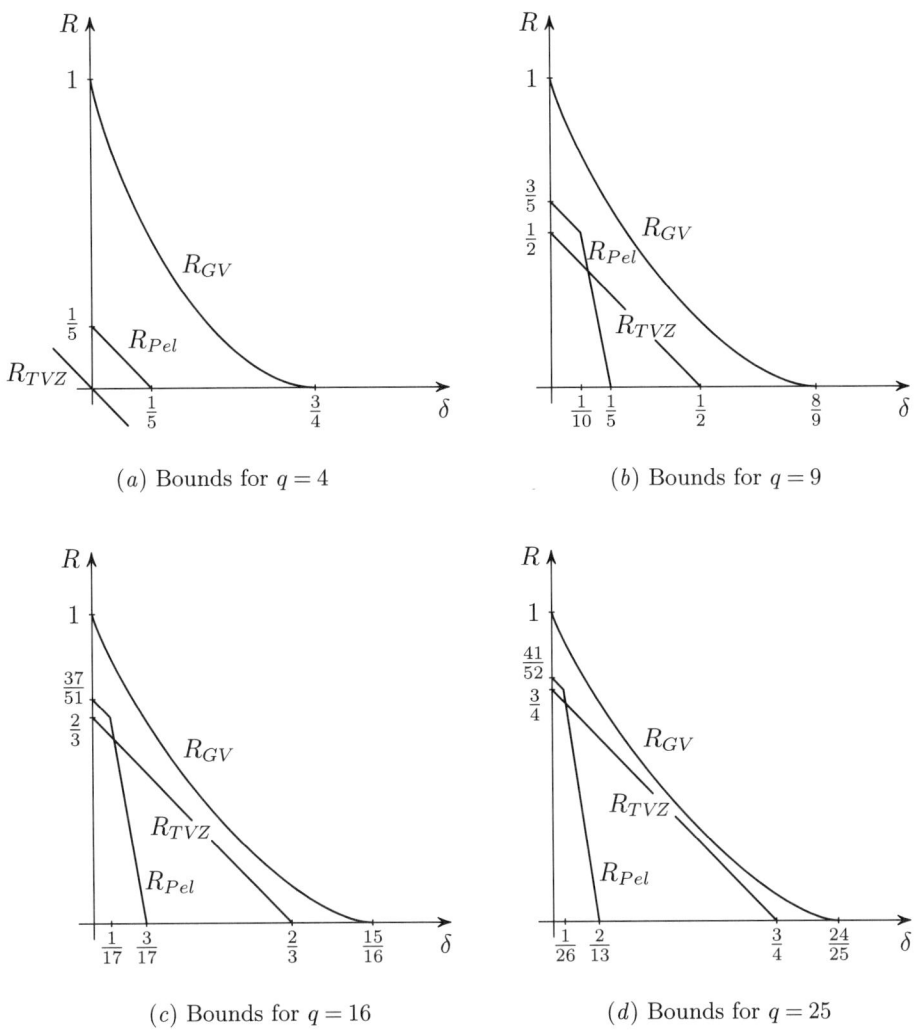

FIGURE 4.2. Lower bounds for $q = p^{2m} < 49$.

Figure 4.2 shows this bound for the cases $q = 4, 9, 16$, and 25. The Pellikaan bound is always below the Gilbert–Varshamov bound, but in the interval $0 \leq \delta < (q - \sqrt{q} + 2)^{-1}$ it is above the basic algebraic geometry bound.

4.5.2. Expurgation Bound and Codes with Many Light Vectors

The Gilbert–Varshamov bound (see Sec. 1.1.4 and 1.3.2) is proved by the expurgation method; i.e., desired objects (in this case, linear codes) are constructed step by step (say, choosing basis vectors one by one), applying random choice whenever possible. Quite to the contrary, the basic algebraic geometry bound is proved using direct algebraic geometry constructions.

Here we combine the two ideas. An algebraic geometry code is constructed from a curve X and a line bundle $\mathscr{L}$ on it. If the curve is fixed, the above estimate for the parameters (Theorem 4.1.13) depends only on $a = \deg \mathscr{L}$. It turns out that among linear bundles of a given degree one can often choose those for which the minimum distance of the corresponding code is much better than the a priori estimate given by Theorem 4.1.13. This is due to the fact that the number of bundles defining codes with a bounded minimum distance can be bounded from above, while the total number of bundles of a given degree is bounded from below using Proposition 3.2.14 (the expurgation method!).

In this subsection q is an even power of a prime. We consider codes constructed from all $\mathbb{F}_q$-points of X using the H-construction.

Let $\mathcal{P} \subseteq X(\mathbb{F}_q)$. Define
$$C_a(\mathcal{P}) = \{D \in \mathrm{Div}^+(X) \mid \deg D = a, \; \mathrm{Supp}\, D \cap X(\mathbb{F}_q) \subseteq \mathcal{P}\};$$
$$C_{a,r} = \{D \in \mathrm{Div}^+(X) \mid \deg D = a, \; |\mathrm{Supp}\, D \cap X(\mathbb{F}_q)| = r\}.$$

By $J_{a,d}$ we denote the set of line bundles $\mathscr{L} \in \mathrm{Pic}(X)$ defined over $\mathbb{F}_q$ such that $\deg \mathscr{L} = a$ and the minimum distance of $C = (X, X(\mathbb{F}_q), \mathscr{L})_H$ is precisely d.

LEMMA 4.5.9. *There exists an embedding of sets*
$$\alpha \colon J_{a,d} \hookrightarrow C_{a,n-d}.$$
In particular,
$$|J_{a,d}| \leq |C_{a,n-d}|.$$

PROOF. Let $\mathscr{L} \in J_{a,d}$. Then there is a section $s \in H^0(\mathscr{L})$ such that the divisor of its zeroes satisfies
$$|\mathrm{Supp}\, D_s \cap X(\mathbb{F}_q)| = n - d,$$
i.e., $D_s \in C_{a,n-d}$. Choose any s with this property and put $\alpha(\mathscr{L}) = D_s$. The divisor class D_s corresponds to the bundle $\mathscr{L}$; hence, the map α never glues nonisomorphic bundles together. □

By $J_a = J_{X,a}$ we denote the set of linear equivalence classes of $\mathbb{F}_q$-divisors of degree a on X.

LEMMA 4.5.10. *Consider a family of curves X_i with $g_i \to \infty$, $\lim(N_i/g_i) = A$. Then for any a we have*
$$\log_q |J_{X_i,a}| \geq g_i \left(1 + A \log_q \frac{q}{q-1}\right) - o(g_i).$$

PROOF. On any curve X having an $\mathbb{F}_q$-point there exist linear $\mathbb{F}_q$-bundles of any degree a. Fix one such bundle, $\mathscr{L}_0$. Then any other line bundle of degree a can be obtained as $\mathscr{L}_0 \otimes \mathscr{L}$, where $\deg \mathscr{L} = 0$; i.e., $\mathscr{L} \in J_X(\mathbb{F}_q)$, which defines an isomorphism $J_{X,a} \cong J_X(\mathbb{F}_q)$. It remains to apply Proposition 3.2.14. □

REMARK 4.5.11. It can be proved that for an asymptotically optimal family with $A = \sqrt{q} - 1$ the assertion of Lemma 4.5.10 holds with equality. This fact follows from much more precise formulae, which we are going to discuss in *Advanced Chapters*.

LEMMA 4.5.12. *Consider a family of asymptotically optimal curves X_i with $g_i \to \infty$, and let sequences of integers $\{a_i\}$ and $\{\ell_i\}$ be such that*
$$\liminf \frac{a_i - \ell_i}{g_i} > 0$$

and that there exists the limit

$$\lambda = \lim \frac{\ell_i}{g_i} > 0.$$

Then

$$\log_2 |C_{a_i,\ell_i}| \le N_i H_2(\ell_i/N_i) + f_{a_i,\ell_i} + o(g_i),$$

where

$$f_{a_i,\ell_i} = \begin{cases} a_i H_2(\ell_i/a_i) & \text{for } \ell_i/a_i > (q-1)/q, \\ a_i \log_2 q - \ell_i \log_2(q-1) & \text{for } \ell_i/a_i \le (q-1)/q. \end{cases}$$

PROOF. A divisor from $C_{a,\ell}$ is determined by a choice of a set of ℓ points $\{P_1, \ldots, P_\ell\}$ such that $\operatorname{Supp} D \cap X(\mathbb{F}_q) = \{P_1, \ldots, P_\ell\}$ and a choice of an effective divisor $D' = D - (P_1 + \ldots + P_\ell)$ of degree $a - \ell$; the divisor D' can be any one from the set $C_{a-\ell}(\{P_1, \ldots, P_\ell\})$. Therefore,

$$|C_{a,\ell}| = \binom{N}{\ell} d_{a,\ell}, \tag{4.5.2}$$

where the number $d_{a,\ell} = |C_{a,\ell}(\{P_1, \ldots, P_\ell\})|$ depends on a and ℓ only and does not depend on the set $\{P_1, \ldots, P_\ell\}$.

A divisor of degree $a - \ell$ is a sum of b_1 points of degree 1, b_2 points of degree 2, etc.; here,

$$b_1 + 2b_2 + 3b_3 + \ldots = a - \ell$$

(see Sec. 2.5.3). Points of degree 1 are chosen from the set $\{P_1, \ldots, P_\ell\}$; other points are taken arbitrarily.

Thus, if we fix a sequence $(b_1, b_2, \ldots, b_{a-\ell})$, then the number of divisors from $C_{a-\ell}(\{P_1, \ldots, P_\ell\})$ with given $(b_1, b_2, \ldots, b_{a-\ell})$ equals

$$\prod_{s=1}^{a-\ell} \binom{B_s + b_s - 1}{b_s}, \tag{4.5.3}$$

where for $s \ge 2$ by B_s we denote (as in Sec. 3.1.1) the number of points of degree s on X and $B_1 = \ell$. To estimate (4.5.3) from above, we can replace B_s with any number $t_s \ge B_s$. Take $t_1 = \ell$, and for $s = 2, \ldots, c$, where $c = \lfloor 2 \log_q g \rfloor$, take

$$t_s = \frac{8gq^{s/2}}{s(c+1-s)};$$

here, $t_s \ge B_s$ by Proposition 3.2.7. Finally, for $s \ge c+1$ take

$$t_s = \frac{4q^s}{s};$$

then

$$sB_s \le N_s = q^s + 1 - \sum_{i=1}^{2g} \omega_i^s \le q^s + 1 + 2gq^{s/2} < 4q^s = st_s$$

since $2\log_q g \le s$.

Thus,

$$d_{a,\ell} \le \sum_{(b_1,b_2,\ldots)} \prod_{s=1}^{a-\ell} \binom{t_s + b_s - 1}{b_s},$$

where the sum is taken over all sequences $(b_1, b_2, \ldots)$ with $b_1 + 2b_2 + 3b_3 + \ldots = a - \ell$. Hence, we have

$$\log_2(d_{a,\ell}) \leq \log_2(p(a-\ell)) + \max\left\{\sum_{s=1}^{a-\ell} \log_2 \binom{t_s + b_s - 1}{b_s}\right\}; \quad (4.5.4)$$

here the maximum is taken over the same sequences $(b_1, b_2, \ldots)$, and $p(n)$ is the *partition function* (number of partitions), the well-known number-theoretical function, satisfying the classical estimate

$$p(n) < e^{\pi\sqrt{2n/3}}.$$

Since $a \leq \text{const} \cdot g$, this implies the asymptotic estimate

$$\log_2(p(a-\ell)) = o(g).$$

Now let us estimate the maximum of the sum in (4.5.4).
By the Stirling formula (for $b_s \neq 0$), we have

$$\log_2 \binom{t_s + b_s - 1}{b_s} \leq \log_2 \binom{t_s + b_s}{b_s} \leq (t_s + b_s) H_2\left(\frac{b_s}{t_s + b_s}\right)$$

$$= t_s \log_2\left(1 + \frac{b_s}{t_s}\right) + b_s \log_2\left(1 + \frac{t_s}{b_s}\right).$$

Since all the zero b_s's add at most $a - \ell$ times 1, which amounts to $o(g)$, we get

$$\log_2 d_{a,\ell} \leq \max\left\{\sum_{s=1}^{a-\ell}\left(t_s \log_2\left(1 + \frac{b_s}{t_s}\right) + b_s \log_2\left(1 + \frac{t_s}{b_s}\right)\right)\right\} + o(g).$$

Disregarding the condition that b_s are integers, we obtain

$$\log_2 d_{a,\ell} \leq \max\left\{\sum_{s=1}^{a-\ell}\left(t_s \log_2\left(1 + \frac{x_s}{t_s}\right)\right.\right.$$

$$\left.\left. + x_s \log_2\left(1 + \frac{t_s}{x_s}\right)\right) \,\Big|\, \sum sx_s = a - \ell\right\} + o(g),$$

where x_s are continuous variables.

This expression can be estimated using the Lagrange method. Since the derivative of the maximized function with respect to x_s equals $\log_2\left(1 + \frac{t_s}{x_s}\right)$, we see that the maximum is attained for $x_s = \dfrac{t_s}{\mu^s - 1}$, where μ is given by the equation

$$\sum_{s=1}^{a-\ell} \frac{st_s}{\mu^s - 1} = a - \ell.$$

Note that $\mu > 1$; otherwise, the sum would be negative. The maximum value is

$$f_{a,\ell} = \sum_{s=1}^{a-\ell} t_s \log_2 \frac{\mu^s}{\mu^s - 1} + \sum_{s=1}^{a-\ell} x_s \log_2(\mu^s) = \sum_{s=1}^{a-\ell} t_s \log_2 \frac{\mu^s}{\mu^s - 1} + (a - \ell) \log_2 \mu.$$

On the other hand,

$$\frac{(a-\ell)t_{a-\ell}}{\mu^{a-\ell} - 1} = (a-\ell)x_{a-\ell} \leq a - \ell,$$

i.e., $t_{a-\ell} \leq \mu^{a-\ell} - 1$; hence,
$$\log_2 \mu \geq \log_2 q - o(1)$$
(we use the fact that $t_{a-\ell} = 4q^{a-\ell}/(a-\ell)$ since $a - \ell \geq \text{const} \cdot g > c$ by the condition of the lemma). Let us estimate the sum

$$\sum_{s=2}^{c} t_s \log_2 \frac{\mu^s}{\mu^s - 1} \leq \sum_{s=2}^{c} \frac{t_s}{(\ln 2)(\mu^s - 1)} = \frac{1}{\ln 2} \sum_{s=2}^{c} \frac{8gq^{s/2}}{s(c+1-s)(\mu^s - 1)}$$
$$\leq \frac{8g}{\ln 2} \sum_{s=2}^{c} \frac{q^{s/2}}{s(c+1-s)q^s} + o(g) \leq \frac{8g}{c \ln 2} \sum_{s=2}^{c} q^{-s/2} + o(g) \leq o(g)$$

and the sum

$$\sum_{s=c+1}^{a-\ell} t_s \log_2 \frac{\mu^s}{\mu^s - 1} \leq \sum_{s=c+1}^{a-\ell} \frac{4q^s}{s(\ln 2)(\mu^s - 1)}$$
$$\leq \frac{4}{c \ln 2} \sum_{s=c+1}^{a-\ell} (q/\mu)^s + o(g) \leq \text{const} \cdot \frac{a-\ell}{c} + o(g).$$

Substituting these estimates into the formula for $f_{a,\ell}$, we get

$$f_{a,\ell} = (a - \ell) \log_2 \mu + \ell \log_2 \frac{\mu}{\mu - 1} + o(g).$$

We know that $\log_2 \mu \geq \log_2 q - o(1)$. However, to compute $f_{a,\ell}$ we need more precise information on the value of μ. Consider two cases.

(a) Assume that $\mu \geq q(1 + \varepsilon)$ for some $\varepsilon > 0$. Then

$$\sum_{s=2}^{c} \frac{st_s}{\mu^s - 1} \leq \sum_{s=2}^{c} \frac{8gq^{s/2}s}{s(c+1-s)} \leq \frac{8g}{c} \sum_{s=2}^{c} s(\sqrt{q}/\mu)^s + o(g) = o(g)$$

since the series $\sum_{s=1}^{\infty} (\sqrt{q}/\mu)^s$ converges. We also have

$$\sum_{s=c+1}^{a-\ell} \frac{st_s}{\mu^s - 1} \leq \sum_{s=c+1}^{a-\ell} \frac{4q^s}{\mu^s - 1} = o(g).$$

Therefore,

$$\frac{\ell}{\mu - 1} = (a - \ell) - \sum_{s=2}^{a-\ell} \frac{st_s}{\mu^s - 1} = a - \ell - o(g),$$

i.e.,

$$\mu = \frac{a}{a - \ell} + o(g),$$

and we obtain

$$f_{a,\ell} = -(a - \ell) \log_2 \frac{a - \ell}{a} - \ell \log_2 \left(1 - \frac{a - \ell}{a}\right) + o(g) = aH_2(\ell/a) + o(g).$$

Note that in this case $\frac{a}{a - \ell} = \mu > q$, i.e.,

$$\frac{\ell}{a} > 1 - \frac{1}{q}.$$

(b) Now let $\mu \leq q$. In this case, since $\log_2 \mu \geq \log_2 q - o(1)$, we have
$$f_{a,\ell} = (a-\ell)\log_2 q + \ell \log_2 \frac{q}{q-1} + o(g) = a\log_2 q - \ell \log_2(q-1) + o(g).$$
Then, since $a - \ell \geq \dfrac{\ell}{\mu-1} = \dfrac{\ell}{q-1} + o(g)$, we get
$$\ell/a \leq 1 - \frac{1}{q}.$$
Note also that for $\ell/a = 1 - 1/q$, the formulae for cases (a) and (b) coincide.

Since (4.5.2) and the Stirling formula yield
$$\log_2 |C_{a,\ell}| = \log_2 \binom{N}{\ell} + \log_2 d_{a,\ell} \leq N H_2(\ell/N) + f_{a,\ell} + o(g),$$
the desired result follows. □

Based on the estimate for the total number of $\mathbb{F}_q$-divisors (Lemma 4.5.10) and the estimate for the number of divisors whose support includes a given number of $\mathbb{F}_q$-points (Lemmas 4.5.9 and 4.5.12), we are now able to prove the following lower bound theorem.

THEOREM 4.5.13. *Let q be an even power of a prime. Fix a family $\{X_i\}$ of curves of genus $g_i \to \infty$ over $\mathbb{F}_q$ such that $\lim \dfrac{|X_i(\mathbb{F}_q)|}{g_i} = \sqrt{q}-1$, and let $\beta_q(\delta)$ be the lower asymptotic bound for linear codes of the form $C = (X_i, X_i(\mathbb{F}_q), \mathscr{L})_H$.*

(a) *If $q < 49$, then*
$$\alpha_q^{\lin}(\delta) \geq \beta_q(\delta) \geq R_{GV}(\delta).$$

(b) *If $q \geq 49$, put $\gamma = (\sqrt{q}-1)^{-1}$, and let δ_1' and δ_4' be the two roots of the equation*
$$H_q(\delta) + \frac{q}{q-1}(1-\delta) = 1 + \gamma$$
and δ_2' and δ_3' be the two roots of
$$H_q(\delta) + (1-\delta)\log_q(q-1) = 1 + \gamma,$$
$$0 < \delta_1' < \delta_2' < \delta_3' < \delta_4' < \frac{q-1}{q}.$$
Then
$$\alpha_q^{\lin}(\delta) \geq \beta_q(\delta) \geq R_V(\delta),$$
where
$$R_V(\delta) = R_{GV}(\delta) \qquad \text{for } \delta \in [0, \delta_1'] \cup \left[\delta_4', \frac{q-1}{q}\right],$$
$$R_V(\delta) = R_{TVZ}(\delta) = 1 - \gamma - \delta \quad \text{for } \delta \in [\delta_2', \delta_3'],$$
and for $\delta \in [\delta_1', \delta_2'] \cup [\delta_3', \delta_4']$ the function $R_V(\delta) = R_V^0(\delta)$ is given by the implicit equation
$$(R_V^0(\delta) + \gamma) H_q\left(\frac{1-\delta}{R_V^0(\delta)+\gamma}\right) + H_q(\delta) = 1 + \gamma. \tag{4.5.5}$$

EXERCISE 4.5.14. Check that $R_V(\delta)$ is everywhere continuously differentiable, and the second derivative is discontinuous at δ_1', δ_2', δ_3', and δ_4'.

4.5. ASYMPTOTIC RESULTS

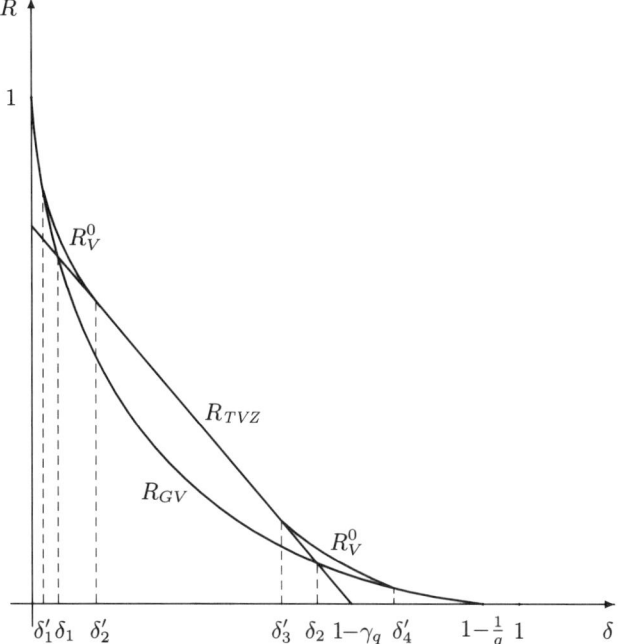

FIGURE 4.3. Lower asymptotic bounds for $q = p^{2m} \geq 49$: R_{GV} is the Gilbert–Varshamov bound, R_{TVZ} is the basic algebraic geometry bound, and R_V^0 is the expurgation bound (where it differs from R_{GV} and R_{TVZ}). For $q = 49$ we have $\delta_1' \approx 0.28$, $\delta_1 \approx 0.37$, $\delta_2' \approx 0.39$, $\delta_3' \approx 0.61$, $\delta_2 \approx 0.62$, $\delta_4' \approx 0.67$, $1 - \gamma_q \approx 0.83$, and $1 - 1/q \approx 0.98$. (The figure is not to scale.)

The function R_V is known as the *expurgation bound*, or the *Vlăduţ bound*. Figure 4.3 shows its behaviour for $q = p^{2m} \geq 49$; see also diagrams (Sec. A.2.2) and tables (Sec. A.2.4) in the Appendix.

PROOF OF THEOREM 4.5.13. As usual, $\gamma = (\sqrt{q} - 1)^{-1}$. Fix some $R \in (0,1)$ and a curve X of genus g from the family under consideration. Put

$$a = \lfloor (\gamma^{-1} R + 1) g \rfloor.$$

Then for the parameters of the code $(X, X(\mathbb{F}_q), \mathscr{L})_H$, where $\mathscr{L}$ is any line bundle of degree a, we have

$$k \geq a - g + 1 \geq \gamma^{-1} R g,$$
$$d \geq n - a \geq n - \gamma^{-1} R g - g.$$

Since the code length n, which is the number of $\mathbb{F}_q$-points of X, behaves (as $g \to \infty$) as $g(\sqrt{q} - 1)$, these inequalities asymptotically turn into

$$k/n \gtrsim R, \qquad \delta = d/n \gtrsim 1 - R - \gamma.$$

By the spoiling lemma (Lemma 1.1.39), we may assume that $k/n \sim R$, and in what follows we consider only the relative minimum distance $\delta = d/n$. Of course, for an arbitrary $\mathscr{L}$ we can obtain $\delta \sim 1 - R - \gamma$ (as in Corollary 4.5.3). Let us prove that it is possible to choose a line bundle $\mathscr{L}_0$ of degree a such that δ asymptotically meets the bound of the theorem. For $\delta \in (\delta_2', \delta_3')$, this is obvious.

For other values of δ we shall show that, for any $\varepsilon > 0$ and any value of g large enough, we have the inequality

$$\sum_{d \leq d_\varepsilon} |C_{a,n-d}| < |J_a|, \tag{4.5.6}$$

where $d_\varepsilon = n(\delta - \varepsilon)$, $\delta = R_V^{-1}(R)$. Indeed, by this inequality and Lemma 4.5.9, the number of line bundles of degree a such that the minimum distance of the corresponding code is at most d_ε is less than the total number of $\mathbb{F}_q$-bundles of degree a:

$$\sum_{d \leq d_\varepsilon} |J_{a,d}| \leq \sum_{d \leq d_\varepsilon} |C_{a,n-d}| < |J_a|.$$

Hence, there exists a bundle $\mathscr{L}_0$ with minimum distance greater than d_ε.

Note that $d_\varepsilon \lesssim \text{const} \cdot g$, and therefore

$$\log_q \left(\sum_{d \leq d_\varepsilon} |C_{a,n-d}| \right) \leq \log_q \left(\max_{d \leq d_\varepsilon} \{|C_{a,n-d}|\} \right) + o(g).$$

To prove inequalities (4.5.6), it suffices to show (use Lemma 4.5.10 and take the logarithm of (4.5.6)) that

$$\limsup \left(\frac{1}{g} \log_q \left(\max\{|C_{a,n-d}|\} \right) \right) < 1 + \gamma^{-1} \log_q \frac{q}{q-1}. \tag{4.5.7}$$

By Lemma 4.5.12, the left-hand side of (4.5.7) is at most

$$\max_{d \leq d_\varepsilon} \left\{ (\sqrt{q} - 1) H_2 \left(1 - \frac{d}{n} \right) + f_{a,n-d}/g \right\}.$$

Put

$$y = d/n, \quad f(y) = (\sqrt{q} - 1) H_2(y) + f_{a,n-d}/g.$$

Furthermore, since we are interested in the asymptotics, we may assume that $a/g = \gamma^{-1} R + 1$ (which holds up to $o(1)$). Then

$$f(y) = \gamma^{-1} H_2(y) + (\gamma^{-1} R + 1) H_2 \left(\frac{1-y}{R+\gamma} \right),$$

$$1 \geq \frac{n-d}{a} = \frac{1-y}{R+\gamma} \geq 1 - \frac{1}{q},$$

and

$$f(y) = \gamma^{-1} H_2(y) + (\gamma^{-1} R + 1) \log_2 q - \gamma^{-1}(1-y) \log_2(q-1),$$

$$\frac{1-y}{R+\gamma} \leq 1 - \frac{1}{q}.$$

Taking the derivative (recall that $H_2'(x) = \log_2 \frac{1-x}{x}$), we get

$$\gamma f'(y) = \log_2 \frac{1-y}{y} - \log_2 \frac{R+\gamma-1+y}{1-y}, \quad 1 \geq \frac{1-y}{R+\gamma} \geq 1 - \frac{1}{q},$$

and

$$\gamma f'(y) = \log_2 \frac{1-y}{y} + \log_2(q-1), \quad \frac{1-y}{R+\gamma} \leq 1 - \frac{1}{q}.$$

Note that in the first case $f'(y) > 0$ for
$$(1-y)^2 > y(R+\gamma-1+y),$$
i.e., for $y < \dfrac{1}{1+R+\gamma}$, but in this case we have
$$y \leq 1 - (R+\gamma)\frac{q-1}{q} < \frac{1}{1+R+\gamma}$$
(since $R+\gamma \geq \gamma > \dfrac{1}{q-1}$). In the second case, $f'(y) > 0$ for $y \leq \dfrac{q-1}{q}$. Hence, $f(y)$ is increasing on the whole interval $(0,(q-1)/q)$. Therefore,
$$\max_{d \leq d_\varepsilon} f(d/n) = f(d_\varepsilon/n) < f(\delta).$$
Thus, inequality (4.5.7) holds for
$$f(\delta) \leq 1 + \gamma^{-1}(1 - \log_2(q-1)). \tag{4.5.8}$$
For $1 \geq \dfrac{1-\delta}{R+\gamma} \geq 1 - \dfrac{1}{q}$, we have
$$f(\delta) - \left(1 + \gamma^{-1}(1 - \log_2(q-1))\right)$$
$$= \gamma^{-1}\left(H_2(\delta) + (R+\gamma)\log_2 q - (1-\delta)\log_2(q-1) - \gamma - 1 + \log_2(q-1)\right)$$
$$= \gamma^{-1}(\log_2 q)\left(H_q(\delta) + (R+\gamma)H_q\left(\frac{1-\delta}{R+\gamma}\right) - (1+\gamma)\right).$$

Thus, since $f(\delta)$ is monotone increasing, inequality (4.5.8) (and hence also (4.5.6)) is satisfied for all δ less than or equal to the solution of
$$H_q(\delta) + (R+\gamma)H_q\left(\frac{1-\delta}{R+\gamma}\right) = 1+\gamma \tag{4.5.9}$$
subject to the auxiliary condition
$$1 - (R+\gamma) \leq \delta \leq 1 - \frac{q-1}{q}(R+\gamma). \tag{4.5.10}$$
For $\dfrac{1-\delta}{R+\gamma} \leq 1 - \dfrac{1}{q}$, we get
$$f(\delta) - \left(1 + \gamma^{-1}(1 - \log_2(q-1))\right)$$
$$= \gamma^{-1}\left(H_2(\delta) + (R+\gamma)\log_2 q - (1-\delta)\log_2(q-1) - \gamma - 1 + \log_2(q-1)\right)$$
$$= \gamma^{-1}(\log_2 q)(H_q(\delta) + R - 1).$$
Thus, inequality (4.5.8) holds for all δ less than or equal to the solution of
$$R = 1 - H_q(\delta) \tag{4.5.11}$$
subject to the condition
$$\delta \geq 1 - \frac{q-1}{q}(R+\gamma). \tag{4.5.12}$$

Equation (4.5.11) defines the Gilbert–Varshamov bound.

Let us sum up our reasoning. First, we have shown that there exist line bundles of the desired degree on curves of our family that define codes with the given R and $\delta \geq 1 - (R+\gamma)$. Next, if the minimum distance is less than the bounds given by (4.5.9) and (4.5.10) and by (4.5.11) and (4.5.12), then we can choose a bundle such that the corresponding code has a better minimum distance. □

The curve (4.5.9) intersect the line $\delta = 1 - (R+\gamma)$ at δ_2' and δ_3' and intersects $\delta = 1 - \dfrac{q-1}{q}(R+\gamma)$ at δ_1' and δ_4' (the points δ_1', δ_2', δ_3', and δ_4' are defined in the statement of the theorem). Bound (4.5.5) takes place between these lines (then condition (4.5.10) is satisfied), i.e., for $\delta \in [\delta_1', \delta_2'] \cup [\delta_3', \delta_4']$. Above the second line, the Gilbert–Varshamov bound (4.5.11) takes place (condition (4.5.12) holds). One can easily check that bounds (4.5.9) and (4.5.11) meet at δ_1' and δ_4'.

For $q < 49$, i.e., for $q = 4, 9, 16, 25$, the line $\delta = 1 - \dfrac{q-1}{q}(R+\gamma)$ is everywhere below the curve (4.5.5) (see Theorem 4.5.4), and δ can be "pulled up" to the Gilbert–Varshamov bound for any R. For $q \geq 49$ both equations giving the values of δ_2' and δ_3' and δ_1' and δ_4' are solvable, and the bound is assembled from five pieces (see Fig. 4.3).

COROLLARY 4.5.15. *Let q be an even power of a prime. Then*
$$\alpha_q(\delta) \geq \alpha_q^{\mathrm{lin}}(\delta) \geq R_V(\delta). \qquad \square$$

The bound R_V is the best presently known lower bound for α_q^{lin} (but not for α_q; cf. Exercise 4.5.38 and Sec. 4.6.2).

EXERCISE 4.5.16. Prove that under the conditions of Theorem 4.5.13, parameters of the code $C = (X_i, X_i(\mathbb{F}_q), \mathscr{L})_H$ with probability 1 meet the bound $R_V(\delta)$ (it is useful to compare this result with Remark 1.3.22).

REMARK 4.5.17. Obviously, in Theorem 4.5.13 we may omit the assumption that q is an even power of a prime, replacing $\gamma = (\sqrt{q}-1)^{-1}$ with $\gamma = A(q)^{-1}$.

Codes with many light vectors. Methods used in deriving the expurgation bound can also be applied to prove the following important coding theory result (Theorem 4.5.18).

Define the function
$$E_q(\delta) = H_2(\delta) - \frac{\log_2 q}{\sqrt{q}-1} - \log_2 \frac{q}{q-1},$$
where $H_2(\delta)$ is the binary entropy function. The function $E_q(\delta)$ has two zeroes, $0 < \bar{\delta}_1(q) < \bar{\delta}_2(q) < (q-1)/q$, and is positive for $\delta \in (\bar{\delta}_1(q), \bar{\delta}_2(q))$.

Recall that by $A_r = A_r(C)$ we denote the number of vectors of weight r in a code C.

THEOREM 4.5.18. *For any $q = 2^{2s}$, $s = 3, 4, \ldots$, there exists a sequence of binary linear codes $\{C_n\}$ of length $n = qN$, $N \to \infty$, with minimum distance $d_n = n\delta/2$, $\delta \in \bigl(\bar{\delta}_1(2^{2s}), \bar{\delta}_2(2^{2s})\bigr)$, such that*
$$\log A_{d_n} \geq N E_q(\delta) - o(N).$$

To prove the theorem, let us first construct a sequence of linear q-ary (algebraic geometry) codes.

Let $q \geq 49$ be an even power of a prime; consider a sequence of curves X with $N \geq g(\sqrt{q}-1)$. Let $C = C(D) = (X, X(\mathbb{F}_q), \mathscr{O}(D))_H$.

THEOREM 4.5.19. *Let $\delta = (N-a)/N$ lie in the interval $(\bar{\delta}_1(q), \bar{\delta}_2(q))$. Then there exists a divisor $D \in \mathrm{Div}_a^+(X)$ such that the corresponding code $C = C(D)$ has*

minimum distance $d = N - a = \delta N$, and for the number A_d of vectors of weight d we have
$$\log A_d \geq N E_q(\delta) - o(N).$$

PROOF. Consider the canonical projection
$$\pi_a \colon \operatorname{Div}_a^+(X) \longrightarrow J_a(X).$$

Its fibres, i.e., inverse images of points of $J_a(X)$, are complete linear systems $|D| = \mathbb{P}(L(D))$ (cf. the proof of Lemma 3.1.8b). Therefore, for any $D \in \operatorname{Div}_a^+(X)$ the number of vectors of weight $d = N - a$ in $C(D)$ equals
$$A_d(D) = (q-1)\left|\pi_a^{-1}(\pi_a(D)) \cap C_{a,a}\right|;$$

here, as above,
$$C_{a,a} = \left\{ D \in \operatorname{Div}^+(X) \mid \deg D = a,\ |\operatorname{Supp} D \cap X(\mathbb{F}_q)| = a \right\}.$$

Hence,
$$\max\{A_d(D) \mid D \in \operatorname{Div}_a^+(X)\} \geq (q-1)\frac{|C_{a,a}|}{|J_a|}.$$

It is obvious that $|C_{a,a}| = \binom{N}{a}$, and therefore
$$\log |C_{a,a}| = N H_2\left(\frac{a}{N}\right) - o(g).$$

Furthermore (see Remark 4.5.11), we have the asymptotic estimate
$$\log_q |J_a| = g\left(1 + (\sqrt{q}-1)\log_q \frac{q}{q-1}\right) - o(g),$$

and the theorem follows. □

PROOF OF THEOREM 4.5.18. It remains to pass to binary codes. For $q = 2^{2s}$, take the binary linear $[n = q-1, n-2s, 3]$ Hamming code and consider its dual code. For simplicity, let us augment each vector with a zero coordinate; this results in a binary linear code S of length q, dimension $2s$, and minimum distance $q/2$ in which every nonzero vector is of weight $q/2$. Now let us establish a one-to-one correspondence between elements of $\mathbb{F}_q$ and vectors in S, and for each vector in a q-ary code $C = C(D)$ let us replace each coordinate by its image under this correspondence. We obtain a linear binary code C_n of length $n = qN$ with minimum distance $d_n = qN\delta/2$. Note that, upon passing to C_n, pairwise distances between vectors of C change by a factor of $q/2$, and thus vectors of weight d_n in C_n are obtained from vectors of weight d in C and only from them. It remains to apply Theorem 4.5.19. □

REMARK 4.5.20. The union of intervals $\left(\bar{\delta}_1(2^{2s}), \bar{\delta}_2(2^{2s})\right)$ of admissible values of δ from Theorem 4.5.18 over all s entirely covers the interval $(0, 1/2)$.

4.5.3. Constructive Bounds

Algebraic geometry codes make it possible to significantly improve the lower bounds for α_q^{pol} and $\alpha_q^{\text{pol,lin}}$ established in Sec. 1.3.4.

THEOREM 4.5.21. *If q is an even power of a prime, then*
$$\alpha_q^{\text{pol,lin}}(\delta) \geq R_{TVZ}(\delta) = 1 - (\sqrt{q}-1)^{-1} - \delta.$$
□

To prove the theorem, it suffices to present a polynomial algorithm for construction of algebraic geometry codes from a family of curves X_i of genus $g_i \to \infty$ over $\mathbb{F}_q$ with
$$\limsup \frac{|X_i(\mathbb{F}_q)|}{g_i} = \sqrt{q}-1.$$
Such algorithms will be described in *Advanced Chapters*. Note that since R_{TVZ} for large q partly lies above R_{GV}, Theorem 4.5.21 obviously improves the polynomial bound R_{BZ} for some values of δ.

Let us now apply some constructions of Sec. 1.2.3.

Concatenation bound. Applying Corollary 1.3.50 to the bound R_{TVZ}, we obtain the following result, which is valid for any q.

THEOREM 4.5.22. *Define*
$$R_{KTV(\mathrm{lin})}(\delta) = \max\left\{\left(1-(q^{k/2}-1)^{-1}\right)\frac{k}{n} - \frac{k}{d}\delta\right\},$$
where the maximum is taken over all linear q-ary $[n,k,d]_q$ codes such that if q is an odd power of a prime, then k is even. Then
$$\alpha_q^{\mathrm{pol,lin}}(\delta) \geq R_{KTV(\mathrm{lin})}(\delta). \qquad \square$$

EXERCISE 4.5.23. Prove the theorem.

Unfortunately, we do not know the precise value of $R_{KTV(\mathrm{lin})}(\delta)$, the *Katsman–Tsfasman–Vlăduţ bound*. The reason is that we do not know the parameters of all good q-ary codes. However, each particular code from the set described in the theorem gives a line segment
$$R = \left(1-(q^{k/2}-1)^{-1}\right)\frac{k}{n} - \frac{k}{d}\delta, \qquad (4.5.13)$$
which is a lower bound for $R_{KTV(\mathrm{lin})}$ and $\alpha_q^{\mathrm{pol,lin}}$.

If we have a set of such codes, then the envelope line of the corresponding segments becomes a lower estimate. Let us study the behaviour of the bound $R_{KTV(\mathrm{lin})}$ at the endpoints of the interval.

PROPOSITION 4.5.24. *For $\delta \to 0$, the bound $R_{KTV(\mathrm{lin})}(\delta)$ behaves as follows:*
$$1 - R \sim -2\delta \log_q \delta.$$
For $x = \left(\frac{q-1}{q}-\delta\right) \to 0$, the bound behaves as follows:
$$R \sim \begin{cases} -2(\sqrt{q}-1)^4(\sqrt{q}^3-1)^{-3}x^3\log_q x & \text{for } q=p^{2m}, \\ -2q^6(q+1)^2(q^3-1)^{-3}x^3\log_q x & \text{for } q=p^{2m+1}. \end{cases} \qquad \square$$

EXERCISE 4.5.25. Prove the proposition. (*Hint*: Estimate the asymptotic behaviour of nodes of the polygonal line defined as the maximum of segments (4.5.13) that correspond to trivial $[\ell,\ell,1]_q$ or $[2\ell,2\ell,1]_q$ codes and to affine $[q^m,m+1,q^m-q^{m-1}]_q$ or $[q^{2m-1},2m,q^{2m-1}-q^{2m-2}]_q$ first-order Reed–Muller codes. The difficult part is to show that no set of codes gives better asymptotics.)

REMARK 4.5.26. For comparison, recall the behaviour of the Gilbert–Varshamov bound: at the upper endpoint (for $\delta \to 0$), we have
$$1 - R_{GV}(\delta) \sim -\delta \log_q \delta,$$
and at the lower endpoint (for $x \to 0$),
$$R_{GV}\left(\frac{q-1}{q} - x\right) \sim \frac{q^2}{2(q-1)\ln q} x^2;$$
and the behaviour of the Blokh–Zyablov bound:
$$1 - R_{BZ}(\delta) \sim \frac{\ln q}{2} \delta (\log_q \delta)^2$$
for $\delta \to 0$, and
$$R_{BZ}\left(\frac{q-1}{q} - x\right) \sim \frac{q^3}{6(q-1)^2 \ln q} x^3$$
for $x \to 0$ (see Exercises 1.3.16 and 1.3.48 and tables in Sec. A.2.4).

THEOREM 4.5.27. *For any q (which is a power of a prime) and for any δ in the interval $0 < \delta < \dfrac{q-1}{q}$, we have*
$$R_{KTV(\mathrm{lin})}(\delta) > R_{BZ}(\delta).$$

SKETCH OF THE PROOF. We do not give a detailed proof here since it involves cumbersome computations. The idea is as follows.

Consider the polygonal line composed of segments (4.5.13) for codes indicated in Exercise 4.5.25. Careful estimation of the difference between the nodes of this polygon from the asymptotics of Proposition 4.5.24, as well as the difference between R_{BZ} from the asymptotics of Remark 4.5.26, proves the theorem in the intervals $(0, \alpha)$ and $(\beta, (q-1)/q)$ for some explicitly computable α and β. For q large enough, we get $\beta < \alpha$. For small q, we should add segments (4.5.13) for several other good codes. To check that the polygonal line lies above the Blokh–Zyablov curve R_{BZ}, it suffices to show that all nodes of the polygon are above R_{BZ} (since R_{BZ} is concave). Far from the endpoints of the interval $\left(0, \dfrac{q-1}{q}\right)$ this can be verified by direct computation using monotonicity of R_{BZ} and tables of its values. The constructed polygon lies "far above" the bound R_{BZ}, so estimates can be rather rough. □

EXERCISE 4.5.28. Give a detailed proof.

For $\delta \to 0$, the bound $R_{KTV(\mathrm{lin})}$ is half as good as R_{GV}. Now we are going to construct another bound, which behaves nicely for small values of δ.

Restriction bound. Let us now make use of field restriction.

THEOREM 4.5.29. *Define*
$$R_{KT}(\delta) = \max_m \left\{ 1 - \frac{2m(q-1)}{q(q^{m/2} - 1)} - \frac{m(q-1)}{q}\delta \right\},$$
where the maximum is taken over all integers $m \geq 1$ such that q^m is a square. Then
$$\alpha_q^{\mathrm{pol,lin}}(\delta) \geq R_{KT}(\delta).$$

This bound is known as the *restriction bound*, or the *Katsman–Tsfasman bound*.

SKETCH OF THE PROOF. Let q^m be an even power of a prime. Consider a family of curves X of growing genus g over $\mathbb{F}_{q^m}$ such that the ratio of the number of $\mathbb{F}_{q^m}$-points to the genus is asymptotically optimal, i.e., $\limsup \dfrac{N}{g} = q^{m/2} - 1$.

We shall need a point over the extension field $\mathbb{F}_{q^{mb}}$ of $\mathbb{F}_{q^m}$ which is not defined over any smaller field:

EXERCISE 4.5.30. Prove that for $b \geq g$ there always exists an $\mathbb{F}_{q^{mb}}$-point on X which is not defined over any smaller field. (*Hint*: Use the fact that
$$\left||X(\mathbb{F}_{q^{mb}})| - q^{mb} - 1\right| \leq 2gq^{mb/2};$$
see inequality (3.1.6), p. 141.)

Let $b \geq g$. Fix such a point, and let $Q_1, \ldots, Q_b$ be the set of its conjugates over $\mathbb{F}_{q^m}$. Set $D_1 = \sum Q_i$ and $\mathcal{P} = X(\mathbb{F}_{q^m})$. Up to spoiling, Theorem 4.2.40 yields an
$$[N, N - m(q-1)b - 1, qb - 2g + 2]_q \text{ code.}$$
Asymptotically, we get a family of codes with parameters
$$R \sim 1 - m(q-1)\beta, \qquad \delta \sim q\beta - 2\gamma,$$
where $\gamma = (q^{m/2} - 1)^{-1}$ and $\beta = \liminf(b/N) \geq \gamma$. The latter inequality is not essential: for $b < g$, the a priori estimate for the code dimension, $k \geq N - m(qb - g + 1)$, is better than that of Theorem 4.2.40.

Appropriately choosing β (depending on δ) and noting that restriction is a polynomial operation and that we restrict algebraic geometry codes with polynomial construction, we obtain
$$\alpha_q^{\text{pol,lin}}(\delta) \geq 1 - \frac{2m(q-1)}{q(q^{m/2}-1)} - \frac{m(q-1)}{q}\delta. \qquad \square$$

THEOREM 4.5.31. *For $\delta \to 0$, the bound $R_{KT}(\delta)$ behaves as*
$$R \sim 1 + 2\frac{q-1}{q}\delta \log_q \delta.$$
In particular, for $q = 2$ this bound behaves as $\delta \to 0$ precisely in the same way as the Gilbert–Varshamov bound: $R \sim 1 + \delta \log_2 \delta$. $\square$

EXERCISE 4.5.32. Give a detailed proof of Theorem 4.5.29 and check the statement of Theorem 4.5.31.

4.5.4. Other Bounds

Algebraic geometry codes lead to improvements for a number of other asymptotic bounds.

Nonlinear codes. Since concatenation can be applied to nonlinear codes as well, Theorem 4.5.22 has the following obvious analogue.

THEOREM 4.5.33. *Define*
$$R_{KTV}(\delta) = \max\left\{\left(1 - (q^{k/2} - 1)^{-1}\right)\frac{k}{n} - \frac{k}{d}\delta\right\},$$

where the maximum is taken over all (not necessarily linear) $[n,k,d]_q$ codes such that k is a natural number and q^k is a square. Then

$$\alpha_q^{\mathrm{pol}} \geq R_{KTV}(\delta).\qquad\square$$

This theorem makes sense since (for a fixed finite length) there do exist nonlinear codes whose parameters are better than those of linear codes. For instance, as one can check, for $q=2$ the inequality $R_{KTV}(\delta) > R_{KTV(\mathrm{lin})}(\delta)$ holds in the interval $\delta \in \left(\frac{197}{1225}; \frac{7}{25}\right) \approx (0.1608; 0.2800)$. Moreover, outside this interval, most probably, we have $R_{KTV}(\delta) = R_{KTV(\mathrm{lin})}(\delta)$, though we cannot prove this strictly. Numerical values for R_{KTV} are given in tables of Sec. A.2.4.

Now we are going to apply nonlinear constructions of Sec. 1.2.3; below, in Theorems and Exercises 4.5.34–4.5.38, the alphabet cardinality q is arbitrary (not necessarily a power of a prime).

THEOREM 4.5.34. *Define*

$$R_{LT}^0(\delta) = \max\left\{ \left(\log_M p^{2m}\right) \left(\frac{(p^m-2)2m}{(p^m-1)n} - \frac{2m}{d}\delta \right) \right\},$$

where the maximum is taken over all $[n,k,d]_{p^{2m}}$ codes, and a prime p and natural m satisfy $M = q^k \geq p^{2m}$. Then

$$\alpha_q(\delta) \geq \alpha_q^{\mathrm{pol}}(\delta) \geq R_{LT}^0(\delta).\qquad\square$$

EXERCISE 4.5.35. *Prove the theorem.* (*Hint*: Apply Exercise 1.3.52 on the alphabet extension to R_{KTV} for $\alpha_{p^{2m}}^{\mathrm{pol}}$.)

THEOREM 4.5.36. *Let $q \geq 46$. Then for δ ranging in some interval, we have the strict inequality*

$$\alpha_q^{\mathrm{pol}}(\delta) > R_{GV}(\delta).$$

PROOF. Apply the part of Theorem 1.3.24 concerning alphabet restriction to the basic algebraic geometry bound

$$\alpha_{q'}(\delta) \geq 1 - \frac{1}{\sqrt{q'}-1} - \delta,$$

where $q' \geq q$, and $q' = p^{2m}$ is an even power of a prime. Then we get

$$\alpha_q^{\mathrm{pol}}(\delta) \geq R_{LZ}(\delta) = 1 - \left((\sqrt{q'}-1)^{-1} + \delta\right)\log_q(q').$$

The obtained *Litsyn–Zinoviev bound* is a line segment, and we can compare it with the Gilbert–Varshamov bound precisely as in the proof of Theorem 4.5.4.

According to Exercise 1.3.19, the tangent to the curve $R_{GV}(\delta)$ passing at the angle of $-\log_q(q')$ touches $R_{GV}(\delta)$ at $\delta_0 = (q-1)/(q'+q-1)$. At this point we have

$$R_{GV}(\delta_0) = 1 + \frac{q'}{q'+q-1}\log_q(q') - \log_q(q'+q-1),$$

$$R_{LZ}(\delta_0) = 1 - \left(\frac{1}{\sqrt{q'}-1} + \frac{q-1}{q'+q-1}\right)\log_q(q').$$

The curve $R_{GV}(\delta)$ and the line $R_{LZ}(\delta)$ intersect at two points if and only if $R_{LZ}(\delta_0) > R_{GV}(\delta_0)$, which is easily seen to be equivalent to

$$p^{2m} + q - 1 > p^{2mp^m/(p^m-1)};$$

i.e.,
$$q' = p^{2m} \geq q > p^{2mp^m/(p^m-1)} - p^{2m} + 1. \qquad (4.5.14)$$

To prove the theorem for a given q, it suffices to show that q belongs to one of the intervals (4.5.14).

Let $p = 2$; then (4.5.14) takes the form
$$2^{2m} \geq q > 2^{2m2^m/(2^m-1)} - 2^{2m} + 1 = 2^{2m}(2^{2m/(2^m-1)} - 1) + 1,$$
and the intervals corresponding to m and $m+1$ overlap if
$$2^{2(m+1)/(2^{m+1}-1)} < \frac{5}{4},$$
i.e., for $m \geq 5$. The left-hand endpoint of the interval corresponding to $m = 5$ is less than 258; i.e., for $q \geq 258$ the theorem is proved.

Now let us successively put $p = 17, 13, 11$; for $m = 1$ we cover the intervals $[124, 289]$, $[91, 169]$, and $[76, 121]$. For $p = 3$ and $m = 2$, $p = 2$ and $m = 3$, and $p = 7$ and $m = 1$, we get the intervals $[61, 81]$, $[53, 64]$, and $[46, 49]$.

It remains to prove the theorem for $q = 50, 51$, and 52. Here we apply the alphabet restriction theorem (Theorem 1.3.24) to the linear algebraic geometry bound (for $p^{2m} \leq q$), which results in
$$\alpha_q^{\text{pol}} > \overline{R}_{LZ}(\delta) = \left(1 - \frac{1}{p^m - 1} - \delta\right) \log_q(p^{2m}).$$

The bounds R_{GV} and $\overline{R}_{LZ}$ are compared similarly. By Exercise 1.3.19, the tangent at the angle of $-\log_q p^{2m}$ touches R_{GV} at $\delta_0 = \dfrac{q-1}{p^{2m} + q - 1}$, and
$$R_{GV}(\delta_0) = 1 + \frac{p^{2m} \log_q(p^{2m})}{p^{2m} + q - 1} - \log_q(p^{2m} + q - 1),$$
$$\overline{R}_{LZ}(\delta_0) = \left(1 - \frac{1}{p^m - 1} - \frac{q-1}{p^{2m} + q - 1}\right) \log_q(p^{2m}).$$

The condition $\overline{R}_{LZ}(\delta_0) > R_{GV}(\delta_0)$ is equivalent to
$$\frac{p^{2m} - 1}{p^{2m/(p^m-1)} - 1} > q \geq p^{2m}.$$

With $p = 7$ and $m = 1$, this inequality is satisfied for $q = 50, 51$, and 52. $\square$

Finally, let us present one more simple theorem, bounding $\alpha_q(\delta)$ in the nonlinear case.

THEOREM 4.5.37. *For any alphabet cardinality q, we have*
$$\alpha_q(\delta) \geq R_{LT}(\delta) = \max\{R'_{LT}(\delta), R''_{LT}(\delta)\},$$
where
$$R'_{LT}(\delta) = \max\left\{1 - \left(1 - R_V^{(q')}(\delta)\right) \log_q(q')\right\};$$
here the maximum is taken over all even powers of primes $q' = p^{2m} \geq q$ (here $R_V^{(q')}(\delta)$ is the expurgation bound for q'-ary codes), and
$$R''_{LT}(\delta) = \max\left\{R_V^{(q')}(\delta) \log_q(q')\right\},$$
where the maximum is taken over $q' = p^{2m} \leq q$.

PROOF. It suffices to apply Theorem 1.3.24 to the expurgation bound. □

This result, the *Litsyn–Tsfasman bound*, is of interest even in the case where q is itself an even power of a prime (this is by no means obvious, since if in the statement of the theorem we change the expurgation bound by the Gilbert–Varshamov bound, we get the same Gilbert–Varshamov bound).

EXERCISE 4.5.38. Check that for $q = 256^2$ there exists a value δ_0 such that $R_{LT}(\delta_0) > R_V(\delta_0)$. (*Hint:* Take $\delta_0 = 0.0146$, compute $R_{TVZ}(\delta_0)$, then estimate $R_{LT}(\delta_0)$ noting that
$$R_{LT}(\delta_0) \geq 1 - (1 - R_{TVZ}(\delta_0))\log_{256} 263,$$
and show that $\delta_0 > \delta_2$, where δ_2 is the smallest root of
$$H_q(\delta) + (1-\delta)\log_q(q-1) = 1 + \gamma_q$$
for $q = 256^2$.)

It is likely that such a δ_0 exists for any q large enough. On the other hand, the bounds $R_{LT}(\delta)$ and $R_V(\delta)$ most likely coincide for small q (say, $q \leq 256$).

Self-dual codes. Let us consider the asymptotic problem for self-dual (or quasi-self-dual) codes. For such codes, we always have $R = 1/2$, and the question is on the value of
$$\delta_q^{\mathrm{sd}} = \limsup \delta$$
taken over all such codes (respectively, δ_q^{qsd} for quasi-self-dual codes, and also $\delta_q^{\mathrm{sd,pol}}$ and $\delta_q^{\mathrm{qsd,pol}}$ for polynomial families).

It is known that there are self-dual codes meeting the Gilbert–Varshamov bound. Therefore, there are questions on asymptotic improvement of this bound and on its polynomiality. For $q \geq 49$, we would like to have an answer to the following question:

PROBLEM 4.5.39. Do there exist quasi-self-dual algebraic geometry codes meeting the basic algebraic geometry bound? (If this is the case, then $\delta_q^{\mathrm{qsd,pol}} \geq \frac{1}{2} - \frac{1}{\sqrt{q}-1}$).

Here is a somewhat weaker bound:

THEOREM 4.5.40. *If q is an even power of a prime, then*
$$\delta_q^{\mathrm{qsd}} \geq \delta_{Sch} = \frac{1}{2} - \frac{1}{\sqrt{q}-3}.$$
If q is an even power of 2, then
$$\delta_q^{\mathrm{sd}} \geq \delta_{Sch}. \qquad \square$$

EXERCISE 4.5.41. Prove the theorem. (*Hint:* Apply Corollary 4.1.32 for a family of curves with $N/g \sim \sqrt{q}-1$).

EXERCISE 4.5.42. Show that for $q = p^{2m} \geq 121$ (respectively, for $q = 2^{2m} \geq 256$) the obtained *Scharlau bound* for δ_q^{qsd} (respectively, for δ_q^{sd}) is better than the Gilbert–Varshamov bound.

Note that the method of Theorem 4.1.31 and Corollary 4.1.32 says nothing about $\delta^{\mathrm{qsd,pol}}$ and $\delta^{\mathrm{sd,pol}}$.

Constant-weight codes. Recall that a nonlinear code $C \subset \mathbb{F}_q^n$ is called a *constant-weight* code of weight w if $\|v\| = w$ for any $v \in C$.

EXERCISE 4.5.43. Let $n \to \infty$, $w/n \to \omega$. Define $\alpha_q(\omega, \delta)$. Show that
$$\alpha_2(\omega, \delta) \geq R_G(\omega, \delta) = H_2(\omega) - \omega H_2(\delta/2\omega) - (1-\omega)H_2(\delta/(2-2\omega)),$$
and extend this inequality (the "Gilbert bound") to the case of $\alpha_q(\omega, \delta)$.

Here is a simple construction of constant-weight codes. Let C be an arbitrary $[n, k, d]_q$ code. Enumerate the elements of $\mathbb{F}_q$ by integers $1, 2, \ldots, q$, and map the ith element to the binary vector with 1 at the ith position and zeroes at the other positions. Then map each vector $x \in \mathbb{F}_q^n$ to the vector $\varphi(x) \in \mathbb{F}_2^{qn}$ obtained from x by the above change of each coordinate.

EXERCISE 4.5.44. Prove that the image $\varphi(C) \subseteq \mathbb{F}_2^{qn}$ has the same cardinality q^k and minimum distance $2d$ and that it is a constant-weight code of weight n.

THEOREM 4.5.45. *Let $\omega = p^{-2m}$. Then*
$$\alpha_2(\omega, \delta) \geq R_{EZ}(\omega, \delta) = -\left(\omega - \frac{\delta}{2} - \frac{\omega\sqrt{\omega}}{1 - \sqrt{\omega}}\right)\log_2 \omega. \qquad \square$$

EXERCISE 4.5.46. Prove the theorem. (*Hint*: Apply Exercise 4.5.44 to algebraic geometry codes). Show that, for ω small enough, the bound of Theorem 4.5.45 is better than that of Exercise 4.5.43 for values of δ in some interval.

PROBLEM 4.5.47. Generalize the *Ericson–Zinoviev bound* of Theorem 4.5.45 to the case of $\alpha_q(\omega, \delta)$ for arbitrary q.

4.6. Nonlinear Algebraic Geometry Constructions

In this section we describe two recent wonderful constructions of nonlinear codes from algebraic curves. The codes thus obtained in a certain (well defined) sense are strictly better than algebraic geometry codes we studied above (except for the fact that they are nonlinear). The Elkies construction instead of struggling against the fact that the value of a function can be infinite—what we did before—uses the infinity to its advantage as an extra symbol of the alphabet. Xing's construction cleverly combines two maps, that of "evaluation at a point" with that of "evaluation of the derivative". For nonlinear codes this permits the basic algebraic geometry bound to slightly improve on the whole interval.

4.6.1. Elkies Codes

In this subsection we describe a recent new construction of nonlinear codes on algebraic curves due to N. Elkies. First we present the construction and some simple properties of these codes, then discuss bounds on their parameters, compare the Elkies codes with algebraic geometry codes, and finally consider the case of Elkies codes on $\mathbb{P}^1$.

Throughout this subsection, X denotes a smooth absolutely irreducible projective curve over $\mathbb{F}_q$.

Construction. The idea of the construction is most easily understood in the case $X = \mathbb{P}^1$, i.e., the case where algebraic geometry codes are merely Reed–Solomon codes (see Example 4.1.5). Recall that in this case a code is the image of the map

$$\text{Ev}_{\mathcal{P}} \colon L(m) \to \mathbb{F}_q^N, \quad f \mapsto (f(P_1), \ldots, f(P_N)),$$

where $\mathcal{P} = \{P_1, \ldots, P_N\} \subseteq \mathbb{P}^1 \setminus \{\infty\}$, and $L(m)$ is the space of polynomials of degree at most m. Now, instead of the space $L(m)$, consider the space $L_0(m)$ of rational functions of degree at most m (the degree of a rational function $f = p/q$ equals $\max\{\deg p, \deg q\}$ for an irreducible fraction p/q). Then $\text{Ev}_{\mathcal{P}}(f)$ is indefinite for a function f having a pole at any point P_i. Let us assume that a function having a pole at a point takes the value ∞ at this point; in other words, let us extend the alphabet and consider the map

$$\widetilde{\text{Ev}}_{\mathcal{P}} \colon L_0(m) \longrightarrow (\mathbb{F}_q \cup \{\infty\})^N,$$

where $\mathcal{P} \subseteq \mathbb{P}^1$. The image of this map is a (nonlinear) code $E_0(m)$ of length N over the alphabet of $q+1$ symbols. This code is called the *Elkies code on a projective line*. Its minimum distance is easily estimated: since $\deg(f_1 - f_2) \leq \deg f_1 + \deg f_2 \leq 2m$, the equality $f_1(P) = f_2(P)$ can hold for at most $2m$ points P, whence $d \geq N - 2m$. In particular, for $m < N/2$ the map $\widetilde{\text{Ev}}_{\mathcal{P}}$ is injective. In the following we always assume that $m < N/2$.

Now let us describe the construction for an arbitrary curve X. For a given divisor D of degree zero and a number $m < N/2$, define the set $L_D(m)$ of rational functions as follows:

$$L_D(m) = \{0\} \cup \{f \in \mathbb{F}_q(X)^* \mid (f) + D \text{ can be represented as}$$
$$\text{the difference of two effective divisors of degree} \leq m\}.$$

We define the map $\widetilde{\mathrm{Ev}}_{\mathcal{P}}$, $\mathcal{P} \subset X$, on the set $L_D(m)$ as follows: for each point $P_i \in \mathcal{P}$ we choose a rational function φ_i whose order P_i equals the order of D at this point, and set

$$\widetilde{\mathrm{Ev}}_{\mathcal{P}}(f) = \big((\varphi_1 f)(P_1), \ldots, (\varphi_N f)(P_N)\big) \in (\mathbb{F}_q \cup \{\infty\})^N. \qquad (4.6.1)$$

The image of this map is a nonlinear *Elkies code* $E_D(m)$. In particular, for $D = 0$ and $\varphi_i = 1$ we obtain the code $E_0(m)$ described above.

Note that in fact the construction (up to equivalence) does not depend on the choice of a function φ_i:

LEMMA 4.6.1. *Different choices of the functions φ_i in (4.6.1) yield equivalent codes.*

PROOF. Let ψ_i be any other choice. Put $\theta_i = \psi_i/\varphi_i$. Then θ_i is a rational function on X with neither pole nor zero at P_i. Hence, if we use ψ_i instead of φ_i in (4.6.1), each coordinate will be multiplied by the nonzero constant $\theta_i(P_i)$, which corresponds to a permutation of the alphabet in each coordinate; i.e., we obtain an equivalent code. □

Moreover, in fact the construction does not depend on the divisor D but depends only on its linear equivalence class:

LEMMA 4.6.2. *If D and D' are linearly equivalent divisors of degree zero, then the codes $E_D(m)$ and $E_{D'}(m)$ are isomorphic.*

PROOF. Let D and D' differ by a divisor of a function g; i.e., $D' - D = (g)$. It is easily seen that $f \in L_{D'}(m)$ if and only if $fg \in L_D(m)$. This defines a one-to-one correspondence between the sets $L_{D'}(m)$ and $L_D(m)$. Next, if φ_i are chosen for D in (4.6.1), we may choose the functions $\varphi_i' = g\varphi_i$ for D'. Then the image of f as an element of $L_{D'}(m)$ and the image of fg as an element of $L_D(m)$ under a map $\widetilde{\mathrm{Ev}}_{\mathcal{P}}$ coincide. This defines an isomorphism of the codes. □

Thus, we may consider D as an element of the Jacobian $J_X(\mathbb{F}_q)$ of the curve X.

Automorphisms. Let us make some remarks on automorphisms of Elkies codes.

First, any nonzero $\theta \in \mathbb{F}_q^*$ defines an automorphism $f \mapsto \theta f$ of the set $L_D(m)$. Thus, the multiplicative group $\mathbb{F}_q^*$ acts on $E_D(m)$. For arbitrary X, D, and m we expect that this is the full automorphism group. Quite the contrary, algebraic geometry codes (as any linear codes) have many more automorphisms (translation by any codeword, as well as multiplication by a scalar). However, as well as algebraic geometry codes, Elkies codes can inherit more symmetries from automorphisms of X and/or $\mathbb{F}_q$. Thus, for instance, if X has an automorphism taking D to a linearly equivalent divisor, then by Lemma 4.6.2 it induces an automorphism of $E_D(m)$. In particular, any automorphism of a curve acts on $E_0(m)$. Likewise, if X and D are defined over a subfield $\Bbbk \subset \mathbb{F}_q$, then the Galois group $\mathrm{Gal}(\mathbb{F}_q/\Bbbk)$ acts on $E_D(m)$.

Finally, the group $\mathrm{PGL}(2, \mathbb{F}_q)$ of fractional linear transformations of the projective line $\mathbb{P}^1 = \mathbb{F}_q \cup \{\infty\}$ also acts on $E_0(m)$. Indeed, any element $\gamma \in \mathrm{PGL}(2, \mathbb{F}_q)$ defines the automorphism $f \mapsto \gamma \circ f$ of $L_0(m)$. These automorphisms have no algebraic geometry analogues.

Minimum distance and cardinality. It is easy to bound the minimum distance:

PROPOSITION 4.6.3. *Let D be a divisor of degree zero on a curve X over $\mathbb{F}_q$. Let f_1 and f_2 be distinct elements of $L_D(m)$ of degrees m_1 and m_2 respectively. Then the codewords corresponding to f_1 and f_2 coincide at no more than m_1+m_2 positions. In particular, the minimum distance of $E_D(m)$ is at least $N-2m$.*

PROOF. We may assume that both f_j are nonzero. Consider the divisors $E_j = (f_j) + D$ and their decompositions $E_j = E_j^+ - E_j^-$ (where $E_j^+, E_j^- \geq 0$). By the definition of $L_D(m)$, the degrees $\deg E_j^+$ and $\deg E_j^-$ are at most m_j. Consider the nonzero rational function $f = f_1 - f_2$ on X. If the codewords corresponding to f_1 and f_2 coincide at the ith coordinate, then P_i is either a zero of $\varphi_i f$ or a pole of both $\varphi_i f_1$ and $\varphi_i f_2$. Consider the set of coordinate positions

$$S = \{i \mid 1 \leq i \leq N,\ (\varphi_i f_1)(P_i) = (\varphi_i f_2)(P_i) = \infty\};$$

let $|S| = m$. Then the negative part of the degree-zero divisor $D+(f)$ is bounded from above by the divisor $E_1^- + E_2^- - \sum_{i \in S} P_i$, and therefore its degree is at most $m_1 + m_2 - m$. Thus the positive part of $D+(f)$ is of degree at most $m_1 + m_2 - m$. Hence, there are at most $m_1 + m_2 - m$ positions i for which $(\varphi_i f)(P_i) = 0$. Since in total there are m common poles, we conclude that the codewords corresponding to f_1 and f_2 have at most $(m_1 + m_2 - m) + m = m_1 + m_2$ common coordinates. $\square$

Estimation of the cardinality of codes $E_D(m)$ requires more effort. We can obtain it in the case where X is an *asymptotically optimal curve*; i.e., X runs over a family of curves of genus $g \to \infty$ with

$$\frac{N(X)}{g} \to \sqrt{q}-1.$$

Let $M(m, X)$ be the average cardinality of $E_D(m)$ over all elements of the Jacobian J_X:

$$M(m,X) = \frac{1}{|J_X|} \sum_{D \in J_X} |E_D(m)|.$$

THEOREM 4.6.4. *If X is an asymptotically optimal curve, then for each*

$$\rho > \frac{2q}{q^2-1}$$

the estimate

$$M(m,X) = \left(\frac{q+1}{q}\right)^{N+o_\rho(N)} q^{2m-g}$$

holds if $2m/N > \rho$ (i.e., if for each curve in the family we choose $m = m(X)$, and $\inf(m/N) > q/(q^2-1)$).

PROOF. Let us first count the elements of $L_D(m) \setminus L_D(m-1)$, i.e., the functions with divisors of the form $E^+ - E^- - D$, where E^+ and E^- are effective divisors of degree exactly m with disjoint supports. In this case the divisor $E^+ - E^-$ is necessarily linearly equivalent to D. Conversely, for each unordered pair (E^+, E^-) of effective divisors of degree m with disjoint supports such that $E^+ - E^- \sim D$, there are $q-1$ rational functions f whose divisor is $E^+ - E^- - D$. Thus, $|L_D(m) \setminus L_D(m-1)|$

is $q-1$ times the number of such ordered pairs (E^+, E^-). Averaging over all $D \in J_X$, we may ignore the condition $E^+ - E^- \sim D$.

So, let us denote by A_m the number of pairs (E^+, E^-) of effective divisors such that $\deg E^+ = \deg E^- = m$ and $\operatorname{Supp} E^+ \cap \operatorname{Supp} E^- = \varnothing$. Then

$$\sum_{D \in J_X} |L_D(m)| = \sum_{n=0}^{m} \sum_{D \in J_X} |(L_D(n) \setminus L_D(n-1))| = \sum_{n=0}^{m} (q-1) A_n.$$

Thus,
$$M = \frac{q-1}{h} \sum_{n=0}^{m} A_n;$$

here $h = |J_X|$ is the number of classes of effective divisors.

To compute the numbers A_m, note that each pair (D^+, D^-) of effective divisors of degree n has a unique representation $(E + E^+, E + E^-)$, where E, E^+, E^- are effective divisors with the condition $\operatorname{Supp} E^+ \cap \operatorname{Supp} E^- = \varnothing$. The number of pairs (D^+, D^-) is M_n^2, where M_n is the number of effective divisors of degree n on X (see Sec. 3.1). Hence,

$$M_n^2 = \sum_{m=0}^{n} M_{n-m} A_m. \tag{4.6.2}$$

Consider the formal power series $\sum_{m=0}^{\infty} A_m t^m$. Then (4.6.2) states that

$$\sum_{n=0}^{\infty} M_n^2 t^n = \left(\sum_{n=0}^{\infty} M_n t^n \right) \left(\sum_{m=0}^{\infty} A_m t^m \right);$$

i.e.,

$$\sum_{m=0}^{\infty} A_m t^m = \frac{Z_{(2)}(t)}{Z(t)},$$

where $Z(t)$ is the zeta function of X, and $Z_{(2)}(t) = \sum_{n=0}^{\infty} M_n^2 t^n$. Thus, our goal is to study the function $Z_{(2)}(t)/Z(t)$.

It follows immediately from the explicit form of the coefficients M_n for $n \geq 2g-1$ (see Exercise 3.1.14) that $Z_{(2)}(t)$ has simple poles at the points $t = q^{-2}$, $t = q^{-1}$, and $t = 1$ and has no other poles, and its residue at q^{-2} is

$$-\left(\frac{q}{q-1} \frac{h}{q^g} \right)^2.$$

Hence, $Z_{(2)}(t)/Z(t)$ has a simple pole at q^{-2} and no other poles with $|t| < 1/\sqrt{q}$, and its residue at q^{-2} is

$$-\frac{\left(\frac{q}{q-1} \frac{h}{q^g} \right)^2}{\frac{P(q^{-2})}{(1-q^2)(1-q)}} = -\frac{q+1}{q} \frac{P^2(q^{-1})}{P(q^{-2})};$$

here we used the equality $h = q^g P(q^{-1})$, which follows from the functional equation for $P(t)$ (Proposition 3.1.16b).

This means that the coefficients A_m have the following asymptotic behaviour as $m \to \infty$:
$$A_m = \frac{q+1}{q} \frac{P^2(q^{-1})}{P(q^{-2})} q^{2m} + O_\varepsilon\left(q^{(\frac{1}{2}+\varepsilon)m}\right), \tag{4.6.3}$$
whence
$$\frac{1}{h} \sum_{n=0}^{m} A_n = \frac{q}{q-1} \frac{P(q^{-1})}{P(q^{-2})} q^{2m-g} + O_\varepsilon\left(q^{(\frac{1}{2}+\varepsilon)m}\right)$$
as $m \to \infty$. However, we are interested in estimates on $A_0 + \ldots + A_m$ for $m < N/2$, not as $m \to \infty$. Nevertheless, below we shall show that, under the conditions of the theorem, the "main term" in (4.6.3) is exponentially greater than the "error term" for $m < N/2$ too, which will imply the theorem. But let us first study the behaviour of the function $\frac{P(q^{-1})}{P(q^{-2})}$.

LEMMA 4.6.5. *For an asymptotically optimal curve, we have*
$$\frac{P(q^{-1})}{P(q^{-2})} = \left(\frac{q+1}{q}\right)^{N+o(N)}.$$

PROOF. Apply Remark 3.2.16:
$$\frac{1}{g} \log P(q^{-s}) \to -(\sqrt{q}-1)\log(1-q^{-s}) \tag{4.6.4}$$
for all s with $\{\operatorname{Re} s > 1/2\}$, uniformly in any half-plane $\operatorname{Re} s > \sigma$, where $\sigma > 1/2$. In particular, since $N/g \to \sqrt{q}-1$, we thus obtain
$$\frac{1}{N} \log \frac{P(q^{-1})}{P(q^{-2})} \to -\log \frac{1-q^{-1}}{1-q^{-2}} = \log \frac{q+1}{q},$$
as required. $\square$

Let us return to expression (4.6.3). Our goal is to estimate
$$A_m - \frac{q+1}{q} \frac{P^2(q^{-1})}{P(q^{-2})} q^{2m} = \frac{1}{2\pi i} \oint_{|t|=r} \frac{Z_{(2)}(t)}{Z(t)} \frac{dt}{t^{m+1}}, \tag{4.6.5}$$
where $r \in (q^{-2}, q^{-1/2})$. Consider only contours with $r \in (q^{-2}, q^{-3/2})$, and let us estimate the integrand. By (4.6.4), on the circle $|t| = r$ we have
$$\log|Z(t)| = -\log|1-t| = o(N). \tag{4.6.6}$$
To estimate $|Z_{(2)}(t)|$, express $Z_{(2)}(t)$ in terms of $Z(t)$ as follows:

LEMMA 4.6.6. *For all $t \neq q^{-1}$ with $q^{-2} < |t| < 1$ we have the representation*
$$Z_{(2)}(t) = \frac{1}{2\pi i} \oint_{|z|=|t|^{1/2}} Z(z) Z\left(\frac{t}{z}\right) \frac{dz}{z} + 2 \frac{q^2}{q-1} P(q^{-1}) Z(qt).$$

PROOF. First, consider t with $0 < |t| < q^{-2}$. For such t, by integrating termwise the product of absolutely convergent series $Z(z)$ and $Z(t/z)$, we obtain
$$Z_{(2)}(t) = \frac{1}{2\pi i} \oint_{|z|=|t|^{1/2}} Z(z) Z\left(\frac{t}{z}\right) \frac{dz}{z}.$$

For any t other than $0, q^{-2}, q^{-1}, 1$, the integrand extends to a meromorphic function on $\mathbb{C}$ with simple poles at $z = t, qt, 1/q, 1$ and a multiple pole at $z = 0$. The contour in the last expression encloses the poles $0, t, qt$ but not the poles $1/q$ and 1. Thus, analytic continuation gives

$$Z_{(2)}(t) = \frac{1}{2\pi i} \oint Z(z) Z\left(\frac{t}{z}\right) \frac{dz}{z}$$

for all $t \notin \{q^{-2}, 1\}$ and any contour that encloses $0, t, qt$ but not $1/q$ and 1. At the same time, the contour in the assertion of the lemma encloses $0, t, 1/q$ but neither qt nor 1. Therefore, starting from the last expression, we should add the residue at $1/q$ and subtract the residue at qt. The former residue is

$$-\frac{q^2}{q-1} P(q^{-1}) Z(qt),$$

and the latter is

$$+\frac{q^2}{q-1} P(q^{-1}) Z(qt).$$

This proves the lemma. □

Thus, the right-hand side in (4.6.5) equals

$$\frac{1}{2\pi i} \oint_{|t|=r} \frac{1}{Z(t)} \left(\frac{2q^2}{q-1} P(q^{-1}) Z(qt) + \frac{1}{2\pi i} \oint_{|z|=|t|^{1/2}} Z(z) Z\left(\frac{t}{z}\right) \frac{dz}{z} \right) \frac{dt}{t^{m+1}}.$$

We use (4.6.4) and (4.6.6) to estimate both terms. For the single integral, we find

$$\log \left| \frac{2q^2}{q-1} \frac{Z(qt)}{Z(t)} \frac{P(q^{-1})}{t^m} \right| = -m \log r + N \log \frac{|1-t|}{(1-q^{-1})|1-qt|} + o(N)$$

$$\leq -m \log r + N \log \frac{1-r}{(1-q^{-1})(1-qr)} + o(N).$$

Let us show that the double integral is exponentially smaller than the last expression. Denote $z' = t/z$ and estimate the integrand of the double integral:

$$\log \left| \frac{2q^2}{q-1} \frac{Z(z) Z(z')}{Z(zz')} \frac{1}{t^m} \right| = -m \log r + N \log \left| \frac{(1-z)(1-z')}{1-zz'} \right| + o(N).$$

Here $|z| = |z'| = \sqrt{r}$, and therefore

$$\left| \frac{(1-z)(1-z')}{1-zz'} \right| = \left| 1 + \frac{z}{1-z} + \frac{z'}{1-z'} \right| \leq 1 + 2 \frac{\sqrt{r}}{1-\sqrt{r}} = \frac{1+\sqrt{r}}{1-\sqrt{r}}.$$

It remains to show that

$$\frac{1-r}{(1-q^{-1})(1-qr)} \geq \frac{1+\sqrt{r}}{1-\sqrt{r}},$$

i.e.,

$$(1-\sqrt{r})^2 \geq (1-q^{-1})(1-qr).$$

Indeed,

$$(1-\sqrt{r})^2 - (1-q^{-1})(1-qr) = q(\sqrt{r} - q^{-1})^2.$$

Thus, we have proved that (4.6.5) can be represented in the following form:

$$A_m = \frac{q+1}{q}\frac{P^2(q^{-1})}{P(q^{-2})}q^{2m} + O\left(B\left(\frac{2m}{N}\right)^N \exp(o(N))\right),$$

where

$$B(\rho) = \min_{q^{-2} \leq r \leq q^{-3/2}} r^{-\rho/2}\frac{1-r}{(1-q^{-1})(1-qr)}.$$

It remains to estimate $B(\rho)$ for $\rho > 0$ and to show that the "error term" is exponentially smaller than the main term, thus completing the proof of the theorem.

By (4.6.4), the main term is

$$q^{2m}\left(\frac{q+1}{q-1}\right)^N \exp(o(N)).$$

Let us show that

$$B(\rho) \leq q^\rho \left(\frac{q+1}{q-1}\right)$$

for all $\rho > 0$, and for $\rho > 2q/(q^2-1)$ the inequality is strict.

Indeed,

$$q^{-\rho}\frac{q-1}{q+1}B(\rho) = \frac{q}{q+1}\min_{q^{-1/2} \leq r \leq q^{-3/2}}(q^2 r)^{-\rho/2}\frac{1-r}{1-qr}.$$

Substituting $r = q^{-2}$, we get

$$q^{-\rho}\frac{q-1}{q+1}B(\rho) \leq \frac{q}{q+1}\frac{1-q^{-2}}{1-q^{-1}} = 1$$

for all ρ. Moreover, the inequality is strict if $r^{-\rho/2}(1-r)/(1-qr)$ is a decreasing function of r at $r = q^{-2}$. One can easily check that the logarithmic derivative of this function at $r = q^{-2}$ is

$$-\frac{q^2}{q^2-1}\left((q^2-1)\frac{\rho}{2} - q\right).$$

This is negative whenever $\rho > 2q/(q^2-1)$. Thus, Theorem 4.6.4 is proved. $\square$

COROLLARY 4.6.7. *Let q be an even power of a prime. Then for codes over an alphabet of cardinality $q+1$ we have the bound*

$$\alpha_{q+1}(\delta) \geq R_E(\delta) = 1 - \left(\frac{1}{\sqrt{q}-1} + \delta\right)\log_{q+1}q. \qquad \square$$

EXERCISE 4.6.8. Prove this bound (the *Elkies bound*).

REMARK 4.6.9. Somewhat subtler arguments allow one to derive an estimate for the cardinality of an individual code $E_D(m)$, i.e., a code with a fixed divisor D. It turns out that for an asymptotically optimal curve, the asymptotics of the cardinality of an individual code is the same as that averaged over the Jacobian but with larger m:

$$|E_D(m)| = \left(\frac{q+1}{q}\right)^{N+o(N)} q^{2m-g}$$

if $m = m(X)$ is chosen such that $\inf(2m/N) > \rho_1$, where $\rho_1 = \rho_1(q) > 0$ is defined by the following condition. Consider the function

$$B_1(\rho) = \frac{q+1}{q} q^{\varkappa} \min_{r \leq q^{-1/2}} r^{-\rho/2}(1+r)(1+\sqrt{qr})^{4\varkappa},$$

where $\varkappa = 1/(\sqrt{q}-1)$. Then there exists a unique value of ρ_1 for which

$$B_1(\rho_1) = q^{\rho_1} \frac{q+1}{q-1};$$

here, $B_1(\rho) < q^\rho(q+1)/(q-1)$ for all $\rho > \rho_1$.

Comparison with algebraic geometry codes. The above results show in particular that for $2m \geq g$, the Elkies codes on the average have $(q+1/q)^{N+o(N)}$ times as many codewords as what algebraic geometry codes of the same length and with the same designed distance should have according to the basic algebraic geometry bound. This, however, does not mean that the rate of these codes is asymptotically $\log((q+1)/q)$ times greater than that of algebraic geometry codes, since the alphabet cardinality became larger by one. We should rather compare the Elkies codes with algebraic geometry codes over a field of $q+1$ elements, but situations where both q and $q+1$ are powers of primes are quite rare, and of course q and $q+1$ cannot both be squares.

REMARK 4.6.10. If we nevertheless disregard this fact and consider the hypothetical situation where for both q and $q+1$ there are asymptotically optimal families of curves, results of the comparison are quite impressive. Indeed, the existence bound for algebraic geometry codes over an alphabet of cardinality $q+1$ would in this case be

$$R + \delta \geq 1 - \frac{1}{\sqrt{q+1}-1} - o(1),$$

whereas by Corollary 4.6.7 the parameters of the Elkies codes satisfy

$$\frac{\log(q+1)}{\log q} R + \delta \geq 1 - \frac{1}{\sqrt{q}-1} + \frac{\log \frac{q+1}{q}}{\log q} - o(1),$$

which improves the preceding bound provided that

$$1 - R > \frac{\frac{1}{\sqrt{q}-1} - \frac{1}{\sqrt{q-1}-1}}{\log\left(\frac{q+1}{q}\right)/\log q} = \frac{\frac{1}{2}q^{-1/2} + O(q^{-1})}{\log q}.$$

It can easily be checked that the latter condition is satisfied for all values of (R, δ) for which the parameters of algebraic geometry codes are above the Gilbert–Varshamov bound.

Another way to compare the codes is to spoil an Elkies code by artificially reducing the alphabet cardinality to q. To do this, we choose for each coordinate position $i = 1, \ldots, N$ a forbidden symbol a_i and consider only words $w \in E_D(m)$ such that $w_i \neq a_i$ for every i. If the forbidden symbols a_i are chosen independently at random, this gives $\left(\frac{q}{q+1}\right)^N |E_D(m)|$ words, which constitute a code of the same length N with minimum distance $\geq N - 2m$ over an alphabet of cardinality q. By Theorem 4.6.4, the cardinality of the obtained code is within a subexponential factor $\exp o(N)$ of q^{2m-g}, i.e., equals the cardinality of an algebraic geometry code with the same parameters. Thus, such an average spoiling of the Elkies code asymptotically gives as good a code as the algebraic geometry construction! So we may justifiably claim that the code $E_D(m)$ itself is better than algebraic geometry codes from the point of view of parameters.

However, these codes have two essential drawbacks: first, they are not linear; second, the above construction, of course, is not polynomial (in a sense, we have not presented a code but have only proved its existence).

Message transmission. The described construction of codes automatically raises related decoding problems. Since the codes come from algebraic curves, these problems can be stated in terms of the geometry of the curves. For example, for the codes $E_0(m)$, the problem of maximum-likelihood decoding is a particular case of the following problem:

PROBLEM 4.6.11. Given an algebraic curve X of genus g over a field $\Bbbk$, a list $(P_1,\ldots,P_N)$ of $\Bbbk$-rational points of X, an N-tuple $(c_1,\ldots,c_N) \in (\Bbbk \cup \{\infty\})^N$, and integers $m, e \geq 0$, find a rational function f of degree at most m on X such that $f(P_i) = c_i$ for each i with at most e exceptions, assuming that at least one such f exists.

Similarly for the codes $E_D(m)$:

PROBLEM 4.6.12. Given an algebraic curve X of genus g over $\Bbbk$, a divisor D of degree m on X, a list $(P_1,\ldots,P_N)$ of $\Bbbk$-points of X and functions φ_i whose divisors have the same order at P_i as D, an N-tuple of symbols $(c_1,\ldots,c_N) \in (\Bbbk \cup \{\infty\})^N$, and integers $m, e \geq 0$, find a rational function $f \in L_D(m)$ on X such that $(\varphi_i f)(P_i) = c_i$ for each i with at most e exceptions, assuming that at least one such f exists.

The special case $e = 0$ of Problems 4.6.11 and 4.6.12 is the error detection, or recognition, problem: Does a given word belong to the code? For an algebraic geometry code, as well as for any linear code, this problem is solved in time polynomial in the length of the code. But for nonlinear codes, an efficient error-detection algorithm is not obvious.

Another problem from the message-transmission point of view is enumeration of codewords: one must have efficient ways to generate the jth codeword given its number j and, vice versa, given a codeword, to recover its number j (the transmitted message). For a linear code with a known basis, this is no harder than error recognition, but for the Elkies codes the problem seems nontrivial. In fact, it is not necessary to enumerate every codeword: it suffices to be able to efficiently enumerate most codewords (i.e., to use a subcode of sufficiently large cardinality M for message transmission). Thus, the following problem arises:

PROBLEM 4.6.13. Construct an injection $\mu \colon \{1,\ldots,M\} \hookrightarrow E_D(m)$ with $M = |E_D(m)|^{1-o(1)}$ and such that both μ and the inverse function $\mu^{-1} \colon \mu(\{1,\ldots,M\}) \to \{1,\ldots,M\}$ are efficiently computable.

REMARK 4.6.14. Even for $X = \mathbb{P}^1$, Problems 4.6.11 and 4.6.13 are nontrivial (Problem 4.6.12 in this case coincides with Problem 4.6.11); for this case, polynomial-time solutions are found.

4.6.2. Xing Codes

While we were writing this book, another remarkable construction of nonlinear algebraic geometry codes appeared, due to C. Xing. Asymptotic parameters of codes obtained by this construction are always above the basic algebraic geometry bound!

Construction. Us usual, let D be a divisor of degree $\deg D = a$ on a curve X, and let $\mathcal{P} = \{P_1, \ldots, P_n\}$ be a system of points such that $\mathcal{P} \cap \operatorname{Supp} D = \varnothing$. In the space $L(D)$, consider only functions f whose images under $\operatorname{Ev}_\mathcal{P}$ lie inside a fixed ball B_r of radius r. Among all balls $B_r \subset \mathbb{F}_q^n$, choose the one with the maximum possible number of functions $f \in L(D)$ mapped into this ball, i.e., the ball whose inverse image $\mathcal{M}_r(D) = \operatorname{Ev}_\mathcal{P}^{-1}(B_r) \subset L(D)$ is maximal (if there are many such balls, take any of them).

EXERCISE 4.6.15. Show that the cardinality of this set $\mathcal{M}_r(D)$ satisfies
$$|\mathcal{M}_r(D)| \geq \frac{|L(D)| \cdot |B_r|}{q^n}.$$

(*Hint*: Each point in $\mathbb{F}_q^n$ belongs to precisely $|B_r|$ distinct balls of radius r.)

Next, let t be a local parameter at P, and let $f \in \mathbb{F}_q(X)$ be a function regular at P. Define the *derivative of f at P* as
$$f'_t(P) = \frac{f - f(P)}{t}(P) \in \mathbb{F}_q.$$

Now, on the set $\mathcal{M}_r(D)$, consider the map
$$\operatorname{Ev}'_\mathcal{P} : \mathcal{M}_r(D) \to \mathbb{F}_q^n, \quad f \mapsto \left(f'_{t_1}(P_1), \ldots, f'_{t_n}(P_n) \right),$$
where t_i is an arbitrary local parameter at P_i. The image of this map is a nonlinear code
$$C_r = C_r(X, \mathcal{P}, D)$$
(which, generally, depends on the choice of local parameters t_i). Let us estimate its minimum distance:

PROPOSITION 4.6.16. *The minimum distance d of C_r satisfies*
$$d \geq 2n - a - 4r.$$

PROOF. Let $f, h \in \mathcal{M}_r(D)$ be two distinct functions. Since $\operatorname{Ev}_\mathcal{P}(f)$ and $\operatorname{Ev}_\mathcal{P}(h)$ belong to the same ball B_r, the weight of the word $\operatorname{Ev}_\mathcal{P}(f - h)$ is not greater than $2r$. In other words, the function $f - h$ has at least $n - 2r$ zeroes in the set $\mathcal{P}$. Let $\mathcal{Q}$ be the set of zeroes of $f - h$ in $\mathcal{P}$, and let $\mathcal{Q}'$ be the set of zeroes of $(f - h)'$ (the set of zeroes of the derivative) in $\mathcal{P}$. Then we have
$$|\mathcal{Q}| \geq n - 2r, \qquad |\mathcal{Q}'| = n - w,$$
where w is the weight of $\operatorname{Ev}'_\mathcal{P}(f - h) = \operatorname{Ev}'_\mathcal{P}(f) - \operatorname{Ev}'_\mathcal{P}(h)$, i.e., the distance between the codewords $\operatorname{Ev}'_\mathcal{P}(f)$ and $\operatorname{Ev}'_\mathcal{P}(h)$ of C_r. Here,
$$|\mathcal{Q} \cap \mathcal{Q}'| = |\mathcal{Q}| + |\mathcal{Q}'| - |\mathcal{Q} \cup \mathcal{Q}'| \geq |\mathcal{Q}| + |\mathcal{Q}'| - n \geq n - 2r + n - w - n = n - 2r - w.$$

Furthermore, since $\mathcal{Q}\setminus\mathcal{Q}'$ is the set of single zeroes of $f-h$, and $\mathcal{Q}\cap\mathcal{Q}'$ is the set of its (at least) double zeroes, we have

$$f - h \in L\Big(D - \sum_{P\in\mathcal{Q}\setminus\mathcal{Q}'} P - 2\sum_{P\in\mathcal{Q}\cap\mathcal{Q}'} P\Big).$$

Since $f-h$ is a nonzero function, this implies (cf. Remark 4.1.3)

$$\begin{aligned} 0 \le \deg\Big(D - \sum_{P\in\mathcal{Q}\setminus\mathcal{Q}'} P - 2\sum_{P\in\mathcal{Q}\cap\mathcal{Q}'} P\Big) \\ = a - |\mathcal{Q}\setminus\mathcal{Q}'| - 2|\mathcal{Q}\cap\mathcal{Q}'| \\ = a - |\mathcal{Q}| - |\mathcal{Q}\cap\mathcal{Q}'| \\ \le a - (n-2r) - (n-2r-w) \\ = a - 2n + 4r + w, \end{aligned}$$

i.e., $w \ge 2n - a - 4r$. □

This immediately yields the following result.

THEOREM 4.6.17. *For $a < 2n-4r$, the map $\mathrm{Ev}'_\mathcal{P}$ on the set $\mathcal{M}_r(D)$ is injective, and the cardinality of the q-ary code C_r of length n with distance $d \ge 2n-a-4r$ satisfies*

$$|C_r| \ge \frac{|L(D)|\cdot|B_r|}{q^n} = \frac{|L(D)|}{q^n}\left(\sum_{i=0}^{r}(q-1)^i\binom{n}{i}\right).$$

PROOF. The inequality for the cardinality follows from Exercise 4.6.15 (and Lemma 1.1.48). □

Asymptotic parameters. Now let $\{X\}$ be a family of curves over $\mathbb{F}_q$ of growing genus $g = g(X)$ such that

$$\lim_{g\to\infty} \frac{N(X)}{g} = A(q) = \gamma_q^{-1}.$$

For each curve X of the family, consider the corresponding code $C_r = C_r(X,\mathcal{P},D)$ of length n, where $n = N(X)$ is the number of $\mathbb{F}_q$-points on X; the value of $r = \lfloor \sigma n \rfloor$ will be chosen in the optimal way.

By Theorem 4.6.17, using the Riemann–Roch formula, we obtain

$$\begin{aligned} \log_q |C_r| + d &\ge \log_q\left(\frac{|L(D)|}{q^n}\sum_{i=0}^{r}(q-1)^i\binom{n}{i}\right) + 2n - a - 4r \\ &= \ell(D) + \log_q\left(\sum_{i=0}^{r}(q-1)^i\binom{n}{i}\right) + n - a - 4r \\ &\ge a - g(X) + 1 + \log_q\left(\sum_{i=0}^{r}(q-1)^i\binom{n}{i}\right) + n - a - 4r \\ &= n - g(X) + 1 - 4r + \log_q\left(\sum_{i=0}^{r}(q-1)^i\binom{n}{i}\right). \end{aligned}$$

Dividing the resulting inequality by n and using the fact (Exercise 1.3.10) that the ratio of the volume of a ball of radius r in $\mathbb{F}_q^n$ to n is asymptotically equivalent to

the q-ary entropy $H_q(r/n)$, in the limit as $n \to \infty$ we get
$$\alpha_q(\delta) + \delta \geq 1 - \gamma_q - 4\sigma + H_q(\sigma).$$
It remains to find the maximum of the function
$$-4\sigma + H_q(\sigma) = -4\sigma + \sigma \log_q(q-1) - \sigma \log_q \sigma - (1-\sigma) \log_q(1-\sigma).$$
This is easily done by direct computation.

EXERCISE 4.6.18. Check that the maximum of the latter expression is attained at
$$\sigma = \frac{q-1}{q^4 + q - 1}$$
and equals
$$\log_q \left(1 + \frac{q-1}{q^4}\right).$$
Thus, we have obtained the following bound.

THEOREM 4.6.19. *Let q be a power of a prime. Then for any $\delta \in (0,1)$ we have*
$$\alpha_q(\delta) \geq 1 - \gamma_q - \delta + \log_q \left(1 + \frac{q-1}{q^4}\right).$$
In particular, if q is an even power of a prime, then
$$\alpha_q(\delta) \geq 1 - (\sqrt{q}-1)^{-1} - \delta + \log_q \left(1 + \frac{q-1}{q^4}\right). \qquad \square$$

Generalization. Thus, the basic algebraic geometry bound R_{TVZ} is improved for nonlinear codes by $\log_q(1 + (q-1)/q^4)$ for all values of δ. Using a natural generalization of the above construction, this result can be slightly refined:

THEOREM 4.6.20 (Xing bound). *Let q be a power of a prime. Then for any $\delta \in (0,1)$ we have*
$$\alpha_q(\delta) \geq 1 - \gamma_q - \delta + \sum_{i=2}^{\infty} \log_q \left(1 + \frac{q-1}{q^{2i}}\right).$$
In particular, if q is an even power of a prime, then
$$\alpha_q(\delta) \geq R_X(\delta) = 1 - (\sqrt{q}-1)^{-1} - \delta + \sum_{i=2}^{\infty} \log_q \left(1 + \frac{q-1}{q^{2i}}\right).$$

SKETCH OF THE PROOF. Define
$$f_t^{(2)}(P) = \frac{f - f(P) - f_t'(P)t}{t^2}(P),$$
where t is a local parameter at P, and then inductively
$$f_t^{(i)}(P) = \frac{f - f(P) - f_t'(P)t - \ldots - f_t^{(i-1)}(P)t^{i-1}}{t^i}(P).$$
Consider the maps
$$\operatorname{Ev}_{\mathcal{P}}^{(i)} : L(D) \to \mathbb{F}_q^n, \quad f \mapsto \left(f_{t_1}^{(i)}(P_1), \ldots, f_{t_n}^{(i)}(P_n)\right)$$
(of course, we assume that $f^{(0)} = f$ and $f^{(1)} = f'$.)

For each i, fix a radius r_i, $0 < r_i < n$, and consider the ball $B_{r_i} \subset \mathbb{F}_q^n$ whose inverse image $\mathcal{M}_{r_i}(D) = \left(\mathrm{Ev}_{\mathcal{P}}^{(i)}\right)^{-1}(B_{r_i})$ has the largest cardinality. Then, for the cardinality of the set

$$\mathcal{M}_m = \bigcap_{i=0}^{m-1} \mathcal{M}_{r_i}(D),$$

we obtain the estimate

$$|\mathcal{M}_m| = |L(D)| \prod_{i=0}^{m-1} \frac{|B_{r_i}|}{q^n}.$$

Consider the *Xing code*

$$C_m = \mathrm{Ev}_{\mathcal{P}}^{(m)}(\mathcal{M}_m).$$

Its minimum distance d_m is estimated as follows:

$$d_m \geq (m+1)n - a - 2\sum_{i=0}^{m-1}(m+1-i)r_i;$$

in particular, for $a < (m+1)n - 2\sum_{i=0}^{m-1}(m+1-i)r_i$, the map $\mathrm{Ev}_{\mathcal{P}}^{(m)}$ on the set $\mathcal{M}_m$ is injective, and the cardinality of C_m satisfies

$$|C_m| \geq \frac{|L(D)|}{q^{mn}} \prod_{i=0}^{m-1}\left(\sum_{j=0}^{r_i}(q-1)^j \binom{n}{j}\right).$$

Next, for a sequence $\{X\}$ of curves of growing genus with $N(X)/g \to A(q)$, consider the code C_m of length $n = N(X)$, choosing $r_i = \lfloor \sigma_i n \rfloor$, where

$$\sigma_i = \frac{q-1}{q^{2(m+1-i)} + q - 1}.$$

Then the code C_m yields the bound

$$\alpha_q(\delta) + \delta \geq 1 - \gamma_q + \sum_{i=2}^{m+1} \log_q\left(1 + \frac{q-1}{q^{2i}}\right). \qquad \square$$

EXERCISE 4.6.21. Give a detailed proof.

REMARK 4.6.22. The Xing bound improves the basic algebraic geometry bound by a constant term on the whole interval. Numerically, this constant is very small: for $q = 49$ it is of the order 10^{-6}, whereas the difference between R_{TVZ} and R_{GV} amounts to 10^{-2}, and the difference between R_{TVZ} and upper bounds (alas!) amounts to $2 \cdot 10^{-1}$. The Xing bound, as well as the Elkies bound, is not polynomial and (as yet) has no analogue for linear codes.

Historical and Bibliographic Notes

Algebraic geometry codes were discovered by V.D. Goppa in 1980 [Gop81] (see also [Gop82, Gop84]). This discovery crowned many years of his reflections on how to generalize constructions of Reed–Solomon codes, BCH codes, and classical Goppa codes. In his seminal work [Gop81] he proposed what we call the Ω-construction (generalizing classical Goppa codes) and proved Theorem 4.1.8.

This wonderful discovery being made, it was not a big deal to find a more natural L-construction. It was known to V.D. Goppa, Yu.I. Manin, and to many other mathematicians. The H-construction is due to Yu.I. Manin; it was first published in [VM84]. Its interpretation in terms of projective systems is given in [TV91].

The first mention of duality for algebraic geometry codes is that of Goppa's paper [Gop82]. Self-dual algebraic geometry codes were studied by Y. Driencourt and J-F. Michon [DM86], and by H. Stichtenoth [Sti88b]. Theorem 4.1.28, Corollary 4.1.29, and Theorem 4.5.40 are due to W. Scharlau [Sch89]; Remark 4.1.33, Theorem 4.1.35 and Remark 4.1.36 to Y. Driencourt and H. Stichtenoth [DS89]. Theorem 4.1.37 was obtained by G.L. Katsman and M.A. Tsfasman [KT87].

The first idea of a decoding algorithm (Sec. 4.1.3) was put forth by J. Justesen [JLJ$^+$89], and its precise formulation and proofs are due to S.G. Vlăduţ and A.N. Skorobogatov [SV90]. The basic decoding algorithm was independently discovered by V.Yu. Krachkovsky [Kra88]. Theorem 4.1.45 will be proved in *Advanced Chapters*.

Proposition 4.2.5 is due to Y. Driencourt and J-F. Michon [DM85].

Improvement of parameters of one-point codes (Theorems 4.2.11 and 4.2.14) by using Weierstrass points is given in [GL92, GKL93]. Transfer of this technique to two-point codes (Theorems 4.2.15 and 4.2.16) is contained in [Mat01, HK01]. Proposition 4.2.22 was proved in [Pel92b]; the fact stated in Proposition 4.2.25 was noted by M.Yu. Rosenbloom and M.A. Tsfasman [RT90], and independently by R. Pellikaan [Pel92b]. The latter paper contains the definition of abundant codes and Theorem 4.2.27. Theorem 4.2.31 was obtained in [PT99]; its idea is based on the decoding algorithm for algebraic geometry codes due to G-L. Feng and T.R.N. Rao (to be described in *Advanced Chapters*). In that paper improved algebraic geometry codes were introduced (Proposition 4.2.45, Theorem 4.2.47 and Example 4.4.45). Theorem 4.2.40 was proved in the paper of G.L. Katsman and M.A. Tsfasman [KT89]. Remark 4.2.42 is based on H. Stichtenoth's paper [Sti90c].

The construction of generalized algebraic geometry codes is given in [XNL99a, XNL99b]; the examples of Sec. 4.4.5 are also given there. Partial algebraic geometry codes (Sec. 4.2.3) were invented by O. Pretzel [Pre99].

The characterization of algebraic geometry codes (presented in Sec. 4.3) is given in [PSvW91].

Codes of genus 1, 2, and 3 and their binary "descendants" (Sec. 4.4.1) were studied by A.M. Barg, G.L. Katsman, S.N. Litsyn, and M.A. Tsfasman [BKT87]; the study was based on the work of J-P. Serre [Ser83a, Ser83b]. Elliptic codes (Sec. 4.4.2) were studied by Y. Driencourt and J-F. Michon [DM87]. The calculation of B_k for these codes and the absence of elliptic MDS codes of length greater than $q+1$ is due to G.L. Katsman and M.A. Tsfasman [KT87].

Hermitian codes were first introduced by V.D. Goppa [Gop82] and studied by H.J. Tiersma [Tie87] and H. Stichtenoth [Sti88a]. Here we follow [Sti88a].

The results of Sec. 4.5.1 (the basic algebraic geometry bound) are due to M.A. Tsfasman, S.G. Vlăduţ, and T. Zink [TVZ82]. Theorem 4.5.13 (the expurgation bound) was obtained by S.G. Vlăduţ in [Vlă87]. Theorems 4.5.18 and 4.5.19 were proved by A. Ashikhmin, A.M. Barg, and S.G. Vlăduţ in [ABV01]. Theorem 4.5.21 was proved by S.G. Vlăduţ [VM84]; Theorems 4.5.22 and 4.5.27 by S.G. Vlăduţ, G.L. Katsman, and M.A. Tsfasman [KTV84, VKT84]; Theorems 4.5.29 and 4.5.31 by G.L. Katsman and M.A. Tsfasman [KT89]. Theorems 4.5.34 and 4.5.36 are due to V.A. Zinoviev and S.N. Litsyn [ZL85]; Theorem 4.5.37 and Remark 4.5.38 to S.N. Litsyn and M.A. Tsfasman [LT86]. Theorem 4.5.40 is taken from [Sch89]; Theorem 4.5.45 from [EZ87].

All results of Sec. 4.6.1 are due to N. Elkies [Elk01a], and those of Sec. 4.6.2 to C. Xing [Xin03].

APPENDIX

Summary of Results and Tables

In the appendix we give a short summary of the main formulae of coding theory. It is naturally divided into three parts: the first contains prelimit formulae, i.e., expressions for parameters of codes of a finite length; in Sec. A.2 we list and compare bounds on asymptotic parameters of codes; and in Sec. A.3 we give bounds on constant-weight and self-dual codes.

A.1. Codes of Finite Length

Here we briefly present results on parameters of q-ary codes. We start with bounds on the cardinality of a code, then give a list of parameters of particular families of codes, and finally describe parameters of codes obtained by various code constructions.

A.1.1. Bounds

Upper bounds. If there exists an $[n,k,d]_q$ code C, then:

(1)
$$k+d \leq n+1 \qquad \text{(Singleton bound)};$$

(2)
$$\log_q \sum_{i=1}^{\lfloor (d-1)/2 \rfloor} \binom{n}{i}(q-1)^i \leq n-k \qquad \text{(Hamming bound)};$$

(3)
$$d \leq n \frac{q^k}{q^k-1} \frac{q-1}{q} \qquad \text{(Plotkin bound)};$$

(4) for any integer w such that $1 \leq w \leq n$ and
$$A = d - 2w + \frac{qw^2}{(q-1)n} > 0,$$
we have
$$n - k \geq \log_q \binom{n}{w} + w\log_q(q-1) - \log_q d + \log_q A$$
$$\text{(Bassalygo–Elias bound)};$$

(5) let
$$P_i(j) = \sum_{v=0}^{i}(-1)^v(q-1)^{i-v}\binom{j}{v}\binom{n-j}{i-v};$$

then for any $a_1, \ldots, a_n \in \mathbb{R}$, $a_i \geq 0$, such that
$$1 + \sum_{i=1}^{n} a_i P_i(j) \leq 0 \quad \text{for any } j = d, d+1, \ldots, n,$$
we have
$$q^k \leq 1 + \sum_{i=1}^{n} a_i \binom{n}{i} (q-1)^i$$

(linear programming bound);

(6) if C is a linear code, then
$$n \geq \sum_{i=0}^{k-1} \left\lceil \frac{d}{q^i} \right\rceil \quad \text{(Griesmer bound)}.$$

Lower bound. On the contrary, if
$$q^{n-k} > \sum_{i=0}^{d-2} \binom{n-1}{i} (q-1)^i$$

(Gilbert–Varshamov bound),

then there always exists an $[n, k, d]_q$ code.

A.1.2. Parameters of Some Codes

Here we give parameters of some particular code families. In fact, in some cases we cannot give exact values of parameters since we often do not know them; it is only known that the exact values are not worse than those given below. If a family depends on certain parameters, then it is assumed that they run over all possible values such that $1 \leq d \leq n$ and $1 \leq k \leq n$.

(1) Trivial codes (n is arbitrary):
   (a) $[n, n, 1]_q$ codes;
   (b) parity-check codes: $[n, n-1, 2]_q$;
   (c) repetition codes: $[n, 1, n]_q$.
(2) Hamming codes:
$$\left[\frac{q^m - 1}{q - 1}, \frac{q^m - 1}{q - 1} - m, 3 \right]_q.$$

(3) Reed–Solomon codes:
$$[q + 1, k, q + 2 - k]_q.$$

(4) First-order projective Reed–Muller codes (simplex codes):
$$\left[\frac{q^m - 1}{q - 1}, m, q^{m-1} \right]_q.$$

(5) Affine Reed–Muller codes (of order r):
$$\left[q^m, \sum_{i=0}^{r} \sum_{j=0}^{\lfloor i/q \rfloor} (-1)^j \binom{m}{j} \binom{m-1+i-qj}{m-1}, (q-b)q^{m-a-1} \right]_q,$$

where a and b are defined by
$$r = a(q-1) + b, \quad 1 \le b \le q-1.$$
In particular, for $q = 2$:
$$\left[2^m, \sum_{i=0}^{r}\binom{m}{i}, 2^{m-r}\right]_2.$$

(6) BCH codes (primitive):
$$\left[q^m - 1, q^m - 1 - m\left(d - 1 - \left\lfloor\frac{d-1}{q}\right\rfloor\right), d\right]_q;$$
$$\left[q^m - 1, q^m - 1 - m\left(d - 1 - \left\lfloor\frac{d-2}{q}\right\rfloor\right) - 1, d\right]_q.$$

(7) Algebraic geometry codes: if there exists a curve $X/\mathbb{F}_q$ of genus g such that $|X(\mathbb{F}_q)| = n$, then
$$[n, k, n+1-g-k]_q$$
(note that we always have $|X(\mathbb{F}_q)| \le q + 1 + 2g\sqrt{q}$).

(8) Golay codes:
$$[24, 12, 8]_2, \qquad [12, 6, 6]_3.$$

(9) Goppa codes:
$$[q^m + 1, q^m + 1 - m(d-1), d]_q;$$
in particular, for $q = 2$:
$$\left[2^m + 1, 2^m + 1 - m\left\lfloor\frac{d}{2}\right\rfloor, d\right]_2.$$

(10) Justesen codes:
$$\left[2m(q^m + 1), mr, \sum_{i=1}^{\ell} i\binom{2m}{i}(q-1)^i\right]_q,$$
where ℓ is the largest integer such that
$$\sum_{i=1}^{\ell} \binom{2m}{i}(q-1)^i \le n - r + 1.$$

(11) Quadratic-residue codes ($q = p$ is a prime, ℓ is another prime, and p is a quadratic residue modulo ℓ):
$$\left[\ell, \frac{\ell+1}{2}, \lceil\sqrt{\ell}\rceil\right]_p.$$

A.1.3. Parameters of Some Constructions

Here we give parameters of codes obtained from codes with known parameters by applying certain constructions; by ℓ and m we denote arbitrary integers such that for codes obtained by these constructions we have $0 \le k \le n$ and $1 \le d \le n$. If we write the symbol $\subseteq$ between parameters of codes, we mean that the corresponding codes are embedded.

Construction	Parameters of original codes	Parameters of new codes
Lengthening by zeroes	$[n,k,d]_q$	$[n+\ell,k,d]_q$
Omitting parity checks	$[n,k,d]_q$	$[n-\ell,k-\ell,d]_q$
Projection	$[n,k,d]_q$	$[n-\ell,k,d-\ell]_q$
Direct sum	$[n_1,k_1,d_1]_q$ $[n_2,k_2,d_2]_q$	$[n_1+n_2,k_1+k_2,d]_q$ $d=\min\{d_1,d_2\}$
Power	$[n,k,d]_q$	$[\ell n,\ell k,d]_q$
Tensor product	$[n_1,k_1,d_1]_q$ $[n_2,k_2,d_2]_q$	$[n_1 n_2,k_1 k_2,d_1 d_2]_q$
Tensor power	$[n,k,d]_q$	$[n^\ell,k^\ell,d^\ell]_q$
Pasting	$[n_1,k,d_1]_q$ $[n_2,k,d_2]_q$	$[n_1+n_2,k,\geq d_1+d_2]_q$
Repetition	$[n,k,d]_q$	$[\ell n,k,\ell d]_q$
Code from an embedded pair	$[n,k,d]_q \subseteq [n,k+1,d-1]_q$	$[n+1,k+1,d]_q$
$(u\,|\,u+v)$ construction	$[n,k_1,d_1]_q$ $[n,k_2,d_2]_q$	$[2n,k_1+k_2,d]_q$ $d=\min\{2d_1,d_2\}$
Suffix construction	$[n_1,k_1,d_1]_q \subseteq [n_1,k_2,d_2]_q$ $[n_3,k_2-k_1,d_3]_q$	$[n_1+n_3,k_2,d]_q$ $d\geq \min\{d_1,d_2+d_3\}$
Combination of four codes	$[n_1,k_1,d_1]_q \subseteq [n_1,k_2,d_2]_q$ $[n_3,k_3,d_3]_q \subseteq [n_3,k_4,d_4]_q$ $d_3\geq d_1,\ k_2-k_1=k_4-k_3$	$[n_1+n_3,k_2+k_3,d]_q$ $d\geq \min\{d_1,d_2+d_4\}$
Shortening by distance	$[n,k,d]_q$	$\left[n-d,k-1,\geq \left\lceil\dfrac{d}{q}\right\rceil\right]_q$
Shortening by dual distance	$[n,k,d]_q$	$[n-d^\perp,k-d^\perp+1,\geq d]_q$
Parity check	$[n,k,2t+1]_2$	$[n+1,k,2t+2]_2$
Subfield restriction	$[n,k,d]_{q^m}$	$[n,\geq n-m(n-k),\geq d]_q$
Concatenation	$[n,k,d]_{q^m}$ $[N,m,D]_q$	$[nN,km,\geq dD]_q$
Field descent	$[n,k,d]_{q^m}$	$[mn,mk,d]_q$

Construction	Parameters of original codes	Parameters of new codes
Generalized concatenation	$[n, k_i, d_i]_{q^{m_i}}$ $\left[N, \sum_{j=1}^{i} m_j, D_i\right]_q,$ $i = 1, \ldots, \ell$ $\left[N, \sum_{j=1}^{i+1} m_j, D_{i+1}\right]_q$ $\supseteq \left[N, \sum_{j=1}^{i} m_j, D_i\right]_q$	$[Nn, K, D]_q,$ $K = \sum_{i=1}^{\ell} m_i k_i,$ $D \geq \min_{1 \leq i \leq \ell} \{d_i D_i\}$
Generalized iterative construction	$C': [N, k, d]_q$ $[n, k_i, d_i]_q, \ i = 1, \ldots, k$ $C'_1 \subset \ldots \subset C'_k = C'$	$[Nn, K, D]_q,$ $K = \sum_{i=1}^{k} k_i,$ $D \geq \min_{1 \leq i \leq k} \{d_i d(C'_i)\}$
Alphabet extension	$[n, k, d]_r, \quad r \leq q$	$[n, k \log_q r, d]_q$
Alphabet restriction	$[n, k, d]_r, \quad r \geq q$	$[n, k_r, \geq d]_q$ $k_r \geq n - (n - k) \log_q r$

Parameters of constructions for higher weights. Given a linear code with parameters $[n, k, d_1, \ldots, d_{k-1}]$, one can construct codes with the following parameters:

- $[n-1, k, \geq d_1 - 1, \ldots, \geq d_{k-1} - 1]$ (if $d_1 \geq 2$);
- $[n, k-1, \geq d_1, \ldots, \geq d_{k-1}]$ (if $k \geq 2$);
- $[n-1, k-1, \geq d_1, \ldots, \geq d_{k-1}]$ (if $k \geq 2$);
- $[n+1, k, \geq d_1, \ldots, \geq d_{k-1}]$;
- $[n - d_s, k - s, d'_1 \geq d_{s+1} - d_s, \ldots, d'_{k-s-1} \geq d_{k-1} - d_s]$ and $[d_s, s, d''_1 \geq d_1, \ldots, d''_{s-1} \geq d_{s-1}]$, for any $s = 1, \ldots, k-1$.

A.2. Asymptotic Bounds

Algebraic geometry codes provide many new asymptotic lower bounds for the possible cardinality of a code. Here we give diagrams illustrating interrelations between these bounds; we also give tables for the asymptotic behaviour of the bounds as $\delta \to 0$ and $\delta \to 1 - 1/q$ and tables of numerical values of the bounds. For simplicity, the alphabet cardinality q is assumed to be a power of a prime. Recall that
$$H_q(x) = x \log_q(q-1) - x \log_q x - (1-x) \log_q(1-x).$$

A.2.1. List of Bounds

"True" bounds.

α_q: the "true" bound for q-ary codes (Theorem 1.3.1);
α_q^{lin}: the "true" bound for linear q-ary codes (Exercise 1.3.2);
α_q^{pol}: the "true" bound for polynomial families of q-ary codes (Theorem 1.3.39);
$\alpha_q^{\mathrm{pol,lin}}$: the "true" bound for polynomial families of linear q-ary codes (Theorem 1.3.39).

Upper bounds.

R_S: the Singleton bound,
$$R_S(\delta) = 1 - \delta;$$

R_P: the Plotkin bound (Theorem 1.3.7),
$$R_P(\delta) = 1 - \frac{q}{q-1}\delta;$$

R_H: the Hamming bound (Theorem 1.3.9),
$$R_H(\delta) = 1 - H_q(\delta/2);$$

R_{BE}: the Bassalygo–Elias bound (Theorem 1.3.11),
$$R_{BE}(\delta) = 1 - H_q\left(\frac{q-1}{q} - \frac{q-1}{q}\sqrt{1 - \frac{q\delta}{q-1}}\right);$$

R_4: the MRRW bound (the McEliece–Rodemich–Rumsey–Welch bound; see Theorem 1.3.12),
$$R_4(\delta) = H_q\left(\frac{q-1-\delta(q-2) - 2\sqrt{(q-1)\delta(1-\delta)}}{q}\right);$$

$R_{4(2)}$: the second MRRW bound for $q = 2$ (Theorem 1.3.13),
$$R_{4(2)}(\delta) = \min_{0 \leq u \leq 1 - 2\delta} \{1 + h(u^2) - h(u^2 + 2\delta u + 2\delta)\},$$

where
$$h(x) = H_2\left(\frac{1 - \sqrt{1-x}}{2}\right);$$

R'_A: the bound obtained by the tangent line to R_4 (Theorem 1.3.14) for $q > 2$,
$$R'_A(\delta) = \begin{cases} 1 - \dfrac{\delta q}{q-2} \log_q(q-1), & 0 \leq \delta \leq \delta'_A = \left(\dfrac{q-2}{q}\right)^2, \\ R_4(\delta), & \delta > \delta'_A; \end{cases}$$

R_A: the Aaltonen bound for $q > 2$ (see Theorem 1.3.16);

R_{LL}: the Laihonen–Litsyn bound (Theorem 1.3.17) for $q \geq 7$,

$$R_{LL}(\delta) \leq \min\left\{ R_4(y) + \left(1 - H_q(x/2) - R_4(y)\right)\frac{\delta - y}{x - y} \right\},$$

where

$$x = \frac{2\tau}{\mu}, \qquad y = \frac{\delta - 2\tau}{1 - \mu}$$

and the minimum is taken over all values of μ and τ such that

$$0 \leq \delta - 2\tau \leq 1 - \mu, \qquad 0 \leq \tau \leq \frac{q-1}{q}\mu.$$

Lower bounds.

R_{GV}: the Gilbert–Varshamov bound (Theorem 1.3.20),

$$R_{GV}(\delta) = 1 - H_q(\delta);$$

R_{BZ}: the Blokh–Zyablov bound (Theorem 1.3.47),

$$R_{BZ}(\delta) = R_{GV}(\delta) - \delta \int_0^{R_{GV}(\delta)} \frac{dx}{R_{GV}^{-1}(x)},$$

where R_{GV}^{-1} is the inverse function of R_{GV};

R_{TVZ}: the Tsfasman–Vlăduţ–Zink bound (basic algebraic geometry bound), q is an even power of a prime (Corollary 4.5.3),

$$R_{TVZ}(\delta) = 1 - (\sqrt{q} - 1)^{-1} - \delta;$$

R_{Pel}: the Pellikaan bound, q is an even power of a prime (Proposition 4.5.8),

$$R_{Pel}(\delta) = \begin{cases} 1 + \dfrac{1}{q+1} - \gamma_q - \delta & \text{for } 0 \leq \delta \leq \dfrac{1}{q+1}, \\ 1 - (q+1)\gamma_q \delta & \text{for } \dfrac{1}{q+1} \leq \delta \leq \dfrac{\sqrt{q}-1}{q+1}; \end{cases}$$

R_V^0: the Vlăduţ bound (expurgation bound, Theorem 4.5.13), q is an even power of a prime; the bound is given by the relation

$$\left(R_V^0(\delta) + (\sqrt{q} - 1)^{-1}\right) H_q\left(\frac{1 - \delta}{R_V^0(\delta) + (\sqrt{q}-1)^{-1}}\right) + H_q(\delta) = 1 + (\sqrt{q}-1)^{-1}$$

and is defined for $\delta \in [\delta_1', \delta_2'] \cup [\delta_3', \delta_4']$, where $\delta_1' < \delta_4'$ are the roots of the equation

$$H_q(\delta) + \frac{q}{q-1}(1 - \delta) = 1 + (\sqrt{q}-1)^{-1},$$

and $\delta_2' < \delta_3'$ are the roots of

$$H_q(\delta) + (1 - \delta)\log_q(q - 1) = 1 + (\sqrt{q}-1)^{-1};$$

$$R_V(\delta) = \begin{cases} R_V^0(\delta) & \text{for } \delta \in [\delta_1', \delta_2'] \cup [\delta_3', \delta_4'], \\ R_{GV}(\delta) & \text{for } \delta \in [0, \delta_1'] \cup [\delta_4', 1 - 1/q], \\ R_{TVZ}(\delta) & \text{for } \delta \in [\delta_2', \delta_3']; \end{cases}$$

R_E: the Elkies bound, q is greater by 1 than an even power of a prime (Corollary 4.6.7, the codes are nonlinear),

$$R_E(\delta) = 1 - \left(\frac{1}{\sqrt{q-1}-1} + \delta\right) \log_q(q-1);$$

R_X: the Xing bound, q is an even power of a prime (Theorem 4.6.20, the codes are nonlinear),

$$R_X(\delta) = 1 - (\sqrt{q}-1)^{-1} - \delta + \sum_{i=2}^{\infty} \log_q\left(1 + \frac{q-1}{q^{2i}}\right);$$

R_{KTV}, $R_{KTV(\text{lin})}$: the Katsman–Tsfasman–Vlăduţ bounds (concatenation bounds, Theorem 4.5.22),

$$R_{KTV}(\delta) = \max\left\{\left(1 - \left(q^{k/2}\right)^{-1}\right)\frac{k}{n} - \frac{k}{n}\delta\right\},$$

where the maximum is taken over all $[n,k,d]_q$ codes, and q^k is a square; $R_{KTV(\text{lin})}$ is defined similarly, but the maximum is taken over linear codes;

R_{KT}: the Katsman–Tsfasman bound (restriction bound, Theorem 4.5.29),

$$R_{KT}(\delta) = \max_m\left\{1 - \frac{2m(q-1)}{q(q^{m/2}-1)} - \frac{m(q-1)}{q}\delta\right\},$$

where the maximum is taken over all natural m such that q^m is a square;

$R_{KTV,KT}(\delta) = \max\{R_{KTV}(\delta), R_{KT}(\delta)\};$

$R_{KTV(\text{lin}),KT}(\delta) = \max\{R_{KTV(\text{lin})}(\delta), R_{KT}(\delta)\};$

R_{LT}: the Litsyn–Tsfasman bound (Theorem 4.5.37),

$$R_{LT}(\delta) = \max_m\{R'_{LT}(\delta), R''_{LT}(\delta)\};$$

here

$$R'_{LT}(\delta) = \max_{q'}\left\{1 - \left(1 - R_V^{(q')}(\delta)\right)\log_q(q')\right\},$$

where the minimum is taken over all even powers of primes $q' \geq q$, and $R_V^{(q')}$ is the expurgation bound for q'-ary codes, and

$$R''_{LT}(\delta) = \max_{q'}\left\{R_V^{(q')}(\delta)\log_q(q')\right\},$$

where the maximum is taken over all even powers of primes $q' \leq q$.

A.2.2. Comparison Diagrams

In the diagrams (Fig. A.1–A.4) we compare various bounds.

An arrow $A \longrightarrow B$ means that $A(\delta) \geq B(\delta)$ for any $\delta \in \left[0, \frac{q-1}{q}\right]$ and that we can prove this inequality.

A dashed arrow $A \dashrightarrow B$ means that $A(\delta) \geq B(\delta)$ for any $\delta \in \left[0, \frac{q-1}{q}\right]$ such that values of $A(\delta)$ and $B(\delta)$ are known, but we do not know any general proof.

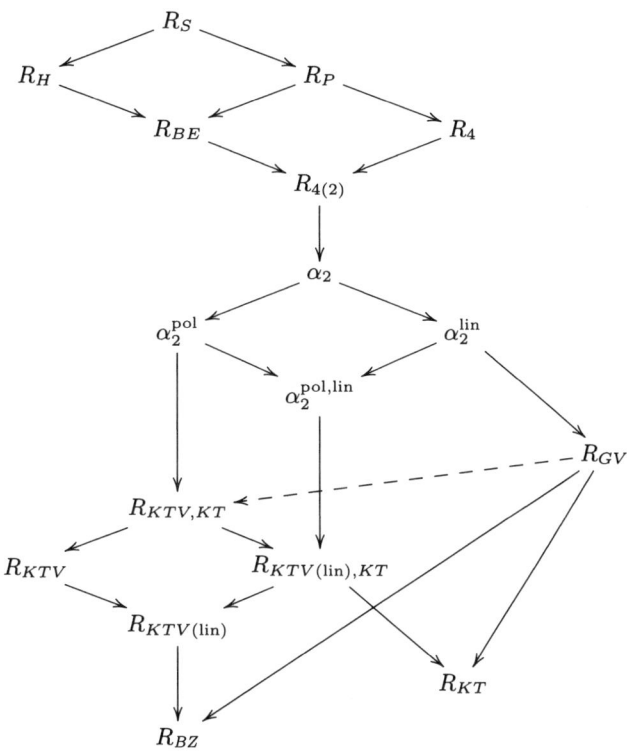

FIGURE A.1. Comparison of bounds for $q = 2$.

Note: It can be shown that $R_{GV}(\delta) > R_{KTV,KT}(\delta)$ for all δ except for a very small interval.

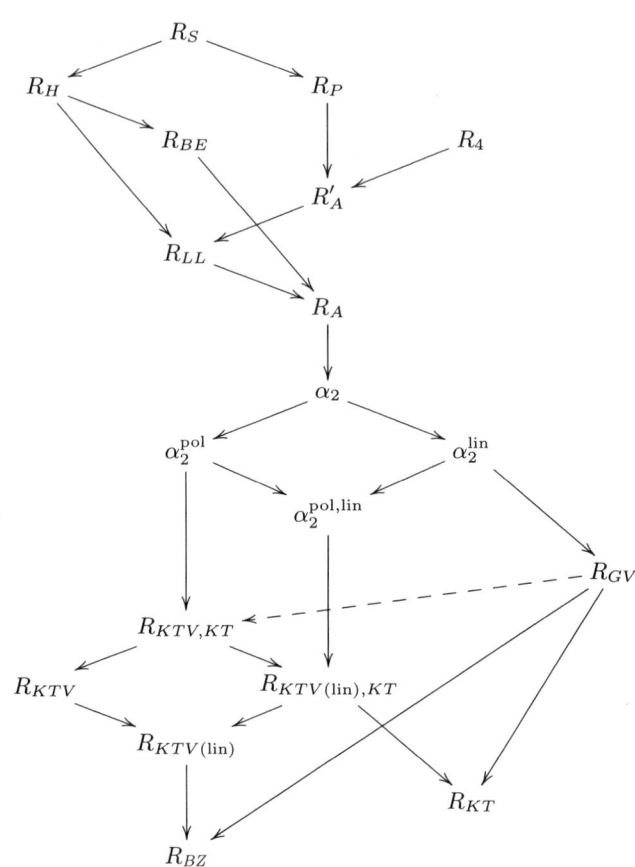

FIGURE A.2. Comparison of bounds for $q = 3$ and $q = 4$.

Note: In the case $q = 3$ we also have $R_P \longrightarrow R_{BE}$.

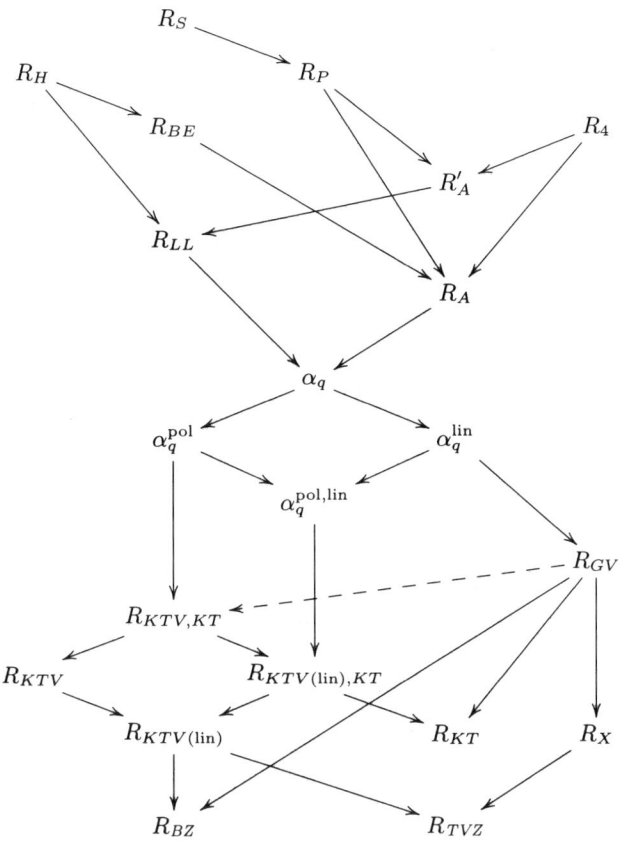

FIGURE A.3. Comparison of bounds for $4 < q = p^{2m} < 49$.

Notes: 1. For $q < 16$ we have $R'_A \longrightarrow R_A$; for $q \leq 16$, we have $R_S \longrightarrow R_{BE}$ (in the case $q = 5$ we also have $R_{LL} \longrightarrow R_A$).

2. If $2 < q = p^m < 49$ and m is odd, the diagram should be changed slightly. First, in this case $R_{TVZ}(\delta) = 1 - \gamma_q - \delta$, where γ_q is unknown. Second, we should also consider the bound R_{LT}, which is of importance in this case; for instance, for $q = 47$ it intersects R_{GV}.

3. One may also consider the case $2 < q < 49$, $q \neq p^m$. In this case there is no notion of a linear bound, and nonlinear constructions leading to the bound R_{LT} play a crucial role.

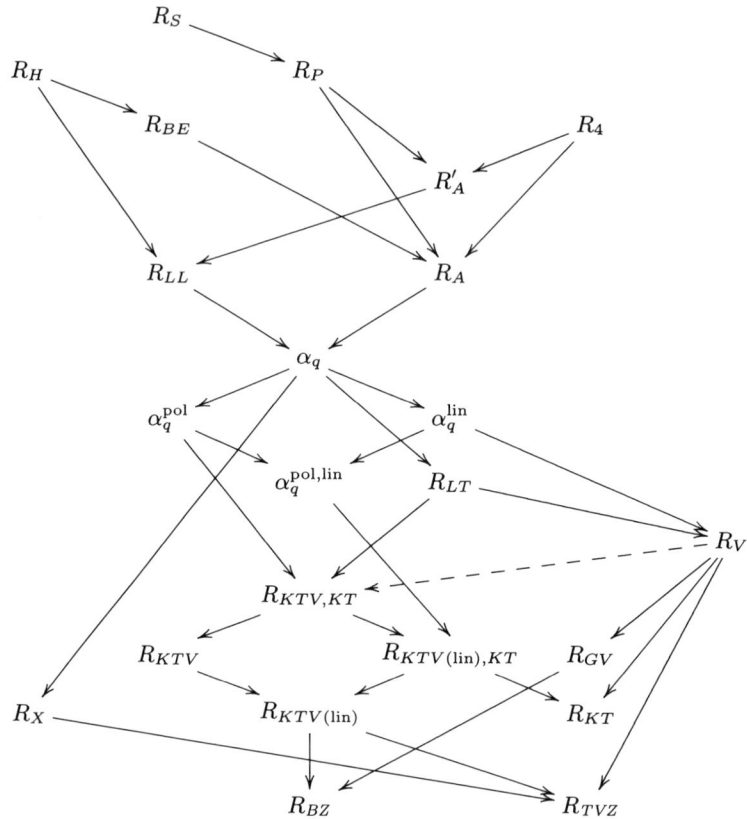

FIGURE A.4. Comparison of bounds for $q = p^{2m} \geq 49$.

Note: In the case $49 \leq q = p^{2m} \leq 289$ we also have $R_{GV} \longrightarrow R_{KT}$.

A.2.3. Behaviour at the Endpoints

When analyzing bounds, it is useful to know their behaviour as $\delta \to 0$ and $\delta \to 1 - 1/q$. In Tables 1 and 2 this behaviour is represented as follows: for a bound $R(\delta)$ we give the asymptotics of $1 - R(\delta)$ as $\delta \to 0$ and of $R\left(\frac{q-1}{q} - x\right)$ as $x \to 0$.

Bound	$1 - R(\delta)$ as $\delta \to 0$	$R\left(\frac{1}{2} - x\right)$ as $x \to 0$
R_{BZ}	$\dfrac{\ln 2}{2}\delta(\log_2 \delta)^2$	$\dfrac{4}{3\ln 2}x^3$
$\left.\begin{array}{l}R_{KTV}\\R_{KTV(\text{lin})}\end{array}\right\}$	$-2\delta\log_2\delta$	$-\dfrac{1152}{343}x^3\log_2 x$
R_{KT}	$-\delta\log_2\delta$	<0
R_{GV}	$-\delta\log_2\delta$	$-\dfrac{2}{\ln 2}x^2$
$R_{4(2)}$	$-\dfrac{1}{2}\delta\log_2\delta$	$-2x^2\log_2 x$
R_4	$\dfrac{2}{\ln 2}\delta$	$-2x^2\log_2 x$
R_{BE}	$-\dfrac{1}{2}\delta\log_2\delta$	$\dfrac{1}{\ln 2}x$
R_H	$-\dfrac{1}{2}\delta\log_2\delta$	>0
R_P	2δ	$2x$
R_S	δ	>0

TABLE 1. Behaviour of bounds for binary codes at 0 and 1/2

A.2.4. Numerical Values

In the following tables (Tables 3–7) we give numerical values of some bounds.

We present values of four upper bounds (R_P, R_{BE}, R_4, and also $R_{4(2)}$ for $q = 2$ and R_A for $q > 2$); for $q = 16$, 64, and 256, we also give values of R_{LL}. Here, the values of R_{LL} are shown only at points where this bound improves upon R_A, and values of R_{BE} and R_4 are shown at points where they are better than R_P (for $q = 16$, 64, and 256, values of R_{BE} and R_4 are placed in one column of the table: R_{BE} is at the top, and R_4 at the bottom).

Also, for $q = 2$, 4, and 16 we give values of four lower bounds (R_{BZ}, R_{KTV}, R_{KT}, and R_{GV}), and for $q = 64$ and 256, of six lower bounds (R_{BZ}, R_{TVZ}, R_{KTV}, R_{KT}, R_V^0, and R_{GV}). Note that the presented values of R_{KTV} are lower estimates

Bound	$1-R(\delta)$ as $\delta \to 0$	$R\left(\dfrac{q-1}{q}-x\right)$ as $x \to 0$
R_{BZ}	$\dfrac{\ln q}{2}\delta(\log_q \delta)^2$	$\dfrac{q^3}{6(q-1)^2 \ln q}x^3$
R_{KTV}, $R_{KTV(\text{lin})}$	$-2\delta \log_q \delta$	$\begin{cases} -2\dfrac{(\sqrt{q}-1)q^4}{(\sqrt{q}^3-1)^3}x^3 \log_q x \\ \quad \text{for } q=p^{2m} \\ -2\dfrac{(q+1)^2 q^6}{(q^3-1)^3}x^3 \log_q x \\ \quad \text{for } q=p^{2m+1} \end{cases}$
R_{KT}	$-2\dfrac{q-1}{q}\delta \log_q \delta$	<0
R_{GV}	$-\delta \log_q \delta$	$\dfrac{q^2}{2(q-1)\ln q}x^2$
R_4	$\dfrac{2}{\ln q}\delta$	$-2x^2 \log_2 x$
R_{BE}	$-\dfrac{1}{2}\delta \log_q \delta$	$\dfrac{q}{2\ln q}x$
R_H	$-\dfrac{1}{2}\delta \log_q \delta$	>0
R_P	$\dfrac{q}{q-1}\delta$	$\dfrac{q}{q-1}x$
R_S	δ	>0

TABLE 2. Behaviour of bounds for q-ary codes at 0 and $\dfrac{q-1}{q}$

but not precise values, whose evaluation involves maximization over all codes of a certain type and which are therefore unknown.

The difference between a precise value of a bound and a value given in the table does not exceed 10^{-4}. Note that the bound R_{LT} in not given in the tables since it always differs (if at all) from R_V by at most 10^{-4}. Similarly, we do not present R_X; being nonconstructive, for $q<49$ it is worse than R_{GV}, and for $q \geq 49$ the difference between R_X and R_{TVZ} is of order at most 10^{-6}.

For very small values of the bounds we give the first significant digit.

A.2. ASYMPTOTIC BOUNDS

δ	Lower bounds				Upper bounds			
	R_{BZ}	R_{KTV}	R_{KT}	R_{GV}	R_P	R_{BE}	R_4	$R_{4(2)}$
0.0001	0.9896	0.9973	0.9980	0.9985	0.9998	0.9992	0.9997	0.9992
0.001	0.9546	0.9801	0.9842	0.9886	0.998	0.9938	0.9971	0.9938
0.01	0.7778	0.8677	0.8804	0.9192	0.98	0.9544	0.9712	0.9542
0.02	0.6609	0.7884	0.7847	0.8586	0.96	0.9185	0.9427	0.9180
0.03	0.5738	0.7298	0.6973	0.8056	0.94	0.8862	0.9143	0.8851
0.04	0.5039	0.6798	0.6173	0.7577	0.92	0.8562	0.8862	0.8544
0.05	0.4456	0.6298	0.5398	0.7136	0.90	0.8279	0.8582	0.8251
0.06	0.3959	0.5896	0.4698	0.6726	0.88	0.8008	0.8305	0.7971
0.07	0.3529	0.5496	0.3998	0.6341	0.86	0.7748	0.8030	0.7699
0.08	0.3152	0.5096	0.3298	0.5978	0.84	0.7498	0.7758	0.7436
0.09	0.2819	0.4696	0.2695	0.5635	0.82	0.7255	0.7487	0.7179
0.10	0.2524	0.4347	0.2095	0.5310	0.80	0.7019	0.7219	0.6927
0.11	0.2261	0.4047	0.1495	0.5001	0.78	0.6789	0.6954	0.6681
0.12	0.2025	0.3747	0.0895	0.4706	0.76	0.6565	0.6691	0.6439
0.13	0.1813	0.3447	0.0295	0.4426	0.74	0.6345	0.6431	0.6201
0.14	0.1622	0.3147		0.4158	0.72	0.6130	0.6173	0.5966
0.15	0.1450	0.2847		0.3902	0.70	0.5920	0.5919	0.5734
0.16	0.1294	0.2547		0.3657	0.68	0.5713	0.5667	0.5506
0.18	0.1027	0.2267		0.3199	0.64	0.5310	0.5172	0.5055
0.20	0.0808	0.2000		0.2781	0.60	0.4920	0.4690	0.4614
0.22	0.0630	0.1733		0.2398	0.56	0.4541	0.4221	0.4179
0.24	0.0485	0.1467		0.2050	0.52	0.4172	0.3767	0.3750
0.26	0.0367	0.1200		0.1733	0.48	0.3812	0.3328	0.3326
0.28	0.0273	0.0933		0.1445	0.44	0.3461	0.2906	0.2906
0.30	0.0198	0.0733		0.1187	0.40	0.3117	0.2502	0.2502
0.32	0.0140	0.0533		0.0956	0.36	0.2781	0.2118	0.2118
0.34	0.0096	0.0388		0.0752	0.32	0.2451	0.1754	0.1754
0.36	0.0062	0.0302		0.0573	0.28	0.2127	0.1414	0.1414
0.38	0.0038	0.0217		0.0420	0.24	0.1808	0.1100	0.1100
0.39	0.0029	0.0174		0.0352	0.22	0.1651	0.0954	0.0954
0.40	0.0022	0.0132		0.0290	0.20	0.1495	0.0815	0.0815
0.41	0.0015	0.0097		0.0235	0.18	0.1340	0.0684	0.0684
0.42	0.0011	0.0058		0.0185	0.16	0.1187	0.0561	0.0561
0.43	0.0007	0.0046		0.0142	0.14	0.1035	0.0448	0.0448
0.44	0.0004	0.0033		0.0104	0.12	0.0884	0.0345	0.0345
0.45	0.0003	0.0021		0.0072	0.10	0.0734	0.0253	0.0253
0.46	0.0001	0.0009		0.0046	0.08	0.0585	0.0172	0.0172
0.47	$5 \cdot 10^{-5}$	0.0005		0.0026	0.06	0.0437	0.0104	0.0104
0.48	$2 \cdot 10^{-5}$	0.0002		0.0012	0.04	0.0290	0.0051	0.0051
0.49	$2 \cdot 10^{-6}$	$2 \cdot 10^{-5}$		0.0003	0.02	0.0145	0.0015	0.0015

TABLE 3. $q = 2$

δ	Lower bounds				Upper bounds			
	R_{BZ}	R_{KTV}	R_{KT}	R_{GV}	R_P	R_{BE}	R_4	R_A
0.0001	0.9920	0.9986	0.9985	0.9992	0.9999	0.9996	0.9999	0.9996
0.001	0.9706	0.9890	0.9882	0.9935	0.9987	0.9965	0.9985	0.9965
0.01	0.8481	0.9241	0.9103	0.9517	0.9867	0.9733	0.9851	0.9733
0.02	0.7599	0.8677	0.8386	0.9134	0.9733	0.9514	0.9698	0.9512
0.04	0.6330	0.7733	0.7129	0.8472	0.9467	0.9124	0.9388	0.9118
0.06	0.5388	0.6933	0.6023	0.7887	0.9800	0.8770	0.9073	0.8756
0.08	0.4637	0.6171	0.4973	0.7355	0.8933	0.8437	0.8756	0.8414
0.10	0.4015	0.5571	0.4071	0.6863	0.8667	0.8120	0.8437	0.8085
0.12	0.3489	0.5067	0.3171	0.6402	0.8400	0.7815	0.8117	0.7767
0.14	0.3039	0.4667	0.2331	0.5969	0.8133	0.7520	0.7796	0.7456
0.16	0.2648	0.4267	0.1581	0.5560	0.7867	0.7234	0.7475	0.7152
0.18	0.2308	0.3867	0.0831	0.5173	0.7600	0.6954	0.7154	0.6852
0.20	0.2009	0.3467	0.0081	0.4805	0.7333	0.6681	0.6834	0.6557
0.22	0.1746	0.3129		0.4456	0.7067	0.6413	0.6515	0.6264
0.24	0.1515	0.2829		0.4123	0.6800	0.6149	0.6196	0.5975
0.26	0.1311	0.2543		0.3806	0.6533	0.5890	0.5879	0.5687
0.28	0.1130	0.2343		0.3504	0.6267	0.5635	0.5564	0.5402
0.30	0.0971	0.2143		0.3216	0.6000	0.5383	0.5251	0.5117
0.32	0.0830	0.1943		0.2942	0.5733	0.5133	0.4940	0.4834
0.34	0.0706	0.1743		0.2681	0.5467	0.4887	0.4631	0.4552
0.36	0.0597	0.1543		0.2434	0.5200	0.4643	0.4325	0.4270
0.38	0.0501	0.1343		0.2198	0.4933	0.4401	0.4023	0.3988
0.40	0.0418	0.1143		0.1975	0.4667	0.4161	0.3725	0.3706
0.42	0.0345	0.0943		0.1764	0.4400	0.3923	0.3430	0.3424
0.44	0.0282	0.0743		0.1565	0.4133	0.3686	0.3140	0.3140
0.46	0.0228	0.0663		0.1378	0.3867	0.3451	0.2855	0.2856
0.48	0.0182	0.0596		0.1202	0.3600	0.3216	0.2576	0.2576
0.50	0.0143	0.0529		0.1038	0.3333	0.2983	0.2304	0.2304
0.52	0.0110	0.0463		0.0885	0.3067	0.2750	0.2038	0.2038
0.54	0.0083	0.0396		0.0744	0.2800	0.2517	0.1781	0.1781
0.56	0.0061	0.0329		0.0614	0.2533	0.2285	0.1532	0.1532
0.58	0.0043	0.0263		0.0496	0.2267	0.2053	0.1294	0.1294
0.60	0.0030	0.0196		0.0390	0.2000	0.1820	0.1067	0.1067
0.62	0.0019	0.0129		0.0296	0.1733	0.1587	0.0853	0.0853
0.64	0.0012	0.0063		0.0215	0.1467	0.1353	0.0654	0.0654
0.66	0.0006	0.0033		0.0146	0.1200	0.1117	0.0473	0.0473
0.68	0.0003	0.0017		0.0089	0.0933	0.0878	0.0313	0.0313
0.70	0.0001	0.0007		0.0046	0.0667	0.0637	0.0178	0.0178
0.72	$2 \cdot 10^{-5}$	0.0002		0.0017	0.0400	0.0390	0.0074	0.0074
0.74	$9 \cdot 10^{-7}$	$1 \cdot 10^{-5}$		0.0002	0.0133		0.0010	0.0010

TABLE 4. $q = 4$

A.2. ASYMPTOTIC BOUNDS

δ	Lower bounds				Upper bounds			
	R_{BZ}	R_{KTV}	R_{KT}	R_{GV}	R_P	R_{BE}/R_4	R_A	R_{LL}
0.0001	0.9932	0.9992	0.9991	0.9995	0.9999	0.9998	0.9998	
0.001	0.9796	0.9940	0.9926	0.9962	0.9989	0.9980	0.9980	
0.01	0.8922	0.9561	0.9440	0.9700	0.9893	0.9837	0.9837	
0.02	0.8246	0.9241	0.8971	0.9451	0.9787	0.9699	0.9698	
0.04	0.7220	0.8641	0.8033	0.9004	0.9573	0.9446	0.9443	
0.06	0.6418	0.8133	0.7420	0.8595	0.9360	0.9210	0.9203	
0.08	0.5751	0.7733	0.6857	0.8213	0.9147	0.8985	0.8972	
0.10	0.5178	0.7333	0.6295	0.7851	0.8933	0.8767	0.8747	
0.12	0.4676	0.6933	0.5732	0.7505	0.8720	0.8555	0.8526	
0.14	0.4232	0.6533	0.5170	0.7172	0.8507	0.8348	0.8308	
0.16	0.3834	0.6133	0.4607	0.6851	0.8293	0.8144	0.8092	
0.18	0.3475	0.5733	0.4125	0.6542	0.8080	0.7943	0.7877	
0.20	0.3151	0.5333	0.3750	0.6242	0.7867	0.7745	0.7662	
0.24	0.2587	0.4533	0.3000	0.5668	0.7440	0.7356	0.7234	
0.28	0.2117	0.3867	0.2250	0.5127	0.7013	0.6972	0.6806	0.6799
0.32	0.1722	0.3467	0.1500	0.4614	0.6587		0.6376	0.6359
0.36	0.1390	0.3067	0.0750	0.4127	0.6160		0.5944	0.5918
0.40	0.1111	0.2667		0.3666	0.5733		0.5510	0.5478
0.44	0.0877	0.2267		0.3228	0.5307	0.5288	0.5074	0.5038
0.48	0.0681	0.1867		0.2815	0.4880	0.4816	0.4637	0.4597
0.50	0.0596	0.1667		0.2616	0.4667	0.4578	0.4418	0.4377
0.54	0.0450	0.1267		0.2237	0.4240	0.4102	0.3979	0.3937
0.58	0.0330	0.0867		0.1881	0.3813	0.3625	0.3539	0.3496
0.62	0.0235	0.0572		0.1549	0.3387	0.3149	0.3096	0.3056
0.66	0.0160	0.0458		0.1242	0.2960	0.2674	0.2651	0.2616
0.70	0.0103	0.0344		0.0960	0.2533	0.2205	0.2202	0.2175
0.72	0.0080	0.0297		0.0829	0.2320	0.1973	0.1973	0.1955
0.74	0.0061	0.0257		0.0705	0.2106	0.1743	0.1743	0.1735
0.75	0.0052	0.0237		0.0646	0.2000	0.1629	0.1629	0.1625
0.76	0.0045	0.0217		0.0589	0.1893	0.1517	0.1517	0.1515
0.78	0.0032	0.0177		0.0481	0.1680	0.1294	0.1294	
0.80	0.0022	0.0137		0.0381	0.1467	0.1078	0.1078	
0.82	0.0014	0.0097		0.0291	0.1253	0.0867	0.0867	
0.84	0.0008	0.0057		0.0210	0.1040	0.0666	0.0666	
0.86	0.0004	0.0023		0.0140	0.0827	0.0477	0.0477	
0.88	0.0002	0.0005		0.0081	0.0613	0.0304	0.0304	
0.90	$5\cdot 10^{-5}$	0.0003		0.0037	0.0400	0.0156	0.0156	
0.92	$6\cdot 10^{-6}$	$3\cdot 10^{-5}$		0.0009	0.0187	0.0044	0.0044	
0.93	$5\cdot 10^{-7}$	$4\cdot 10^{-6}$		0.0002	0.0080	0.0010	0.0010	

TABLE 5. $q = 16$

	Lower bounds						Upper bounds			
δ	R_{BZ}	R_{TVZ}	R_{KTV}	R_{KT}	R_V^0	R_{GV}	R_P	R_{BE}/R_4	R_A	R_{LL}
0.0001	0.9936	0.8570	0.9995	0.9994		0.9997	0.9999	0.9998	0.9998	
0.001	0.9828	0.8561	0.9958	0.9948		0.9971	0.9990	0.9985	0.9985	
0.01	0.9087	0.8471	0.9680	0.9589		0.9766	0.9898	0.9874	0.9874	
0.02	0.8493	0.8371	0.9441	0.9294		0.9565	0.9797	0.9765	0.9764	
0.04	0.7569	0.8171	0.9041	0.8703		0.9198	0.9594	0.9561	0.9558	
0.06	0.6830	0.7971	0.8641	0.8203		0.8857	0.9390	0.9369	0.9360	0.9356
0.08	0.6203	0.7771	0.8241	0.7809		0.8533	0.9187	0.9183	0.9167	0.9151
0.10	0.5658	0.7571	0.7841	0.7416		0.8222	0.8984		0.8974	0.8946
0.12	0.5173	0.7371	0.7441	0.7022		0.7922	0.8781		0.8770	0.8741
0.14	0.4738	0.7171	0.7171	0.6628		0.7632	0.8578		0.8565	0.8535
0.16	0.4344	0.6971	0.6971	0.6234		0.7349	0.8375		0.8360	0.8330
0.18	0.3985	0.6771	0.6771	0.5841		0.7073	0.8171		0.8155	0.8125
0.20	0.3657	0.6571	0.6571	0.5447		0.6804	0.7968		0.7950	0.7920
0.24	0.3077	0.6171	0.6171	0.4825	0.6284	0.6284	0.7561		0.7540	0.7510
0.28	0.2582	0.5771	0.5771	0.4431	0.5797	0.5785	0.7156		0.7130	0.7100
0.32	0.2157	0.5371	0.5371	0.4038		0.5305	0.6749		0.6720	0.6690
0.36	0.1792	0.4971	0.4971	0.3644		0.4842	0.6343		0.6310	0.6280
0.40	0.1477	0.4571	0.4571	0.3250		0.4397	0.5937		0.5900	0.5870
0.44	0.1206	0.4171	0.4171	0.2856		0.3967	0.5530		0.5490	0.5460
0.48	0.0973	0.3771	0.3771	0.2462		0.3553	0.5124		0.5081	0.5050
0.50	0.0870	0.3571	0.3571	0.2266		0.3352	0.4921		0.4876	0.4845
0.54	0.0686	0.3171	0.3171	0.1872		0.2961	0.4514		0.4466	0.4435
0.58	0.0531	0.2771	0.2771	0.1478		0.2586	0.4108		0.4056	0.4024
0.62	0.0402	0.2371	0.2371	0.1084		0.2227	0.3702		0.3646	0.3614
0.66	0.0295	0.1971	0.1971	0.0691		0.1884	0.3295		0.3236	0.3204
0.70	0.0208	0.1571	0.1571	0.0297		0.1558	0.2889		0.2826	0.2794
0.72	0.0172	0.1371	0.1371	0.0100	0.1402	0.1402	0.2686		0.2621	0.2589
0.74	0.0140	0.1171	0.1171			0.1250	0.2483		0.2416	0.2384
0.76	0.0112	0.0971	0.0971			0.1104	0.2279		0.2211	0.2179
0.78	0.0088	0.0771	0.0771			0.0963	0.2076		0.2006	0.1974
0.80	0.0067	0.0571	0.0571			0.0827	0.1873		0.1801	0.1769
0.82	0.0050	0.0371	0.0371			0.0698	0.1670		0.1596	0.1564
0.84	0.0035	0.0171	0.0171			0.0575	0.1467	0.1462	0.1391	0.1359
0.86	0.0024		0.0065			0.0459	0.1263	0.1229	0.1186	0.1154
0.88	0.0015		0.0048			0.0351	0.1060	0.0997	0.0977	0.0949
0.90	0.0009		0.0035			0.0252	0.0857	0.0769	0.0769	0.0744
0.92	0.0004		0.0023			0.0165	0.0654	0.0546	0.0546	0.0539
0.94	0.0002		0.0013			0.0090	0.0451	0.0334	0.0334	
0.96	$3 \cdot 10^{-5}$		0.0003			0.0033	0.0248	0.0143	0.0143	
0.98	$2 \cdot 10^{-7}$		$2 \cdot 10^{-6}$			0.0001	0.0044	0.0009	0.0009	

TABLE 6. $q = 64$

δ	Lower bounds						Upper bounds			
	R_{BZ}	R_{TVZ}	R_{KTV}	R_{KT}	R_V^0	R_{GV}	R_P	R_{BE}/R_4	R_A	R_{LL}
0.0001	0.9938	0.9932	0.9996	0.9995		0.9997	0.9999	0.9999	0.9999	
0.001	0.9844	0.9323	0.9968	0.9956		0.9976	0.9990	0.9987	0.9987	
0.01	0.9173	0.9233	0.9761	0.9687		0.9799	0.9900	0.9893	0.9893	
0.02	0.8622	0.9133	0.9561	0.9445		0.9623	0.9800	0.9798	0.9797	
0.04	0.7751	0.8933	0.9161	0.9047		0.9297	0.9598		0.9597	0.9597
0.06	0.7046	0.8733	0.8761	0.8649		0.8991	0.9398		0.9396	0.9396
0.08	0.6443	0.8533	0.8533	0.8250		0.8698	0.9197		0.9195	0.9194
0.10	0.5912	0.8333	0.8333	0.7852		0.8414	0.8996		0.8994	0.8993
0.12	0.5438	0.8133	0.8133	0.7477	0.8142	0.8139	0.8795		0.8792	0.8792
0.14	0.5010	0.7933	0.7933	0.7277		0.7871	0.8595		0.8591	0.8590
0.16	0.4619	0.7733	0.7733	0.7078		0.7608	0.8394		0.8390	0.8389
0.18	0.4261	0.7533	0.7533	0.6879		0.7351	0.8193		0.8188	0.8187
0.20	0.3931	0.7333	0.7333	0.6680		0.7099	0.7992		0.7987	0.7986
0.24	0.3344	0.6933	0.6933	0.6281		0.6608	0.7591		0.7584	0.7583
0.28	0.2838	0.6533	0.6533	0.5883		0.6133	0.7189		0.7182	0.7180
0.32	0.2399	0.6133	0.6133	0.5484		0.5672	0.6787		0.6779	0.6778
0.36	0.2016	0.5733	0.5733	0.5086		0.5224	0.6386		0.6377	0.6375
0.40	0.1683	0.5333	0.5333	0.4688		0.4789	0.5984		0.5974	0.5972
0.44	0.1393	0.4933	0.4933	0.4289		0.4366	0.5583		0.5572	0.5569
0.50	0.1028	0.4333	0.4333	0.3691		0.3754	0.4980		0.4968	0.4965
0.54	0.0826	0.3933	0.3933	0.3293		0.3360	0.4579		0.4565	0.4562
0.58	0.0652	0.3533	0.3533	0.2895		0.2977	0.4177		0.4162	0.4160
0.62	0.0504	0.3133	0.3133	0.2496		0.2607	0.3776		0.3760	0.3757
0.66	0.0380	0.2733	0.2733	0.2098		0.2249	0.3374		0.3357	0.3354
0.70	0.0277	0.2333	0.2333	0.1699		0.1903	0.2973		0.2955	0.2951
0.72	0.0233	0.2133	0.2133	0.1500		0.1736	0.2771		0.2753	0.2750
0.74	0.0193	0.1933	0.1933	0.1301		0.1572	0.2571		0.2552	0.2548
0.76	0.0158	0.1733	0.1733	0.1102		0.1412	0.2370		0.2351	0.2347
0.78	0.0127	0.1533	0.1533	0.0902		0.1255	0.2169		0.2150	0.2146
0.80	0.0100	0.1333	0.1333	0.0703		0.1103	0.1969		0.1948	0.1944
0.82	0.0077	0.1133	0.1133	0.0504		0.0956	0.1768		0.1747	0.1743
0.84	0.0057	0.0933	0.0933	0.0305		0.0813	0.1567		0.1546	0.1541
0.86	0.0041	0.0733	0.0733	0.0105		0.0676	0.1366		0.1344	0.1340
0.88	0.0028	0.0533	0.0533			0.0545	0.1165		0.1143	0.1139
0.90	0.0017	0.0333	0.0333			0.0420	0.0965		0.0942	0.0937
0.92	0.0010	0.0133	0.0133			0.0304	0.0764		0.0740	0.0736
0.94	0.0005		0.0010			0.0197	0.0563		0.0539	0.0534
0.96	0.0002		0.0005			0.0104	0.0362	0.0362	0.0338	0.0333
0.98	$2 \cdot 10^{-5}$		0.0001			0.0030	0.0162	0.0132	0.0132	
0.99	$1 \cdot 10^{-6}$		$2 \cdot 10^{-5}$			0.0006	0.0061	0.0033	0.0033	

TABLE 7. $q = 256$

A.3. Additional Bounds

Here we present some numerical values for constant-weight and self-dual codes (see Sec. 4.5.4) and list bounds on higher weights of linear codes.

A.3.1. Constant-Weight Codes

This table (borrowed from [EZ87]) compares two bounds for constant-weight codes:

$R_G(\omega, \delta)$: the "Gilbert bound", i.e., the random choice bound,

$$R_G(\omega, \delta) = H_2(\omega) - \omega H_2\left(\frac{\delta}{2\omega}\right) - (1-\omega) H_2\left(\frac{\delta}{2-2\omega}\right),$$

and

$R_{EZ}(\omega, \delta)$: the Ericson–Zinoviev bound, obtained from R_{TVZ} for $\omega = p^{-2m}$,

$$R_{EZ}(\omega, \delta) = -\left(\omega - \frac{\delta}{2} - \frac{\omega\sqrt{\omega}}{1-\sqrt{\omega}}\right)\log_2 \omega.$$

For $\omega = p^{-2m} \leq 1/81$ we give values of $\varkappa = 1 - \delta/2\omega$ such that for any $\delta \in (2\omega(1-\varkappa_2), 2\omega(1-\varkappa_1))$ we have the inequality

$$R_{EZ}(\omega, \delta) > R_G(\omega, \delta)$$

and the difference $\Delta(\omega, \delta) = R_{EZ}(\omega, \delta) - R_G(\omega, \delta)$ is maximal for $\delta = \delta_0 = 2\omega(1-\varkappa_0)$; also, values of $R_{EZ}(\omega, \delta_0)$, $R_G(\omega, \delta_0)$, $\Delta(\omega, \delta_0)$, and $(\Delta(\omega, \delta_0)R_G(\omega, \delta_0)) \cdot 100\%$ are given.

ω^{-1}	$\varkappa_1$	$\varkappa_2$	$\varkappa_0$	$R_{EZ}(\omega, \delta)$	$R_G(\omega, \delta)$	$\Delta(\omega, \delta)$	%
81	0.290	0.482	0.386	0.02043	0.01994	0.00049	2.4
121	0.250	0.567	0.385	0.01630	0.01512	0.00117	7.7
169	0.164	0.609	0.384	0.01317	0.01188	0.00129	10.8
256	0.127	0.647	0.383	0.00989	0.00870	0.00118	13.6
289	0.119	0.657	0.383	0.00907	0.00794	0.00112	14.1
361	0.105	0.672	0.383	0.00771	0.00670	0.00101	15.0
529	0.085	0.694	0.383	0.00576	0.00495	0.00080	16.2
625	0.078	0.702	0.382	0.00506	0.00434	0.00072	16.6
729	0.071	0.709	0.382	0.00448	0.00384	0.00065	16.9
841	0.066	0.715	0.382	0.00400	0.00342	0.00058	17.0

A.3.2. Self-dual Codes

Since a self-dual or quasi-self-dual code always has $R = 0.5$, asymptotic behaviour of such codes is determined by the value of δ. We present (for some particular values of q) numerical values of the Scharlau bound

$$\delta_{Sch} = \frac{1}{2} - \frac{1}{\sqrt{q}-3}$$

and (for comparison) values of the basic algebraic geometry bound and the Gilbert–Varshamov bound, i.e., values δ_{TVZ} and δ_{GV} for which

$$R_{TVZ}(\delta_{TVZ}) = R_{GV}(\delta_{GV}) = \frac{1}{2}.$$

Recall that δ_{Sch} is a lower bound for quasi-self-dual codes, and if $q = 2^m$, for self-dual codes.

q	25	49	64	81	121	169	256	65536
δ_{Sch}	0	0.25	0.3	0.3333	0.375	0.4	0.4231	0.4960
δ_{GV}	0.3113	0.3375	0.3462	0.3532	0.3639	0.3718	0.3805	0.4382
δ_{TVZ}	0.25	0.3333	0.3571	0.375	0.4	0.4167	0.4333	0.4961

A.3.3. Bounds for Higher Weights

Here we list some bounds for generalized weights of linear codes; recall that existence bounds are formulated in Theorem 1.1.82 in the form of a "scheme of theorems". Generalized Hamming and Bassalygo–Elias bounds are valid as well for nonlinear codes (with an appropriate definition of generalized weights).

The generalized Singleton bound (Corollary 1.1.65):
$$d_r \leq n - k + r.$$

The generalized Plotkin bound (Theorem 1.1.68):
$$d_r \leq \left\lfloor \frac{n(q^r - 1)q^{k-r}}{q^k - 1} \right\rfloor.$$

The generalized Griesmer bound (Corollary 1.1.70):
$$d_r \geq \sum_{i=0}^{r-1} \left\lceil \frac{d_1}{q^i} \right\rceil,$$

and its corollaries: for any $r \leq s$ (Corollary 1.1.73),
$$d_s \geq d_r + \sum_{i=1}^{s-r} \left\lceil \frac{(q-1)d_r}{(q^r - 1)q^i} \right\rceil;$$

in particular, for any $r \leq k$,
$$n \geq d_r + \sum_{i=1}^{k-r} \left\lceil \frac{(q-1)d_r}{(q^r - 1)q^i} \right\rceil;$$

for $r \leq s$ (Corollary 1.1.74),
$$d_r \leq \left\lfloor \frac{d_s(q^r - 1)q^{s-r}}{q^s - 1} \right\rfloor;$$

for any $\ell \leq s - r$ (Corollary 1.1.75),
$$d_r \leq \left\lfloor \frac{(d_s - \ell)(q^r - 1)q^{s-\ell-r}}{q^{s-\ell} - 1} \right\rfloor.$$

Lower bound (Theorem 1.1.76): for $r \leq s < k$,
$$d_r \geq n - \left\lfloor \frac{(q^{k-r} - 1)(n - d_s)}{q^{k-s} - 1} \right\rfloor.$$

The generalized Hamming bound (Theorem 1.1.77):
$$\sum_{i=0}^{t}\binom{n}{i}(q-1)^i \leq q^{n-k+r-1},$$
where $t = \lfloor (d_r-1)/(r+1) \rfloor$.

The generalized Bassalygo–Elias bound (Theorem 1.1.78):
$$d_r \leq \min_{w}\left\{ n - w\prod_{i=0}^{r-1}\left(\frac{w}{n(q-1)} - \frac{q^{n-k+i}}{\binom{n}{w}(q-1)^w}\right) \right.$$
$$\left. -(n-w)\prod_{i=0}^{r-1}\left(\frac{n-w}{n} - \frac{q^{n-k+i}}{\binom{n}{w}(q-1)^w}\right)\right\},$$
where the minimum is taken over all integers w such that $0 < w < n$ and
$$\binom{n}{w}(q-1)^w \min\left\{\frac{w}{n(q-1)}, \frac{n-w}{n}\right\} \geq q^{n-k+r-1},$$
and its corollary (Corollary 1.1.79): let $\omega = w/n$ and $\delta_r = d_r/n$; then
$$1 - \delta_r \geq \max_{w}\left\{\omega\left(\frac{\omega}{q-1} - \frac{q^{n-k+r-1}}{\binom{n}{w}(q-1)^w}\right)^r + (1-\omega)\left(1 - \omega - \frac{q^{n-k+r-1}}{\binom{n}{w}(q-1)^w}\right)^r\right\},$$
where the maximum is taken over all integers w such that $0 < w < n$ and
$$\binom{n}{w}(q-1)^w \min\left\{\frac{w}{n(q-1)}, \frac{n-w}{n}\right\} \geq q^{n-k+r-1}.$$

Bibliography

[Aal84] M.J. Aaltonen, "Notes on the asymptotic behavior of the information rate of block codes," *IEEE Trans. Inform. Theory*, vol. 30, no. 1, pp. 84–85, 1984.

[Aal87] M.J. Aaltonen, "A new upper bound on nonbinary block codes," in *The very knowledge of coding*, pp. 7–39, Turku: Univ. Turku, 1987.

[Aal90] M. Aaltonen, "A new upper bound on nonbinary block codes," *Discrete Math.*, vol. 83, no. 2–3, pp. 139–160, 1990.

[AB99] A. Ashikhmin and A. Barg, "Binomial moments of the distance distribution: Bounds and applications," *IEEE Trans. Inform. Theory*, vol. 45, no. 2, pp. 438–452, 1999.

[ABL99] A. Ashikhmin, A. Barg, and S. Litsyn, "New upper bounds on generalized weights," *IEEE Trans. Inform. Theory*, vol. 45, no. 4, pp. 1258–1263, 1999.

[ABL01a] A. Ashikhmin, A. Barg, and S. Litsyn, "Estimates of the distance distribution of codes and designs," *IEEE Trans. Inform. Theory*, vol. 47, no. 3, pp. 1050–1061, 2001.

[ABL01b] A. Ashikhmin, A. Barg, and S. Litsyn, "Estimates of the distance distribution of nonbinary codes, with applications," in *Codes and Association Schemes* (A. Barg and S. Litsyn, eds.), pp. 287–303, Providence, RI: Amer. Math. Soc., 2001.

[ABV01] A. Ashikhmin, A. Barg, and S. Vlăduţ, "Linear codes with exponentially many light vectors," *J. Combin. Theory Ser. A*, vol. 96, no. 2, pp. 396–399, 2001.

[AHU74] A.V. Aho, J.E. Hopcroft, and J.D. Ullman, *The design and analysis of computer algorithms*, Reading, Mass.: Addison-Wesley, 1974.

[AM02] B. Anglès and C. Maire, "A note on tamely ramified towers of global function fields," *Finite Fields Appl.*, vol. 8, no. 2, pp. 207–215, 2002.

[AP95] Y. Aubry and M. Perret, "Coverings of singular curves over finite fields," *Manuscripta Math.*, vol. 88, no. 4, pp. 467–478, 1995.

[Atk67] A.O.L. Atkin, "Weierstrass points at cusps of $\Gamma_0(n)$," *Ann. of Math.*, vol. 85, pp. 42–45, 1967.

[Bar97] A. Barg, "The matroid of supports of a linear code," *Appl. Algebra Engrg. Comm. Comput.*, vol. 8, no. 2, pp. 165–172, 1997.

[Bar98] A. Barg, "Complexity issues in coding theory," in *Handbook of Coding Theory* (V. Pless and W.C. Huffman, eds.), vol. 1, pp. 649–754, Amsterdam: Elsevier, 1998.

[Bar02] A. Barg, "Extremal problems of coding theory," in *Coding Theory and Cryptology* (H. Niederreiter, ed.), pp. 11–48, River Edge, NJ: World Scientific, 2002.

[Bas94] L.A. Bassalygo, 1994. Private communication.

[BCS82] A. Bos, J.H. Conway, and N.J.A. Sloane, "Further lattice packings in high dimensions," *Mathematika*, vol. 29, no. 2, pp. 171–180, 1982.

[BDN03] F. Baldassarri, C. Deninger, and N. Naumann, "A motivic version of Pellikaan's two variable zeta function." Mathematics ArXiv, `math.AG/0302121` (electronic), 2003.

[Ber68] E.R. Berlekamp, *Algebraic coding theory*, New York: McGraw-Hill, 1968.

[Ber74] E.R. Berlekamp, ed., *Key papers in the development of coding theory*, New York: IEEE Press, 1974. IEEE Press Selected Reprint Series.

[Bet82] T. Beth, "Some aspects of coding theory between probability, algebra, combinatorics and complexity theory," in *Combinatorial theory (Schloss Rauischholzhausen, 1982)*, vol. 969 of Lecture Notes in Math., pp. 12–29, Berlin: Springer, 1982.

[BF02]   A. Barg and G.D. Forney, Jr., "Random codes: Minimum distances and error exponents," *IEEE Trans. Inform. Theory*, vol. 48, no. 9, pp. 2568–2573, 2002.

[BHSTR00]   J. Buchmann, T. Høholdt, H. Stichtenoth, and H. Tapia-Recillas, eds., *Coding theory, cryptography and related areas*, Berlin: Springer, 2000.

[BK75]   B.J. Birch and W. Kuyk, eds., *Modular functions of one variable. IV*, vol. 476 of Lecture Notes in Math., Berlin: Springer, 1975.

[BKT87]   A.M. Barg, G.L. Katsman, and M.A. Tsfasman, "Algebraic-geometric codes from curves of small genus," *Problemy Peredachi Informatsii*, vol. 23, no. 1, pp. 42–46, 1987.

[Bli05]   V.M. Blinovskii, "Random sphere packing," *Problemy Peredachi Informatsii*, vol. 41, no. 4, pp. 23–35, 2005.

[Bom74]   E. Bombieri, "Counting points on curves over finite fields (d'après S.A. Stepanov)," in *Séminaire Bourbaki, 25ème année (1972/1973), exp. no. 430*, vol. 383 of Lecture Notes in Math., pp. 234–241, Berlin: Springer, 1974.

[Bos00]   W. Bosma, ed., *Algorithmic number theory*, vol. 1838 of Lecture Notes in Comput. Sci., Berlin: Springer, 2000.

[Bro77]   M. Broué, "Codes correcteurs d'erreurs auto-orthogonaux sur le corps à deux éléments et formes quadratiques entières définies positives à discriminant +1," *Discrete Math.*, vol. 17, no. 3, pp. 247–269, 1977.

[Bro]   A.E. Brouwer, Tables of linear codes [electronic]. Available online at http://www.win.tue.nl/~aeb/voorlincod.html.

[BS66]   Z.I. Borevich and I.R. Shafarevich, *Number theory*, vol. 20 of Pure and Applied Mathematics, New York: Academic, 1966.

[BS85]   Z.I. Borevich and I.R. Shafarevich, *Teoriya chisel*, Moscow: Nauka, 3rd ed., 1985. English translation of the 1st ed.: see [BS66].

[BT98]   D.A. Buell and J.T. Teitelbaum, eds., *Computational perspectives on number theory*, vol. 7 of AMS/IP Studies in Advanced Mathematics, Providence, RI: Amer. Math. Soc., 1998.

[BZ82]   E.L. Blokh and V.V. Zyablov, *Lineinye kaskadnye kody* [Linear concatenated codes], Moscow: Nauka, 1982 (in Russian).

[Cal86]   J. Calmet, ed., *Algebraic algorithms and error-correcting codes*, vol. 229 of Lecture Notes in Comput. Sci., Berlin: Springer, 1986.

[Cal95]   R. Calderbank, ed., *Different aspects of coding theory*, vol. 50 of Proceedings of Symposia in Applied Mathematics, Providence, RI: Amer. Math. Soc., 1995.

[Car57]   P. Cartier, "Une nouvelle opération sur les formes différentielles," *C. R. Acad. Sci. Paris*, vol. 244, pp. 426–428, 1957.

[Cas59]   J.W.S. Cassels, *An introduction to the geometry of numbers*, Die Grundlehren der mathematischen Wissenschaften in Einzeldarstellungen mit besonderer Berücksichtigung der Anwendungsgebiete, Bd. 99, Berlin–Göttingen–Heidelberg: Springer, 1959.

[CGM95]   G. Cohen, M. Giusti, and T. Mora, eds., *Applied algebra, algebraic algorithms and error-correcting codes*, vol. 948 of Lecture Notes in Comput. Sci., Berlin: Springer, 1995.

[Che48]   N.G. Chebotarev, *Teoriya algebraicheskikh funktsii* [Theory of algebraic functions], Moscow–Leningrad: OGIZ, 1948 (in Russian).

[Che51]   C. Chevalley, *Introduction to the theory of algebraic functions of one variable*, Mathematical Surveys, no. VI, New York: Amer. Math. Soc., 1951.

[Che92]   V.V. Chepyzhov, "New lower bounds for minimum distance of linear quasi-cyclic and almost linear cyclic codes," *Problemy Peredachi Informatsii*, vol. 28, no. 1, pp. 39–51, 1992.

[CHKT00]   A. Cossidente, J.W.P. Hirschfeld, G. Korchmáros, and F. Torres, "On plane maximal curves," *Compositio Math.*, vol. 121, no. 2, pp. 163–181, 2000.

[CHMT01]   C. Ciliberto, F. Hirzebruch, R. Miranda, and M. Teicher, eds., *Applications of algebraic geometry to coding theory, physics and computation*, vol. 36 of NATO Science Series II: Mathematics, Physics and Chemistry, Dordrecht: Kluwer, 2001.

[CHZ94]   G. Cohen, L. Huguet, and G. Zémor, "Bounds on generalized weights," in *Algebraic coding (Paris, 1993)*, vol. 781 of Lecture Notes in Comput. Sci., pp. 270–277, Berlin: Springer, 1994.

[CK86]   R. Calderbank and W.M. Kantor, "The geometry of two-weight codes," *Bull. London Math. Soc.*, vol. 18, no. 2, pp. 97–122, 1986.

[CK01]   W. Chen and T. Kløve, "Weight hierarchies of binary linear codes of dimension 4," *Discrete Math.*, vol. 238, no. 1–3, pp. 27–34, 2001. Designs, codes and finite geometries (Shanghai, 1999).

[CKdR99]   N.J. Calkin, J.D. Key, and M.J. de Resmini, "Minimum weight and dimension formulae for some geometric codes," *Des. Codes Cryptogr.*, vol. 17, no. 1–3, pp. 105–120, 1999.

[CKL03]   G. Cohen, M. Krivelevich, and S. Litsyn, "Bounds on the distance distribution of codes of a given size," in *Communications, Information, and Network Security* (V.K. Bhargava et al., eds.), pp. 33–41, Boston: Kluwer, 2003.

[CL99]   C-L. Chai and W-C.W. Li, "Function fields: arithmetic and applications," in *Applications of curves over finite fields (Seattle, WA, 1997)*, vol. 245 of Contemp. Math., pp. 189–199, Providence, RI: Amer. Math. Soc., 1999.

[CLY98]   H. Chen, H.S. Luk, and S. Yau, "Explicit computation of generalized Hamming weights for some algebraic geometric codes," *Adv. in Appl. Math.*, vol. 21, no. 1, pp. 124–145, 1998.

[CLZ94]   G. Cohen, S. Litsyn, and G. Zémor, "Upper bounds on generalized distances," *IEEE Trans. Inform. Theory*, vol. 40, no. 6, pp. 2090–2092, 1994.

[CMM93]   G. Cohen, T. Mora, and O. Moreno, eds., *Applied algebra, algebraic algorithms and error-correcting codes*, vol. 673 of Lecture Notes in Comput. Sci., Berlin: Springer, 1993.

[Coh96]   H. Cohen, ed., *Algorithmic number theory*, vol. 1122 of Lecture Notes in Comput. Sci., Berlin: Springer, 1996.

[CRTL89]   M. Carral, D. Rotillon, and A. Thiong Ly, "Codes from some Artin–Schreier curves," in *Coding theory and applications (Toulon, 1988)*, vol. 388 of Lecture Notes in Comput. Sci., pp. 13–27, New York: Springer, 1989.

[CS88]   J.H. Conway and N.J.A. Sloane, *Sphere packings, lattices and groups*, vol. 290 of Grundlehren der Mathematischen Wissenschaften [Fundamental Principles of Mathematical Sciences], New York: Springer, 1988. With contributions by E. Bannai, J. Leech, S.P. Norton, A.M. Odlyzko, R.A. Parker, L. Queen, and B.B. Venkov. See also [CS99].

[CS99]   J.H. Conway and N.J.A. Sloane, *Sphere packings, lattices and groups*, vol. 290 of Grundlehren der Mathematischen Wissenschaften [Fundamental Principles of Mathematical Sciences], New York: Springer, 3rd ed., 1999. With additional contributions by E. Bannai, R.E. Borcherds, J. Leech, S.P. Norton, A.M. Odlyzko, R.A. Parker, L. Queen, and B.B. Venkov.

[Del75]   P. Deligne, "Courbes elliptiques: formulaire d'après J. Tate," in *Modular functions of one variable, IV (Proc. Internat. Summer School, Univ. Antwerp, 1972)*, vol. 476 of Lecture Notes in Math., pp. 53–73, Berlin: Springer, 1975.

[Den03]   C. Deninger, "Two-variable zeta functions and regularized products," *Doc. Math.*, Extra Vol., pp 227–259 (electronic), 2003.

[Deu41]   M. Deuring, "Die Typen der Multiplikatorenringe elliptischer Funktionenkörper," *Abh. Math. Sem. Hansischen Univ.*, vol. 14, pp. 197–272, 1941.

[DH87]   P. Deligne and D. Husemoller, "Survey of Drinfel'd modules," in *Current trends in arithmetical algebraic geometry (Arcata, Calif., 1985)*, vol. 67 of Contemp. Math., pp. 25–91, Providence, RI: Amer. Math. Soc., 1987.

[Dic15]   L.E. Dickson, "Geometric and invariantive theory of quartic curves modulo 2," *Amer. J. Math.*, vol. 37, pp. 337–354, 1915.

[Die74]   J. Dieudonné, *Cours de géométrie algébrique, I, Aperçu Historique sur le Développement de la Géométrie Algébrique*, Collection Sup., Press Univ. France, 1974.

[Die78]   J. Dieudonné, ed., *Abrégé d'histoire des mathématiques 1700–1900*, Paris: Hermann, 1978. Tome I: Algèbre, analyse classique, théorie des nombres. Tome II: Fonctions elliptiques, analyse fonctionnelle, topologie, géométrie différentielle, probabilités, logique mathématique.

[DK73]   P. Deligne and W. Kuyk, eds., *Modular functions of one variable. II*, vol. 349 of Lecture Notes in Math., Berlin: Springer, 1973.

[DM85]   Y. Driencourt and J-F. Michon, "Remarques sur les codes géométriques," *C. R. Acad. Sci. Paris Sér. I Math.*, vol. 301, no. 1, pp. 15–17, 1985.

[DM86]   Y. Driencourt and J-F. Michon, "Rapport sur les codes géométriques." Preprint, 1986.

[DM87]   Y. Driencourt and J-F. Michon, "Elliptic codes over fields of characteristic 2," *J. Pure Appl. Algebra*, vol. 45, no. 1, pp. 15–39, 1987.

[DM88]   Y. Driencourt and J-F. Michon, "Amelioration of the McWilliams–Sloane tables using geometric codes from curves with genus 1, 2 or 3 (following A.M. Barg, G.L. Katsman and M.A. Tsfasman)," in *Coding theory and applications (Cachan, 1986)*, vol. 311 of Lecture Notes in Comput. Sci., pp. 96–105, Berlin: Springer, 1988.

[DNX00]   C. Ding, H. Niederreiter, and C. Xing, "Some new codes from algebraic curves," *IEEE Trans. Inform. Theory*, vol. 46, no. 7, pp. 2638–2642, 2000.

[DR73]   P. Deligne and M. Rapoport, "Les schémas de modules de courbes elliptiques," in *Modular functions of one variable, II (Proc. Internat. Summer School, Univ. Antwerp, Antwerp, 1972)*, vol. 349 of Lecture Notes in Math., pp. 143–316, Berlin: Springer, 1973.

[Dri74]   V.G. Drinfel'd, "Elliptic modules," *Mat. Sb. (N.S.)*, vol. 94 (136), pp. 594–627, 1974.

[Dri77]   V.G. Drinfel'd, "Elliptic modules. II," *Mat. Sb. (N.S.)*, vol. 102 (144), no. 2, pp. 182–194, 1977.

[Dri85]   V.G. Drinfel'd, Unpublished manuscript on the Langlands conjecture for $GL(2)$, 1985.

[Dri86]   Y. Driencourt, "Some properties of elliptic codes over a field of characteristic 2," in *Algebraic algorithms and error correcting codes (Grenoble, 1985)*, vol. 229 of Lecture Notes in Comput. Sci., pp. 185–193, Berlin: Springer, 1986.

[Dri87]   Y. Driencourt, "Un exemple de codes géométriques: les codes elliptiques," *Traitement Signal*, vol. 4, no. 2, pp. 147–153, 1987.

[DS89]   Y. Driencourt and H. Stichtenoth, "A criterion for self-duality of geometric codes," *Comm. Algebra*, vol. 17, no. 4, pp. 885–898, 1989.

[DS98]   S. Dodunekov and J. Simonis, "Codes and projective multisets," *Electron. J. Combin.*, vol. 5, no. 1, 1998. Research Paper 37, 23 pp. (electronic).

[Duu99]   I.M. Duursma, "Weight distributions of geometric Goppa codes," *Trans. Amer. Math. Soc.*, vol. 351, no. 9, pp. 3609–3639, 1999.

[Elk94]   N.D. Elkies, "Mordell–Weil lattices in characteristic 2. I. Construction and first properties," *Internat. Math. Res. Notices*, no. 8, pp. 343 ff. (electronic), 1994.

[Elk97a] N.D. Elkies, "Mordell–Weil lattices in characteristic 2. II. The Leech lattice as a Mordell–Weil lattice," *Invent. Math.*, vol. 128, no. 1, pp. 1–8, 1997.

[Elk97b] N. Elkies, 1997. Private communication.

[Elk01a] N.D. Elkies, "Excellent codes from modular curves," in *Proceedings of the 33rd Annual ACM Symposium on Theory of Computing*, pp. 200–208 (electronic), New York: ACM, 2001.

[Elk01b] N.D. Elkies, "Explicit towers of Drinfeld modular curves," in *European Congress of Mathematics, Vol. II (Barcelona, 2000)*, vol. 202 of Progr. Math., pp. 189–198, Basel: Birkhäuser, 2001.

[Elk01c] N.D. Elkies, "Mordell–Weil lattices in characteristic 2. III. A Mordell–Weil lattice of rank 128," *Experiment. Math.*, vol. 10, no. 3, pp. 467–473, 2001.

[ES01]   B. Engquist and W. Schmid, eds., *Mathematics unlimited—2001 and beyond. Parts I, II*, Berlin: Springer, 2001.

[EZ87]   T. Ericson and V.A. Zinoviev, "An improvement of the Gilbert bound for constant weight codes," *IEEE Trans. Inform. Theory*, vol. 33, no. 5, pp. 721–723, 1987.

[FGT97] R. Fuhrmann, A. García, and F. Torres, "On maximal curves," *J. Number Theory*, vol. 67, no. 1, pp. 29–51, 1997.

[For66] G.D. Forney, Jr., *Concatenated codes*, Cambridge, Mass.: MIT Press, 1966. MIT Research Monograph, no. 37.

[FR95] G-L. Feng and T.R.N. Rao, "Improved geometric Goppa codes. I. Basic theory," *IEEE Trans. Inform. Theory*, vol. 41, no. 6, part 1, pp. 1678–1693, 1995. Special issue on algebraic geometry codes.

[Fri99a] M.D. Fried, "Applications of curves over finite fields," in *Applications of curves over finite fields (Seattle, WA, 1997)*, vol. 245 of Contemp. Math., pp. ix–xxiii, Providence, RI: Amer. Math. Soc., 1999.

[Fri99b] M.D. Fried, ed., *Applications of curves over finite fields*, vol. 245 of Contemp. Math., Providence, RI: Amer. Math. Soc., 1999.

[FT72] L. Fejes Tóth, *Lagerungen in der Ebene auf der Kugel und im Raum*, Berlin: Springer, 1972. Zweite verbesserte und erweiterte Auflage, Die Grundlehren der mathematischen Wissenschaften, Bd. 65.

[FT96] R. Fuhrmann and F. Torres, "The genus of curves over finite fields with many rational points," *Manuscripta Math.*, vol. 89, no. 1, pp. 103–106, 1996.

[Ful69] W. Fulton, *Algebraic curves. An introduction to algebraic geometry*, New York–Amsterdam: Benjamin, 1969. Notes written with the collaboration of R. Weiss, Mathematics Lecture Notes Series.

[Gar92] A. García, "On Goppa codes and Artin–Schreier extensions," *Comm. Algebra*, vol. 20, no. 12, pp. 3683–3689, 1992.

[Gar01] A. García, "Curves over finite fields attaining the Hasse–Weil upper bound," in *European Congress of Mathematics, Vol. II (Barcelona, 2000)*, vol. 202 of Progr. Math., pp. 199–205, Basel: Birkhäuser, 2001.

[Gar02] A. García, "On curves with many rational points over finite fields," in *Finite fields with applications to coding theory, cryptography, and related areas (Oaxaca, México, 2001)* (G.L. Mullen, H. Stichtenoth, and H. Tapia-Recillas, eds.), pp. 152–163, Berlin: Springer, 2002.

[Gar03] A. Garzón, "Euclidean algorithm and Kummer covers with many points," *Rev. Colombiana Mat.*, vol. 37, no. 1, pp. 37–50, 2003.

[GG03] A. García and A. Garzon, "On Kummer covers with many rational points over finite fields," *J. Pure Appl. Algebra*, vol. 185, no. 1–3, pp. 177–192, 2003.

[GH78] P. Griffiths and J. Harris, *Principles of algebraic geometry*, Pure and Applied Mathematics, New York: Wiley, 1978. See also [GH94b].

[GH94a] A. García and M. Homma, "Frobenius order-sequences of curves," in *Algebra and number theory (Essen, 1992)*, pp. 27–41, Berlin: Gruyter, 1994.

[GH94b] P. Griffiths and J. Harris, *Principles of algebraic geometry*, Wiley Classics Library, New York: Wiley, 1994. Reprint of the 1978 original.

[GH95] D.G. Glynn and J.W.P. Hirschfeld, "On the classification of geometric codes by polynomial functions," *Des. Codes Cryptogr.*, vol. 6, no. 3, pp. 189–204, 1995.

[GKL93] A. García, S.J. Kim, and R.F. Lax, "Consecutive Weierstrass gaps and minimum distance of Goppa codes," *J. Pure Appl. Algebra*, vol. 84, no. 2, pp. 199–207, 1993.

[GL92] A. García and R.F. Lax, "Goppa codes and Weierstrass gaps," in *Coding theory and algebraic geometry (Luminy, 1991)*, vol. 1518 of Lecture Notes in Math., pp. 33–42, Berlin: Springer, 1992.

[Gop81] V.D. Goppa, "Codes on algebraic curves," *Dokl. Akad. Nauk SSSR*, vol. 259, no. 6, pp. 1289–1290, 1981.

[Gop82] V.D. Goppa, "Algebraic-geometric codes," *Izv. Akad. Nauk SSSR Ser. Mat.*, vol. 46, no. 4, pp. 762–781, 1982.

[Gop84] V.D. Goppa, "Codes and information," *Uspekhi Mat. Nauk*, vol. 39, no. 1, pp. 77–120, 1984.

[Gop88] V.D. Goppa, *Geometry and codes*, vol. 24 of Mathematics and Its Applications (Soviet Series), Dordrecht: Kluwer, 1988.

[Gos80] D. Goss, "The algebraist's upper half-plane," *Bull. Amer. Math. Soc. (N.S.)*, vol. 2, no. 3, pp. 391–415, 1980.

[GPS98]   K. Győry, A. Pethő, and V.T. Sós, eds., *Number theory*, Berlin: Gruyter, 1998. Diophantine, computational and algebraic aspects.

[GQ01]   A. García and L. Quoos, "A construction of curves over finite fields," *Acta Arith.*, vol. 98, no. 2, pp. 181–195, 2001.

[GR06]   V. Guruswami and A. Rudra, "Limits to list decoding Reed–Solomon codes," *IEEE Trans. Inform. Theory*, vol. 52, no. 8, pp. 3642–3649, 2006.

[Gro60]   A. Grothendieck, "Éléments de géométrie algébrique. I. Le langage des schémas," *Inst. Hautes Études Sci. Publ. Math.*, no. 4, p. 228, 1960.

[Gro61a] A. Grothendieck, "Éléments de géométrie algébrique. II. Étude globale élémentaire de quelques classes de morphismes," *Inst. Hautes Études Sci. Publ. Math.*, no. 8, p. 222, 1961.

[Gro61b] A. Grothendieck, "Éléments de géométrie algébrique. III. Étude cohomologique des faisceaux cohérents. I," *Inst. Hautes Études Sci. Publ. Math.*, no. 11, p. 167, 1961.

[Gro63]   A. Grothendieck, "Éléments de géométrie algébrique. III. Étude cohomologique des faisceaux cohérents. II," *Inst. Hautes Études Sci. Publ. Math.*, no. 17, p. 91, 1963.

[Gro64]   A. Grothendieck, "Éléments de géométrie algébrique. IV. Étude locale des schémas et des morphismes de schémas. I," *Inst. Hautes Études Sci. Publ. Math.*, no. 20, p. 259, 1964.

[Gro65]   A. Grothendieck, "Éléments de géométrie algébrique. IV. Étude locale des schémas et des morphismes de schémas. II," *Inst. Hautes Études Sci. Publ. Math.*, no. 24, p. 231, 1965.

[Gro66]   A. Grothendieck, "Éléments de géométrie algébrique. IV. Étude locale des schémas et des morphismes de schémas. III," *Inst. Hautes Études Sci. Publ. Math.*, no. 28, p. 255, 1966.

[Gro67]   A. Grothendieck, "Éléments de géométrie algébrique. IV. Étude locale des schémas et des morphismes de schémas. IV," *Inst. Hautes Études Sci. Publ. Math.*, no. 32, p. 361, 1967.

[Gro90]   B.H. Gross, "Group representations and lattices," *J. Amer. Math. Soc.*, vol. 3, no. 4, pp. 929–960, 1990.

[GS64]   E.S. Golod and I.R. Shafarevich, "On the class field tower," *Izv. Akad. Nauk SSSR Ser. Mat.*, vol. 28, pp. 261–272, 1964.

[GS95a]   A. García and H. Stichtenoth, "Algebraic function fields over finite fields with many rational places," *IEEE Trans. Inform. Theory*, vol. 41, no. 6, part 1, pp. 1548–1563, 1995. Special issue on algebraic geometry codes.

[GS95b]   A. García and H. Stichtenoth, "A tower of Artin–Schreier extensions of function fields attaining the Drinfel'd–Vlăduţ bound," *Invent. Math.*, vol. 121, no. 1, pp. 211–222, 1995.

[GS96a]   A. García and H. Stichtenoth, "Asymptotically good towers of function fields over finite fields," *C. R. Acad. Sci. Paris Sér. I Math.*, vol. 322, no. 11, pp. 1067–1070, 1996.

[GS96b]   A. García and H. Stichtenoth, "On the asymptotic behaviour of some towers of function fields over finite fields," *J. Number Theory*, vol. 61, no. 2, pp. 248–273, 1996.

[GS99a]   A. García and H. Stichtenoth, "A class of polynomials over finite fields," *Finite Fields Appl.*, vol. 5, no. 4, pp. 424–435, 1999.

[GS99b]   V. Guruswami and M. Sudan, "Improved decoding of Reed–Solomon and algebraic-geometry codes," *IEEE Trans. Inform. Theory*, vol. 45, no. 6, pp. 1757–1767, 1999.

[GST97]   A. García, H. Stichtenoth, and M. Thomas, "On towers and composita of towers of function fields over finite fields," *Finite Fields Appl.*, vol. 3, no. 3, pp. 257–274, 1997.

[GSX00]   A. García, H. Stichtenoth, and C-P. Xing, "On subfields of the Hermitian function field," *Compositio Math.*, vol. 120, no. 2, pp. 137–170, 2000.

[GT99]   A. García and F. Torres, "On maximal curves having classical Weierstrass gaps," in *Applications of curves over finite fields (Seattle, WA, 1997)*, vol. 245 of Contemp. Math., pp. 49–59, Providence, RI: Amer. Math. Soc., 1999.

[GV88]   A. García and J.F. Voloch, "Fermat curves over finite fields," *J. Number Theory*, vol. 30, no. 3, pp. 345–356, 1988.

[Hac95]   G. Haché, "Computation in algebraic function fields for effective construction of algebraic-geometric codes," in *Applied algebra, algebraic algorithms and error-correcting codes (Paris, 1995)*, vol. 948 of Lecture Notes in Comput. Sci., pp. 262–278, Berlin: Springer, 1995.

[Ham93]   N. Hamada, "A characterization of some $[n, k, d; q]$-codes meeting the Griesmer bound using a minihyper in a finite projective geometry," *Discrete Math.*, vol. 116, no. 1–3, pp. 229–268, 1993.

[Han87a] J.P. Hansen, "Codes on the Klein quartic, ideals, and decoding," *IEEE Trans. Inform. Theory*, vol. 33, no. 6, pp. 923–925, 1987.

[Han87b] J.P. Hansen, "Group codes and algebraic curves." Mathematica Göttingensis, Schriftenreihe SFB Geometrie und Analysis, Heft 9, Feb. 1987.

[Han95]   S.H. Hansen, *Rational points on curves over finite fields*, vol. 64 of Lecture Notes Series, Aarhus: Aarhus Universitet Matematisk Institut, 1995.

[Har77]   R. Hartshorne, *Algebraic geometry*, Graduate Texts in Mathematics, no. 52, New York: Springer, 1977.

[Hir83]   J.W.P. Hirschfeld, "Maximum sets in finite projective spaces," in *Surveys in combinatorics (Southampton, 1983)*, vol. 82 of London Math. Soc. Lecture Note Ser., pp. 55–76, Cambridge: Cambridge Univ. Press, 1983.

[Hir84]   J.W.P. Hirschfeld, "Linear codes and algebraic curves," in *Geometrical combinatorics (Milton Keynes, 1984)*, vol. 114 of Res. Notes in Math., pp. 35–53, Boston, MA: Pitman, 1984.

[Hir90]   J.W.P. Hirschfeld, "Codes and curves," in *Finite geometries, buildings, and related topics (Pingree Park, CO, 1988)*, Oxford Sci. Publ., pp. 129–144, New York: Oxford Univ. Press, 1990.

[Hir98]   J.W.P. Hirschfeld, "Codes on curves and their geometry," *Rend. Circ. Mat. Palermo (2) Suppl.*, no. 51, pp. 123–137, 1998.

[HK91]   A. Hefez and N. Kakuta, "New bounds for Fermat curves over finite fields," in *Enumerative algebraic geometry (Copenhagen, 1989)*, vol. 123 of Contemp. Math., pp. 89–97, Providence, RI: Amer. Math. Soc., 1991.

[HK01]   M. Homma and S.J. Kim, "Goppa codes with Weierstrass pairs," *J. Pure Appl. Algebra*, vol. 162, no. 2–3, pp. 273–290, 2001.

[HKLY95] T. Helleseth, T. Kløve, V.I. Levenshtein, and Ø. Ytrehus, "Bounds on the minimum support weights," *IEEE Trans. Inform. Theory*, vol. 41, no. 2, pp. 432–440, 1995.

[HKM77]   T. Helleseth, T. Kløve, and J. Mykkeltveit, "The weight distribution of irreducible cyclic codes with block length $n_1((q^l - 1)/N)$," *Discrete Math.*, vol. 18, no. 2, pp. 179–211, 1977.

[HKY92]   T. Helleseth, T. Kløve, and Ø. Ytrehus, "Generalized Hamming weights of linear codes," *IEEE Trans. Inform. Theory*, vol. 38, no. 3, pp. 1133–1140, 1992.

[HKY94]   T. Helleseth, T. Kløve, and Ø. Ytrehus, "Generalizations of the Griesmer bound," in *Error control, cryptology, and speech compression (Moscow, 1993)*, vol. 829 of Lecture Notes in Comput. Sci., pp. 41–52, Berlin: Springer, 1994.

[HL03]   E.W. Howe and K.E. Lauter, "Improved upper bounds for the number of points on curves over finite fields," *Ann. Inst. Fourier*, vol. 53, no. 6, pp. 1677–1737, 2003.

[HLB95]   G. Haché and D. Le Brigand, "Effective construction of algebraic geometry codes," *IEEE Trans. Inform. Theory*, vol. 41, no. 6, part 1, pp. 1615–1628, 1995. Special issue on algebraic geometry codes.

[HS90]   J.P. Hansen and H. Stichtenoth, "Group codes on certain algebraic curves with many rational points," *Appl. Algebra Engrg. Comm. Comput.*, vol. 1, no. 1, pp. 67–77, 1990.

[HS98]   J.W.P. Hirschfeld and L. Storme, "The packing problem in statistics, coding theory and finite projective spaces," *J. Statist. Plann. Inference*, vol. 72, no. 1–2, pp. 355–380, 1998. R.C. Bose Memorial Conference (Fort Collins, CO, 1995).

[HS01]   J.W.P. Hirschfeld and L. Storme, "The packing problem in statistics, coding theory and finite projective spaces: update 2001," in *Finite Geometries* (A. Blokhuis, J.W. Hirschfeld, J. Dieter, and J.A. Thas, eds.), vol. 3 of Developments in Mathematics, pp. 201–246, Dordrecht: Kluwer, 2001. Available from http://www.maths.susx.ac.uk/Staff/JWPH/CONF/IOT/book13.ps.

[HT78] N. Hamada and F. Tamari, "On a geometrical method of construction of maximal t-linearly independent sets," *J. Combin. Theory Ser. A*, vol. 25, no. 1, pp. 14–28, 1978.

[HTV94] J.W.P. Hirschfeld, M.A. Tsfasman, and S.G. Vlăduţ, "The weight hierarchy of higher-dimensional Hermitian codes," *IEEE Trans. Inform. Theory*, vol. 40, no. 1, pp. 275–278, 1994.

[HV88] J.W.P. Hirschfeld and J.F. Voloch, "The characterization of elliptic curves over finite fields," *J. Austral. Math. Soc. Ser. A*, vol. 45, no. 2, pp. 275–286, 1988.

[Ibu93] T. Ibukiyama, "On rational points of curves of genus 3 over finite fields," *T'hoku Math. J.*, vol. 45, pp. 311–329, 1993.

[Iha75] Y. Ihara, "On modular curves over finite fields," in *Discrete subgroups of Lie groups and applications to moduli (Internat. Colloq., Bombay, 1973)*, pp. 161–202, Bombay: Oxford Univ. Press, 1975.

[Iha81] Y. Ihara, "Some remarks on the number of rational points of algebraic curves over finite fields," *J. Fac. Sci. Univ. Tokyo Sect. IA Math.*, vol. 28, no. 3, pp. 721–724, 1981.

[Iha82] Y. Ihara, "Congruence relations and fundamental groups," *J. Algebra*, vol. 75, no. 2, pp. 445–451, 1982.

[Iha99] Y. Ihara, "Shimura curves over finite fields and their rational points," in *Applications of curves over finite fields (Seattle, WA, 1997)*, vol. 245 of Contemp. Math., pp. 15–23, Providence, RI: Amer. Math. Soc., 1999.

[ILP+89] M. Isaacs, A. Lichtman, D. Passman, S. Sehgal, N.J.A. Sloane, and H. Zassenhaus, eds., *Representation theory, group rings, and coding theory*, vol. 93 of Contemp. Math., Providence, RI: Amer. Math. Soc., 1989. Papers in honour of S.D. Berman (1922–1987).

[Jan90] H. Janwa, "Some optimal codes from algebraic geometry and their covering radii," *European J. Combin.*, vol. 11, no. 3, pp. 249–266, 1990.

[Jan91] H. Janwa, "On the parameters of algebraic geometric codes," in *Applied algebra, algebraic algorithms and error-correcting codes (New Orleans, LA, 1991)*, vol. 539 of Lecture Notes in Comput. Sci., pp. 19–28, Berlin: Springer, 1991.

[JLJ+89] J. Justesen, K.J. Larsen, H.E. Jensen, A. Havemose, and T. Høholdt, "Construction and decoding of a class of algebraic geometry codes," *IEEE Trans. Inform. Theory*, vol. 35, no. 4, pp. 811–821, 1989.

[Kaw03] M.Q. Kawakita, "Kummer curves and their fibre products with many rational points," *Appl. Algebra Engrg. Comm. Comput.*, vol. 14, no. 1, pp. 55–64, 2003.

[Kel01] A. Keller, "Cyclotomic function fields with many rational places," in *Finite fields and applications (Augsburg, 1999)*, pp. 293–302, Berlin: Springer, 2001.

[KFR00] M.S. Kolluru, G-L. Feng, and T.R.N. Rao, "Construction of improved geometric Goppa codes from Klein curves and Klein-like curves," *Appl. Algebra Engrg. Comm. Comput.*, vol. 10, no. 6, pp. 433–464, 2000.

[Kie94] H. Kiechle, "Points on Fermat curves over finite fields," in *Finite fields: theory, applications, and algorithms (Las Vegas, NV, 1993)*, vol. 168 of Contemp. Math., pp. 181–183, Providence, RI: Amer. Math. Soc., 1994.

[Kir96] C. Kirfel, "On the Clifford defect for special curves," in *Arithmetic, geometry and coding theory (Luminy, 1993)*, pp. 67–75, Berlin: Gruyter, 1996.

[KL78] G.A. Kabatiansky and V.I. Levenshtein, "On bounds for packings on a sphere and in space," *Problemy Peredachi Informatsii*, vol. 14, no. 1, pp. 3–25, 1978.

[KL89] N. Katz and R. Livné, "Sommes de Kloosterman et courbes elliptiques universelles en caractéristiques 2 et 3," *C. R. Acad. Sci. Paris Sér. I Math.*, vol. 309, no. 11, pp. 723–726, 1989.

[Klø78] T. Kløve, "The weight distribution of linear codes over $GF(q^l)$ having generator matrices over $GF(q)$," *Discrete Math.*, vol. 23, pp. 159–168, 1978.

[Klø92] T. Kløve, "Support weight distribution of linear codes," *Discrete Math.*, vol. 106/107, pp. 311–316, 1992. A collection of contributions in honour of J. van Lint.

[Klø93] T. Kløve, "Minimum support weights of binary codes," *IEEE Trans. Inform. Theory*, vol. 39, no. 2, pp. 648–654, 1993.

[KM85]   N.M. Katz and B. Mazur, *Arithmetic moduli of elliptic curves*, vol. 108 of Annals of Mathematics Studies, Princeton, NJ: Princeton Univ. Press, 1985.

[Kob93]   N. Koblitz, *Introduction to elliptic curves and modular forms*, vol. 97 of Graduate Texts in Mathematics, New York: Springer, 2nd ed., 1993.

[Koc87]   H.V. Koch, "Unimodular lattices and self-dual codes," in *Proceedings of the International Congress of Mathematicians, Vols. 1, 2 (Berkeley, Calif., 1986)*, pp. 457–465, Providence, RI: Amer. Math. Soc., 1987.

[KP95]   C. Kirfel and R. Pellikaan, "The minimum distance of codes in an array coming from telescopic semigroups," *IEEE Trans. Inform. Theory*, vol. 41, no. 6, part 1, pp. 1720–1732, 1995. Special issue on algebraic geometry codes.

[Kra88]   V.Yu. Krachkovsky, "Decoding of codes on algebraic curves." Preprint, 1988 (in Russian).

[KS73]   W. Kuyk and J-P. Serre, eds., *Modular functions of one variable. III*, vol. 350 of Lecture Notes in Math., Berlin: Springer, 1973.

[KT87]   G.L. Katsman and M.A. Tsfasman, "Spectra of algebraic-geometric codes," *Problemy Peredachi Informatsii*, vol. 23, no. 4, pp. 19–34, 1987.

[KT89]   G.L. Katsman and M.A. Tsfasman, "A remark on algebraic geometric codes," in *Representation theory, group rings, and coding theory*, vol. 93 of Contemp. Math., pp. 197–199, Providence, RI: Amer. Math. Soc., 1989.

[KT02]   G. Korchmáros and F. Torres, "On the genus of a maximal curve," *Math. Ann.*, vol. 323, no. 3, pp. 589–608, 2002.

[KTII78]   T. Kasami, N. Tokura, Y. Iwadare, and Y. Inagaki, *Teoriya kodirovaniya* [Coding theory], Moscow: Mir, 1978 (in Russian; translated from Japanese).

[KTV84]   G.L. Katsman, M.A. Tsfasman, and S.G. Vlăduţ, "Modular curves and codes with a polynomial construction," *IEEE Trans. Inform. Theory*, vol. 30, no. 2, pp. 353–355, 1984.

[KTV92]   G.L. Katsman, M.A. Tsfasman, and S.G. Vladuţ, "Spectra of linear codes and error probability of decoding," in *Coding theory and algebraic geometry (Luminy, 1991)*, vol. 1518 of Lecture Notes in Math., pp. 82–98, Berlin: Springer, 1992.

[Kuy73]   W. Kuyk, ed., *Modular functions of one variable. I*, vol. 320 of Lecture Notes in Math., Berlin: Springer, 1973.

[Lac86]   G. Lachaud, "Les codes géométriques de Goppa," *Astérisque*, no. 133–134, pp. 189–207, 1986. Seminar Bourbaki, Vol. 1984/85.

[Lac87]   G. Lachaud, "Sommes d'Eisenstein et nombre de points de certaines courbes algébriques sur les corps finis," *C. R. Acad. Sci. Paris Sér. I Math.*, vol. 305, no. 16, pp. 729–732, 1987.

[Lac89a]   G. Lachaud, "Distribution of the weights of the dual of the Melas code," *Discrete Math.*, vol. 79, no. 1, pp. 103–106, 1989.

[Lac89b]   G. Lachaud, "Exponential sums and the Carlitz–Uchiyama bound," in *Coding theory and applications (Toulon, 1988)*, vol. 388 of Lecture Notes in Comput. Sci., pp. 63–75, New York: Springer, 1989.

[Lac90]   G. Lachaud, "The parameters of projective Reed–Muller codes," *Discrete Math.*, vol. 81, no. 2, pp. 217–221, 1990.

[Lac91]   G. Lachaud, "Artin–Schreier curves, exponential sums, and the Carlitz–Uchiyama bound for geometric codes," *J. Number Theory*, vol. 39, no. 1, pp. 18–40, 1991.

[Lan65]   S. Lang, *Algebra*, Reading, Mass.: Addison-Wesley, 1965.

[Lan70]   S. Lang, *Algebraic number theory*, Reading, Mass.–London–Don Mills, Ont.: Addison-Wesley, 1970.

[Lan73]   S. Lang, *Elliptic functions*, Reading, Mass.–London–Amsterdam: Addison-Wesley, 1973. With an appendix by J. Tate.

[Lan76]   S. Lang, *Introduction to modular forms*, Berlin: Springer, 1976. Grundlehren der mathematischen Wissenschaften, no. 222. See also [Lan95].

[Lan87]   S. Lang, *Elliptic functions*, vol. 112 of Graduate Texts in Mathematics, New York: Springer, 2nd ed., 1987. With an appendix by J. Tate.

[Lan94]   S. Lang, *Algebraic number theory*, vol. 110 of Graduate Texts in Mathematics, New York: Springer, 2nd ed., 1994.

[Lan95]   S. Lang, *Introduction to modular forms*, vol. 222 of Grundlehren der Mathematischen Wissenschaften [Fundamental Principles of Mathematical Sciences], Berlin: Springer, 1995. With appendixes by D. Zagier and W. Feit, Corrected reprint of the 1976 original.

[Lan02]   S. Lang, *Algebra*, vol. 211 of Graduate Texts in Mathematics, New York: Springer, 3rd ed., 2002.

[Lau99a]   K. Lauter, "A formula for constructing curves over finite fields with many rational points," *J. Number Theory*, vol. 74, no. 1, pp. 56–72, 1999.

[Lau99b]   K. Lauter, "Improved upper bounds for the number of rational points on algebraic curves over finite fields," *C. R. Acad. Sci. Paris Sér. I Math.*, vol. 328, no. 12, pp. 1181–1185, 1999.

[Lau00]   K. Lauter, "Zeta functions of curves over finite fields with many rational points," in *Coding theory, cryptography and related areas (Guanajuato, 1998)*, pp. 167–174, Berlin: Springer, 2000.

[Lau01]   K. Lauter, "Geometric methods for improving the upper bounds on the number of rational points on algebraic curves over finite fields," *J. Algebraic Geom.*, vol. 10, no. 1, pp. 19–36, 2001. With an appendix by J-P. Serre.

[LB86]   D. Le Brigand, "On computational complexity of some algebraic curves over finite fields," in *Algebraic algorithms and error correcting codes (Grenoble, 1985)*, vol. 229 of Lecture Notes in Comput. Sci., pp. 223–227, Berlin: Springer, 1986.

[LB88]   D. Le Brigand, "A [32, 17, 14]-geometric code coming from a singular curve," in *Coding theory and applications (Cachan, 1986)*, vol. 311 of Lecture Notes in Comput. Sci., pp. 106–115, Berlin: Springer, 1988.

[LBR88]   D. Le Brigand and J-J. Risler, "Algorithme de Brill–Noether et codes de Goppa," *Bull. Soc. Math. France*, vol. 116, no. 2, pp. 231–253, 1988.

[Lee64]   J. Leech, "Some sphere packings in higher space," *Canad. J. Math.*, vol. 16, pp. 657–682, 1964.

[Lee67]   J. Leech, "Notes on sphere packings," *Canad. J. Math.*, vol. 19, pp. 251–267, 1967.

[Len86]   H.W. Lenstra, Jr., "Codes from algebraic number fields," in *Mathematics and computer science, II (Amsterdam, 1986)*, vol. 4 of CWI Monogr., pp. 95–104, Amsterdam: North-Holland, 1986.

[Lev83]   V.I. Levenshtein, "Bounds for packings of metric spaces and some of their applications," *Problemy Kibernet.*, no. 40, pp. 43–110, 1983.

[Li96]   W-C.W. Li, *Number theory with applications*, vol. 7 of Series on Univ. Mathematics, River Edge, NJ: World Scientific, 1996.

[Li00]   W-C.W. Li, "Various constructions of good codes," in *Combinatorial and computational algebra (Hong Kong, 1999)*, vol. 264 of Contemp. Math., pp. 271–285, Providence, RI: Amer. Math. Soc., 2000.

[LL98]   T. Laihonen and S. Litsyn, "On upper bounds for minimum distance and covering radius of non-binary codes," *Des. Codes Cryptogr.*, vol. 14, no. 1, pp. 71–80, 1998.

[LL04]   I. Luengo and B. López, "Algebraic curves over $\mathbb{F}_3$ with many rational points," in *Algebra, arithmetic and geometry with applications (West Lafayette, IN, 2000)* (C. Christensen, G. Sundaram, A. Sathaye, and C. Bajaj, eds.), pp. 619–626, Berlin: Springer, 2004.

[LMD90]   G. Lachaud and M. Martin-Deschamps, "Nombre de points des jacobiennes sur un corps fini," *Acta Arith.*, vol. 56, no. 4, pp. 329–340, 1990.

[Lom02]   C. Lomont, "Yet more projective curves over $\mathbb{F}_2$," *Experiment. Math.*, vol. 11, no. 4, pp. 547–554, 2002.

[LR03]   J.C. Lagarias and E. Rains, "On a two-variable zeta function for number fields," *Ann. Inst. Fourier (Grenoble)*, vol. 53, no. 1, pp. 1–68, 2003.

[LSH97]   J. Little, K. Saints, and C. Heegard, "On the structure of Hermitian codes," *J. Pure Appl. Algebra*, vol. 121, no. 3, pp. 293–314, 1997.

[LT85]   S.N. Litsyn and M.A. Tsfasman, "Algebraic-geometric and number-theoretic packings of balls in $\mathbb{R}^N$," *Uspekhi Mat. Nauk*, vol. 40, no. 2, pp. 185–186, 1985.

[LT86]   S.N. Litsyn and M.A. Tsfasman, "A note on lower bounds," *IEEE Trans. Inform. Theory*, vol. 32, no. 5, pp. 705–706, 1986.

[LT87]   S.N. Litsyn and M.A. Tsfasman, "Constructive high-dimensional sphere packings," *Duke Math. J.*, vol. 54, no. 1, pp. 147–161, 1987.

[LT97]   G. Lachaud and M.A. Tsfasman, "Formules explicites pour le nombre de points des variétés sur un corps fini," *J. Reine Angew. Math.*, vol. 493, pp. 1–60, 1997.

[LTJW95]   G. Lachaud, M.A. Tsfasman, J. Justesen, and V.K. Wei, eds., Special issue on algebraic geometry codes: *IEEE Trans. Inform. Theory*, vol. 41, no. 6, part 1, 1995.

[LW87]   G. Lachaud and J. Wolfmann, "Sommes de Kloosterman, courbes elliptiques et codes cycliques en caractéristique 2," *C. R. Acad. Sci. Paris Sér. I Math.*, vol. 305, no. 20, pp. 881–883, 1987.

[LW90]   G. Lachaud and J. Wolfmann, "The weights of the orthogonals of the extended quadratic binary Goppa codes," *IEEE Trans. Inform. Theory*, vol. 36, no. 3, pp. 686–692, 1990.

[Man81]   Yu.I. Manin, "What is the maximum number of points on a curve over $\mathbb{F}_2$?," *J. Fac. Sci. Univ. Tokyo Sect. IA Math.*, vol. 28, no. 3, pp. 715–720, 1981.

[Mar78]   J. Martinet, "Tours de corps de classes et estimations de discriminants," *Invent. Math.*, vol. 44, no. 1, pp. 65–73, 1978.

[Mat01]   G.L. Matthews, "Weierstrass pairs and minimum distance of Goppa codes," *Des. Codes Cryptogr.*, vol. 22, no. 2, pp. 107–121, 2001.

[MH73]   J. Milnor and D. Husemoller, *Symmetric bilinear forms*, Ergebnisse der Mathematik und ihrer Grenzgebiete, Bd. 73, New York: Springer, 1973.

[Mic84]   J-F. Michon, "Codes de Goppa," in *Seminar on number theory, 1983–1984 (Talence, 1983/1984)*, vol. 13, exp. no. 7, pp. 1–17, Talence: Univ. Bordeaux I, 1984.

[Mic85]   J-F. Michon, "Les codes BCH comme codes géométriques." Preprint, 1985.

[Mic86]   J-F. Michon, "Amélioration de paraméters de codes de Goppa." Preprint, 1986.

[Mil76]   J. Milnor, "Hilbert's problem 18: on crystallographic groups, fundamental domains, and on sphere packing," in *Mathematical developments arising from Hilbert problems (Proc. Sympos. Pure Math., Northern Illinois Univ., De Kalb, Ill., 1974)* (F.E. Browder, ed.), vol. XXVIII of Proc. Symposia Pure Math., pp. 491–506, Providence, R.I.: Amer. Math. Soc., 1976.

[MM91]   C.J. Moreno and O. Moreno, "Exponential sums and Goppa codes. I," *Proc. Amer. Math. Soc.*, vol. 111, no. 2, pp. 523–531, 1991.

[MM92a]   C.J. Moreno and O. Moreno, "Exponential sums and Goppa codes. II," *IEEE Trans. Inform. Theory*, vol. 38, no. 4, pp. 1222–1229, 1992.

[MM92b]   C.J. Moreno and O. Moreno, "An improved Bombieri–Weil bound and applications to coding theory," *J. Number Theory*, vol. 42, no. 1, pp. 32–46, 1992.

[Mor91]   C.J. Moreno, *Algebraic curves over finite fields*, vol. 97 of Cambridge Tracts in Mathematics, Cambridge: Cambridge Univ. Press, 1991.

[MP93a]   C. Munuera and R. Pellikaan, "Equality of geometric Goppa codes and equivalence of divisors," *J. Pure Appl. Algebra*, vol. 90, no. 3, pp. 229–252, 1993.

[MP93b]   C. Munuera and R. Pellikaan, "Self-dual and decomposable geometric Goppa codes," in *Eurocode'92 (Udine, 1992)*, vol. 339 of CISM Courses and Lectures, pp. 77–87, Vienna: Springer, 1993.

[MS77]   F.J. MacWilliams and N.J.A. Sloane, *The theory of error-correcting codes. I, II*, North-Holland Mathematical Library, Vol. 16, Amsterdam: North-Holland, 1977.

[MS94]   G.L. Mullen and P.J-S. Shiue, eds., *Finite fields: theory, applications, and algorithms*, vol. 168 of Contemp. Math., Providence, RI: Amer. Math. Soc., 1994.

[Mun94]   C. Munuera, "On the generalized Hamming weights of geometric Goppa codes," *IEEE Trans. Inform. Theory*, vol. 40, no. 6, pp. 2092–2099, 1994.

[MZK95]   O. Moreno, V.A. Zinoviev, and P.V. Kumar, "An extension of the Weil–Carlitz–Uchiyama bound," *Finite Fields Appl.*, vol. 1, no. 3, pp. 360–371, 1995.

[MZZ95]   O. Moreno, D. Zinoviev, and V. Zinoviev, "On several new projective curves over $\mathbb{F}_2$ of genus 3, 4, and 5," *IEEE Trans. Inform. Theory*, vol. 41, no. 6, part 1, pp. 1643–1648, 1995. Special issue on algebraic geometry codes.

[NX96]   H. Niederreiter and C. Xing, "Explicit global function fields over the binary field with many rational places," *Acta Arith.*, vol. 75, no. 4, pp. 383–396, 1996.

[NX98] H. Niederreiter and C. Xing, "A general method of constructing global function fields with many rational places," in *Algorithmic number theory (Portland, OR, 1998)*, vol. 1423 of Lecture Notes in Comput. Sci., pp. 555–566, Berlin: Springer, 1998.

[NX98] H. Niederreiter and C. Xing, "Nets, (t,s)-sequences, and algebraic geometry," in *Random and quasi-random point sets*, vol. 138 of Lecture Notes in Statist., pp. 267–302, New York: Springer, 1998.

[NX99a] H. Niederreiter and C. Xing, "Algebraic curves with many rational points over finite fields of characteristic 2," in *Number theory in progress, Vol. 1 (Zakopane-Kościelisko, 1997)*, pp. 359–380, Berlin: Gruyter, 1999.

[NX99b] H. Niederreiter and C. Xing, "Curve sequences with asymptotically many rational points," in *Applications of curves over finite fields (Seattle, WA, 1997)*, vol. 245 of Contemp. Math., pp. 3–14, Providence, RI: Amer. Math. Soc., 1999.

[NX99c] H. Niederreiter and C. Xing, "Global function fields with many rational places and their applications," in *Finite fields: theory, applications, and algorithms (Waterloo, ON, 1997)*, vol. 225 of Contemp. Math., pp. 87–111, Providence, RI: Amer. Math. Soc., 1999.

[NX00] H. Niederreiter and C. Xing, "A propagation rule for linear codes," *Appl. Algebra Engrg. Commun. Comput.*, vol. 10, no. 6, pp. 425–432, 2000.

[NX01] H. Niederreiter and C. Xing, *Rational points on curves over finite fields: theory and applications*, vol. 285 of London Mathematical Society Lecture Note Series, Cambridge: Cambridge Univ. Press, 2001.

[ÖG98] F. Özbudak and M. Glukhov, "Codes on superelliptic curves," *Turkish J. Math.*, vol. 22, no. 2, pp. 223–234, 1998.

[Ogg74] A.P. Ogg, "Hyperelliptic modular curves," *Bull. Soc. Math. France*, vol. 102, pp. 449–462, 1974.

[ÖS99a] F. Özbudak and H. Stichtenoth, "Constructing codes from algebraic curves," *IEEE Trans. Inform. Theory*, vol. 45, no. 7, pp. 2502–2505, 1999.

[ÖS99b] F. Özbudak and H. Stichtenoth, "Curves with many points and configurations of hyperplanes over finite fields," *Finite Fields Appl.*, vol. 5, no. 4, pp. 436–449, 1999.

[ÖS02] F. Özbudak and H. Stichtenoth, "Note on Niederreiter–Xing's propagation rule for linear codes," *Appl. Algebra Engrg. Comm. Comput.*, vol. 13, no. 1, pp. 53–56, 2002.

[ÖT05] F. Özbudak and B. G. Temur, "Fibre products of Kummer covers and curves with many points." Preprint, Middle East Technical University, Ankara, 2005.

[Özb99] F. Özbudak, "Codes on fibre products of some Kummer coverings," *Finite Fields Appl.*, vol. 5, no. 2, pp. 188–205, 1999.

[Pel88] R. Pellikaan, "On decoding linear codes by error correcting pairs." Preprint, 1988.

[Pel89] R. Pellikaan, "On a decoding algorithm for codes on maximal curves," *IEEE Trans. Inform. Theory*, vol. 35, no. 6, pp. 1228–1232, 1989.

[Pel92a] R. Pellikaan, "On decoding by error location and dependent sets of error positions," *Discrete Math.*, vol. 106/107, pp. 369–381, 1992. A collection of contributions in honour of J. van Lint.

[Pel92b] R. Pellikaan, "On the gonality of curves, abundant codes and decoding," in *Coding theory and algebraic geometry (Luminy, 1991)*, vol. 1518 of Lecture Notes in Math., pp. 132–144, Berlin: Springer, 1992.

[Pel96a] R. Pellikaan, "On special divisors and the two variable zeta function of algebraic curves over finite fields," in *Arithmetic, geometry and coding theory (Luminy, 1993)*, pp. 175–184, Berlin: Gruyter, 1996.

[Pel96b] R. Pellikaan, "The shift bound for cyclic, Reed–Muller and geometric Goppa codes," in *Arithmetic, geometry and coding theory (Luminy, 1993)*, pp. 155–174, Berlin: Gruyter, 1996.

[Per89a] M. Perret, "Families of codes exceeding the Varshamov–Gilbert bound," in *Coding theory and applications (Toulon, 1988)*, vol. 388 of Lecture Notes in Comput. Sci., pp. 28–36, New York: Springer, 1989.

[Per89b] M. Perret, "Sur le nombre de points d'une courbe sur un corps fini; application aux codes correcteurs d'erreurs," *C. R. Acad. Sci. Paris Sér. I Math.*, vol. 309, no. 3, pp. 177–182, 1989.

[Per91] M. Perret, "Tours ramifiées infinies de corps de classes," *J. Number Theory*, vol. 38, no. 3, pp. 300–322, 1991.

[Pet61] W.W. Peterson, *Error-correcting codes*, Cambridge, Mass.: MIT Press, 1961.

[Poi77] G. Poitou, "Minorations de discriminants (d'après A.M. Odlyzko)," in *Séminaire Bourbaki, Vol. 1975/76, 28ème année, exp. no. 479*, vol. 567 of Lecture Notes in Math., pp. 136–153, Berlin: Springer, 1977.

[PPV96] R. Pellikaan, M. Perret, and S.G. Vlăduţ, eds., *Arithmetic, geometry and coding theory*, Berlin: Gruyter, 1996.

[Pre92] O. Pretzel, *Error-correcting codes and finite fields*, Oxford Applied Mathematics and Computing Science Series, Oxford: Clarendon; New York: Oxford Univ. Press, 1992.

[Pre98] O. Pretzel, *Codes and algebraic curves*, vol. 8 of Oxford Lecture Series in Mathematics and Its Applications, Oxford: Clarendon; New York: Oxford Univ. Press, 1998.

[Pre99] O. Pretzel, "Partial geometric codes," *IEEE Trans. Inform. Theory*, vol. 45, no. 7, pp. 2506–2512, 1999.

[PS94] D. Polemi and T. Sakkalis, "Singular algebraic curves over finite fields," in *Information theory and applications (Rockland, ON, 1993)*, vol. 793 of Lecture Notes in Comput. Sci., pp. 24–37, Berlin: Springer, 1994.

[PSP92] S.C. Porter, B-Z. Shen, and R. Pellikaan, "Decoding geometric Goppa codes using an extra place," *IEEE Trans. Inform. Theory*, vol. 38, no. 6, pp. 1663–1676, 1992.

[PSvW91] R. Pellikaan, B-Z. Shen, and G.J.M. van Wee, "Which linear codes are algebraic-geometric?," *IEEE Trans. Inform. Theory*, vol. 37, no. 3, part 1, pp. 583–602, 1991.

[PT99] R. Pellikaan and F. Torres, "On Weierstrass semigroups and the redundancy of improved geometric Goppa codes," *IEEE Trans. Inform. Theory*, vol. 45, no. 7, pp. 2512–2519, 1999.

[PW72] W.W. Peterson and E.J. Weldon, Jr., *Error-correcting codes*, Cambridge, Mass.–London: MIT Press, 2nd ed., 1972.

[Que88] H.G. Quebbemann, "Cyclotomic Goppa codes," *IEEE Trans. Inform. Theory*, vol. 34, no. 5, pp. 1317–1320, 1988. Coding techniques and coding theory.

[Que91a] H.G. Quebbemann, "Estimates of regulators and class numbers in function fields," *J. Reine Angew. Math.*, vol. 419, pp. 79–87, 1991.

[Que91b] H.G. Quebbemann, "On even codes," *Discrete Math.*, vol. 98, no. 1, pp. 29–34, 1991.

[RB79] S.S. Ryshkov and E.P. Baranovskii, "Classical methods of the theory of lattice packings," *Uspekhi Mat. Nauk*, vol. 34, no. 4, pp. 3–63, 256, 1979.

[RB99] I. Reuven and Y. Be'ery, "Generalized Hamming weights of nonlinear codes and the relation to the Z_4-linear representation," *IEEE Trans. Inform. Theory*, vol. 45, no. 2, pp. 713–720, 1999.

[Rog64] C.A. Rogers, *Packing and covering*, Cambridge Tracts in Mathematics and Mathematical Physics, no. 54, New York: Cambridge Univ. Press, 1964.

[Roq70] P. Roquette, "Abschätzung der Automorphismenanzahl von Funktionenkörpern bei Primzahlcharakteristik," *Math. Z.*, vol. 117, pp. 157–163, 1970.

[RS87] J.A. Rush and N.J.A. Sloane, "An improvement to the Minkowski–Hlawka bound for packing superballs," *Mathematika*, vol. 34, no. 1, pp. 8–18, 1987.

[RS94] H.G. Rück and H. Stichtenoth, "A characterization of Hermitian function fields over finite fields," *J. Reine Angew. Math.*, vol. 457, pp. 185–188, 1994.

[RT90] M.Yu. Rosenbloom and M.A. Tsfasman, "Multiplicative lattices in global fields," *Invent. Math.*, vol. 101, no. 3, pp. 687–696, 1990.

[RT91] C. Rentería and H. Tapia, "A geometric code on an Artin–Schreier curve," in *Proceedings of the 22nd Southeastern Conference on Combinatorics, Graph Theory, and Computing (Baton Rouge, LA, 1991)*, vol. 84, pp. 3–7, 1991.

[RT92]　C. Rentería and H. Tapia, "Some geometric Goppa codes over a special curve," in *International Seminar on Algebra and Its Applications (Spanish) (México City, 1991)*, vol. 6 of Aportaciones Mat. Notas Investigación, pp. 49–57, México: Soc. Mat. Mexicana, 1992.

[Rüc87]　H.G. Rück, "A note on elliptic curves over finite fields," *Math. Comp.*, vol. 49, no. 179, pp. 301–304, 1987.

[Rüc90]　H.G. Rück, "On Goppa codes defined by Kummer and Artin–Schreier extensions," *J. Pure Appl. Algebra*, vol. 64, no. 2, pp. 163–169, 1990.

[Rus89]　J.A. Rush, "A lower bound on packing density," *Invent. Math.*, vol. 98, no. 3, pp. 499–509, 1989.

[SAK+01]　K.W. Shum, I. Aleshnikov, P.V. Kumar, H. Stichtenoth, and V. Deolaikar, "A low-complexity algorithm for the construction of algebraic-geometric codes better than the Gilbert–Varshamov bound," *IEEE Trans. Inform. Theory*, vol. 47, no. 6, pp. 2225–2241, 2001.

[Sch51]　B. Schoeneberg, "Über die Weierstrass-Punkte in den Körpern der elliptischen Modulfunktionen," *Abh. Math. Sem. Univ. Hamburg*, vol. 17, pp. 104–111, 1951.

[Sch85]　R. Schoof, "Elliptic curves over finite fields and the computation of square roots mod p," *Math. Comp.*, vol. 44, no. 170, pp. 483–494, 1985.

[Sch87]　R. Schoof, "Nonsingular plane cubic curves over finite fields," *J. Combin. Theory Ser. A*, vol. 46, no. 2, pp. 183–211, 1987.

[Sch89]　W. Scharlau, "Selbstduale Goppa-Codes," *Math. Nachr.*, vol. 143, pp. 119–122, 1989.

[Sch90]　R. Schoof, "Algebraic curves and coding theory," Tech. Rep. UTM 336, Univ. of Trento, 1990.

[Sch95a]　R. Schoof, "Counting points on elliptic curves over finite fields," *J. Théor. Nombres Bordeaux*, vol. 7, no. 1, pp. 219–254, 1995. Les Dix-huitièmes Journées Arithmétiques (Bordeaux, 1993).

[Sch95b]　R. Schoof, "Families of curves and weight distributions of codes," *Bull. Amer. Math. Soc. (N.S.)*, vol. 32, no. 2, pp. 171–183, 1995.

[Seg67]　B. Segre, "Introduction to Galois geometries," *Atti Accad. Naz. Lincei Mem. Cl. Sci. Fis. Mat. Natur. Sez. I*, vol. 8, pp. 133–236, 1967.

[Ser59]　J-P. Serre, *Groupes algébriques et corps de classes*, Publications de l'institut de mathématique de l'université de Nancago, VII. Hermann, Paris, 1959.

[Ser83a]　J-P. Serre, "Nombres de points des courbes algébriques sur $\mathbb{F}_q$," in *Seminar on number theory, 1982–1983 (Talence, 1982/1983)*, exp. no. 22, pp. 1–8, Talence: Univ. Bordeaux I, 1983.

[Ser83b]　J-P. Serre, "Sur le nombre des points rationnels d'une courbe algébrique sur un corps fini," *C. R. Acad. Sci. Paris Sér. I Math.*, vol. 296, no. 9, pp. 397–402, 1983.

[Ser83c]　J-P. Serre, "The number of rational points on curves over finite fields." Princeton lectures, Fall 1983. Notes by E. Bayer.

[Ser84]　J-P. Serre, *Groupes algébriques et corps de classes*, vol. 7 of Publications de l'Institut Mathématique de l'Université de Nancago, Paris: Hermann, 2nd ed., 1984. Actualités Scientifiques et Industrielles, 1264.

[Ser85a]　J-P. Serre, "Quel est le nombre maximum de points rationnels que peut avoir une courbe algébrique de genre g sur un corps fini $\mathbb{F}_q$?." Résumé des cours de 1983–1984, Annuaire du Collége de France, Paris, 1985.

[Ser85b]　J-P. Serre, "Rational points on curves over finite fields," 1985. Notes of lectures at Harvard Univ.

[Ser91]　J-P. Serre, "Lettre à M. Tsfasman," *Astérisque*, no. 198–200, pp. 351–353, 1991. Journées Arithmétiques, 1989 (Luminy, 1989).

[SH95]　K. Saints and C. Heegard, "Algebraic-geometric codes and multidimensional cyclic codes: a unified theory and algorithms for decoding using Gröbner bases," *IEEE Trans. Inform. Theory*, vol. 41, no. 6, part 1, pp. 1733–1751, 1995. Special issue on algebraic geometry codes.

[Sha72]　I.R. Shafarevich, *Osnovy algebraicheskoi geometrii* [Basic algebraic geometry], Moscow: Nauka, 1972 (in Russian); see also [Sha88a, Sha88b].

[Sha88a] I.R. Shafarevich, *Osnovy algebraicheskoi geometrii. T. 1: Algebraicheskie mnogoobraziya v proektivnom prostranstve*, Moscow: Nauka, 2nd ed., 1988. English translation: see [Sha94a].

[Sha88b] I.R. Shafarevich, *Osnovy algebraicheskoi geometrii. T. 2: Skhemy. Kompleksnye mnogoobraziya*, Moscow: Nauka, 2nd ed., 1988. English translation: see [Sha94b].

[Sha94a] I.R. Shafarevich, *Basic algebraic geometry. Vol. 1: Varieties in projective space*, Berlin: Springer, 2nd ed., 1994.

[Sha94b] I.R. Shafarevich, *Basic algebraic geometry. Vol. 2: Schemes and complex manifolds*, Berlin: Springer, 2nd ed., 1994.

[She92] B.Z. Shen, "Solving a congruence on a graded algebra by a subresultant sequence and its application," *J. Symbolic Comput.*, vol. 14, no. 5, pp. 505–522, 1992.

[Shi71] G. Shimura, *Introduction to the arithmetic theory of automorphic functions*, Kanô Memorial Lectures, 1, Publications of the Mathematical Society of Japan, no. 11, Tokyo: Iwanami Shoten, 1971.

[Shi89a] T. Shioda, "Mordell–Weil lattices and Galois representation. I," *Proc. Japan Acad. Ser. A Math. Sci.*, vol. 65, no. 7, pp. 268–271, 1989.

[Shi89b] T. Shioda, "Mordell–Weil lattices and Galois representation. II, III," *Proc. Japan Acad. Ser. A Math. Sci.*, vol. 65, no. 8, pp. 296–303, 1989.

[Shi91] T. Shioda, "Mordell–Weil lattices and sphere packings," *Amer. J. Math.*, vol. 113, no. 5, pp. 931–948, 1991.

[Shi94] G. Shimura, *Introduction to the arithmetic theory of automorphic functions*, vol. 11 of Publications of the Mathematical Society of Japan, Princeton, NJ: Princeton Univ. Press, 1994. Reprint of the 1971 original, Kanô Memorial Lectures, 1.

[Sil86] J.H. Silverman, *The arithmetic of elliptic curves*, vol. 106 of Graduate Texts in Mathematics, New York: Springer, 1986.

[Sim94] J. Simonis, "The effective length of subcodes," *Appl. Algebra Engrg. Comm. Comput.*, vol. 5, no. 6, pp. 371–377, 1994.

[Sko91] A.N. Skorobogatov, "The parameters of subcodes of algebraic-geometric codes over prime subfields," *Discrete Appl. Math.*, vol. 33, no. 1–3, pp. 205–214, 1991. Applied algebra, algebraic algorithms, and error-correcting codes (Toulouse, 1989).

[Sle56] D. Slepian, "A class of binary signaling alphabets," *Bell System Tech. J.*, vol. 35, pp. 203–234, 1956.

[Slo72] N.J.A. Sloane, "Sphere packings constructed from BCH and Justesen codes," *Mathematika*, vol. 19, pp. 183–190, 1972.

[Slo84] N.J.A. Sloane, "The packing of spheres," *Scient. Amer.*, vol. 250, no. 1, pp. 116–125, 1984.

[SÖ97] S.A. Stepanov and F. Özbudak, "Stratified products of hyperelliptic curves, and geometric Goppa codes," *Diskret. Mat.*, vol. 9, no. 3, pp. 36–42, 1997.

[Sør91] A.B. Sørensen, "Projective Reed–Muller codes," *IEEE Trans. Inform. Theory*, vol. 37, no. 6, pp. 1567–1576, 1991.

[Spr57] G. Springer, *Introduction to Riemann surfaces*, Reading, Mass.: Addison-Wesley, 1957.

[SS01] S.A. Stepanov and M.K. Shalalfekh, "Codes on fiber products of Artin-Schreier curves," *Diskret. Mat.*, vol. 13, no. 2, pp. 3–13, 2001.

[ST92] H. Stichtenoth and M.A. Tsfasman, eds., *Coding theory and algebraic geometry*, vol. 1518 of Lecture Notes in Math., Berlin: Springer, 1992.

[ST95] B-Z. Shen and K.K. Tzeng, "Generation of matrices for determining minimum distance and decoding of algebraic-geometric codes," *IEEE Trans. Inform. Theory*, vol. 41, no. 6, part 1, pp. 1703–1708, 1995. Special issue on algebraic geometry codes.

[ST01] S. Shlosman and M.A. Tsfasman, "Random lattices and random sphere packings: typical properties," *Mosc. Math. J.*, vol. 1, no. 1, pp. 73–89, 2001.

[Ste74] S.A. Stepanov, "Rational points of algebraic curves over finite fields," in *Current problems of analytic number theory (Proc. Summer School, Minsk, 1972)* (in Russian), pp. 223–243, Minsk: Nauka i Tehnika, 1974.

[Ste75]   S.A. Stepanov, "An elementary method in the theory of equations over a finite field," in *Proceedings of the International Congress of Mathematicians (Vancouver, B.C., 1974)*, Vol. 1, pp. 383–391, Montreal, Que.: Canad. Math. Congress, 1975.

[Ste91]   S.A. Stepanov, *Arifmetika algebraicheskikh krivykh*, Moscow: Nauka, 1991. English translation: see [Ste94].

[Ste94]   S.A. Stepanov, *Arithmetic of algebraic curves*, Monographs in Contemporary Mathematics, New York: Consultants Bureau, 1994.

[Ste96]   S.A. Stepanov, "Character sums and coding theory," in *Finite fields and applications (Glasgow, 1995)*, vol. 233 of London Math. Soc. Lecture Note Ser., pp. 355–378, Cambridge: Cambridge Univ. Press, 1996.

[Ste97a] S.A. Stepanov, "Character sums, algebraic curves and Goppa codes," in *Algebraic geometry (Ankara, 1995)*, vol. 193 of Lecture Notes in Pure and Appl. Math., pp. 313–345, New York: Dekker, 1997.

[Ste97b] S.A. Stepanov, "Codes on fiber products of hyperelliptic curves," *Diskret. Mat.*, vol. 9, no. 1, pp. 83–94, 1997.

[Ste99a] S.A. Stepanov, *Codes on algebraic curves*, New York: Kluwer/Plenum, 1999.

[Ste99b] S.A. Stepanov, "Fibre products, character sums, and geometric Goppa codes," in *Number theory and its applications (Ankara, 1996)*, vol. 204 of Lecture Notes in Pure and Appl. Math., pp. 227–259, New York: Dekker, 1999.

[Sti73]   H. Stichtenoth, "Über die Automorphismengruppe eines algebraischen Funktionenkörpers von Primzahlcharakteristik. I, II.," *Arch. Math. (Basel)*, vol. 24, pp. 527–544; 615–631, 1973.

[Sti88a] H. Stichtenoth, "A note on Hermitian codes over GF(q^2)," *IEEE Trans. Inform. Theory*, vol. 34, no. 5, part 2, pp. 1345–1348, 1988. Coding techniques and coding theory.

[Sti88b] H. Stichtenoth, "Self-dual Goppa codes," *J. Pure Appl. Algebra*, vol. 55, no. 1–2, pp. 199–211, 1988.

[Sti90a] H. Stichtenoth, "Algebraic-geometric codes associated to Artin–Schreier extensions of $\mathbb{F}_q(z)$," in *Proc. 2nd Int. Workshop on Algebraic and Combinatorial Coding Theory (Leningrad, 1990)*, pp. 203–206, 1990.

[Sti90b] H. Stichtenoth, "On automorphisms of geometric Goppa codes," *J. Algebra*, vol. 130, no. 1, pp. 113–121, 1990.

[Sti90c] H. Stichtenoth, "On the dimension of subfield subcodes," *IEEE Trans. Inform. Theory*, vol. 36, no. 1, pp. 90–93, 1990.

[Sti93]   H. Stichtenoth, *Algebraic function fields and codes*, Universitext, Berlin: Springer, 1993.

[Sti95]   H. Stichtenoth, "Algebraic geometric codes," in *Different aspects of coding theory (San Francisco, CA, 1995)*, vol. 50 of Proc. Sympos. Appl. Math., pp. 139–152, Providence, RI: Amer. Math. Soc., 1995.

[Sti01]   H. Stichtenoth, "Explicit constructions of towers of function fields with many rational places," in *European Congress of Mathematics, Vol. II (Barcelona, 2000)*, vol. 202 of Progr. Math., pp. 219–224, Basel: Birkhäuser, 2001.

[SV86]   K.O. Stöhr and J.F. Voloch, "Weierstrass points and curves over finite fields," *Proc. London Math. Soc. (3)*, vol. 52, no. 1, pp. 1–19, 1986.

[SV90]   A.N. Skorobogatov and S.G. Vlăduţ, "On the decoding of algebraic-geometric codes," *IEEE Trans. Inform. Theory*, vol. 36, no. 5, pp. 1051–1060, 1990.

[SvdV91] R. Schoof and M. van der Vlugt, "Hecke operators and the weight distributions of certain codes," *J. Combin. Theory Ser. A*, vol. 57, no. 2, pp. 163–186, 1991.

[SX95]   H. Stichtenoth and C.P. Xing, "On the structure of the divisor class group of a class of curves over finite fields," *Arch. Math. (Basel)*, vol. 65, no. 2, pp. 141–150, 1995.

[SZ77a]   J-P. Serre and D.B. Zagier, eds., *Modular functions of one variable. V*, vol. 601 of Lecture Notes in Math., Berlin: Springer, 1977.

[SZ77b]   J-P. Serre and D.B. Zagier, eds., *Modular functions of one variable. VI*, vol. 627 of Lecture Notes in Math., Berlin: Springer, 1977.

[Tan96]   L. Tang, "Consecutive Weierstrass gaps and weight hierarchy of geometric Goppa codes," *Algebra Colloq.*, vol. 3, no. 1, pp. 1–10, 1996.

[Tat74]   J.T. Tate, "The arithmetic of elliptic curves," *Invent. Math.*, vol. 23, pp. 179–206, 1974.

[Tem01]   A. Temkine, "Hilbert class field towers of function fields over finite fields and lower bounds for $A(q)$," *J. Number Theory*, vol. 87, no. 2, pp. 189–210, 2001.

[Tie87]   H.J. Tiersma, "Remarks on codes from Hermitian curves," *IEEE Trans. Inform. Theory*, vol. 33, no. 4, pp. 605–609, 1987.

[Top03]   J. Top, "Curves of genus 3 over small finite fields," *Indag. Math. (N.S.)*, vol. 14, no. 2, pp. 275–283, 2003.

[Tsf82]   M.A. Tsfasman, "Goppa codes that are better than the Varshamov–Gilbert bound," *Problemy Peredachi Informatsii*, vol. 18, no. 3, pp. 3–6, 1982.

[Tsf85a]   M.A. Tsfasman, "Group of points of an elliptic curve over a finite field," in *Theory of numbers and its applications (Tbilisi, 1985)*, pp. 286–287, 1985.

[Tsf85b]   M.A. Tsfasman, "Group of points of an elliptic curve over a finite field." Preprint, 1985.

[Tsf91a]   M.A. Tsfasman, "Algebraic-geometric codes and asymptotic problems," *Discrete Appl. Math.*, vol. 33, no. 1–3, pp. 241–256, 1991. Applied algebra, algebraic algorithms, and error-correcting codes (Toulouse, 1989).

[Tsf91b]   M.A. Tsfasman, "Global fields, codes and sphere packings," *Astérisque*, no. 198–200, pp. 373–396, 1991. Journées Arithmétiques, 1989 (Luminy, 1989).

[Tsf92a]   M.A. Tsfasman, "Some remarks on the asymptotic number of points," in *Coding theory and algebraic geometry (Luminy, 1991)*, vol. 1518 of Lecture Notes in Math., pp. 178–192, Berlin: Springer, 1992.

[Tsf92b]   M.A. Tsfasman, "Notes on bounds for higher weights." Handwritten notes in limited circulation, Brighton, 1992.

[Tsf99]   M.A. Tsfasman, "Lectures on global fields, lattices and codes," in *Methods of discrete mathematics (Braunschweig, 1999)*, vol. 5 of Quad. Mat., pp. 145–183, Aracne, Rome, 1999.

[Tsf02]   M.A. Tsfasman, "Asymptotic properties of global fields," in *Finite fields with applications to coding theory, cryptography, and related areas (Oaxaca, México, 2001)* (G.L. Mullen, H. Stichtenoth, and H. Tapia-Recillas, eds.), pp. 328–334, Berlin: Springer, 2002.

[TV91]   M.A. Tsfasman and S.G. Vlăduţ, *Algebraic-geometric codes*, vol. 58 of Mathematics and Its Applications (Soviet Series), Dordrecht: Kluwer, 1991.

[TV95]   M.A. Tsfasman and S.G. Vlăduţ, "Geometric approach to higher weights," *IEEE Trans. Inform. Theory*, vol. 41, no. 6, part 1, pp. 1564–1588, 1995. Special issue on algebraic geometry codes.

[TV97]   M.A. Tsfasman and S.G. Vlăduţ, "Asymptotic properties of zeta-functions," *J. Math. Sci. (New York)*, vol. 84, no. 5, pp. 1445–1467, 1997. Algebraic geometry, 7.

[TVZ82]   M.A. Tsfasman, S.G. Vlăduţ, and T. Zink, "Modular curves, Shimura curves, and Goppa codes, better than Varshamov–Gilbert bound," *Math. Nachr.*, vol. 109, pp. 21–28, 1982.

[Ueh98]   T. Uehara, "Introduction to the theory of error-correcting codes," *Sūrikaisekikenkyūsho Kōkyūroku*, no. 1060, pp. 137–144, 1998. Number theory and its applications (Japanese) (Kyoto, 1997).

[Ueh99]   T. Uehara, "On minimum distances of algebraic-geometric codes," *Adv. Stud. Contemp. Math. (Pusan)*, vol. 1, pp. 1–15, 1999. Algebraic number theory (Hapcheon/Saga, 1996).

[Vai89]   I. Vainsencher, "On Stöhr–Voloch's proof of Weil's theorem," *Manuscripta Math.*, vol. 64, no. 1, pp. 121–126, 1989.

[VD83]   S.G. Vlăduţ and V.G. Drinfel'd, "The number of points of an algebraic curve," *Funktsional. Anal. i Prilozhen.*, vol. 17, no. 1, pp. 68–69, 1983.

[vdG91]   G. van der Geer, "Codes and elliptic curves," in *Effective methods in algebraic geometry (Castiglioncello, 1990)*, vol. 94 of Progr. Math., pp. 159–168, Boston, MA: Birkhäuser, 1991.

[vdG01a]   G. van der Geer, "Curves over finite fields and codes," in *European Congress of Mathematics, Vol. II (Barcelona, 2000)*, vol. 202 of Progr. Math., pp. 225–238, Basel: Birkhäuser, 2001.

[vdG01b]   G. van der Geer, "Error-correcting codes and curves over finite fields," in *Mathematics unlimited—2001 and beyond*, pp. 1115–1138, Berlin: Springer, 2001.

[vdG01c]   G. van der Geer, "Coding theory and algebraic curves over finite fields: a survey and questions," in *Applications of algebraic geometry to coding theory, physics and computation (Eilat, 2001)*, vol. 36 of NATO Sci. Ser. II Math. Phys. Chem., pp. 139–159, Dordrecht: Kluwer, 2001.

[vdGS00]   G. van der Geer and R. Schoof, "Effectivity of Arakelov divisors and the theta divisor of a number field," *Selecta Math. (N.S.)*, vol. 6, no. 4, pp. 377–398, 2000.

[vdGSvdV92]   G. van der Geer, R. Schoof, and M. van der Vlugt, "Weight formulae for ternary Melas codes," *Math. Comp.*, vol. 58, no. 198, pp. 781–792, 1992.

[vdGvdV91]   G. van der Geer and M. van der Vlugt, "Artin–Schreier curves and codes," *J. Algebra*, vol. 139, no. 1, pp. 256–272, 1991.

[vdGvdV92]   G. van der Geer and M. van der Vlugt, "Reed–Muller codes and supersingular curves. I," *Compositio Math.*, vol. 84, no. 3, pp. 333–367, 1992.

[vdGvdV93]   G. van der Geer and M. van der Vlugt, "Curves over finite fields of characteristic 2 with many rational points," *C. R. Acad. Sci. Paris Sér. I Math.*, vol. 317, no. 6, pp. 593–597, 1993.

[vdGvdV94]   G. van der Geer and M. van der Vlugt, "Generalized Hamming weights of Melas codes and dual Melas codes," *SIAM J. Discrete Math.*, vol. 7, no. 4, pp. 554–559, 1994.

[vdGvdV96a]   G. van der Geer and M. van der Vlugt, "Quadratic forms, generalized Hamming weights of codes and curves with many points," *J. Number Theory*, vol. 59, no. 1, pp. 20–36, 1996.

[vdGvdV96b]   G. van der Geer and M. van der Vlugt, "Generalized Hamming weights of codes and curves over finite fields with many points," in *Proceedings of the Hirzebruch 65 Conference on Algebraic Geometry (Ramat Gan, 1993)*, vol. 9 of Israel Math. Conf. Proc., pp. 417–432, Ramat Gan: Bar-Ilan Univ., 1996.

[vdGvdV97]   G. van der Geer and M. van der Vlugt, "How to construct curves over finite fields with many points," in *Arithmetic geometry (Cortona, 1994)*, Sympos. Math., XXXVII, pp. 169–189, Cambridge: Cambridge Univ. Press, 1997.

[vdGvdV98]   G. van der Geer and M. van der Vlugt, "Generalized Reed–Muller codes and curves with many points," *J. Number Theory*, vol. 72, no. 2, pp. 257–268, 1998.

[vdGvdV99]   G. van der Geer and M. van der Vlugt, "Constructing curves over finite fields with many points by solving linear equations," in *Applications of curves over finite fields (Seattle, WA, 1997)*, vol. 245 of Contemp. Math., pp. 41–47, Providence, RI: Amer. Math. Soc., 1999.

[vdGvdV00a]   G. van der Geer and M. van der Vlugt, "Kummer covers with many points," *Finite Fields Appl.*, vol. 6, no. 4, pp. 327–341, 2000.

[vdGvdV00b]   G. van der Geer and M. van der Vlugt, "Tables of curves with many points," *Math. Comp.*, vol. 69, no. 230, pp. 797–810, 2000.

[vdGvdV02]   G. van der Geer and M. van der Vlugt, "An asymptotically good tower of curves over the field with eight elements," *Bull. London Math. Soc.*, vol. 34, no. 3, pp. 291–300, 2002.

[vdGvdV]   G. van der Geer and M. van der Vlugt, "Tables of curves with a large number of points." Updated version available at http://www.science.uva.nl/~geer.

[vdV91]   M. van der Vlugt, "A new upper bound for the dimension of trace codes," *Bull. London Math. Soc.*, vol. 23, no. 4, pp. 395–400, 1991.

[VKT84]   S.G. Vlăduţ, G.L. Katsman, and M.A. Tsfasman, "Modular curves and codes with polynomial construction complexity," *Problemy Peredachi Informatsii*, vol. 20, no. 1, pp. 47–55, 1984.

[vL82]   J.H. van Lint, *Introduction to coding theory*, vol. 86 of Graduate Texts in Mathematics, New York: Springer, 1982.

[vL90]   J.H. van Lint, "Algebraic geometric codes," in *Coding theory and design theory, Part I*, vol. 20 of IMA Vol. Math. Appl., pp. 137–162, New York: Springer, 1990.

[vL99]   J.H. van Lint, *Introduction to coding theory*, vol. 86 of Graduate Texts in Mathematics, Berlin: Springer, 3rd ed., 1999.

[Vlă87]   S.G. Vlăduţ, "An exhaustion bound for algebraic-geometric 'modular' codes," *Problemy Peredachi Informatsii*, vol. 23, no. 1, pp. 28–41, 1987.

[Vlă89]   S.G. Vlăduţ, "Algebraic-geometric 'modular' codes as group codes." Preprint, 1989.

[Vlă90]   S. Vlăduţ, "On the decoding of algebraic-geometric codes over $\mathbb{F}_q$ for $q \geq 16$," *IEEE Trans. Inform. Theory*, vol. 36, no. 6, pp. 1461–1463, 1990.

[Vlă91]   S.G. Vlăduţ, *Kronecker's Jugendtraum and modular functions*, vol. 2 of Studies in the Development of Modern Mathematics, New York: Gordon and Breach, 1991.

[Vlă96]   S.G. Vlăduţ, "An upper bound for generalized Hamming weights," in *Arithmetic, geometry and coding theory (Luminy, 1993)*, pp. 263–267, Berlin: Gruyter, 1996.

[vLS87]   J.H. van Lint and T.A. Springer, "Generalized Reed–Solomon codes from algebraic geometry," *IEEE Trans. Inform. Theory*, vol. 33, no. 3, pp. 305–309, 1987.

[vLvdG88] J.H. van Lint and G. van der Geer, *Introduction to coding theory and algebraic geometry*, vol. 12 of DMV Seminar, Basel: Birkhäuser, 1988.

[VM84]   S.G. Vlăduţ and Yu.I. Manin, "Linear codes and modular curves," in *Current problems in mathematics, Vol. 25*, Itogi Nauki i Tekhniki, pp. 209–257, Moscow: Akad. Nauk SSSR, Vsesoyuz. Inst. Nauchn. i Tekhn. Inform., 1984 (in Russian).

[Vol83]   J.F. Voloch, "Codes and curves," *Eureka*, vol. 43, pp. 53–61, 1983.

[Vol88]   J.F. Voloch, "A note on elliptic curves over finite fields," *Bull. Soc. Math. France*, vol. 116, no. 4, pp. 455–458, 1988.

[Vol01]   J.F. Voloch, "On the duals of binary BCH codes," *IEEE Trans. Inform. Theory*, vol. 47, no. 5, pp. 2050–2051, 2001.

[Vos91]   C. Voss, "Asymptotically good families of geometric Goppa codes and the Gilbert–Varshamov bound," in *Eurocode'90 (Udine, 1990)*, vol. 514 of Lecture Notes in Comput. Sci., pp. 150–157, Berlin: Springer, 1991.

[VS90]   C. Voss and H. Stichtenoth, "Asymptotically good families of subfield subcodes of geometric Goppa codes," *Geom. Dedicata*, vol. 33, no. 1, pp. 111–116, 1990.

[VS91]   S.G. Vlăduţ and A.N. Skorobogatov, "Weight distributions of subfield subcodes of algebraic-geometric codes," *Problemy Peredachi Informatsii*, vol. 27, no. 1, pp. 24–36, 1991.

[Wat69]   W.C. Waterhouse, "Abelian varieties over finite fields," *Ann. Sci. École Norm. Sup. (4)*, vol. 2, pp. 521–560, 1969.

[Wei48a]   A. Weil, *Sur les courbes algébriques et les variétés qui s'en déduisent*, Actualités Sci. Ind., no. 1041 = Publ. Inst. Math. Univ. Strasbourg 7 (1945), Paris: Hermann & Cie., 1948.

[Wei48b]   A. Weil, *Variétés abéliennes et courbes algébriques*, Actualités Sci. Ind., no. 1064 = Publ. Inst. Math. Univ. Strasbourg 8 (1946), Paris: Hermann & Cie., 1948.

[Wei52]   A. Weil, "Sur les 'formules explicites' de la théorie des nombres premiers," *Comm. Sém. Math. Univ. Lund* [Medd. Lunds Univ. Mat. Sem], Tome Supplementaire, pp. 252–265, 1952.

[Wei67]   A. Weil, *Basic number theory*, Die Grundlehren der mathematischen Wissenschaften, Bd. 144, New York: Springer, 1967.

[Wei72]   A. Weil, "Sur les formules explicites de la théorie des nombres," *Izv. Akad. Nauk SSSR Ser. Mat.*, vol. 36, pp. 3–18, 1972.

[Wei84]   A. Weil, *Number theory: an approach through history. From Hammurapi to Legendre*, Boston, MA: Birkhäuser, 1984.

[Wei91]   V.K. Wei, "Generalized Hamming weights for linear codes," *IEEE Trans. Inform. Theory*, vol. 37, no. 5, pp. 1412–1418, 1991.

[Wei95]   A. Weil, *Basic number theory*, Classics in Mathematics, Berlin: Springer, 1995. Reprint of the 2nd (1973) edition.

[Wei96]   V.K. Wei, "Generalized Hamming weights; fundamental open problems in coding theory," in *Arithmetic, geometry and coding theory (Luminy, 1993)*, pp. 269–281, Berlin: Gruyter, 1996.

[Wir88]   M. Wirtz, "On the parameters of Goppa codes," *IEEE Trans. Inform. Theory*, vol. 34, no. 5, pp. 1341–1343, 1988. Coding techniques and coding theory.

[Wol86] J. Wolfmann, "Recent results on coding and algebraic geometry," in *Algebraic algorithms and error correcting codes (Grenoble, 1985)*, vol. 229 of Lecture Notes in Comput. Sci., pp. 167–184, Berlin: Springer, 1986.

[Wol87] J. Wolfmann, "Nombre de points rationnels de courbes algébriques sur des corps finis associées à des codes cycliques," *C. R. Acad. Sci. Paris Sér. I Math.*, vol. 305, no. 8, pp. 345–348, 1987.

[Wol89a] J. Wolfmann, "New bounds on cyclic codes from algebraic curves," in *Coding theory and applications (Toulon, 1988)*, vol. 388 of Lecture Notes in Comput. Sci., pp. 47–62, New York: Springer, 1989.

[Wol89b] J. Wolfmann, "The number of points on certain algebraic curves over finite fields," *Comm. Algebra*, vol. 17, no. 8, pp. 2055–2060, 1989.

[Wol89c] J. Wolfmann, "The weights of the dual code of the Melas code over GF(3)," *Discrete Math.*, vol. 74, no. 3, pp. 327–329, 1989.

[Xin91] C.P. Xing, "Remarks on self-dual elliptic codes," *Chinese Sci. Bull.*, vol. 36, no. 8, pp. 629–631, 1991.

[Xin92] C.P. Xing, "When are two geometric Goppa codes equal?," *IEEE Trans. Inform. Theory*, vol. 38, no. 3, pp. 1140–1142, 1992.

[Xin95a] C. Xing, "On automorphism groups of the Hermitian codes," *IEEE Trans. Inform. Theory*, vol. 41, no. 6, part 1, pp. 1629–1635, 1995. Special issue on algebraic geometry codes.

[Xin95b] C.P. Xing, "Automorphism group of elliptic codes," *Comm. Algebra*, vol. 23, no. 11, pp. 4061–4072, 1995.

[Xin03] C. Xing, "Nonlinear codes from algebraic curves beating the Tsfasman–Vlăduţ–Zink bound," *IEEE Trans. Inform. Theory*, vol. 49, no. 7, pp. 1653–1657, 2003.

[XN98] C. Xing and H. Niederreiter, "Towers of global function fields with asymptotically many rational places and an improvement on the Gilbert–Varshamov bound," *Math. Nachr.*, vol. 195, pp. 171–186, 1998.

[XNL99a] C. Xing, H. Niederreiter, and K.Y. Lam, "Constructions of algebraic-geometry codes," *IEEE Trans. Inform. Theory*, vol. 45, no. 4, pp. 1186–1193, 1999.

[XNL99b] C. Xing, H. Niederreiter, and K.Y. Lam, "A generalization of algebraic-geometry codes," *IEEE Trans. Inform. Theory*, vol. 45, no. 7, pp. 2498–2501, 1999.

[XS95] C.P. Xing and H. Stichtenoth, "The genus of maximal function fields over finite fields," *Manuscripta Math.*, vol. 86, no. 2, pp. 217–224, 1995.

[YB92] T. Yaghoobian and I.F. Blake, "Hermitian codes as generalized Reed–Solomon codes," *Des. Codes Cryptogr.*, vol. 2, no. 1, pp. 5–17, 1992.

[YK92] K. Yang and P.V. Kumar, "On the true minimum distance of Hermitian codes," in *Coding theory and algebraic geometry (Luminy, 1991)*, vol. 1518 of Lecture Notes in Math., pp. 99–107, Berlin: Springer, 1992.

[YKS94] K. Yang, P.V. Kumar, and H. Stichtenoth, "On the weight hierarchy of geometric Goppa codes," *IEEE Trans. Inform. Theory*, vol. 40, no. 3, pp. 913–920, 1994.

[ZE87] V.A. Zinoviev and T. Ericson, "On concatenated constant-weight codes beyond the Varshamov–Gilbert bound," *Problemy Peredachi Informatsii*, vol. 23, no. 1, pp. 110–111, 1987.

[Zin85] T. Zink, "Degeneration of Shimura surfaces and a problem in coding theory," in *Fundamentals of computation theory (Cottbus, 1985)*, vol. 199 of Lecture Notes in Comput. Sci., pp. 503–511, Berlin: Springer, 1985.

[ZL85] V.A. Zinoviev and S.N. Litsyn, "On codes beyond the Gilbert bound," *Problemy Peredachi Informatsii*, vol. 21, no. 1, pp. 109–111, 1985.

List of Names

Aaltonen, M.J., 53, 67, 68
Abel, N.H., 131, 132
Anglès, B., 189
Artin, E., 132, 173, 183, 187, 189
Ashikhmin, A., 285

Barg, A.M., xv, 284, 285
Bassalygo, L.A., xv, 17, 18, 67, 68
Be'ery, Y., 68
Bernoulli, Jacob, 131
Bernoulli, Johann, 131
Blokh, E.L., 63, 67
Bose, R.K., 67
Brouwer, A.E., xv

Carlitz, L., 188
Cartier, P., 83, 104, 132
Castelnuovo, G., 229
Chebotarev, N.G., 132
Chepyzhov, V.V., 68
Chévalley, C., 132
Clebsch, R.F.A., 132
Clifford, W.K., 102
Cohen, G., 68

Dedekind, R., 132, 133
Deligne, P., 189
Descartes, R., 131
Deuring, M., 161, 189
di Fagnano dei Toschi, G.C., 131
Dieudonné, J., 131
Diophantos ($\Delta\iota\acute{o}\varphi\alpha\nu\tau o\varsigma$), 131
Dodunekov, S.M., 67
Driencourt, Y., 284
Drinfeld, V.G., xiv, 146, 188, 189

Elias, P., 17, 67
Elkies, N.D., 188, 189, 271, 272, 285
Ericson, T., 270
Euler, L., 131

Feng, G-L., 213, 214, 284
Fermat, P., 131
Forney, G.D., Jr., 67
Frobenius, F.G., 100
Fulton, W., 132

García, A., 177, 189
Gelfand, S.I., xv
Gilbert, E.N., 20, 21, 67
Gleason, A.M., 67
Golay, M.J.E., 42
Goppa, V.D., ix, 41, 67, 68, 191, 284, 285
Griesmer, J.H., 16
Griffiths, P.A., 132
Grothendieck, A., 132

Hamming, R.V., x, 1, 17, 37, 67
Harris, J., 132
Hartshorne, R., 132
Hasse, H., 132, 162, 187, 189
Helleseth, T., 68
Hocquenghem, A., 67
Houzél, C., 131
Huguet, L., 68
Hurwitz, A., 100

Ibukiyama, T., 189
Ihara, Y., 188

Jacobi, C.G.J., 131
Justesen, J., 41, 67, 284

Kabatiansky, G.A., xv, 68
Katsman, G.L., xv, 68, 264, 266, 284, 285
Klein, F.C., 245
Kløve, T., 68
Krachkovsky, V.Yu., 284
Krawtchouk, M.F., 19
Kummer, E.E., 168

Lachaud, G., xv
Laihonen, T., 54, 68
Landau, E., 188
Lang, S., 132
Legendre, A.M., 112, 131
Leontiev, V.K., 67
Levenshtein, V.I., 68
Litsyn, S.N., xv, 54, 68, 267, 269, 284, 285
Lloyd, S.P., 67

MacWilliams, F.J., 8, 67
Maire, C., 189
Manin, Yu.I., ix, xiv, 68, 284
McEliece, R.J., 52, 67
Michon, J-F., 284
Möbius, A.F., 148
Muller, D.E., 36, 37
Mykkeltveit, J., 68

Newton, I., 131
Niederreiter, H., 188, 189
Noether, M., 100, 110, 132

Oesterlé, J., 188

Pellikaan, R., 144, 188, 252, 284
Perret, M., 189
Picard, R., 76
Plotkin, M., 16
Prange, E., 67
Pretzel, O., 284

Rao, T.R.N., 213, 214, 284
Ray-Chaudhuri, D.K., 67
Reed, I.S., 33, 36, 37, 67
Reuven, I., 68
Riemann, G.F.B., 91, 95, 131–133
Roch, G., 91, 95, 132
Rodemich, E.R., 52, 67
Roquette, P., 132
Rosenbloom, M.Yu., xv, 284
Rück, H.G., 189
Rumsey, H.C., Jr., 52, 67

Scharlau, W., 269, 284
Schmidt, F.K., 132, 137, 187
Schoeneberg, B., 132
Schoof, R., 189
Schreier, O., 173
Serre, J-P., 132, 142, 188, 189, 284
Shafarevich, I.R., 131, 132
Simonis, J., 67, 68
Singleton, R.C., 16

Skorobogatov, A.N., xv, 284
Slepian, D., 67
Sloane, N.J.A., 67
Solomon, G., 33, 67
Springer, J., 132
Stepanov, S.A., 189
Stichtenoth, H., 132, 177, 188, 189, 284, 285

Tate, J., 189
Temkine, A., 189
Tiersma, H.J., 285
Tietäväinen, A., 67
Tsfasman, M.A., 67, 68, 189, 251, 264, 266, 269, 284, 285

van der Geer, G., xv, 68, 189
van der Vlugt, M., xv, 68, 189
van der Waerden, B.L., 132
van Lint, J.H., 67
Varshamov, R.R., 20, 21, 67
Vlăduţ, S.G., ix, 67, 68, 146, 251, 259, 264, 284, 285
Voloch, J.F., 189

Waterhouse, W.C., 161, 189
Weber, H., 132
Wei, V.K., 6, 68
Weierstrass, K., 103, 118
Weil, A., 131, 132, 141, 187–189
Welch, L.R., 52, 67

Xing, C., 188, 189, 280, 282, 283, 285

Ytrehus, Ø., 68

Zariski, O., 71
Zémor, G., 68
Zink, T., ix, 189, 251, 285
Zinoviev, V.A., 67, 267, 270, 285
Zyablov, V.V., 62, 63, 67

Index

$A(q)$, 146, 147, 188, 250
B-gap, 209, 221
B-nongap, 221
g-family of codes, 66
H-construction, 195, 284
H-duality, 198
j-invariant, 114, 115, 158
𝕜-divisor, 129
𝕜-variety, 127
L-construction, 192, 284
$L(D)$, see space associated to a divisor
$\ell(D)$, 80, 130
L-function, 183
L/Ω-duality, 198
n-set, 5
n-torsion point, 115
$[n, k, d]_q$ code, x, 1
  up to spoiling, 16
$[n, k, d]_q$ system, 3, 67
  dual, 5
  projective, 4
Ω-construction, 193, 284
P-abundance of a divisor, 212
P-construction, 196
$P(t)$, see numerator of the zeta function
p-linearized polynomial, 168
q-ary entropy, 52
$(u\,|\,u+v)$ construction, 45, 290

absolute invariant, see j-invariant
absolute norm, 141
adjoint polynomial, 109
admissible change of variables, 159
affine closed set, 70
affine space, 70
algebra of distributions, 96
algebraic function field, see function field
algebraic group, 86
  additive, 86
  multiplicative, 86
algebraic set, 127
alphabet, 1
  extension, 47, 56, 291
  restriction, 47, 56, 291
annihilator, 107
Artin–Schreier cover, 173, 185
atlas, 88
automorphism group
  of a code, 3
  of a curve, 95
  of an elliptic curve, 117, 160

Bassalygo lemma, 18
bound
  Aaltonen, 53, 293
  basic algebraic geometry, 251, 259, 285, 293
  Bassalygo–Elias, 17, 67, 287
    asymptotic, 52, 292
    asymptotic, for generalized weights, 59
    generalized, 25, 308
  Blokh–Zyablov, 63, 293
  Castelnuovo, 229
  concatenation, 264, 266, 294
  Drinfeld–Vlăduţ, 146
  Elkies, 277, 294
  Ericson–Zinoviev, 270, 306
  expurgation, 259, 285, 293
  Feng Rao, 213
  Gilbert, 21, 270, 306
  Gilbert–Varshamov, 20, 67, 251, 259, 288, 293
    asymptotic, 56
    for self-dual codes, 57
  Griesmer, 16, 288
    generalized, 23, 307
  Hamming, 17, 287

asymptotic, 52, 292
asymptotic, for generalized weights, 58
generalized, 25, 308
Katsman–Tsfasman, *see* restriction bound
Katsman–Tsfasman–Vlăduţ, *see* concatenation bound
Laihonen–Litsyn, 54, 293
linear programming, 20, 288
Litsyn–Tsfasman, 269, 294
Litsyn–Zinoviev, 267
McEliece–Rodemich–Rumsey–Welch, 52, 292
second, 53, 292
MRRW, 53, 292
Pellikaan, 252, 293
Plotkin, 16, 287
asymptotic, 51, 292
asymptotic, for generalized weights, 58
generalized, 22, 307
random choice, 306
restriction, 266, 294
Scharlau, 269, 306
Serre, 142
Singleton, 5, 16, 287
asymptotic, 292
generalized, 22, 307
sphere packing, 17
Tsfasman–Vlăduţ–Zink, *see* basic algebraic geometry bound
Vlăduţ, *see* expurgation bound
Weil, 141
Xing, 282, 283, 294
Zyablov, 62

canonical class, 92
canonical class curve over a finite field, 199
canonical embedding, 100
Cartier operator, 104, 105, 132
character
of a field
additive, 11, 182
multiplicative, 181
of a group, 181
index of, 181
induced, 182
order of, 181
trivial, 11, 181
class number, 135, 139, 141, 149, 154
limit formula for, 154

Clifford theorem, 102
closed set, 70
code, x, 1, 67
abundant, 212, 252, 284
AG, 225
minimal representation of, 226
algebraic geometry, 192–197, 284, 289
generalized, 216, 247, 284
generalized, designed distance of, 217
improved, 217, 218, 244, 284
partial, 222, 284
spectrum of, 201
alternant, 41
automorphism group of, 3
BCH, 39, 64, 215, 284, 289
designed distance of, 39
narrow-sense, 39
primitive, 39
cardinality of, 1
concatenated, *see* concatenation
constant-weight, 18, 270, 306
cyclic, 38, 67
generator polynomial of, 38
parity-check polynomial of, 38
degenerate, 4
dimension, 3
distance, *see* minimum distance
domain, 49
dual, 8
Elkies, 272, 278
on a projective line, 271
elliptic, 233, 234, 284
MDS, 238, 284
spectrum of, 234
from an embedded pair, 45, 290
error-correcting, 2, 67
of genus 0, *see* MDS code
of genus at most g, 9
geometric Goppa, *see* algebraic geometry code
Golay, 42, 230, 289
Goppa, 41, 215, 284, 289
group, 43
Hamming, 37, 288
Hermitian, 241, 285
hierarchy of, 7
iterative, 44
Justesen, 41, 63, 289
length, 1
linear, x, 2, 67, 228

maximum distance separable, *see* MDS code
MDS, 5, 10, 33, 238
nondegenerate, 4
nonredundant, 33
one-point, 209, 284
parity-check, 33
perfect, 43
power of, 290
projection of, 290
q-ary, 1
quadratic-residue, 40, 289
quasi-cyclic, 40
quasi-self-dual, 13, 199
rate, 2
redundancy, 2
Reed–Muller, 36, 37, 231, 288
Reed–Solomon, 33, 38, 193, 195, 232, 284, 288
    dual code for, 33
    generalized, 34
    spectrum of, 34
repetition, 33
repetition of, 44, 290
r-MDS, 22
rth rank MDS, 22
SAG, 225
self-dual, 13, 57, 199, 306
    formally, 13
simplex, 288
spectrum of, 7, 68
spherical, 18
strongly algebraic-geometric, *see* SAG code
tensor power of, 44, 290
trivial, 33, 288
two-point, 210, 284
vector, 2
WAG, 225
weakly algebraic-geometric, *see* WAG code
Xing, 280
y-self-dual, 13
codes
    algebraic geometry
        automorphisms of, 197
        basic decoding algorithm for, 205, 284
        decoding of, 202, 284
        duality of, 198, 284
        self-dual, 199, 269, 284
    asymptotically good family of, 40, 49

combination of four, 290
direct sum of, 44, 290
equivalent, 3
g-family of, 66
Kronecker product of, 44
polynomial family of, 61
tensor product of, 44, 290
with many light vectors, 262
codeword, 2
codimension, 71
complex topology, 88
complexity, 60, 67
concatenation, 45, 46, 56, 62, 67, 214, 290
    bound, 264, 266, 294
    generalized, 46, 67, 291
conductor
    of a curve, 108
    of an extension, 107
    of a numerical semigroup, 214
    of an order, 117
constant field, 120
coordinate form, 4
coordinate function, 4
correction of errors and erasures, 48
cotangent space, 74
cover, 79
covering, 124
covering degree, 79
criterion of nonsingularity, 77
curve, 71, 77
    adjoint, 94
    algebraic, 77, 132
    Artin–Schreier, 173, 185
    asymptotically optimal, 273
    automorphism group of, 95
    canonical, 100
    complete, 77
    conductor of, 108
    degree of, 84
    elliptic, 111, 145, 158, 188, 211
        absolute invariant of, *see* j-invariant
        automorphism group of, 117, 160
        automorphisms of, 113
        endomorphism ring of, 117, 162
        group law on, 111–113, 131
        group of points of, 113, 119, 158, 163
        ordinary, 116
        zeta function of, 162
    Fermat, 180
    Hermitian, 165
    hyperelliptic, 99, 104, 167, 211

Kummer, 168, 183
maximal, 142, 166, 167
normal, 106
normalization of, 107
number of points of, 134, 141, 142
plane model of, 108
point of, 78, 128
projective, 77
projectively normal, 100
quasiprojective, 77
reducible, 77
singular, 106, 132
  genus of, 109
smooth, 106
smooth complete, 78
smooth plane, 93
  genus of, 94
smooth projective, 126
sub-Hermitian, 166, 189
supersingular, 116
zeta function of, 187
curves
  asymptotically exact family of, 151
  asymptotically good family of, 150
  asymptotically optimal family of, 148
  cover of, 79
  elliptic
    isogeneous, 117
    isogeny classes of, 161
    isomorphism classes of, 160
cusp, 108

decoding, 2, 48
  of algebraic geometry codes, 202, 284
  of concatenated codes, 48
  of Reed–Solomon codes, 35
decomposition field, 128
degree
  covering, 79
  of a curve, 84
  of a divisor, 79, 126, 129
  of an isogeny, 115
  of a map, 79
  of a place, 123, 129
  of a point, 128
derivative of a function at a point, 280
designed dimension, 192, 194
designed distance, 192, 194
  of a BCH code, 39
  of a generalized algebraic geometry code, 217
Deuring–Waterhouse theorem, 161
differential form, 91, 109
  divisor associated with, 92
  divisor of, 110
  exact, 105
  logarithmic, 105
  rational, 92
  regular, 91
  residue of, 97
differential of a function, 91
dimension
  of a code, 3
  of a variety, 71, 127
direct sum of codes, 44, 290
discrete valuation, 79, 122
divisible group, 113
divisor, 79, 84, 85, 125, 128–130, 188
  absolute norm of, 141
  abundance of, 212
  abundant, 212, 244
  associated with a differential form, 92
  base point free, 226
  Cartier, 83
  degree of, 79, 126, 129
  of a differential form, 110
  double-point, 108
  effective, 80, 126
  of a form, 83
  of a function, 80
  group, 79, 125
  hyperelliptic, 206
  hyperplane section, 84
  inverse image of, 83
  k-rational, 129
  order of, at a point, 209
  P-abundant, 212
  of poles, 80, 126
  positive, 80
  prime, 85, 125, 207
  principal, 80, 85, 126, 128
  ramification, 101, 110, 188
  rational, 129
  simple, 85
  space associated to, 80, 130
  special, 102, 206
  support of, 79, 125
  of zeroes, 80, 83, 126
  of a section, 84
divisor class group, see Picard group
Drinfeld–Vlăduţ theorem, see Drinfeld–Vlăduţ bound
dual hierarchy, 14
dual system, 5
duality theorem, 98

effective length, *see* generalized weight enumerator, 7
equivalent $[n, k, d]_q$ systems, 4
equivalent codes, 3
equivalent divisors, 80
equivalent projective systems, 4, 5
erasure, 48
   locators, 48
   vector, 48
error correction, 2
error locators, 35, 48, 203
error vector, 35, 48
evaluation map, 33, 192
exact sequence, 96
existence bounds, *see* lower bounds
explicit formula, 143, 188
expurgation, 253
extension to points of Supp D, 194

family of lattices, 82
family of vector spaces, 75
   restriction of, 75
   trivial, 75
Feng–Rao designed distance, 214
fibre, 75, 82, 130
   product, 173
field descent, 45, 290
field of rational functions, 72
   on a curve, 126, 128, 132, 187
form, 70
Frobenius eigenvalues, *see* Frobenius roots
Frobenius endomorphism, 162
Frobenius morphism, 100, 115, 116
Frobenius roots, 141, 153, 166, 185
Frobenius trace, 163
function
   (element of a function field), 123
   derivative of, at a point, 280
   differential of, 91
   divisor of, 80
   elliptic, 118, 131, 132
   order of, at a point, 128
   pole of, 124
   rational, 71
   regular, 72, 123
      power series expansion of, 78
   on a variety, 74
   zero of, 124
function field, 120, 123, 132, 187, 188
   field of constants of, 120
   genus of, 128
   valuation ring of, 120

functional equation, 133, 138, 139, 145

Galois group of a field, 126
gap, 103, 219
   number, 219
   at a pair, 210
   pure, 210
García–Stichtenoth towers, 177, 189
Gaussian binomial coefficient, 21
general linear group, 86
generalized iterative construction, 46, 68, 291
generalized spectrum, 14
generalized Weierstrass equation, 158
generalized weight, 6, 68, 291, 307
   of a nonlinear code, 24, 26, 68
generator matrix, 3
genus, 87, 89, 91, 128, 130, 131, 188
   arithmetic, 109
   of a function field, 128
   of a Riemann surface, 89
   of a singular curve, 109
   of a smooth plane curve, 94
Gilbert–Varshamov curve, 54
global field, 187
gonality, 211, 244
ground field, 70
group algebra, 43
group of n-torsion points, 115
group of units, 120

Hamming distance, *see* Hamming metric
Hamming metric, x, 1
Hamming sphere, 17
Hamming weight, 3
Hasse theorem, 162, 187
Hermitian product, 8
higher weight, *see* generalized weight
hoax, 152, 155
Hurwitz formula, 100, 101, 188
hyperplane section divisor, 84

image, 73
inductive semigroup, 218
inductive tower of semigroups, 218
inertia, 124
integral closure, 106
integral domain, 106
integral element over an integral domain, 106
integrally closed domain, 106
inverse Frobenius roots, *see* Frobenius roots

inverse image
  of a divisor, 83
  of a point, 84
irreducible component, 71, 77
irreducible topological space, 71
isogeny, 115, 161
  degree of, 115
  dual, 115
isomorphism
  birational, 73
  biregular, 73
  of varieties, 73

Jacobian, 86, 87, 128, 131, 132, 188
  embedding in, 87

Klein quartic, 245
Krawtchouk polynomial, 19
Kronecker product, 44
Kummer cover, 168, 183

lattice, 82, 118
  family, 82
    asymptotically standard, 82, 130
Legendre parameter, 112
Leibniz rule, 91
lengthening by zeroes, 290
line bundle, 75–77, 84, 130
line vector bundle, see line bundle
linear equivalence, 80, 83
linear system, 81
  complete, 81, 98
  incomplete, 218
local invariants, 107
  fundamental, 107
local parameter, 77, 122
local ring, 74, 120, 128
log-cardinality, 1
lower bounds, xi, 20–21
  asymptotic, 54, 293, 299–305
    for generalized weights, 60
  for generalized weights, 24, 26, 307

MacWilliams identity, 8, 12
MacWilliams-type identities for the generalized spectrum, 29–32
main asymptotic problem, xii
main conjecture on MDS codes, 6
map
  degree of, 79
  dominant, 73
  normalization, 107
  ramified, 85

  tamely, 101
  rational, 72
    image of, 73
  regular, 72
  separable, 94
  unramified, 84, 85
mass formula, 103, 118, 160
maximal ideal, 74, 128
maximum likelihood, 2
MDS conjecture, 6
minimum distance, 1
  relative, 2
Möbius function, 149
Möbius inversion formula, 148
model
  of an algebraic variety, 106
  of a field, 106
  plane, 108
morphism
  of algebraic groups, 116
  of families of vector spaces, 75
  Frobenius, 100, 115, 116
  of varieties, 72
  Verschiebung, 116

node, 108
Noether $A\varphi + B\psi$ theorem, 110
Noether theorem, 100, 110
nongap, 103, 219
  at a pair, 210
nonsingularity criterion, 77
norm, 182
  absolute, 182
  of a place, 187
  relative, 182
normalization, 107
number field, 187
number of B-gaps, 221
number of effective divisors, 134, 138, 144
number of points, 134, 141, 142
  of degree r, 134
numerator of the zeta function, 138–140, 185, 186
numerical semigroup, 214

omitting parity checks, 290
open covering, 76
open set, 70
order in a division algebra, 117

parity check, 45, 290
parity-check matrix, 3

partial affine ring of a point, 219
partial L-space, 220
partition function, 256
pasting, 44, 290
perfect field, 106, 120
period lattice, 118
Picard group, 76
place, 122, 128, 187
   degree of, 123, 129
   inert, 124
   norm of, 187
   totally ramified, 124
   unramified, 124
   valuation ring of, 122
   value at, 123
plane cubic, 111
point, 70, 122, 128
   of a curve, 78, 128
   cuspidal, 108
   decomposition field of, 128
   degree of, 128
   of degree s, 129
   double, 108
   hyperelliptic, 104
   inverse image of, 84
   local invariants of, 107
   local ring of, 74
   n-torsion, 115
   nodal, 108
   nonsingular, 75, 77, 107
   partial affine ring of, 219
   ramification, 84, 85
   regular, 75
   simple, 75
   singular, 75
   smooth, 75
   Weierstrass, 103, 207
pole of a function, 124
pole order, 79, 85, 124
polynomial $P(t)$, *see* numerator of the zeta function
polynomial bounds, 61
polynomially decodable family, 61
position, 2
possibility bounds, *see* upper bounds
power of a code, 290
power series expansion of a regular function, 78
principal ideal, 77
product
   of line bundles, 76
   of varieties, 75

projection of a code, 290
projective $[n, k, d]_q$ system, 4
projective closed set, 70
projective closure, 70
projective space, 70
   embedding in, 98
projective system, x, 4
   dual, 5
puncturing, 30

quadratic residue, 40
quasiprojective set, 71
   irreducible, 71
   reducible, 71
quaternion algebra over $\mathbb{Q}$, 117

ramification, 124
   divisor, 101, 110, 188
   index, 85, 124
   tame, 101
   total, 124
   wild, 101
rate, 2
redundancy, 2
repetition of a code, 290
residue
   of a differential form, 97
   formula, 34, 97
   of a function at a point, 34
   quadratic, 40
residue class field, 123
Riemann existence theorem, 89
Riemann hypothesis, 132, 133, 141, 188
Riemann inequality, 132
Riemann surface, 88, 131, 132
   genus of, 89
Riemann–Roch theorem, 95, 96, 98, 131, 132

Schmidt theorem, 137
section, 76, 82
   divisor of zeroes of, 84
   value of, at a point, 82
   zero set of, 76
Segre embedding, 75
semi-local ring, 107
separable closure, 126
shortening, 29, 30
   by distance, 45, 290
   by dual distance, 45, 290
simplest double point, 108
simplest singularity, 108
space associated to a divisor, 80, 130

space of sections, 76, 82
spectrum, 7, 68
   of an algebraic geometry code, 201
   of an elliptic code, 234
   generalized, 14
sphere
   Hamming, 17
   packing, x
   volume of, 17
splitting, 124
spoiling, 15, 22
   lemma, 15
subfield restriction, 39, 45, 215, 290
suffix construction, 290
super code property, 218
support of a divisor, 79, 125
support weight, *see* generalized weight
surface, 71
syndrome, 35, 203

tangent space, 74
tautological bundle, 76
tensor power of a code, 44, 290
tensor product
   of codes, 44, 290
   of line bundles, 76
threefold, 71
total splitting, 124
trace, 182
   absolute, 182
   relative, 182
transmission rate, 2
triangle inequality, 123
trivialization, 195
true asymptotic bounds, 292
twisted inner product, 8

uniformization, 119
unramified extension, 124
upper bounds, xi, 16–20, 287
   asymptotic, 51–54, 292, 299–305
     for generalized weights, 58–60
   for generalized weights, 22–26

valuation ring, 79
  discrete, 121
  of a function field, 120
  of a place, 122
value
  of a function at a point, 72
  at a place, 123
  of a section at a point, 82
variety, 71

abelian, 86
absolutely irreducible, 127
affine, 74, 127
codimension of, 71
dimension of, 71, 127
function on, 74
nonsingular, 75
normal, 106
projective, 71, 74, 127
quasiprojective, 71, 127
rational, 74
smooth, 75, 127
Varshamov procedure, 21
Verschiebung, 116

weak approximation theorem, 125
Wei weight, *see* generalized weight
Weierstrass equation, 158, 159
Weierstrass normal form, 115
Weierstrass pair, 210
Weierstrass $\mathcal{P}$-function, 118
Weierstrass point, 103, 207
Weierstrass semigroup, 213
Weierstrass weight, 103
weight, 3
  enumerator, 7
  generalized, *see* generalized weight
  hierarchy, 7
weight function on an algebra, 213
Weil theorem, 141, 187

Zariski topology, 71, 127
zero of a function, 124
zero order, 79, 85, 124
zero section, 76
zero set, 76
zeta function, 133
  of a curve, 133, 134, 187
  Dedekind, 133, 141, 188
  of an elliptic curve, 162
  of a number field, 133
  numerator of, 138–140, 185, 186
  Pellikaan, 144, 188
  Riemann, 133, 135, 141

Titles in This Series

140 **Elisabetta Barletta, Sorin Dragomir, and Krishan L. Duggal,** Foliations in Cauchy-Riemann geometry, 2007

139 **Michael Tsfasman, Serge Vlăduţ, and Dmitry Nogin,** Algebraic geometric codes: Basic notions, 2007

138 **Kehe Zhu,** Operator theory in function spaces, 2007

137 **Mikhail G. Katz,** Systolic geometry and topology, 2007

136 **Jean-Michel Coron,** Control and nonlinearity, 2007

135 **Bennett Chow, Sun-Chin Chu, David Glickenstein, Christine Guenther, James Isenberg, Tom Ivey, Dan Knopf, Peng Lu, Feng Luo, and Lei Ni,** The Ricci flow: Techniques and applications, Part I: Geometric aspects, 2007

134 **Dana P. Williams,** Crossed products of C^*-algebras, 2007

133 **Andrew Knightly and Charles Li,** Traces of Hecke operators, 2006

132 **J. P. May and J. Sigurdsson,** Parametrized homotopy theory, 2006

131 **Jin Feng and Thomas G. Kurtz,** Large deviations for stochastic processes, 2006

130 **Qing Han and Jia-Xing Hong,** Isometric embedding of Riemannian manifolds in Euclidean spaces, 2006

129 **William M. Singer,** Steenrod squares in spectral sequences, 2006

128 **Athanassios S. Fokas, Alexander R. Its, Andrei A. Kapaev, and Victor Yu. Novokshenov,** Painlevé transcendents, 2006

127 **Nikolai Chernov and Roberto Markarian,** Chaotic billiards, 2006

126 **Sen-Zhong Huang,** Gradient inequalities, 2006

125 **Joseph A. Cima, Alec L. Matheson, and William T. Ross,** The Cauchy Transform, 2006

124 **Ido Efrat, Editor,** Valuations, orderings, and Milnor K-Theory, 2006

123 **Barbara Fantechi, Lothar Göttsche, Luc Illusie, Steven L. Kleiman, Nitin Nitsure, and Angelo Vistoli,** Fundamental algebraic geometry: Grothendieck's FGA explained, 2005

122 **Antonio Giambruno and Mikhail Zaicev, Editors,** Polynomial identities and asymptotic methods, 2005

121 **Anton Zettl,** Sturm-Liouville theory, 2005

120 **Barry Simon,** Trace ideals and their applications, 2005

119 **Tian Ma and Shouhong Wang,** Geometric theory of incompressible flows with applications to fluid dynamics, 2005

118 **Alexandru Buium,** Arithmetic differential equations, 2005

117 **Volodymyr Nekrashevych,** Self-similar groups, 2005

116 **Alexander Koldobsky,** Fourier analysis in convex geometry, 2005

115 **Carlos Julio Moreno,** Advanced analytic number theory: L-functions, 2005

114 **Gregory F. Lawler,** Conformally invariant processes in the plane, 2005

113 **William G. Dwyer, Philip S. Hirschhorn, Daniel M. Kan, and Jeffrey H. Smith,** Homotopy limit functors on model categories and homotopical categories, 2004

112 **Michael Aschbacher and Stephen D. Smith,** The classification of quasithin groups II. Main theorems: The classification of simple QTKE-groups, 2004

111 **Michael Aschbacher and Stephen D. Smith,** The classification of quasithin groups I. Structure of strongly quasithin K-groups, 2004

110 **Bennett Chow and Dan Knopf,** The Ricci flow: An introduction, 2004

109 **Goro Shimura,** Arithmetic and analytic theories of quadratic forms and Clifford groups, 2004

108 **Michael Farber,** Topology of closed one-forms, 2004

TITLES IN THIS SERIES

107 **Jens Carsten Jantzen,** Representations of algebraic groups, 2003
106 **Hiroyuki Yoshida,** Absolute CM-periods, 2003
105 **Charalambos D. Aliprantis and Owen Burkinshaw,** Locally solid Riesz spaces with applications to economics, second edition, 2003
104 **Graham Everest, Alf van der Poorten, Igor Shparlinski, and Thomas Ward,** Recurrence sequences, 2003
103 **Octav Cornea, Gregory Lupton, John Oprea, and Daniel Tanré,** Lusternik-Schnirelmann category, 2003
102 **Linda Rass and John Radcliffe,** Spatial deterministic epidemics, 2003
101 **Eli Glasner,** Ergodic theory via joinings, 2003
100 **Peter Duren and Alexander Schuster,** Bergman spaces, 2004
99 **Philip S. Hirschhorn,** Model categories and their localizations, 2003
98 **Victor Guillemin, Viktor Ginzburg, and Yael Karshon,** Moment maps, cobordisms, and Hamiltonian group actions, 2002
97 **V. A. Vassiliev,** Applied Picard-Lefschetz theory, 2002
96 **Martin Markl, Steve Shnider, and Jim Stasheff,** Operads in algebra, topology and physics, 2002
95 **Seiichi Kamada,** Braid and knot theory in dimension four, 2002
94 **Mara D. Neusel and Larry Smith,** Invariant theory of finite groups, 2002
93 **Nikolai K. Nikolski,** Operators, functions, and systems: An easy reading. Volume 2: Model operators and systems, 2002
92 **Nikolai K. Nikolski,** Operators, functions, and systems: An easy reading. Volume 1: Hardy, Hankel, and Toeplitz, 2002
91 **Richard Montgomery,** A tour of subriemannian geometries, their geodesics and applications, 2002
90 **Christian Gérard and Izabella Łaba,** Multiparticle quantum scattering in constant magnetic fields, 2002
89 **Michel Ledoux,** The concentration of measure phenomenon, 2001
88 **Edward Frenkel and David Ben-Zvi,** Vertex algebras and algebraic curves, second edition, 2004
87 **Bruno Poizat,** Stable groups, 2001
86 **Stanley N. Burris,** Number theoretic density and logical limit laws, 2001
85 **V. A. Kozlov, V. G. Maz'ya, and J. Rossmann,** Spectral problems associated with corner singularities of solutions to elliptic equations, 2001
84 **László Fuchs and Luigi Salce,** Modules over non-Noetherian domains, 2001
83 **Sigurdur Helgason,** Groups and geometric analysis: Integral geometry, invariant differential operators, and spherical functions, 2000
82 **Goro Shimura,** Arithmeticity in the theory of automorphic forms, 2000
81 **Michael E. Taylor,** Tools for PDE: Pseudodifferential operators, paradifferential operators, and layer potentials, 2000
80 **Lindsay N. Childs,** Taming wild extensions: Hopf algebras and local Galois module theory, 2000
79 **Joseph A. Cima and William T. Ross,** The backward shift on the Hardy space, 2000
78 **Boris A. Kupershmidt,** KP or mKP: Noncommutative mathematics of Lagrangian, Hamiltonian, and integrable systems, 2000

For a complete list of titles in this series, visit the
AMS Bookstore at **www.ams.org/bookstore/**.